D1710385

PACKAGE PRINTING

PACKAGE PRINTING

by
Nelson R. Eldred, Ph.D.

Jelmar Publishing Co., Inc.
P O Box 488
Plainview, NY 11803

Library of Congress Cataloging-in-Publication Data

Eldred, Nelson Richards.
Package printing / by Nelson R. Eldred.
p. cm.
Includes bibliographical references (p.) and index.
ISBN 0-9616302-5-6 : $83.50
1. Package printing. 2. Printing machinery and supplies.
I. Title.
TS196.7.E43 1993 92-30588
658.5'64—dc20 CIP

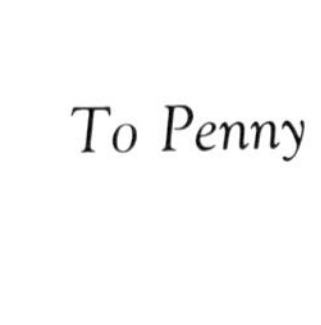

To Penny

Preface

In his 40 years experience in the paper, ink and printing industries, the author has seen the need grow for a book on Package Printing. Prior to World War II, package printing was little more than the adaptation of publication and commercial printing to labels and to a variety of paperboard products sometimes called "cardboard."

Printing on boxes, barrels and most shipping containers was limited to a stencil that marked the address of the consignee. Folding cartons were printed by journeyman printers who used the printing techniques they knew, largely letterpress, although some used the evolving process of offset lithography. Rotogravure has long been used where high quality printing is required as on cereal and detergent boxes.

During the 1940s, corrugated and plastic films became important packaging materials, replacing wooden boxes, barrels and paper sacks as protective and wrapping materials. In 1952, the U. S. Census of Manufacturers still combined folding cartons and corrugated containers. It was not until 1957 that the Census Bureau considered corrugated containers to be sufficiently important to warrant a separate classification.

Cellophane film was the first important packaging film, and printing cellophane was the first important use of flexography. Polyethylene, the first of the myriad plastic packaging films, became an important industrial commodity during "The War." Nylon, first used for ladies' hosiery, became an important film shortly after that.

By the 1980s, flexography had become a true graphic arts process, but it had been developed primarily for package printing where its ability to produce good images economically on virtually all packaging materials made it the leading process. The first printing process developed primarily for packaging, flexography is now finding its way into publication printing, reversing the trend of the adaptation of publication processes for package printing.

It is curious that at this late date, little attention has been paid to the subject of package printing although it comprises a major segment of all printing (35% of all printing inks in the United States are used for package printing).

This is another example where texts and reference books follow technology by many years.

A keyword search of the computerized file at a nearby university library showed 591 entries that included books and printing, 41 including newspapers and printing, 25 involving magazines and printing, but only two involving packaging/packages and printing: one was a census report on printing, the other was Alec Davis's fine book on package and print. Bob Long's 1964 landmark book on "Package Printing" is, unfortunately, not in the collection.

The need for a current text and reference book on the subject of package printing seems obvious.

New Port Richey, FL
May, 1993

Nelson R. Eldred

Acknowledgements

It is impossible to mention and thank all of the many people who have helped by answering questions or submitting relevant information. I especially want to thank Page Crouch of Clemson University who has carefully reviewed the manuscript, making a great many useful suggestions. Special thanks also go to Jim Hagan of W. R. Grace & Co., to Tony Bart of DuPont, and to Joseph Rach and Colleen Larkin of Hercules Inc. for reviewing the chapter on printing plates and making many improvements. Bert Moore of IDAT Consulting & Education and Richard Swartz and Donna Keller of PSC Inc. reviewed the chapter on bar codes. Pictorial and graphic contributions have been acknowledged in the book, but I want to thank Bill Temesy for the original photography. The acknowledgements would not be complete without thanking Joel Shulman, the editor and publisher, who, in addition to being patient, helpful and encouraging, has contributed much information from his long experience in flexographic printing and packaging.

Among those who have made significant contributions are:

Thomas Allison	Allison Systems Corp.
Leonard Anderson	Boise Cascade Corp.
Dennis Andersen	Armotek Industries
Mark Andrews	Mark Andy Inc.
Hank Apfelberg	California Polytechnic State University
Mark Armstrong	Fleishman-Hillard
Peter Basler	Bobst Group

Mike Barisonek	Epic Products, International
Robert Bassemir	Sun Chemical Corp.
Kitty Bernard	Hunter Associates Laboratory, Inc.
Robert Booker	ZED Instruments USA Ltd.
Hank Brandtjen, III	Brandtjen & Kluge
Gary Brogdon	Rexham Corp.
Michael Bruno	Michael Bruno, Consultant
Richard Bucknam	Eastman Kodak Co.
Russell Cannon	Crown Cork & Seal Corp.
Paul Cappa	North American Cerutti Corp.
Linda Casatelli	Flexo Magazine
Richard Chestnut	W. R. Chestnut Co.
John Coburn	Screen Printing Assn. International
George Collins	U. S. Government Printing Office
Rita Conrad	Flint Ink Corp.
Jerry Cozart	Quality Process Consulting
Doug Craig	Misomex North America, Inc.
Charles Dale	LSI/KALA, Div. of Shawk, Inc.
Ed Dalton, Jr.	Photoengraving, Inc.
Ed Dalton, Sr.	Photoengraving, Inc.
Mike Daum	Testing Machines, Inc.
Hans Deamer	Windmoeller & Hoelscher
David DiTallo	Ohio Electronic Engravers
E. L. (Dutch) Drehle	Screen Printing Assn. International
Michael Eamer	InterCité Papers
Georgia Ehrenberg	Harper Corp. of America
Pius Eigemann	Butler, Inc.
Templeton Elliott	Thiele-Engdahl
Robert Engdahl	Thiele-Engdahl
Richard Ernst	Eastman Kodak Co.
E. W. (Ned) Evans	Carnegie-Mellon University
Lucille Fabia	Baldwin Graphic Products
Foster Fargo	Iris Graphics
Milton Field	Consultant
Stanley Field	Flint Ink Corp.
Jim Fitzgerald	Gallus, Inc.
Kurt Flathmann	Fischer & Krecke, Inc.
Russell Fox	Hurletron Corp.
John Friedery	Ball Metal Container Div.
Edmund Funk	Sun Chemical Corp.
Paul Gardner	Paul N. Gardner Co., Inc.
Raymond Gauvin	Oxy-Dry Corp.
Shellee Gero	Pantone Corp.
Cliff Gober	Wikoff Color Corp.
George Goldberg	Scan Newsletter, Ltd.
Robert Gradl	Laminating Arts & Finishing
Eugene Green, Jr.	Wilson Engraving Co., Inc.

Eugene Green, Sr.	Wilson Engraving Co., Inc.
John Hamelink	Boxboard Research & Dev. Assn.
Earle Harley	E. L. Harley, Inc.
Michael Heckaman	Stork Cellramic
Fred Higgins	Graphic Arts Technical Foundation
Kathy von Holtz	Gallus Inc.
Edward Hughes	Flint Ink Corp.
Leslie Hull	Hull & Co.
Dennis Ittner	Klemm of North America Corp.
Dick James	Master Packaging, Inc.
Spence Johnson	Paperboard Packaging Council
Gary Jones	Graphic Arts Technical Foundation
Frank Kanonik	Graphic Arts Technical Foundation
Cheryl Kasunich	Gravure Association of America
Edward Kelly	3M Co.
Peter Kerchner	Mark Andy Inc.
Ken Konrad	James River Corp.
Alan Kotok	Graphic Communications Association
William Lamparter	PrintCom Consulting Group
William Leutwyler	Gallus, Inc.
Tom Lovesheimer	Mead Corp.
Margaret MacBeth	Atlas Electric Devices
John MacPhee	Baldwin Graphic Products
John Magee	Tobias Associates, Inc.
Thomas Manning	Advance Process Supply Co.
Mary Louise McCarthy	Monarch Marking Systems Div., Pitney Bowes
Gerald McLain	Sun Chemical Corp.
Peter Menzian	E. I. duPont de Nemours & Co.
John Mercer	U. S. Bureau of Engraving & Printing
Raymond Moore	Mark Andy, Inc.
Steve Mott	California Polytechnic State University
Kimberly Meyers	Linotype-Hell
Robert Mrak	North American Cerutti Corp.
Gerry Nigg	KoPack International USA
Robert Oppenheimer	Gravure Association of America
Wendy Painter	Stretch Devices, Inc.
Charles Parsons	Weyerhaeuser Co.
Richard Podhajny	Richard Podhajny, Consultant
Raymond Prince	Graphic Arts Technical Foundation
Steven Rach	TEC Systems, W. R. Grace & Co.
Mary Pat Ralston	Bobst Group
Janet Randall	Canadian Standards Office
James Renson	Natl. Assn. of Printing Ink Mfrs. (NAPIM)
Garland Richardson	Ball Corporation
Edward Riggs	Stork Bedford B.V.
Theresa Rossin	Videojet Systems
Timothy Rozgonyi	Graphic Arts Technical Foundation

Robert Savageau	Flint Ink Corp.
Richard Schmiedle	Gravure Association of America
Myron Segal	Premier Mill Corp.
Susan Selke	Michigan State University
Thomas Shaney	Harper Corporation of America
David Share	Stork Cellramic, Inc.
Robert Shutte	MAN Roland Inc.
Dwayne Shrader	Ward Machinery Co.
Benjamin Slatin	American Forest & Paper Association
Richard Smith	Gravure Association of America
Kent Stumpe	MAN Roland Inc.
James Sutphin	Natl. Assn. of Printing Ink Mfrs. (NAPIM)
Gary Syvertson	Flower City Packaging, Inc.
Elaine Tito	Macbeth, Div. of Kollmorgen
Hugh Van Brimer	Trident, Inc.
Gaston Vandermeerssche	Gavarti Associates Ltd.
David VanderZanden	Consolidated Papers Inc.
Donald Voas	James River Corp.
Paul Volpe	Natl. Assn. of Printing Ink Mfrs. (NAPIM)
Kathy VonHolz	Gallus, Inc.
James Votava	American National Can Co.
Garrett Walker	The Wellesley Group
Norman Weiland	Monarch Marking Systems Div., A Pitney Bowes Co.
Steve Woodard	ARC International

Cover photograph courtesy of King Kullen Grocery Co., Inc.

Table of Contents

Guest Introduction

by Robert P. Long

A world of substrate has sped through the presses in the decades since the first book on package printing was published in 1964 and updated in 1973. In those times the burgeoning printing industry was largely unaware, and startled, to learn that printed packaging volume had become a major part of its production.

Today, this totally new book reflects enormous growth in package printing, coupled with extraordinary improvements in concepts and technology. The volume of this category of printing in the U. S. now has reached more than $58 billion annually.

Non-photo digital imaging, just beginning in the 1970s, now is a major factor in the vast advancements in prepress operations. Computer generated graphics and electronic color imaging in plate- and cylinder-making are everyday techniques. Press and ink controls, digital color proofs, press "fingerprinting," "just in time" delivery, robot cylinder changing, ink jet printing, holography, electrostatic methods, and the remarkable alchemy of bar codes—all have found their places in the new technology.

Thorough descriptions and technical information on all of the basic printing processes, as well as the new systems now finding their niches in package printing are given clearly in this book. It is significant that in recent years flexography has become the leading printing process in this field as its technology and quality have emerged.

Environmental considerations, including wide use of waterbase inks, are now at the top of package printing priorities. Recycling of paper and board is rapidly emerging, and already about 70 percent of the recycled stock is used in corrugated products, and 50 percent of folding cartons is made from recycled board. Much plastic packaging is recycled, and the same is true of metal cans and containers. Increasing amounts of these are going back into packaging materials. The trend is growing.

For an overview of the comprehensive scope of this book, look closely at the Table of Contents, and read the author's Introduction. This volume covers every phase of its subject in great depth, for the printer, supplier, the purchaser, and every kind of packaging anyone can think of is included. This book should be the foundation of every serious learning or training program in colleges and universities, as well as in corporate training projects, and it belongs in every serious library of commerce and industry.

Want to see a fantastic exhibition of the best of package printing? Just stroll through the dazzling corridors of your local supermarket, with your eyes and consciousness focused on the vast variety of excellent printing on display!

INTRODUCTION

CONTENTS

LIST OF ILLUSTRATIONS

LIST OF TABLES

INTRODUCTION

THE NEED FOR THIS BOOK

This book is written for the package printer, the package buyer, the artist and designer, the supplier and the student of packaging and printing. One of its goals is to help those in the industry obtain better quality printing for the money and effort spent.

There are many reasons for writing a book on Package Printing. First of all, there is very little written on this technically and economically important subject. In the second place, printing processes used for package printing differ from those used for publication and general commercial printing (figure 1). Publications are printed largely by offset lithography, with some gravure and other processes. Package printing is largely done by flexography, with some offset, gravure and other processes. A third reason is that most package printers consider themselves to be converters, not printers, and because printing is sometimes a secondary consideration, a great deal of package printing is slower, more wasteful and of poorer quality than it needs to be.

This book should help package buyers understand the possibilities and limitations of the printing processes. When a printer or converter tells a package buyer that something cannot be done, or that "my way is better," the package buyer must decide whether the story is likely to be true. The buyer's judgement is affected by his/her knowledge of the process.

Packagers and package buyers need to understand how the printing of the package can best increase its effectiveness as a sales agent, increasing profits

Figure 1. Representative packages printed by a variety of methods. Squirt is printed offset letterpress. Crackerjack is printed gravure. Kentucky Fried Chicken is printed flexo. Others are printed offset, although many cereal boxes and soup labels are printed gravure or flexo.

while keeping costs to a minimum. A well-printed package does not necessarily cost more than a poorly printed one, but the quality of printing is closely related to successful promotion of the product.

Printers, who try to follow the customer's instructions and requests, are often reluctant to tell the customer how to do better. A well-informed customer usually gets the best buy or the best bargain.

Similarly, this book should help artists and designers understand the possibilities and limitations of the printing processes and materials used for packaging so that the printing process can enhance their art and design rather than do battle with it.

Another aim is to help manufacturers and developers of package printing materials and equipment to better understand the customer's needs and difficulties and to help develop new products to meet their needs. The book contains information useful for the film extruder, the paper and paperboard manufacturer, the inkmaker and the manufacturer of presses, prepress hardware/software and instruments as well as manufacturers of rigid plastic, metal and glass containers.

SCOPE OF THIS BOOK

A package is a covering wrapper or container. It should serve four purposes: protection, communication, promotion and convenience. This book is directed principally to promotion and communication although attention is also directed to protection and convenience. Bags, boxes, cans and bottles are clearly packages. We have chosen to include labels within the subject of packaging, but to exclude envelopes, which require much specialized equipment.

Package printing is the application of inks or coating material to a package, directly or with a label, to enhance sales, to convey information or to protect the printing or the surface of the package. It includes printing, coating, embossing and decoration of the package. Package printing includes in-line converting operations in addition to the production of the image.

It would require an encyclopedia to describe all printing processes in detail. Some, such as pad printing and collotype, are little used in package printing. It is not possible in a text such as this to mention every possible application of each printing process, nor to mention every combination of equipment used to print packages. In fact, it is doubtful if any two printing presses are identical in every detail. Those who need additional detail can find it in the references mentioned in "Further Reading" at the end of each chapter.

While technological knowledge is truly international, its application is not. Flexography is growing in Europe as it is in the United States and Canada, but flexo is used more in North America than in Europe, and in Europe more than Japan. Gravure is still the leading package printing process used in Europe, where label printing is still dominated by letterpress although European label printing is swinging toward flexo. There are differences within Europe and the Pacific rim countries. Western Europe makes greater use of flexo than Eastern Europe, and Australia and New Zealand are world leaders in the use of flexography.

The above represents a snapshot taken in 1993. Since technology changes continuously and rapidly and all printing processes compete, this picture will certainly change, possibly in ways we cannot even imagine, as electronic imaging, now in its infancy, develops.

We estimate the 1990 volume of package production in the United States to be around $73 billion of which $58 billion represents packages with printing. If, as we know, the volume of packaging in the United States is half of world volume, the volume of package printing worldwide is in the order of $120 billion.

TECHNOLOGY AND THE MARKETPLACE

Package design and printing are dictated from the source rather than primarily from distribution channels. New technology has created not larger markets but more of them—smaller niche markets, each requiring its own packages. Nowhere is this more evident than on the supermarket shelf. We used to see one type of potato chip on the grocery shelf. Now we have many versions including low sodium, low cholesterol, BBQ flavor and more. Quaker Oats used to sell one type of oatmeal. Today, there are three types and 12 flavors. This requires greater emphasis on short runs, fast makeready, quality control and just-in-time delivery.

Progress in printing technology is usually incremental, but improved technology is introduced at every major printing equipment show. New technology rarely develops rapidly, but such change is seen in the electronic prepress where, in the early 1990s, desktop publishing rapidly replaced the use of photographic film in preparing artwork for printing.

PURPOSES OF PACKAGE PRINTING

Package printing produces corporate imagery and identification of the product and the manufacturer or seller. Most package printing is required by law to meet certain informational or labeling requirements. Package printing involves:

- Eye-catching graphics
- Corporate imagery and identification
- Identification of contents and information about them
- Legal requirements concerning contents and their use or restrictions
- Graphic representation of color, type and appearance of the contents
- Bar codes indicating price, lot number and inventory

STUDIES OF PACKAGE PRINTING

The study of package printing has not developed as rapidly as the study of package structure, layout and design and environmental issues. The gradual shift to supermarket and retail warehouse merchandising should promote the study of package printing. Studies of packaging are directed toward structural, environmental and legal aspects and requirements, but often overlook the ways in which printing can enhance or detract from the purposes of package printing.

On the other hand, direct mail advertisers and publishers of books, magazines and newspapers have long considered printing as the foremost concern in the manufacture of their product.

CONTENTS OF THE BOOK

This book presents the conventional methods of printing common packaging materials: corrugated, folding cartons, labels, flexible packaging, metal and glass. It points out the characteristics of each of the major printing processes and how these characteristics make the process suitable or unsuitable for each kind of packaging product. The book discusses design and preparation of artwork for the press. The book explains color and color printing and describes how printing equipment, printing inks and coatings affect decisions about printing a package.

The book discusses items of current concern: health, safety and the environment, total quality management (TQM) and just-in-time manufacture and delivery (JIT). It includes a discussion of bar codes and tells how to evaluate the printer's capability. The book includes a glossary of printing defects and common terms used in printing and packaging.

The production of a superior package incorporates many considerations and technologies: strength and protection, layout and design, legal and environmental decisions, as well as printing. It is not always possible to ignore one subject while discussing another. Discussions of these items show how they affect the printing requirements.

THE PACKAGING INDUSTRY

Four tables give a glimpse of the packaging industry. Table I lists the latest

TABLE I. PACKAGE PRODUCTS CLASSES REPORTED BY THE U. S. CENSUS OF MANUFACTURES

SIC #	Name	Value Shipped ($Millions)		
		(1987)	(1982)	Growth (%)
2652	Setup paperboard boxes	517.9	485.1	6.8
2653	Corrugated and solid fiber boxes	15,602.2	10,303.1	51.4
2655	Fiber cans, drums, etc	1,495.0	1,490.3	0.3
2656	Sanitary food containers	1,959.6	1,841.1	6.5
2657	Folding paperboard boxes	5,521.8	4,458.6	23.8
2671	Paper coating and laminating for packaging	2,460.7	1,329.2	85.1
2672	Paper coating and laminating (other)	5,497.7	3,574.0	53.8
26723	Pressure sensitive products	3,100.0		
2673	Bags, plastic, laminated and coated	3,936.5	2,762.6	42.4
2674	Bags: uncoated paper and multiwall	2,360.6	2,010.3	17.5
27522	Labels and wrappers (litho)	794.0	677.0	17.3
27542	Labels and wrappers (gravure)	434.4	437.4	− 0.7
27592	Labels and wrappers (letterpress)	422.5	483.8	− 12.7
27597	Flexo printed labels	731.7	405.2	80.6
27598	Screen printed labels	229.1	113.8	101.3
3085	Plastic bottles	2,849.6	1,921.4	48.3
30894	Plastics packaging	4,005.9	(NA)	
3221	Glass containers	4,720.7	5,143.8	− 8.3
3411	Metal cans	10,652.5	10,551.9	0.9
3466	Crown and closures (metal)	811.4	790.2	33.2
	(includes home canning crowns)	**65,003.2**	**48,778.8**	**33.2**

TABLE IA. STANDARD INDUSTRIAL CLASSIFICATIONS FOR PACKAGING MATERIALS (METAL CANS), 1987

34111	Steel cans and tinware	4,852.6
34112	Aluminum cans	5,626.7
34110	Others	173.2

TABLE IB. STANDARD INDUSTRIAL CLASSIFICATIONS FOR PACKAGING MATERIALS (FLEXOGRAPHIC PRINTED LABELS), 1987

27597	14	Paper—Flat (except pressure sensitive)	33.9
27597	16	Paper—Rolls (except pressure sensitive)	61.6
27597	18	Paper—Flat (pressure sensitive)	54.0
27597	20	Paper—Rolls (pressure sensitive)	522.3
27597	30	Other materials except cloth	59.9

Source: U.S. Census of Manufactures, 1987. The census is published every five years.

TABLE II. VALUE OF PRINTED PACKAGING (U. S. SHIPMENTS—1990)

	(Billions of Dollars)				
	Flexo	**Offset**	**Gravure**	**LP**	**Screen & Misc**
Corrugated	12.9	0.5	—	0.5	—
Plastics					
Flexible packaging	7.0	—		—	—
Bags	1.5	0.5	0.5		
Other[1]	2.0				
Folding[2] cartons	1.5	2.3	1.0	0.5	—
Multiwall sacks	0.6	—	0.2	—	—
Paper bags[3]	4.5	1.2	0.3	—	—
Labels[4]	2.0	4.0	0.4	0.5	0.4
Sanitary packaging	2.0	—	—	—	—
Metal cans	0.5	1.0	—	2.0	
All other	2.5	—	—	0.5	1.1
	37.0	**9.5**	**6.2**	**4.0**	**1.5**

Total sales of packages = \$73.0 billion; total of printed packages = \$58.2 billion

Source: Composite of various industrial sources

Note: Figures cited are only approximations, and lack of numbers in some categories does not indicate lack of activity, only that such activity is minimal.

[1] Includes plastic grocery bags.

[2] Includes milk cartons and beverage carriers.

[3] Includes grocery sacks, specialty and boutique bags, etc.

[4] Much label printing is to be found classified as commercial printing.

TABLE III. SHARE OF PACKAGING SALES (U. S. SHIPMENTS—1990)

Corrugated & solid fiber boxes	25.1%
Folding cartons	9.0
Other paper, board and bags	21.4
Metals	20.5
Plastics	13.8
Glass	6.5
Wood & textiles	3.7

Source: Fibre Box Association, 1992

TABLE IV. PRODUCTION OF PAPER AND PAPERBOARD (UNITED STATES, 1992; UNADJUSTED)

	Thousands of Tons	**(%)**
Paper and board	82,383	100.0
Paper*	40,405	49.0
Containerboard	29,318	35.6
Boxboard	12,660	15.4

* Includes newsprint, printing and writing, tissue, packaging and other papers.

Courtesy of American Forest and Paper Association.

(1987) U. S. Bureau of Census figures on the value of various package products. Table II shows the value of packages printed by various packaging processes. Of the $73 billion worth of packages sold in the U. S. in 1990, some $58 billion were printed, of which $37 billion worth were printed by flexo. Table III shows that 25 percent of all packaging consists of corrugated and solid fiber boxes as well as the percentage of other types of packages. About half of all paper fiber produced goes into corrugated and folding cartons (table IV). Differences among tables I, II, III and IV reflect differences in the products defined as packages.

Owing to the methods by which the Bureau of the Census data are collected and reported, they do not always agree well with data from other sources.

Chapter

Packaging: Yesterday, Today and Tomorrow

CONTENTS

LIST OF ILLUSTRATIONS

Chapter

Packaging: Yesterday, Today and Tomorrow

THE PLACE OF PACKAGING IN OUR SOCIETY

Functions of the Package

A package must do three things: it must protect the contents, promote the product and inform the customer. A fourth function, convenience, is closely related to promotion. Convenient packages promote sales. We buy cereal in single-service boxes and insecticides in aerosol cans not because of the economics but for convenience of use. A package that does not protect the contents may result in a loss of the product. A package that does not promote the product may put that product out of business.

Packaging is an essential part of modern life. It reduces waste and the cost of products by protecting goods against damage and spoilage. It promotes public health by reducing diseases transmitted in the handling of foods. Environmental concerns have become a high priority consideration in the design of packaging.

The package must tell the consumer what is in the package, but modern food packages convey far more information than merely a statement of the contents. Through pictures and printed information, the consumer can learn about the benefits resulting from the product's use, the manufacturer or distributor of the product, its color or pattern, how to assemble it, and any special precautions about its use. If it is a food product, printing also tells about the ingredients, the quantity, its storage life, nutritional information and how to prepare or serve.

Printing is not only a means of identifying the contents and a reflection of a corporate identity, but it urges the retail customer to buy the package. Printing often represents the greatest single cost component of a package bought by a packager, but if it fulfills its functions, the cost of printing is fully justified. If the printing does not fulfill its function, both package and contents may be lost. Generic packages, whose appeal is based on the appearance of economy, have had small impact on the food industry, demonstrating the importance of print to purchasing decisions.

Packaging and Waste

There are sound economic reasons for keeping the amount and costs of packaging to a minimum for the protection and sale of goods. There is no point in wasting money on unnecessary packaging. Packages eventually become part of the solid landfill problem, but the real question is their net impact on waste. There is a drive to reduce the amount of packaging or to eliminate it in North America and Europe. Since the forces are not market driven, they are unpredictable. There are many proposals before the United States Congress and various parliaments. Consumers aren't organized, but environmentalists are.

Efforts of both industry and government to correct packaging's alleged deficiencies (particularly in the environmental arena) often suffer from narrow approaches that may be counterproductive in the long run. These efforts, however, will provide incentives to develop improved packages that will harmonize with recycling and avoid other environmental problems. It is unlikely that consumers will return to the inconveniences experienced prior to modern packaging technology.

Packaging is an essential part of modern life. Testin and Vergano* write: "Today, less than three percent of Americans live and work on farms. The fact that these few Americans can feed more than 240 million Americans and millions more overseas is the result of the successful development of agricultural technology and the development of the distribution systems and packaging necessary to avoid spoilage and wastes."

They continue: "Packaging is an essential part of the public health, standard of living and lifestyle in the United States and other societies. Without packaging, life as we know it would quickly grind to a halt; goods would rot on farms, docks and in warehouses, becoming infested with insects and other vermin; food poisoning and disease transmission would increase and product losses would likely result in an increase in total wastes. The cost of packaging is significant—some $70 billion per year, yet as a percentage of the cost of goods it is quite modest, about 7% on average."

The industrialized countries of the world have about six percent of their population engaged in production of foodstuffs, and the less industrialized coun-

* For references, see "Further Reading" at the end of each chapter.

tries, with less packaging, have much larger percentages of their populations engaged in the production of foodstuffs.

Packaging not only reduces spoilage of products, but it often reduces municipal solid waste. Typically, 1,000 chickens produce about 1,650 pounds of waste feathers, viscera, heads and feet. The use of about 15 pounds of packaging for 1,000 chickens allows the 1,650 pounds of waste to be made available for byproduct uses. Because of modern packaging, the butcher does not throw away fat and bone from the beef. They are available for secondary products such as soap, fertilizer and animal feed.

Food and beverage packaging accounts for about two-thirds of the $70 billion packaging industry in the U. S. Food spoilage in the U.S. is less than three percent for processed food and 10–15 percent for fresh food. In lesser developed countries, food spoilage due to inadequate or non-existent packaging can reach 50 percent. Without protective packaging, our modern food distribution system would cease to exist, and our living standards would plummet.

One effect of packaging in modern distribution systems has been its direct linking of the producer and consumer. Products that are packaged, but unbranded, are taking up a smaller and smaller portion of supermarket shelf space. Fresh poultry is now almost universally branded. Branded fresh fruits and vegetables are becoming the norm. A recent innovation has been the wholesale/retail warehouse outlet that sells a vast array of goods ranging from appliances to zucchini. These new approaches to distribution would be impossible without taking full advantage of new technologies for protection and printing for selling.

Historically, promotion was carried out by a sales representative or clerk, but now the package has become the means of promoting the product at the point of sale. The orange crate and the cracker barrel served only as containers for shipping and dispensing the product. They had little, if anything, to do with promoting sales of the product or making it convenient to use. In fact, the product often had to be transferred to another package before it could be used.

A BRIEF HISTORY OF PACKAGING

Gourds and hollow tree trunks, animal bladders and skins were used before mankind learned to make jars or boxes. Leaves are still used as packaging in primitive societies. Ceramic pots for wine and cosmetics have been found in the ruins of ancient Greece and Rome and in sunken ships from Biblical times. They were often labeled by marking the name of the contents or the proprietor in the ceramics before firing.

Packages were originally designed by the merchant who had products to sell. It was not until the late Nineteenth century that an independent packaging industry developed. The earliest package printing was the printing of labels which seem to have incorporated the best available printing. As it does today, package printing through the centuries has always incorporated the best available printing.

Modern packaging can be traced to around 1550 (about the time the printing press was developed) when labels for packages of handmade paper were printed on a sheet of paper made by the manufacturer. A hundred years later, vendors of medicinal powders packaged their products in paper wrappers marked with their signatures.

The end of the Eighteenth century brought two inventions that made possible the modern packaging industry: the paper machine, and lithography. The Nineteenth century saw the birth of the packaging industry, an industry that made and decorated packages for sale to other manufacturers who used them to protect and promote the goods they produced. Growth of packaging was promoted not only by packing and filling machines, but by national corporations made possible by growth of the railroads and steamships. Without modern transportation, large corporations that can produce and distribute large quantities of packaged goods would not exist.

The development of plastic materials for packaging and the invention of flexography together with the evolution of gravure and offset have dramatically changed packaging since the late Nineteenth century (figure 1-1) and early Twentieth century (figure 1-2).

Development of packaging in the U. S. is neatly summarized in four illustrations supplied by Anheuser-Busch Companies, Inc. (figures 1-3 to 1-6). (Figure 1-3 is seen in the color insert, page 169.)

Labels

The earliest known American package is a label or wrapper for paper produced around 1800 at the Luke Bemis paper mill in Watertown, Massachusetts. It was printed from a simple wood block. By 1888, lithographed multicolor cigar labels cost $50 for 1000 labels, and the *New York Sun* commented that the label was

Figure 1-1. Late Nineteenth century packages.

Figure 1-2. Early Twentieth century packages.

often better than the cigar. By this time, packagers began to recognize that a good looking package could actually be influential in selling the product.

Labels from the late Nineteenth century are shown in figure 1-7.

Folding Cartons

What Americans call paperboard or cartonboard is called cardboard by the British. This product is not to be confused with containerboard or corrugated board. Paperboard boxes, in use by 1800, were very flimsy structures. They were made in Philadelphia before 1810, apparently by immigrants from England. They were either labeled or printed upon directly. Robert Gair, in the 1890s, in the United States, invented a boxmaking machine that made possible the mass production of folding cartons. The in-line printing and carton-making machine of Louis Chambon was viewed by many of the millions of visitors to the Universal Exhibition in Paris in 1900. Chambon's machine printed carton board in colors in register, and it slit and creased the board.

Two folding cartons of the late Nineteenth century are illustrated in figure 1-8.

Corrugated Containers

Corrugated containers are made from corrugated board which is also called containerboard. It comprises one or more fluted boards adhered to flat liners. The simplest structure is one fluted layer adhered to one liner or facer. The commonest structure is one fluted layer adhered to two liners.

It was not until 1914 that U.S. railroads allowed goods packaged in corrugated to be shipped by rail, and truckers soon followed. By the end of World War II, corrugated had virtually replaced wooden crates for all but the heaviest products,

Figure 1-4. Beer bottle and label used from 1891 to 1908. (Photo courtesy of Anheuser-Busch Companies, Inc.)

Figure 1-5. Wooden box for transport of beer bottles, used in 1914. (Photo courtesy of Anheuser-Busch Companies, Inc.)

such as large machinery, and for farm produce. About one third of all the paper fiber produced in the United States goes into corrugated board.

In recent years, printers have learned ways to decorate corrugated containers by wrapping them in printed wrappers, or by coating the rough kraft liner and printing it either before (preprinted liner) or after the box is formed. The rough, brown kraft paper that contributes to strength is not suitable for printing good quality color work.

Paper Bags and Flexible Packaging

The first paper bags in England were made by E. S. & A. Robinson in 1844. In 1900, the modern function of flexible packages was mostly fulfilled with paper bags (figure 1-9). The first commercial plastic film was invented in 1912 by Dr. Jacques Edwin Brandenberger and named Cellophane. Aluminum foil was introduced about 1910, some 25 years after Charles M. Hall developed an inexpensive method of separating aluminum from the ore. Improved paper has enabled paper bags to maintain a large market in spite of the excellent packaging qualities provided by film and foil.

Metal and Ceramics

The earliest tinplate was made in Bohemia in the Sixteenth century. Lead snuff boxes are recorded in use in England before 1800. It was easy to adhere paper labels to these cans, but transfer printing (in which the ink was applied to paper and transferred to the metal) and offset lithography were soon used to print the metal. The principal metal, in the early Nineteenth century, as now, was tinplate, although now it is steel coated with tin, whereas the earliest tinplate was iron, coated with tin. The tin coating is now very thin and covered with lacquer. Several metal boxes from the late 1800s are shown in figure 1-10.

An early method for printing metal and glass was transfer printing. For glass, a specially prepared paper was printed from engraved copper plates or by lithography. It was transferred to pottery before it was fired. The colored metallic oxides fused into the pottery when it was fired. The transfer printing could also be applied to metal and fused. By 1850, lids for pharmaceutical bottles were printed by transfer color in as many as four colors.

Figure 1-6. Corrugated boxes for beer bottles, used in 1957. (Photo courtesy of Anheuser-Busch Companies, Inc.)

Figure 1-7. Labels from the late Nineteenth century, from the collection of Container Corporation of America. (Photo courtesy of Robert P. Long.)

PACKAGING TODAY

Every commercial printing method is used today for package printing: flexography, gravure, offset lithography, letterpress, screen, and even nonimpact processes.

Packaging and the Environment

The environment, both inside and outside the printing plant, will continue to demand increasing attention from the package maker and the package buyer. The subject is discussed at length in Chapter 12.

Disposal of packaging involves several problems: litter, recycling, toxic waste and availability of landfill. Unsightly litter causes unfavorable public reaction, and some groups attack the packaging industry for plastic, glass bottles and metal

Figure 1-8. Folding cartons and paperboard and metal boxes of the late Nineteenth century. (Photo courtesy of Robert P. Long.)

cans that litter the nation's roadsides. Consumers and the industry cannot ignore these problems.

Recycling is receiving growing attention, partly because land available for landfill is becoming increasingly scarce and expensive. Packagers will continue to use more recycled materials and materials that are capable of being recycled. This will affect both package design and printing since the nature of recycled materials is often different from the properties of the virgin materials.

Material that is sent to landfill must be free of products that can pollute the groundwater. In packaging, this comes mostly from printing inks and coatings. Although inkmakers are working on formulations that will meet government requirements, the package printer has the ultimate responsibility for any toxic printing inks or other toxic materials that may be buried.

The printer is no longer permitted to discharge evaporated ink solvents to the atmosphere. Disposal of evaporated solvents and products printed with toxic inks are principal environmental concerns of the printer. Change from solventbase to waterbase inks is easing the environmental pressure on the flexographer and gravure printer just as elimination of isopropyl alcohol reduces the environmental impact of offset lithography. However, total elimination of the volatile organic compounds (VOCs) remains a technological challenge.

The safety of workers also affects the design and printing of packages. OSHA has developed a complex group of regulations involving physical, mechanical, electrical and chemical safety within the printing plant. Regulations affect the design of printing presses and the chemistry of printing inks.

Waterbase and Soy Inks

Waterbase inks are impractical in offset lithography, but they are rapidly growing popular in flexography and gravure. Many problems are involved, but the driving force is the attempt to reduce the amount of petroleum products and other VOCs discharged to the atmosphere. New solutions create new challenges: waterbase inks reduce VOCs, but they dry more slowly, creating a different set of challenges.

Figure 1-9. Turn-of-the-century flexible packages made of paper, from the collection of Container Corporation of America. (Photo courtesy of Robert P. Long.)

Flexo easily prints waterbase inks, which are gaining market share. The volatile organic compounds (VOCs) evolved from solventbase inks are potentially explosive and hazardous to health. When discharged into the air they create atmospheric problems so that they must be either recaptured or burned. Waterbase inks have many additional advantages: reduced risk of price fluctuations with changes in oil prices, reduced plant insurance costs by reducing the hazard of fire, washup with water, ease of disposal, little or no swelling of rubber plates, virtual lack of odor. The best waterbase inks nearly match solventbase inks in printability, drying speed, gloss, heat resistance, shelf life and stability.

Use of inks based on soy bean oil also reduces VOCs. While these are not entirely new, there was a great surge of interest in soy inks in the early 1990s. They offer some advantages over petroleum inks. One of the most attractive is that they are readily degraded by organisms in the environment.

Radiation Curing Inks

Inks cured by ultraviolet (UV) or electron beam (EB) radiation contain

Figure 1-10. Metal boxes and a wooden cigar box from the late 1800s, from the collection of Container Corporation of America. (Photo courtesy of Robert P. Long.)

minimal volatile materials or VOCs. They are used in flexo, offset lithography, letterpress and screen printing. With suitable radiation equipment, the inks can be cured in line at high speeds. The cured inks are hard, glossy, resistant to physical and chemical damage, and they are used to make very attractive packages. Their use grew rapidly during the 1980s, and growth seems likely to continue.

These inks require special deinking facilities, and if waste containing UV or EB inks is mixed with waste printed with conventional inks, deinking becomes even more difficult. The volume of drinking of UV/EB printed packaging is currently too small to support commercialization in the U.S. When enough UV/EB volume is generated, it is likely that commercial deinking will become a reality.

Off the Shelf—The Retail Warehouse

Although the largest fraction of packaging is used for food products, the growth of off-the-shelf retailing provides new opportunities for packages. Bicycles and lawn mowers, requiring minimal assembly, are packaged in compact corrugated containers. Small hardware items such as nails and house numbers are sold in blister packs that reduce pilferage and carry bar codes. The optical scanner reads the bar code at the checkout counter and informs the central computer of its removal from stock in addition to adding the item to the customer's bill. The package actually reduces the cost of moving the product from manufacturer to the ultimate customer.

A trip through the retail warehouse will still, in the late Twentieth century, reveal many opportunities to improve the graphics of packages displayed on shelves. The item is often displayed nearby, so that colorful graphics may be considered redundant. Nevertheless, the history of retailing points to the use of increasingly colorful packaging. Statistics on sales of inks show that color printing is growing continually; other studies show that most of the growth involves process color.

Growing Use of Color

The growth of color and graphics on packages has paralleled the growth of the supermarket and off-the-shelf retailing. Even the old corner grocery had some colorful labels and cereal boxes, but as the supermarket grew, so did the use of printed color, often process color. As the discount mart replaced the general merchandise store and the dry goods store, the package replaced the retail clerk as the principal purchasing motivator for the customer.

Spot color (simple, flat colors) is slowly being replaced by process color (full color reproduction of color photography). Although process color allows an almost infinite variation in hues and tones, it also requires strict attention to details of quality control because dozens of variables from photography to final printing and viewing affect the appearance of color. Even worse than failure to match the original is the variability that occurs in process color when production is not carefully controlled.

Desktop Publishing

Techniques of desktop publishing that permit the copy, complete with graphics, to be assembled on the computer, then printed on the final product, have been adapted to the design and printing of the graphics for packaging. Desktop publishing avoids the lengthy processes of photography, stripping of film and proofing that is part of the expensive process of getting an image ready for printing. Driven by the need for just-in-time manufacture, the package printer has put to use computerized and digital techniques for image preparation and for die cutting and finishing functions. The packaging world has totally married the disciplines of desktop publishing with computer assisted design (CAD) and computer assisted manufacturing (CAM) to automate graphic and structural design and manufacturing. The direct printing from digital data is also used to code and mark packages, permitting automation in pricing and inventory control.

The speed of desktop publishing (better called "electronic preparation" in package manufacturing) offers opportunities for quick turnaround for printing of special offers and new prices onto the package. The package printer often places a flexo unit at the end of the gravure press to handle variable information, but, some day, ink jet printing will make it practical to apply variable printing so that each box can be printed individually as lot codes and sequential numbers are now printed.

Faster and Better Proofing

The long, slow, expensive process of preparing and sending press proofs ("wet proofs") to customers to be marked up for further dot etching (changes in the color separations) has been replaced, for most needs, with prepress or off-press proofs. Now, digital or electronic proofs are replacing many of these. The digital proof has many advantages due to the free access to special colors not readily available from offpress systems which were designed for process colors only.

Much of the delay in accepting digital proofs results from reluctance of the industry to change its practices, but there are also some difficult technical problems that remain to be solved. New proofing systems are a feature of every major graphic arts show.

Shorter Runs

Increased market size or market share promote longer runs, and run length does increase for some packages. However, there are far more forces promoting shorter runs: just-in-time packaging operations, price changes, and marketing programs that promote special offers. Improved printing technology and marketing plans and practices conspire to reduce run length.

Marketers like to pinpoint their targets and push for shorter runs to meet the smaller targets. Nowhere is this more evident than on the supermarket shelf. "The mass food market has splintered into many niche markets. Quaker Oats, for example, used to sell one type of oatmeal. Today, it markets three types and 12 flavors of oatmeal, and the types and flavors vary by region of the country.

"The emergence of niche markets for food consumed at home can best be seen at the local supermarkets. More than 10,000 new food products were introduced in 1990, five times the number of new products a decade ago according to the Food Marketing Institute.

"Products aimed at the mass market are now being overtaken by products aimed at specific consumer segments. From Campbell soup to McDonald's hamburgers, food companies are aiming at smaller market niches, a strategy that requires more careful product development and marketing.

"Three forces are behind the recent shift to smaller food market niches: a new emphasis on nutrition, changes in the American life-style, and changes in demographics. Together, these forces translate into strong consumer demands for a greater variety of healthier, more convenient foods."*

These forces drive the development of technology, especially the computer and printing technology, that makes short runs economically practical. This trend especially challenges gravure printing whose strength and special advantage lie in long runs.

DIFFERENCES BETWEEN PACKAGE AND PUBLICATION PRINTING

Flexography vs Offset and Gravure

A striking difference between package and publication or commercial printing is the relative use of flexography and offset lithography. Some 64 percent of package printing is done by flexo with offset and gravure making up most of the rest (see table II). In publication and commercial printing, offset makes up some 80 percent with gravure supplying nearly all of the rest. Only a small portion is done by flexo. Offset and gravure are older processes than flexo and are better understood. Accordingly, flexo technology is changing faster, and the future of flexography depends more upon technical advances than does the future of the better-developed processes.

As flexo has developed, its quality capability has become competitive with both offset lithography and gravure. Combined with cost advantages, its development has allowed flexo to take work from both. The trend shows no sign of letting up.

Role of Printing in Publications and Packaging

Printability of paper is its most important property in commercial printing. Packaging materials must perform many other functions as well. Accordingly, the manufacturer of publication printing paper can concentrate on its performance on the printing press and in the hands of the reader. Magazine or book paper is designed to support the printing.

With packages, protection and promotion of the product are at least equal

---

* Barkema, Drabenshott and Welch: Federal Reserve Bank of Kansas City's *Economic Review* for May/June 1991.

in importance to printing. The design of the package cannot be subservient to the design of the graphics.

The manufacturer of packaging materials must be concerned with many other properties. Although film and glass and metal have many attractive properties, package printing presses have been designed to produce an image on them, while paper has been designed to accept an image from the publication printing press. This presents a challenge to the package printer which has been met with ever-increasing success. If the magnitude of this issue isn't clear, imagine the challenge of printing Saran or polyethylene with a waterbase ink. The materials are essential to many packaging requirements, and their surfaces must be made suitable for printing with water.

Attitude of Printers and Packagers

Package printers usually consider themselves to be converters or package manufacturers, and the printing is only a part of their function. Publication printers think of themselves as printers, and they view printing as their chief concern.

Different packaging materials require different printing techniques, each of which requires specialized expertise. Even though printing of the various packaging materials is in many ways more difficult than the printing of paper, the package printer cannot devote his entire attention to it. Close attention must also be directed to adhesion, slip, rub, creasing, retained solvents, slitting and diecutting, all controlled at the same time and on the same machine.

Number of Package and Publication Printers

There are more than 50,000 publication and commercial printers in the United States but no more than 8,000 package printers. This means there are more opportunities to find a publication printer, but this also means that there are more people experienced in publication printing. The fields of publication and commercial printing have a larger infrastructure: more experts, more books, more seminars and more research.

NEW MANAGEMENT PROBLEMS

The last quarter of the Twentieth century has added some technical challenges to management. Failure to meet challenges of total quality management and just-in-time manufacture or delivery, or failure to conform to health, safety and environmental regulations, can become just as fatal to a business as can failure to manage finances, sales, production and the workforce. Technical problems of quality, health and the environment are similar in all commercial and manufacturing operations, but some aspects are unique to package printing.

These subjects, which are relevant to packaging—today and tomorrow—are discussed in Chapter 12.

PACKAGING TRENDS

Over the centuries, the technology of package printing has closely paralleled the technology of publication printing, and it seems certain it will continue to do so. Major trends are toward better, more efficient manufacturing procedures and greater control by society and government.

Electronic processing of digital data will continue to reduce the time required to move from the original design to the final printed package. Most of the reduction will be in prepress time, the time required to get a plate or cylinder on the press and ready to run, although some printing of bar codes and lot number codes will be done directly from the computer.

New techniques of distribution, such as off-the-shelf merchandising or the retail warehouse, still offer expanding opportunities to the package printer—increased sales resulting from more color, better graphics and coatings at economical prices.

New and developing technology will make possible packages that are more attractive and convenient, that provide better protection, and that cause fewer environmental problems. Printing will continue to grow more colorful and more appealing.

We are quickly learning to reduce the amount of solid waste that we produce and to recycle it more effectively. Recycling of printed packages will become less troublesome as new packaging materials and recycling techniques are developed and present facilities are expanded. The management of environmental problems, especially those of waste disposal, involves not only technology and economics but political processes.

We can expect more legislation and regulations that restrict packaging as people turn to government for solutions to economic and social problems. Improved technology and design will make it possible for packagers to comply with the new regulations.

Evolving technology, better management techniques and training programs make it possible to improve product quality and service while keeping costs low. While the statement: "you can have any two—better quality, faster delivery, lower price—but not all three" applies over the short run, over the long run improved technology, training and management do make all three possible.

FURTHER READING

Packaging in America in the 1990s: Packaging's Role in Contemporary American Society—The Benefits and Challenges. Robert F. Testin and Peter J. Vergano, Clemson, SC. (1990). Available from Council on Plastics and Packaging in the Environment, 1001 Connecticut Ave. NW., Suite 401, Washington, DC 20036.

Package and Print. Alec Davis. Clarkson N. Potter, Inc., New York, NY. (1967).

The Package as a Marketing Tool. Stanley Sacharow. Chilton Book Co., Radnor, PA. (1982).

Package Printing. Robert P. Long. Graphic Magazines, Garden City, NY. (1964).

Chapter

2

Packages, Packaging Materials and How They Are Printed

CONTENTS

LIST OF ILLUSTRATIONS

LIST OF TABLES

Chapter

2

Packages, Packaging Materials and How They Are Printed

INTRODUCTION

Packages are made of wood, paper, paperboard, flexible film, rigid plastics, metal, glass, or textiles and various combinations of them.

Much of the wood used for paper and paperboard manufacture is grown on large farms, particularly in America's Southeast, where wood has replaced cotton as the most important agricultural crop. Some pulp is also made from byproduct sawdust from sawmills, particularly in the Northwest.

Paper can be recycled, and recycled paper has many desirable qualities. It is smoother and often more opaque than paper made from virgin pulp. Recycled paper and board often contain foreign matter, particularly in Europe. For paperboard packages, recycled board that is not deinked can be coated with clay or white paper to produce a fine packaging material. Recycled corrugated can be coated in different ways to provide excellent printing properties. Because recycling reduces the length of the fibers, the board is not as strong as virgin board, but the printing properties may be as good or better.

Most plastics come from petroleum, which is a nonrenewable resource, but supplies are sufficient to last for many decades. Cellophane and glassine, which are sometimes competitive with plastic film, are produced from wood pulp.

Ores for aluminum and steel are plentiful, although as the best ores become exhausted the cost of obtaining pure metal increases. The supply of sand for glass is effectively inexhaustible, but not all sand is suitable for glass.

Every important printing process is used to print packages and packaging materials. Many things dictate which printing process is to be used on any job, but cost is always a major factor. Flexography is the leading process for package printing. It is used in almost every major packaging market: flexible packaging (paper, film and foil), multiwall bags, corrugated and preprinted linerboard, labels and wrappers, folding cartons, beverage carriers and cans. Gravure is also used on many different kinds of packages, especially labels, plastic film and cartonboard. Offset lithography, which is the leading printing process for publications, is not as well suited to plastics and foils so common in packaging, but it is used extensively for printing labels and folding cartons.

Marketing considerations often determine the printing process. If run lengths are short, flexo or litho is favored over gravure. Direct printing of corrugated has an advantage over preprinted liner if monthly repeat runs vary in length, and if print lengths cannot be determined in advance, especially when they tend to be short. Direct print is more responsive to the market needs, while preprint may be more responsive to the desires of the package buyer.

Package printers should make sure that the finished package is compatible with the product it will contain. Labels for liquor bottles must withstand alcohol lest there be breakage during transport. Ink on soap boxes must not be discolored by the soap. Non-skid coatings applied to multiwall bags must be compatible with the inks used for printing the bags.

Printing and coating must be compatible with the plastic, paperboard, foil or other material selected to meet filling, shipping, storage and selling requirements. Some packaging papers, for instance, are acidic. Although this may not hurt the package contents if it is a plastic toy, it may discolor the ink and generate a negative message to the consumer. Similar papers made by different mills may accept ink differently, and inks adhere better to some plastics than to others. Corona treatment of polymer surfaces to promote ink adhesion sometimes fails.

In addition, inks must meet legal requirements for child safety, for human consumption, and for other purposes. The package printer should be aware of these requirements. Although this is the ultimate responsibility of the packager, the package printer's knowledge can avoid aggravation and enhance business. If the printer actually purchases the substrate, he has total responsibility.

Package printers should be able to perform significant tests. The moisture content of paper or paperboard can be critical in ink acceptance, register and the ability of the printed packages to lie flat. The surface tension or wettability of the plastic substrate must be controlled if ink is to wet and adhere to the surface. For paper and paperboard, the porosity of the substrate is vital. Porosity helps set the ink, but it affects dot size or dot gain and can even cause bleed through into the product.

The same rules that relate to the substrate and the ink also relate to laminating adhesives, coatings and all other materials applied to the surface or surfacds of the packaging materials.

Printing presses are discussed in Chapter 8. They are mentioned here only as necessary to support the discussion of the processes by which packages are printed.

Diagrams of the kinds of presses used to print packaging materials are presented in figure 2-1. Consumption of printing ink is a measure of the volume of printing of various types of packaging (table 2-I). Flexible packaging, corrugated and folding cartons comprise about two thirds of all package printing.

Flexography continues to gain market share. More rapid progress is inhibited by the outdated perception that flexo is a low-cost process that does not do high quality work.

FLEXIBLE PACKAGING

Flexible packaging is produced by printing and converting paper, film or foil, alone or in combination, for use in consumer and industrial applications. A more comprehensive definition of flexible packaging includes casting, extruding, metallizing, coating , printing, embossing, slitting, laminating, folding, sheeting, or heat sealing of flexible or semirigid materials, converted from films, foils and paper.

Flexible packaging is the oldest form of packaging known to man. Animal skins, soft barks and large leaves were used by primitive man in ages gone by as they are today. A hundred years ago (and even more recently), old newspapers were used to wrap fish and many other groceries. In fact, British fish-'n'-chips are still customarily wrapped in newspaper.

Social and economic factors have stimulated new technology in the flexible packaging industry. The reduction in the size of the average family unit, the rising number of older people, and the increasing number of single-person households have created a need for convenience foods and single portion servings and small package sizes. These stimulate new product design, new materials and new manufacturing processes.

Trends in Western Europe and Japan vary from those in North America. Flexography has the dominant market share for flexible packaged products in the U.K. On continental Europe, gravure is estimated to have 55 percent of the market, flexography with 40 percent and the remainder going to offset and letterpress. In Japan, gravure dominates over flexography by at least a 2:1 margin. The reasons for the disparities are not related to the technologies but to historical accounting procedures.

The Products

Flexible packaging includes a diverse group of products such as candy wrappers, snack bags, bread wrappers, grocery bags and multiwall bags. For our purposes, we shall also include shipping sacks and single service products such as drinking straws that are sometimes grouped in other classes (figure 2-2).

Flexible packaging firms are classified under several standard industrial classification (SIC) codes (see table I in the Introduction), and there are many trade and technical associations.

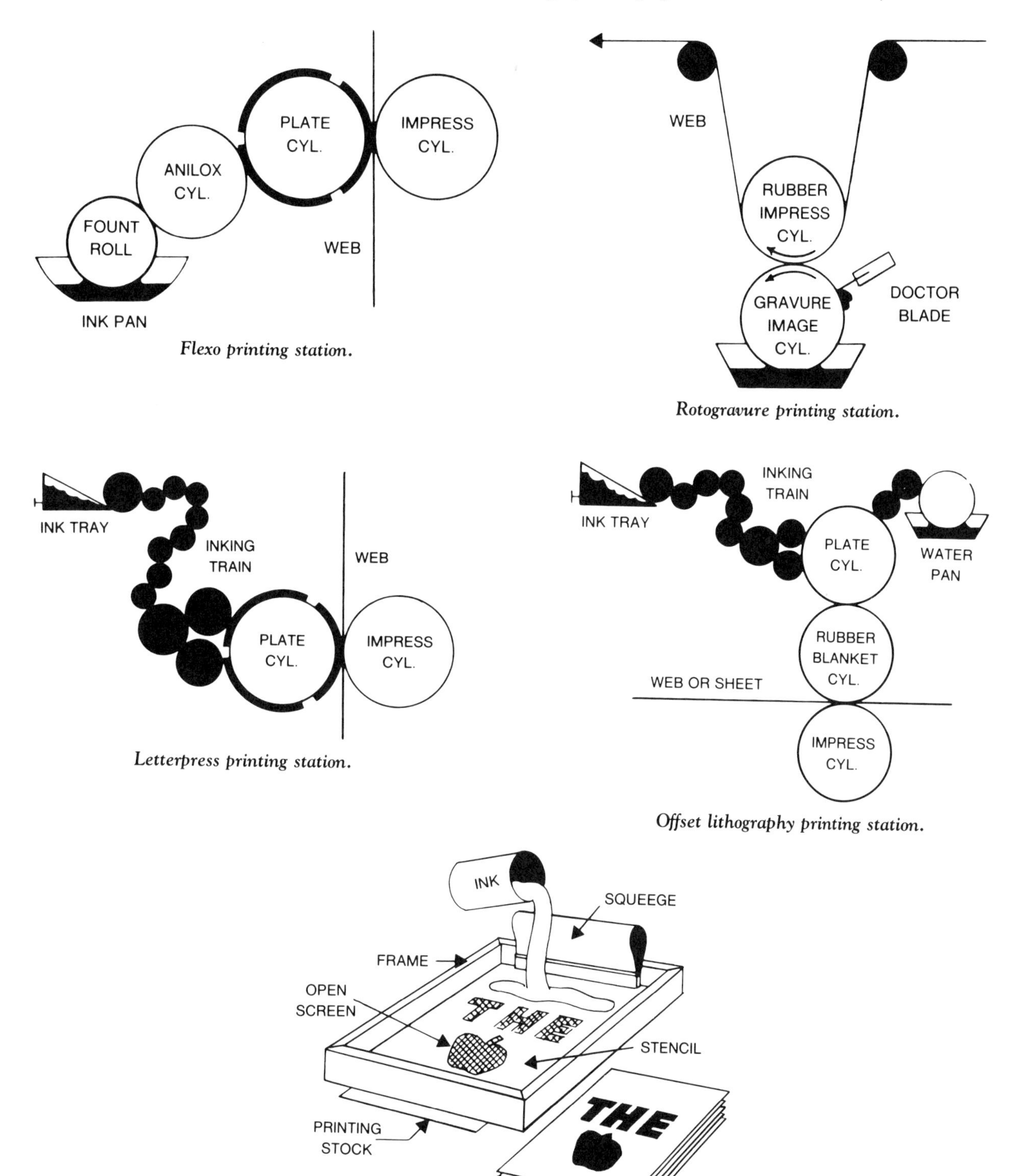

Flexo printing station.

Rotogravure printing station.

Letterpress printing station.

Offset lithography printing station.

A typical screen process.

Figure 2-1. The printing processes. (Source: Flexography: Principles and Practices, Fourth Edition.)

TABLE 2-I. VALUE OF PRINTING INKS IN PACKAGING (U.S. 1990)

	Value	
	($ millions)	(%)
Folding Cartons	134.1	14.5
Corrugated	141.0	15.3
Flexible Packaging	349.9	37.8
Others	300.2	32.4
	925.2	**100.0**

Source: Hull & Co.

In the early 1990s, conventional flexible packaging sales approached $10 billion in the U.S. Most of this is the cost of materials. The value added by manufacture (the actual printing and converting operations) averaged only 15 percent of the total. Many plastic bags and wrappers such as garbage bags and sandwich bags are unprinted. Growth during the 1980s was 10 percent annually, of which inflation represented roughly 6.5 percent and real growth about 3.5 percent. In addition to the above (shopping and grocery), paper bags and shipping sacks accounted for an additional $2 billion.

Materials for Flexible Packaging

Materials used for flexible packaging include many types of paper and plastic film and all sorts of combinations.

Paper. Many types of paper (kraft, waxed, oiled and glassine) are used (table

Figure 2-2. Flexible packaging.

TABLE 2-II. PAPER SUBSTRATES FOR FLEXIBLE PACKAGING

Pouch papers
Glassine and greaseproof
Waxed paper
Parchment
Bleached kraft paper
Bleached sulfite paper

Source: GATF: TechnoEconomic Market Profile No. 13: Flexible Package Printing

TABLE 2-III. PAPER STOCKS USED IN LAMINATIONS

Clay-coated
Tissue
Parchment
Glassine
Supercalendered
Pouch papers

Source: GATF: TechnoEconomic Market Profile No. 13: Flexible Package Printing

2-II). Paper is relatively inexpensive, it has a good strength-to-weight ratio, good opacity and is easily printed. It is made from a renewable resource and is easily recycled. Paper can be printed by all printing processes, but it must be calendered or coated for satisfactory printing by gravure or letterpress which require very smooth printing surfaces.

In spite of the growing variety of plastics, paper usage remains strong, especially multiwall shipping sacks, grocery bags and high quality retail carryout bags.

Waterbase inks are used more extensively for printing of paper than for film. In printing of flexible paper products, flexo uses only waterbase inks, gravure is estimated to use about seven percent waterbase inks, and litho does not used waterbase inks.

Paper is commonly used in combination with other materials (figure 2-3). Various laminating grades of paper have been developed to provide specific characteristics (table 2-III).

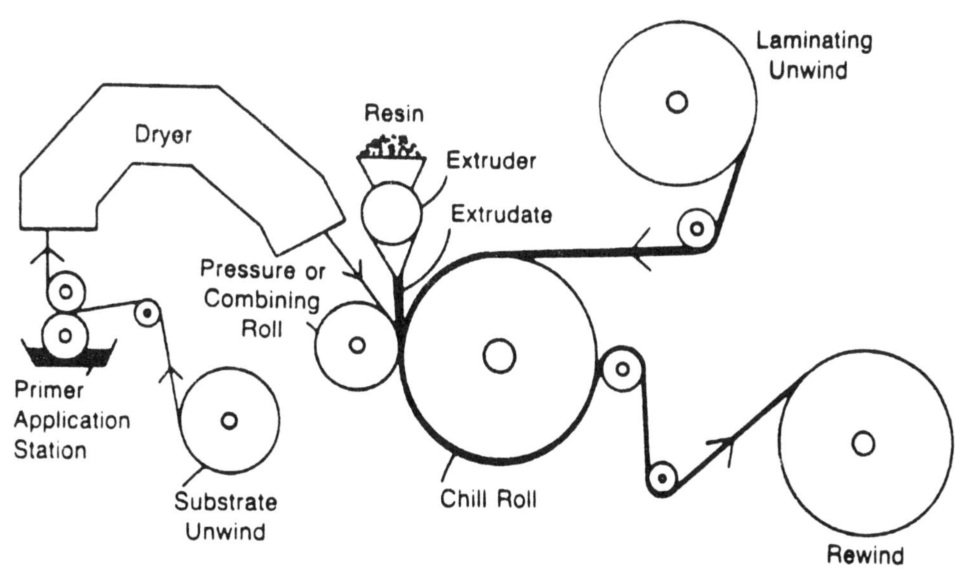

Figure 2-3. Sketch of extrusion laminating process. (Source: Paper, Film and Foil Converter, November, 1986.)

TABLE 2-IV. FILMS COMMONLY USED FOR FLEXIBLE PACKAGING

Cellophane: nitrocellulose coated: PVDC coated, uncoated
Nylon: Uncoated, treated, PVDC coated, metallized
Polyethylene: Linear low density, low, medium and high density
Polypropylene: EVA modified, non-oriented, oriented, oriented metallized, acrylic coated, PVDC coated.
Foil: Uncoated, shellac coated, nitrocellulose coated, vinyl coated.

Source: GATF: TechnoEconomic Market Profile No. 13: Flexible Package Printing)

The primary objective of laminating is to enhance the barrier and mechanical properties of a package for use in specific applications, often in food packaging. Laminations provide barriers against moisture, oxygen, light, odor and flavor. Laminating a printed sheet with a clear film provides locked-in printing with high gloss and scuff resistance.

Films. Paper is slowly giving way to the more cost-effective plastic films that can meet the wide range of properties that package buyers require (table 2-IV).

The first plastic film, cellophane, a nitrocellulose film, has seen a dramatic decline in use, but many millions of pounds are still used annually throughout the world. It is easy to print by all printing processes. Cellophane coated with polyvinylidene chloride has much better barrier properties than the uncoated material.

Since the introduction of polyethylene around 1950, a great variety of plastic films has been developed. Plastic consumption grew in double-digit figures annually during the 1960s and 1970s, but slowed in the 1980s. The polymers most commonly used for making films and other packages are listed in table 2-V.

Coextrusion is the process of combining two or more different polymers into a single multilayered film (figure 2-4). As many as 10 or more different layers may be used to give the required properties to the package: strength, opacity, sealability, printability and a barrier to moisture, air and odors.

Coextrusions have increased rapidly as a share of all flexible packaging materials (paper, film and foil). Even before 1980, bakery products and snack foods were packaged in oriented, coextruded film. Coextrusions amounted to over half of all plastic film consumed in 1985 with trash and other bags, stretch wrapping and medical packaging accounting for roughly 80 percent of coextruded film. This proportion will continue to rise in the foreseeable future.

A typical coextrusion used as the inner layer in snack packaging is a three-layer structure: two layers of high-density polyethylene (HDPE) and one of ethylene vinyl acetate (EVA), using the sequence HDPE/HDPE/EVA. A double layer of HDPE provides good levels of moisture barrier, stiffness and strength. In addition, EVA adds good printability and more than adequate sealant properties.

It is only recently that any material except metal or glass has been used to preserve heat-processed foods because no other material could withstand processing temperatures that range from 212° F for such acidic foods as juices and

TABLE 2-V. SYNTHETIC POLYMERS USED IN PACKAGING

BON	Biaxially oriented nylon
BOPET	Biaxially oriented polyester
BOPP	Biaxially oriented polypropylene
EVA	Ethylene vinyl acetate copolymer
EVOH	Ethylene vinyl alcohol
HDPE	High-density polyethylene
LDPE	Low-density polyethylene
LLDPE	Linear low-density polyethylene
MDPE	Medium-density polyethylene
OPP	Oriented polypropylene
PA	Polyamide (nylon)
PC	Polycarbonate
PE	Polyethylene
PET	Polyethylene terephthalate (polyester)
PP	Polypropylene
PVC	Polyvinyl chloride
PVDC	Polyvinylidene chloride (Saran)

Source: GATF: TechnoEconomic Market Profile No. 13, Flexible Package Printing

fruits to 250° F (120° C) (15 pounds steam pressure) for such low-acid foods as meats and poultry products. Today, film and foil pouches are routinely used for "boil-in" bags for processed foods filled with meats, vegetables and fruits.

Foil. Aluminum foil tends to be very thin gauge, usually in the range of 0.00025–0.002 inch (0.006–0.05 mm). It is used with a variety of paper and film substrates to give desired combinations of strength, barrier properties and appearance. Foil is sometimes coated with heat-sealable coatings.

Foil is susceptible to attack by high-acid foods, and thin films are susceptible to pinholes and flex cracking. Metallized papers (papers with a thin foil or metal coating) are used for labeling beverages and other containers, but they have been largely replaced by paper-foil laminates.

Metallized films are capturing markets in flexible packaging, particularly in the snack food market where the metallized film offers an excellent background for bold graphics that enhance sales. Metallized plastic film has lower cost and improved barrier capabilities and better resistance to tearing and puncture than foil.

Ink. Flexible packaging consumes about 40 percent of the ink used in all packaging (table 2-I). Virtually all of this is fluid ink (flexo and gravure.) Waterbase inks are gaining an increased share of the market for printing flexible packaging. They are generally accepted for printing on paper and gaining share in printing on film. Coatings and adhesives are important specialties for flexible packaging manufacture. They are discussed in Chapter 10.

Dollar figures for ink consumption do not provide a good measure of the value of packaging produced by different printing processes. They record quantities of ink consumed, not the value of the product or the amount or value of the substrates on which they are printed. The ink is usually a small fraction of the cost of the package. Where extensive color is involved, ink consumption rises.

How Flexible Packaging is Printed

Flexography. Flexo and gravure dominated the printing of flexible packaging in the early 1990s, with roughly 80 percent of the output. Offset lithography, letterpress and screen printing will continue to lose market share to the two dominant processes. Flexography is generally preferred for printing on film for two reasons: the central impression press needed for film printing is best suited to flexo, and the cost of gravure preparation demands longer runs and makes changes prohibitively expensive. Flexo process color has become competitive with gravure quality for many applications. Its ability to print on imperfect surfaces is a major factor. With the continuing trend toward reduced run lengths and its ease of handling waterbase inks, flexo is expected to continue to grow faster than gravure.

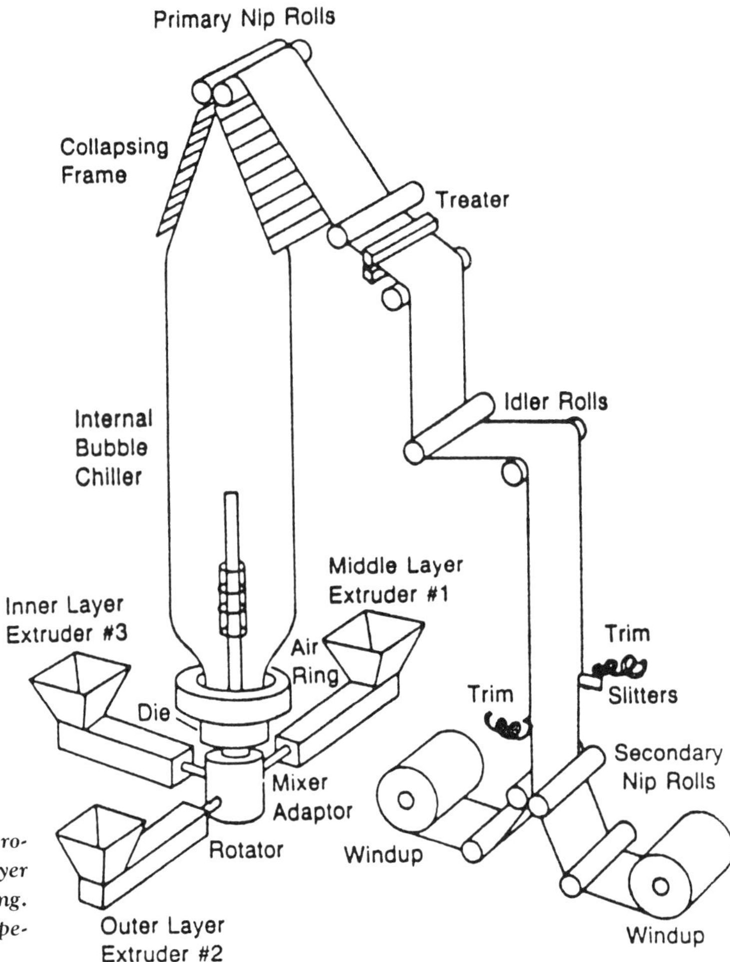

Figure 2-4. Coextrusion process producing a three-layer film for flexible packaging. (Source: Packaging Encyclopedia, 1988.)

Foil is more easily printed by flexo than by gravure. Gravure requires a compressible substrate; and foil laminates are easily printed by gravure.

Flexo can economically print orders from a few hundred packages up to a few million, a range that covers most of the market. Waterbase flexo inks print well on many flexible packaging substrates. Some polymeric films, notably polyolefins (PE and PP), require special treatment if they are to be printed with waterbase inks.

Gravure. Modern developments have helped to keep gravure competitive: electronic engraving (which has replaced carbon tissue), offpress proofing, press instrumentation and automation. However, waterbase inks do not have as great a market share with gravure as they do with flexo.

Gravure is preferred for critical printing of paper and foil. Gravure is also used on film for long runs. Long-run packages such as those for nonwoven diapers are typically printed by gravure, roll-to-roll at one thousand feet per minute on polyethylene film, then laminated to form a sandwich with the print on the inside. Polyester is often printed gravure. Most brick-pack coffee packages are printed by gravure on polypropylene or polyester, then laminated.

Offset and Letterpress. Offset and letterpress have virtually no share of printing of flexible packaging. Their inks are tacky, tending to pull or stretch the substrate, and offset has the added disadvantage of water balance on a nonabsorptive substrate. Sheetfed feeding of the substrates used for flexible packaging is close to impossible.

Printing and Converting Equipment

Flexible converting plants carry out five major operations: extrusion, printing, slitting, lamination and bagmaking, as well as some of the following operations: metallizing, coating, sheeting, embossing and sealing.

State-of-the-art flexo presses are capable of printing jobs with eight or even more colors at 1200 feet per minute while reducing setup time and increasing productivity. The new equipment encourages the switching of jobs from gravure to flexo. Preparation costs are lower because flexo plates can be modified more economically than gravure cylinders to handle such changes as mandatory ingredient copy to meet legal requirements, price changes or copy changes for new marketing programs. Some converters have not yet completely digested all that is available today. It will take some time for the more costly innovations to be fully absorbed into the flexible packaging industry.

In addition to printing, flexible packaging often requires the application of coatings. Coatings enhance not only the appearance of the package, they enhance its performance. Like printing with ink, the coating process involves the application of a liquid and its conversion into a solid. In the last decade, coating technology has developed in three major ways: equipment has given better control of coating thickness, more versatility in coater capabilities, and the introduction of radiation-curable materials.

Adhesives are much like inks and coatings. They are applied as a liquid and must be converted, at least to a semi-solid. Adhesives may be applied over the print, or they may be used to laminate the components of a multiply film together.

CORRUGATED BOARD

Corrugated boxes are made from corrugated board, sometimes called containerboard, a product manufactured by laminating flat sheets of paper to a corrugated inner layer called "corrugating medium." The resulting structure gives the product a high stiffness-to-weight ratio.

In 1990, U.S. production of paper and board was 80 million tons—almost exactly half paper and half board. According to the Fibre Box Association, corrugated production in the U. S. was 24.4 million tons in 1991, equal to some 320 billion square feet, and sales were $15.4 billion.

Corrugated containers and other products have gained an excellent reputation in the industry and among environmentalists as being highly recyclable and therefore less likely to end up in landfills, compared to plastics, which degrade very slowly and some of which are harder to recycle.

Trends include increasing sophistication of buyers, primarily large consumer products companies who have learned the cost/quality/value relationships between the print processes. They are driving the change toward improved graphics. The rest of the market has enormous potential. Market data show that the customer often communicates far less potential than actually exists.

The market for point-of-purchase displays on corrugated is exploding. Offset is used to print labels for corrugated boxes, while letterpress and flexo are used to print corrugated sheets directly.

Leaders in flexo printing of corrugated are working to reduce "washboarding" (the rough surface of containerboard produced by the corrugated medium inside), paying strict attention to substrate requirements and using recycled board when it is suitable. They are learning to use anilox rolls tailored to each print job, and are considering drying between print stations rather than just at the end of the machine line. Corrugated printers are developing specialty inks, especially high-pigmentation inks that improve print quality, developing specialty coatings that do not crack at the fold, and developing high-holdout substrates. They are studying registration changes that accompany changes in press speed and improving computer control of sensing and feedback systems for registration (instead of merely alerting the operator) and control of color. In quality printing, there is a trend toward divorcing the printing press from the diecutter. Perhaps, some day, technology may bring these units back together.

Leaders are investing heavily in CAD-CAM for both structural design and graphics. The marriage of the two allows some interesting "what if" iterations.

One method of increasing productivity and print quality is to print the liner before it is made into corrugated. The product is referred to as preprinted liner. The U.S. produced 6.3 billion square feet of preprinted linerboard in 1990. If the average box consumes 7.5 square feet, that's approaching a billion boxes. Customers are demanding more quality as they learn more about quality/cost/reliability relationships.

The point at which preprinted liner is more economical than direct print on board is probably around three to six rolls, depending on the complexity of the job. It is highly unlikely that direct print will consistently achieve the quality of preprinted liner. High print quality can be produced on webs as wide as 98 inches; most wide webs are around 60–70 inches.

Use of preprinted liner is growing rapidly because the higher level of graphics sells more products or raises the profit margin of the products. The U. S. market for preprint grew 55 percent annually from 1982 to 1990. Preprinted linerboard will continue to grow as printing technology improves and demand for high quality graphics increases.

Much of the pressure for improvements in direct printing has come from market pressures generated by the high quality of preprinting the linerboard. As a result, early estimates that preprint would account for as much as eight percent of the corrugated market have been reduced to six percent—still a very considerable amount of the total current market of about 320 billion square feet. This would be about 20 billion square feet when the preprint market reaches its full potential early in the Twenty-first century.

The Product

To manufacture corrugated board, the corrugating medium is corrugated, adhesive is applied to the tips of the corrugations on one side and the corrugated medium is merged together with the liner. The top liner is supported with a rigid roll to form a reasonably smooth printing surface. This product, called singleface board, is useful for some applications (such as wrappers for light bulbs), but it is usually glued to a second piece of linerboard, the bottom liner, to form the

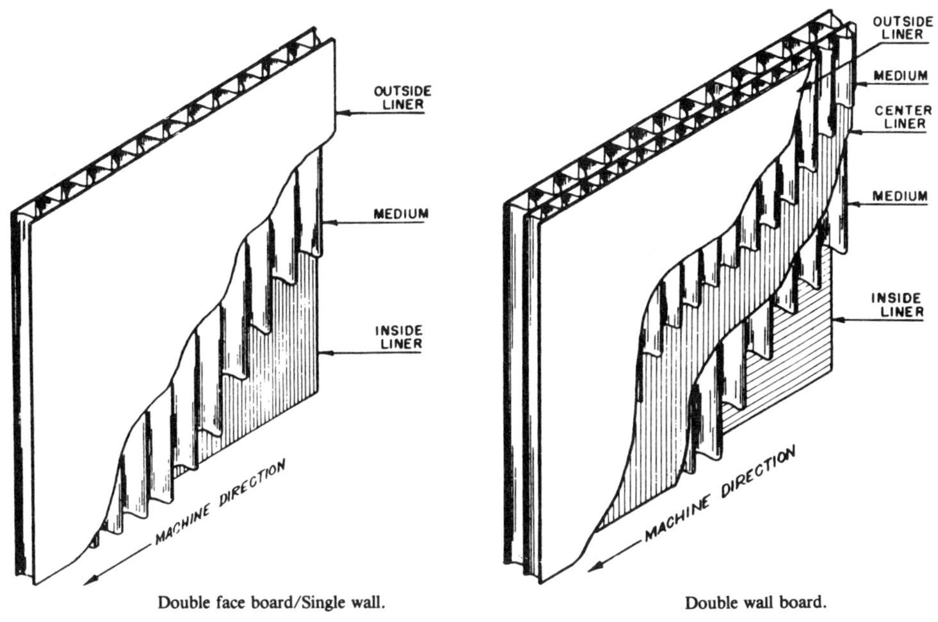

Double face board/Single wall.

Double wall board.

Figure 2-5. Corrugated board.

TABLE 2-VI. DESIGNATION OF CORRUGATED FLUTES

Designation	Height (inches)	Flutes Per Foot
A-Flute	0.184	30–36
B-Flute	0.097	44–50
C-Flute	0.142	36–42
E-Flute	0.062	86–94
F-Flute	0.045	92–100

Height does not include thickness of liners
Source: Fibre Box Handbook

basic corrugated board. The bottom liner must not be forced against the singleface board with enough force to crush the corrugations or the package will lose strength. The bottom liner sometimes winds up with a washboard effect that cannot be easily printed (figure 2-5) (table 2-VI).

Corrugated is printed by several different processes. Flexo and letterpress are used to print directly onto flat corrugated sheets. This is called "direct print." Preprinted liner is printed chiefly by flexo before the board is manufactured, and litho-laminates or labels are printed by offset (figure 2-6).

Average runs are becoming shorter and shorter. The demand for high quality printing is present for small orders that cannot justify preprint. Direct print and labels are the alternative to preprint, with labels offering superior quality at much higher costs.

Figure 2-6. Litho-laminated corrugated box.

Shipping containers are sometimes designed so that they can also serve as P.O.P. (point-of-purchase) displays, and boxes are often opened in order to display the products at the supermarket. The stiff boxes make possible the stacking of virtually any product—light bulbs and power tools, toys and grocery products.

Large "solid fiber" boxes are used for products such as returnable beer bottles. They are generally made by the same companies that make corrugated containers.

Materials for Corrugated Board

The primary raw material used in the corrugated packaging industry is containerboard, a composite wood pulp product made of corrugating medium and linerboard. Linerboard is usually made of unbleached, softwood kraft fiber while corrugating medium is often made of a semichemical pulp.

The outside layers (liners) of corrugated board are generally made from unbleached softwood fibers. The kraft or sulfate process produces a strong brown sheet used to make the liners, or walls, of a corrugated board. Bleached sulfate is used where a good printing surface is required. Mottled white, another mill grade, is produced by putting a thin layer of bleached pulp on top of a layer of brown pulp and compressing them during production. This improves the printability of the board. A clay coating gives a still better printing surface.

Corrugating medium is the fluted material sandwiched between linerboard. It is made from recycled corrugated containers or from virgin hardwood fibers by semichemical pulping.

Color sells, and the corrugated box can be a great selling tool. However, the rough surface and uneven absorptivity make corrugated a difficult surface to print. Furthermore, the brown color interferes with color reproduction, and inks used to print on brown linerboard or boxes must be opaque. Higher grade linerboard, either mottled or white coated, helps to reduce these problems. To produce top-quality printing on corrugated requires a clay-coated, calendered surface, often used for preprinted liners.

The inks for high quality corrugated flexo printing are waterbase acrylics. Heat-resistant inks must be used on preprint to withstand the heat of the corrugating process. In low-cost corrugated, ink is used only for marking and identification of shipping boxes.

Aqueous coatings are generally applied for non-skid purposes, but some are used for decorative purposes as well. The rigors of shipping often damage the shipping container or scratch the graphics unless protected with an overprint coating.

Adhesives used to manufacture corrugated board are mostly corn starch.

How Corrugated is Printed

Corrugated can be printed or decorated in many ways. Ink can be applied directly to corrugated board, linerboard can be printed prior to making the corrugated board, or labels can be applied to the box. Printing processes include flexo, letterpress, litho, gravure, screen and ink jet (table 2-VII).

About 10 percent of all corrugated boxes are unprinted, 50 percent are printed

TABLE 2-VII. METHODS USED TO PRINT CORRUGATED (1990)

	Direct	Preprint	Laminates	Labels	Total
Flexo	73	4	1	*	78
Letterpress	5	*	*	*	5
Litho	0	0	4	1	5
Gravure	0	*	*	*	1
Screen	*	0	*	*	1
Total	78	5	6	1	90

* less than one percent. The total comes to 90 percent because the remainder is unprinted.
Source: Industry Reports

in one color, 35 percent printed in two or more colors and no more than five percent are printed in process color, but these proportions are changing rapidly as multicolor printing grows. Printing satisfies three needs: product identification, sales promotion and description of the contents of the box. Printing to promote brand identification and sales can be simple or very elaborate.

The Campbell's soup logo, for example, is printed on corrugated containers for canned soup in a very stylized manner, even though the containers are seldom seen by consumers. A packaging buyer at a major brewery once said; "We use corrugated for case quantities of beer, but most beer is sold in six packs. Consumers usually do not see the corrugated shipping container, so there is little reason to spend money on its appearance. On the other hand, shipping containers provide a billboard for our name, and there are a lot of them. Since we have to label the carton anyway, we want the graphics to present our brand name as well as possible."

Graphics on corrugated are continually improving as a result of more sophisticated printing presses with improved inking systems and doctor blades, anilox rolls with higher screen counts and better platemaking materials. Thinner photopolymer plates have consistently provided better print quality, longer plate life, predictable dot gain, improved registration, better ink transfer from anilox to plate to substrate, less mounting time and less distortion when mounting larger plates. Laser-engraved rubber plates have the proven characteristics of the molded rubber plate—durability, stability and good ink transfer. The largest format laser-engraved rubber plate currently available is 48x105 inch, available in thicknesses from 0.067–0.250 inch. It is ideal for large one-piece display work. Its accuracy is guaranteed to ±0.001 inch, far better than the molded rubber and equal to photopolymers. Conversion to computer-generated art favors laser plates. The all-electronic flexographic platemaking system enables one to go from the computer or compatible CAD-CAM directly into the laser, eliminating the use of film. Electronic prepress outputs first-generation film, but laser-engraved rubber plates eliminate even this stage.

Most corrugated shipping containers carry the boxmaker's certificate (figure 2-7) as required by the railroad and trucking industries. U.S.D.A. requires a certificate for meat and poultry products that have been inspected. Stamping is not printing, and boxes that carry nothing but the stamped certificate are considered to be unprinted.

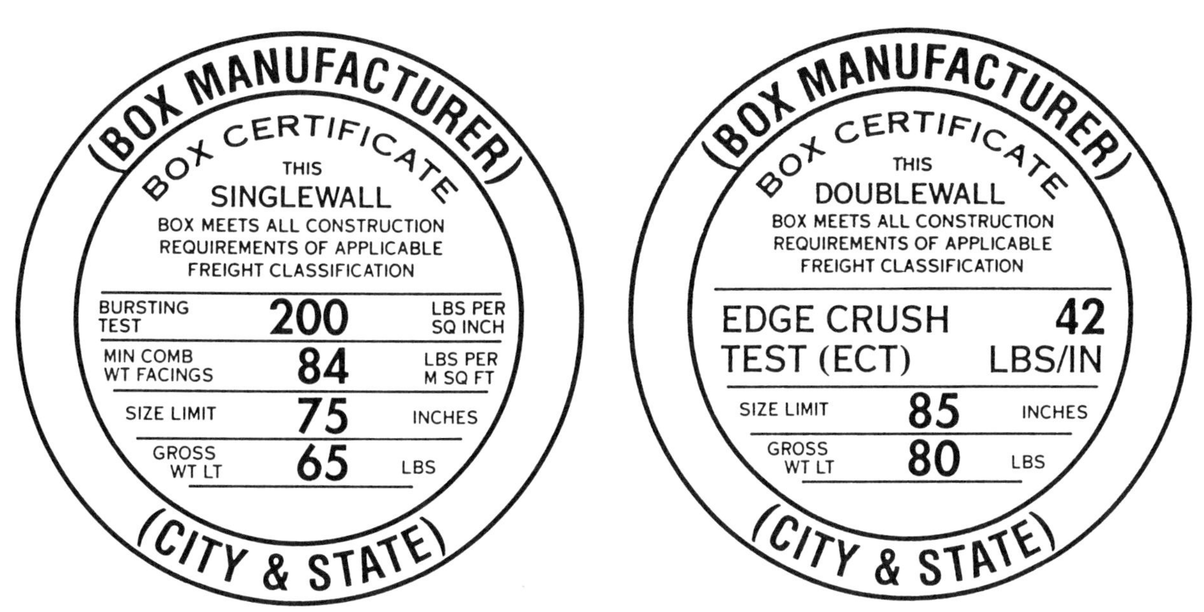

Figure 2-7. Boxmaker's certificate.

In several situations, corrugated shipping containers do not have to have a boxmaker's certificate. Corrugated boxes to be carried by the packager's own trucks need not carry a certificate, and highly decorated boxes that are packed in larger shipping containers likewise need no certificate.

Unprinted corrugated board is frequently used as partitions, slip sheets and other types of separators and dunnage in packaging. The remainder can be divided into four groups: direct printed, preprinted, labeled and litho-laminated.

Many corrugated manufacturers are unable or unwilling to provide superior graphics. They consider themselves to be boxmakers, not printers, and printing increases the cost of their product while decreasing productivity. This is changing rapidly as the value added by printing begins to represent most of the profitability of many corrugating operations. Companies are instituting printing training programs for their personnel, including non-production people. In some plants, the term "pressman" is now applied to the operator of the flexo folder-gluer.

Spot colors on bulk-produced boxes are commonly matched to the Glass Packaging Institute (GCMI) kraft color guide. Wide color variation can no longer be accepted. If ink properties are properly specified, only the viscosity need be altered by the printer. GCMI colors are one of several groups of standard colors used in the packaging and publishing industries. The subject is discussed further in Chapter 5.

Printing of corrugated is changing faster than any other printing industry. Use of thin flexo plates is an important contributing factor. Once 0.250 inch thick, plates are now 0.125 inch thick or thinner and backed with compressible polyurethane blankets. Process color now uses a screen of 45 to 80 lines per inch, but screens of 100 to 110 are being tried. This kind of work is often referred to as "high graphics" or "enhanced graphics."

Figure 2-8. Corrugated printer-slotter. This numerically controlled printer slotter has an inline rollout die cutter. (Courtesy of The Ward Machinery Co.)

Direct Printing. Direct printing refers to the printing of corrugated board that has already been formed. It is applied on boxmaking machines in corrugated plants: printer-slotters or other machines (figure 2-8). About 80 percent of corrugated production is printed in this way. Direct printing is the practical, inexpensive process for short runs.

Direct printing of corrugated boxes is more than 80 percent flexo. The remainder, printed by letterpress, has declined to a low level, mostly for displays. Letterpress printing of shipping containers can be expected to disappear during the next few years.

On the downside, direct printing usually produces a lower print quality than other methods. While tremendous strides have been made toward improving print quality with direct printing, and more improvement is expected, the rough surface of corrugated board precludes truly superior print quality. Even if a better grade of linerboard is used, direct printing cannot match the quality possible with other methods. If the board has a washboard surface, print uniformity is hard to achieve.

Direct printing by flexo will remain cost effective compared to preprinted liner for medium- to small-run quantities for a long time. Direct print cannot consistently achieve the quality levels of preprinted liner.

Direct flexo printing is predominately one-color although some multicolor printing is done. About 60 percent of direct printing is one color, 30 percent two colors and 10 percent three or four colors.

Most new flexo folder-gluers are being sold with three or four printing stations, and some with even five or six stations. These flexo presses always have in-line configurations because corrugated board cannot be bent around a central impression cylinder or stack press.

Some of the stations can be set up while others are running, reducing makeready time. Premounting the plates also saves makeready time. With the cost of a six-color press ranging around two million dollars, makeready time becomes very costly.

Preprinted Liner. One way to get around print quality problems with direct printing is to print the linerboard before it is fed into the corrugator. This provides a smoother surface for printing. Also, the printing is done on presses and in plants dedicated to high-quality printing.

The main disadvantages of preprint are that it often creates more waste—and more expensive waste—and slower production speeds on the corrugator. Also, the setup time and startup waste required to prepare corrugators to run preprint make only very long-run work economical.

Nevertheless, preprint has some cost advantages over labels and litho-laminates. Preprinting is most cost effective for fairly long run work. For top quality graphics, laminates are good: for medium or lower, preprinted is more expensive than direct printing, although preprint is becoming comparable to offset in both quality and line screen counts.

Preprint conventionally prints 120 and 133 line screens, across a press width of 98 inches. Quality is in every way comparable to offset, as is shown by the submissions to the annual FTA Awards Competition. Quality of this sort demands clay-coated board.

Flexography is by far the most common method of preprinting linerboard. The presses are designed for higher quality printing than the printing units normally incorporated into boxmaking equipment. These presses are capable of printing four or more colors in comparison to the one- or two-color printing capability of most boxmaking equipment. Also, the anilox inking system, the web handling systems and the controls on these presses allow greater control of the printing process. With rapid advances in flexo technology, further improvements can be expected.

Gravure is also used to print linerboard, providing superior graphics and color, but run lengths required for economical operation limit it to a few jobs.

Litho-Laminates and Spot Labels. Preprinted liner is a web process. The liners are delivered to the corrugator on a roll. Laminates are usually printed by sheetfed offset, and they are applied to singleface board after the box has been cut. They usually provide higher quality graphics than flexo, but the difference is becoming narrower. The best flexo is comparable to good quality offset.

The use of offset lithography is so pervasive that this method of decorating corrugated is called "litho-lamination" even when another printing process is used.

Occasionally, spot labels are applied to a finished box to provide decoration.

Litho-laminating of corrugated is essentially a short-run operation. It takes advantage of the quick makeready and low cost of offset for runs shorter than can be economically produced on preprint. With the rapid advances in quality of direct printing, litho-laminating should yield market share to the lower cost of printing and converting in a single operation. The lost market share may be compensated for by increased decoration of corrugated packages.

Printing Processes

Flexography. Flexo plates are softer than letterpress plates, permitting them to print on a surface as rough as softwood kraft. Flexo, well established in the '60s in paper bag, labels and flexible packaging, started making inroads into the corrugated market, and in 20 years became the preeminent printing process for corrugated, for both direct and preprint operations.

A major advantage of flexo and the reason it became the dominant method on corrugated board is that the quick drying flexo inks permit inline scoring, slotting, folding and gluing all on one machine—the flexo folder-gluer. All flexo inks used on corrugated are waterbase.

Thin photopolymer plates are helping to further improve corrugated graphics. These plates were at first largely limited to narrow web, preprinted linerboard and flexible packaging markets. They are now used for all levels and grades of flexo printing, and give a cleaner, more consistent printing for premium grades. Plate sleeves of nickel and other synthetic composite materials make it possible to store mounted flexo plates for repeat jobs on preprinted corrugated liner. This greatly aids in meeting demands of JIT customers while permitting much shorter production runs.

Five factors appear responsible for making flexo the dominant printing process for corrugated:

- variable repeat length capabilities
- development of the electronic or direct drive corrugator cutoff knife that cuts precise blank lengths
- photopolymer printing plates
- preprint liner presses with multicolor printing and coating stations
- waterbase flexo inks that can withstand the heat of the corrugating process (350° to 400° F: 175°-200° C) and offer crisp, bright colors demanded of quality graphics.

High print quality can be produced on presses as wide as 98 inches, although most are 60 to 70 inches wide. The fastest operate at speeds up to 1,000 feet per minute. Preprint presses can print up to 10 colors or even more.

Gravure. Gravure is capable of very high printing speeds and print quality. Good quality gravure printing requires a smooth, compressible surface such as clay-coated paper. It is unsuitable for direct printing because the high print pressure would crush the box. It is used for preprinting linerboard on one U. S. press, several presses in Europe and many in Japan.

Gravure is capable of printing an image the size of several large detergent boxes with eight or 10 colors, including the necessary heavy laydown of fluorescent inks, on long runs. The job may include a top coating to protect the ink and produce high gloss and slip resistance.

Offset. Offset lithography is used for litho-laminates and labels. It is capable of the highest print quality. Labels and laminates to be applied to corrugated are printed primarily by sheetfed offset.

Some E-flute is printed directly by offset, generally where this fine flute has replaced folding cartons for packaging a specific product. Offset printing of E-flute is not a major factor in the North American marketplace, but it is widely practiced in Europe and in Japan. Europeans are also using microflute corrugated which has smaller flutes and is still smoother.

Web offset, with its fixed cutoff, is unsuitable for preprinting the wide variety of sizes of corrugated packages that are used. Flexo is making gains, but the best quality graphics on labels or laminates are still produced by lithography.

Other Printing Processes. Letterpress is now printed mostly from photopolymer plates using glycol-base inks. Its use has declined dramatically with the popularization of flexography and the advent of the in-line flexo folder-gluer. However, many small sheet plants have nothing but letterpress equipment. While the number of such plants is declining, it will be many years before all of this equipment is converted to flexography.

Screen printing is slower than the other printing processes, but it is capable of very heavy ink coverage. It can be an economical printing process for short-run work. It is a minor process for corrugated printing but well suited for specialty applications such as display work.

Printing and Converting Equipment

Digital drive technology is replacing mechanical control. Great benefits from electronics come from integrating the electronic systems completely with the mechanics of the press to yield a CNC (computerized numerically controlled) flexo press.

Direct printing is one of the normal steps in creating a corrugated box. While not all boxes are printed, corrugated boxmaking machines usually embody printing units. Printer-slotters are widely used to produce conventional corrugated shipping cases (figure 2-8). The box is printed and then scored and slotted and moved as a flat blank to a separate folding and gluing machine. Printer-slotters are either letterpress or flexo. A more sophisticated system is the flexo folder-gluer (figure 2-9). Here, printing, scoring, slotting, folding, gluing and stacking are handled in a continuous operation. The process is suitable for all run lengths.

A great deal of the printing and converting equipment used to manufacture corrugated containers is combined into sequential operations on a single machine. Such is the flexo folder-gluer, for example, which accepts a blank that has been

Figure 2-9. Flexo folder-gluer. (Courtesy of Bobst Group Inc.)

previously scored on the corrugator, and then prints, slots, creases, die cuts, glues, folds, counts, stacks and even ties counted bundles. Flexo printer-slotters and letterpress printer-slotters, on the other hand, accept previously scored blanks from the corrugator, print, slot and crease them and deliver a flat (but converted) blank to a stacker. Folding and gluing to make the finished box is carried out in a subsequent operation.

For some types of corrugated packaging, (e.g. trays, Bliss boxes), conversion takes place in the field where produce is picked or packed, or in the customer's plant. Trays carrying products such as juice cans can be shrink wrapped and stacked.

FOLDING CARTONS

The Products

Folding cartons are packages made of boxboard that are folded flat during shipment and set up at the packaging line. Typical folding cartons include everything from small toothpaste cartons to jumbo detergent and cereal boxes. Rigid or set-up boxes (discussed in a later section of this chapter) are not folded; they are shipped already set up or are manufactured at the packaging plant. Set-up boxes include the familiar "Whitman's Sampler" candy box.

Most consumers already have a favorable environmental image of paperboard packaging. A survey showed that 83 percent of consumers in 1990 said that paper packaging is easy to recycle. An antimoisture coating is no obstacle to recycling.

The growth of self-service and supermarket merchandising in the consumer marketplace has forced the carton user to place ever-greater emphasis on design and decoration of the package. This has brought about a rising demand for more process color printing in carton decoration.

Advertisers believe that many people equate color usage with high quality. As improving technology has made the costs of color more reasonable, carton users have increased their use of color. Competition among carton end users for shelf space in merchandising outlets will continue to promote use of color (figure 2-10). Today, a carton converter requires five- six- and even seven-color presses to meet the demands of the typical carton buyer. Such presses commonly possess an additional coating station for applying aqueous, acrylic or UV coatings.

Although run lengths are decreasing, gravure still plays a major role in printing of cigarette cartons (a large percent of them for export), cereal boxes and detergent boxes.

Web offset printing of folding cartons is growing. There may be 20 wide webs (30 inches or greater) in the U. S. for printing folding cartons. Flexography also is making inroads into folding carton production, particularly in press widths of 20 inches and less.

Materials for Folding Cartons

The basic raw material of the folding carton industry is paperboard (or, more specifically, boxboard) (table 2-VIII). Three grades of pulp are used to make

TABLE 2-VIII. SUBSTRATES FOR FOLDING CARTONS
Solid bleached kraft or sulfate (SBS)
Unbleached kraft (SUS)
Recycled paperboard
Patent coated

Source: GATF TechnoEconomic Market Profile No. 12: Folding Cartons

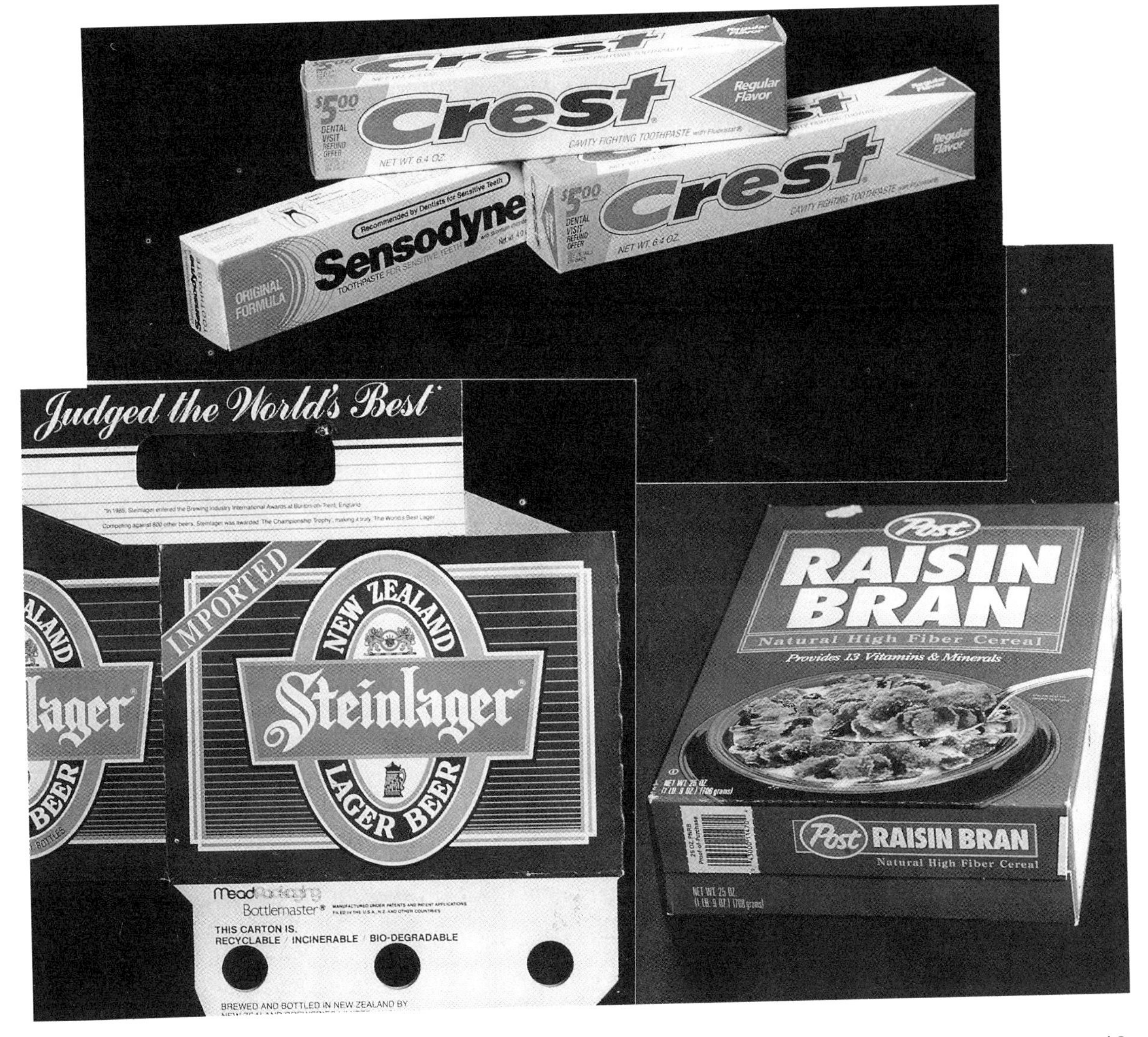

Figure 2-10. Folding cartons.

paperboard: solid bleached sulfate (SBS), solid unbleached sulfate (SUS) and recycled paperboard.

"Basis weight" is the unit of measure used for the weight of the board per unit area in the U. S. For paperboard, this unit is expressed in pounds for either 1,000 or 3,000 square feet of board. Outside of the U. S., basis weight is expressed in grams per square meter, commonly referred to as "grammage." The relationships between basis weight and grammage are noted in table 2-IX.

TABLE 2-IX. CONVERSION FROM BASIS WEIGHT TO GRAMMAGE

Product	Basis	Basis Weight	Grammage
Paperboard*	1000 sq ft	100 pounds	488.2 g/m^2
Paperboard	1000 sq ft	20.5	100
Bag Paper	3000 sq ft	100	162.7
Bag Paper	3000 sq ft	61.4	100.0
Writing Paper	(17 × 22)	20	75.2
Book Paper	(25 × 38)	50	74.0

Source: TAPPI Test Methods

* Paperboard is often referred to in terms of "points," that is, thousands of an inch in thickness.

Some confusion in terminology exists within the folding carton industry with regard to the term "paperboard." Paperboard is the general term for the materials used for folding cartons, corrugated boxes and solid fiber boxes. The material used for folding cartons is boxboard (which is also used for set-up boxes). However, the term "paperboard" is commonly used instead of "boxboard" by carton converters when referring to the board converted into folding cartons, and it is used that way in this chapter. The correct term for board used in corrugated and solid fiber boxes in containerboard. One also hears the term "cardboard" used in Europe for packaging boards. In the U.S. this term is used only by people unfamiliar with the industry.

SBS is usually given a clay coating on one side to create a smooth printing surface. (Since it is made of bleached fiber, the board is white on both sides.) Its white surface provides an excellent medium for high-quality printing. Its single-ply composition lends itself to efficient diecutting. It is less susceptible to damage from humidity than many other grades of paperboard.

Unbleached kraft (SUS) is usually found in single-ply, but it can be of multi-ply construction. It is used in cartons for heavy products that require extra tear and wet strength, such as those for hardware items and bottle or can carriers. Unbleached kraft paperboard includes clay-coated unbleached kraft and bleach-lined unbleached kraft-backed. (Coated SUS is brown on the back side.)

Recycled paperboard is made from fibers from various grades of paper stock. Recycled includes unlined chipboard (which is un-deinked and looks gray), kraft-lined, white-lined and clay-coated boards. Lined or coated boards are easily distinguished by the difference in appearance of the two sides. Bending chipboard is the least expensive grade of paperboard. It is composed of recycled fibers and is used for cartons that require little or no printing quality or board strength. Paperboard is easy to recycle, and coated board has been recycled for years.

Ink consumed in printing folding cartons is 45 percent offset, 46 percent roto, 5 percent flexo and 5 percent letterpress as shown in table 2-X. By volume, this represents about 16 million pounds of lithographic ink and 30–35 million pounds of gravure ink; around five million pounds of flexo ink and two to three million pounds of letterpress ink.

In 1990, the packaging industries purchased $925 million worth of printing inks, or 35 percent of all printing ink sales (table 2-I). More than half of all flexo inks used for folding cartons are waterbase.

TABLE 2-X. INK CONSUMED IN PRINTING FOLDING CARTONS—1990 (BY DOLLAR VALUE)

	%	$
Offset Lithograhy	45	60.3
Rotogravure	46.3	61.6
Flexography	4.3	5.8
Letterpress	4.8	6.4

Source: Hull & Co.

Blister Packs and Carded Packages

One of the most important variations of the folding carton is the blister pack in which the product is packaged under a rigid plastic blister attached to a piece of printed boxboard. Blister packs are used to package a wide variety of small products such as sliced meats, small light bulbs, and screws and washers.

The boxboard is printed by offset litho then run through a special assembly machine that inserts the product and adheres the plastic. Blister packs and carded packages usually refer to the same product, but some carded packages do not have plastic blisters on them.

Printing of Folding Cartons

Folding cartons were once printed largely by sheetfed offset presses, but the trend has long been toward web presses: gravure, flexo and web offset.

Folding carton stock for packaging milk, orange juice and the like is sometimes referred to as sanitary food board. These packages are printed solely by flexo.

Printing and Converting Equipment

Printing Equipment. Presses for printing of cartonboard are not notably different from presses for label or publication printing. They are discussed in further detail in Chapter 8.

Most presses operated by carton converters are equipped with a wide array of accessories. Production-oriented accessories include such items as blanket washers, continuous deliveries and feeds, pin registers and speed clamps. Those that are more market-oriented include UV, EB and IR drying coaters. Some accessories are oriented toward improving quality performance: press mounted densitometers, computer controlled inking systems, and error detection systems for nonstandard products.

Flexography. Folding cartons are one of the few types of packages not dominated by flexo. Although flexo is capable of meeting the high quality graphics required in most folding cartons, offset and gravure are the chief printing processes, except for sanitary containers. While flexo is economically competitive against gravure on short to medium runs, it is making slow progress against offset. Flexo is widely used for inexpensive folding cartons printed in one color.

Flexographic presses currently used in the folding carton industry are similar in size and configuration to rotogravure presses. Flexo has the potential for serving a very large fraction of the present folding carton industry. Flexo is an economical printing process for shorter runs as long as the press size is carefully chosen with regard to the product mix it must produce. There is a trend to narrow web flexo presses that can be made ready quickly.

The preparation of the printing cylinders on a flexo press is far less time-consuming and less expensive than gravure cylinder preparation. This could change should photopolymer cylinder technology for gravure become a reality. Given the generally perceived higher quality of the rotogravure-produced printed product, flexography would then experience new competitive problems in the carton printing field.

Future press innovations in flexography will center upon production improvements, such as automatic registration and monitoring devices that make it easier to set up and run with less waste. Microprocessor technology will simplify press controls; computer controls and sensors will reduce makeready times and spoilage. New systems are being designed to guarantee registration, and rapidly improving printing plates will further enhance the competitiveness of flexo.

Improvements in rotary diecutting eliminate one of the advantages of sheetfed printing. Rotary diecutters on flexo presses are helping these become important machines in the manufacture of folding cartons.

Gravure. As in other printing markets, gravure is favored for long runs such as cigarette boxes and cartons and soap boxes. Gravure colors are bright, and the eight, 10 or even 12 units found on gravure presses can print process colors, several spot color and one or even two coatings. Gravure presses are often equipped with flexo units that are used to print features that may be changed frequently, such as the price and bar codes, and to apply spot coating. This is an example of taking the best from different processes.

Gravure presses used by the folding carton industry range from 15 inches to 56 inches in width. Like sheetfed offset presses, these presses are available in multi-unit configurations for both printing and coating. One of the major markets for folding cartons, the six-pack beverage carrier, has traditionally been printed by gravure, but new flexo presses to print these products have been put into operation in the 1990s.

Innovative design changes are being made to permit rapid job changes on rotogravure presses. These will help adapt this printing process to shorter run orders.

Offset. Sheetfed offset lithography is the most widely used printing process for folding cartons today. Sheetfed presses for cartons differ little from those used for commercial printing. Press size is, perhaps, the single most distinguishing difference. Those most commonly used by the folding carton converting industry today are between 44 inches and 63 inches wide, and some old presses as wide as 74 inches are still in use. Large, sheetfed offset presses are preferred in order to maximize the number of finished carton blanks per sheet (commonly referred to by carton converters as the number "up"). Sheetfed has the ability to economically print cartons of various sizes on combination sheets (sheets that carry different boxes).

Large presses are favored because of their higher productivity. The basis weight and dimensional stability of the board make large sheets more stable than paper of similar size.

Web offset, so favored in publication printing, is beginning to make inroads into the printing of folding cartons in spite of relatively high waste. The high speed of printing and ease of makeready make web offset practical for printing of standard-size cartons (such as half-gallon ice cream cartons or single serve cereal packages). Webfed technology permits printing, coating, diecutting, laminating and delivering a finished carton blank in one in-line operation. In sheetfed systems, in-line operation is difficult to achieve.

While various technological problems for carton converting need to be addressed further (i.e., the gap necessitated by the lockup mechanism of the offset printing plate and problems created by jobs requiring different repeat lengths), web offset appeals to many carton converters. The offset process is familiar to the industry; offset plates can be made easily and quickly in-house, and offset achieves a high level of printing quality on a wide range of surfaces.

In 1985, there were 1364 sheetfed offset presses in use by 404 carton converters in the U. S. In 267 of those plants, offset was the only printing technology used according to a GATF study. Only 300 flexo and gravure presses were used for printing folding cartons.

Since the 1985 figures were published, more web offset and flexo presses have come on stream. Web presses run at higher speed, and accordingly have a higher output than sheetfed presses.

Diecutting Equipment. The diecutting process is crucial in the manufacturing of folding cartons. In order to convert paperboard into a carton, it must be cut, creased and stripped of all waste to create a blank that is ready to be folded, glued and shipped.

Sheetfed systems often hold an advantage over webfed in diecutting. Sheetfed diecutters are capable of easily handling a wide variety of diecut configurations. If care is taken in layout, a webfed diecutter can diecut many of the configurations that can be handled by a sheetfed diecutter.

Web presses possess sophisticated feed units that unwind, pull and feed the web into the printing and coating units continuously. Once the board has passed through the printing and coating units, it can either be rewound, cut into sheets or fed directly into the diecutter. In the latter case, an automatic transitional device usually transforms the continuous web feed into an intermittent feed that is synchronized with the reciprocating movement of a platen diecutter (figure 2-11).

Diecutting machines are termed "presses." Their sizes are comparable with the sizes of printing presses. Diecutting presses used in the carton industry are of two types: platen and cylinder. Platen presses involve two flat platens. One platen (called the male or cutting die) holds the cutting and creasing die rules, while the other (called the female) holds the male die in register on the paperboard placed between them. Actual cutting and creasing is accomplished by pressing the two dies against each other.

The method by which the two platens are pressed defines the type of platen diecutter. In reciprocating diecutters, one platen remains in a fixed position while the other platen moves against it in a reciprocating motion (figure 2-11). In jaw platen diecutters, both the male and the female dies are hinged, and press in and out against each other in a motion that resembles the biting of an alligator's jaw, hence the name. Jaw type platen cutters can reach

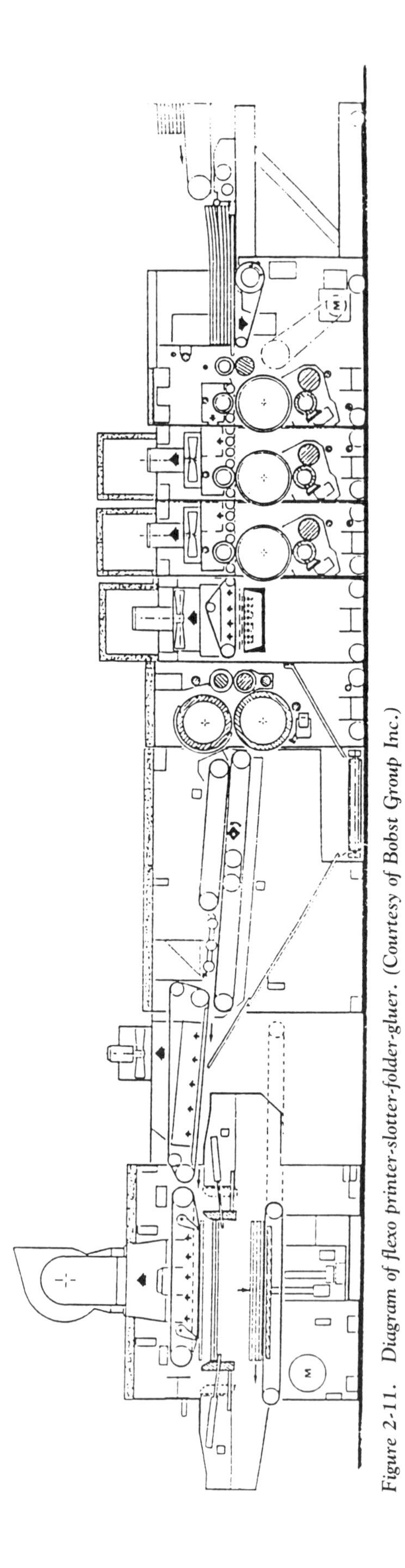

Figure 2-11. Diagram of flexo printer-slotter-folder-gluer. (Courtesy of Bobst Group Inc.)

Figure 2-12. Cylinder diecutting press. The roll-out die cutter is partially rolled out. (Courtesy of The Ward Machinery Co.)

speeds of 6000 sheets per hour. Reciprocating platen diecutters are inherently faster than jaw platen types, reaching speeds of 9,000 sheets per hour.

With cylinder diecutting presses, the cutting die is mounted on a flat bed which moves back and forth in unison with a rotating impression cylinder that carries the female die (figure 2-12). Each rotation of the cylinder produces a diecut sheet.

Platen diecutting presses are the type most commonly used today by the folding carton converting industry. They are mostly sheetfed machines, reflecting the traditional approach to carton converting.

Rotary dies are faster than flatbed types, and the trend is probably toward their use. The can be used for folding cartons and corrugated.

Some folding cartons are finished when diecutting and stripping are done; others require further finishing such as gluing, folding, coating, windowing, stamping and embossing. Gluing is usually performed after the prefolding or at the feed. Liquid glues or hot melts may be used (see Chapter 10).

Strippers and Blankers. In diecutting, the cut edges of the carton are sometimes left attached to the waste portion of the board by a number of small tabs. Following removal from the cutting press, stacks of carton blanks are manually separated from the waste board with an air hammer. This method of waste removal is labor intensive and generates additional waste. New diecutting presses automatically strip the waste board from the carton blanks after cutting and before stacking.

The "blanker" automatically strips the waste board. A blanker not only completely separates a carton blank from the waste board, it also counts and stacks the cut blanks for inplant packing and finishing. It is particularly efficient for producing carton products that do not require further finishing.

The latest available diecutting presses carry an increasing number of auxiliary features designed to reduce downtime and to increase efficiency. These include: premakeready preparation presses, preparation tables for waste strippers and blankers, equipment for automatically removing completed piles from the delivery unit and introducing new pallets without stopping the machinery of the press, and computerized automatic set-up capability that make possible job changes in 10 to 15 minutes (as opposed to the hours required by manual systems).

Off-line Coaters. Esthetic coatings on cartons are in high demand. A shiny, lustrous coating is often applied by coating units on the printing press. Coatings can, however, be applied to individual blanks as part of the finishing operation. Commercial copolymers are replacing the traditional wax as finishing coatings. UV, EB and other coating products are applied off line when production conditions make it practical.

Embossing and Stamping. Embossing and stamping are usually performed immediately before or after diecutting. Embossing is the pressing of a design into the substrate. Stamping is the adhesion of a pattern or stamp onto the packaging material. Converters often combine these with diecutting by using accessories attached to diecutting presses.

Foil stamping is used to decorate cartons that require high-quality visibility, such as perfume and cosmetic cartons. Foil stamping is considered as the ultimate in providing crisp, opaque and visually striking metallic colors. Silver, bronze or colored metallic foil is applied to the board by an embossing plate or die.

METAL CANS

Cans are made of metal, rigid board or plastic. If made of glass, they are referred to as "jars." Cylindrical containers made of plastic or fiberboard are discussed under the section on rigid containers. Those made of glass are discussed as bottles.

The Industry

Food products have been preserved in metal cans for two centuries. There are two types of metal cans: two-piece and three-piece, and they are printed by different processes (figure 2-13).

The three-piece can is made by forming the flat sheet of metal into a cylinder and sealing the side seam. The bottom is sealed to the cylinder, and after the can is filled the packager seals the top onto the can. A gasket in the flange provides a hermetic closure. Three-piece cans can be labeled or formed from printed metal.

Figure 2-13. Two-piece and three-piece cans. Hunt's three-piece cans are printed on flat sheets by offset lithography. Soft drink and beer cans are shaped first then printed by offset letterpress.

Two-piece beverage cans are made from aluminum or steel. A circle is stamped from a thick aluminum sheet, formed into a cup and then forced through a series of tooling dies that stretch and form the cup into a can. Cans made in this way are called "drawn and ironed" (D & I). The can is carefully cleaned and dried. A coat of lacquer is applied to the bottom of the can to protect the metal from abrasion. The can is next given an internal coat to protect the metal from the contents. The can is printed with a coating of titanium dioxide if a white background is required for the printing. The printed can is sent to the cannery, and closed with a single double-seamed end after filling. The process is now applied to steel fabrication for tin-free steel cans. D & I two-piece cans of both aluminum and steel now have over half the beverage can market.

Materials of Construction

"Tin" cans contain very little tin today. Coatings may be as thin as 0.00001 inch (0.00025 mm). This minuscule coating is thick enough to prevent corrosion of the steel underneath. But because tin cannot protect the steel from all products, the insides of many cans are coated with materials that cover slight imperfections in the tin coating and provide barriers for aggressive ingredients in some foods and beverages. Tinless cans made from steel alone are gaining a market share. They are uncoated in some cases or coated with an enamel or lacquer.

Some products, such as nuts and coffee, are provided with a plastic re-closure lid that may be printed.

Unprinted three-piece cans used for soups and other processed foods are often assembled at the cannery.

Figure 2-14. Cookie can printed in process color by offset lithography.

Figure 2-15. Shell of two-piece can, printed by offset letterpress, is ready for filling.

How Cans are Printed

Three-piece cans may be labeled or formed from printed metal sheets. Those that are printed are produced from flat stock that is printed by an offset litho metal decorating process. The printed sheets that once required up to 20 minutes to cure were placed on wickets that carried them slowly through a gas-fired oven. Most of these units have now been replaced by ultraviolet curing that can cure the litho inks in a fraction of a second. The higher cost of the UV ink is compensated for by higher productivity.

Metal decorating is sometimes used for metal food trays or other decorative objects. Screen printing can also be used for metal decorating although this is rarely applied in packaging.

Customer demand for increasing quality affects can printing around the world. Europe and Japan lead in high quality printing of three-piece cans. Cans decorated with mountain scenes or pictures of bakery products have startling reality. In Europe, some two-piece cans are printed in halftone process color, but it is not considered cost effective in the U. S. (figure 2-14).

Two-piece cans are printed by offset letterpress, also called dry offset or letterset. The inks are placed, one on top of the other, on an offset blanket, then transferred to the can. The printing plates are plastic letterpress plates. After printing, the cans are heated briefly at 250° F (120° C) to cure the ink in five to 10 seconds. A thin lacquer is finally applied to protect the print (figure 2-15).

The final printing on most cans is applied by the cannery using ink jet printing to apply lot numbers or control numbers.

Printing and Converting Equipment

Although the printing processes are offset litho or offset letterpress, the presses are very different than letterpress or litho presses for printing publication work. They are built specifically for metal decorating and are found only in can manufacturing facilities. (See Chapter 8.)

LABELS, WRAPPERS AND TAGS

Labels, wrappers and tags are printed products applied to a package and carrying information about its contents: brand identification, contents and information about use. Labels and wrappers are adhered to the package while tags are applied by other means such as wire or thread. Wrappers on corrugated boxes are usually called laminates (or even litho-laminates).

Originally printed on paper, labels are now printed on plastic film and foil laminates as well as paper, while tags are usually printed on thin paperboard.

The Products

Labels, wrappers and tags are closely related. Labels or wrappers for canned foods and other three-piece cans are the largest market for labels. The label on a three-piece can or a stick of chewing gum is a wrapper for the product. The label on a corrugated box technically becomes a wrapper if it covers the entire box. Tags on tea bags serve not only to identify the product, but they also serve as a handle for removing the tea bag from the hot brewed tea (figure 2-16).

Every important printing process is used to print labels, wrappers and tags. No packaging product is printed by a greater variety of printing processes.

There are many types of labels, such as flat labels or plain paper labels that are attached to cans and bottles with adhesive; gummed labels that, like postage stamps, have a moisture-sensitive adhesive; heat-seal labels that carry a heat-seal adhesive on the back, and pressure sensitive labels that have a pressure-sensitive adhesive on them.

Flat labels, gummed labels and heat-seal labels are printed by a variety of processes, but pressure sensitive labels are printed by flexography, rotary letterpress or screen. A few are still printed by sheetfed offset. Gravure and offset, preferred for in-mold labels, are being replaced by flexo. Ink jet is not yet developed to the point where it can apply colored pictures at speeds of most other printing presses.

Spot color is still more predominant than process color, but, as in other areas of printing, process color is on the increase.

In the late 1980s, 90 percent of roll labels in the U.S., 40–45 percent in Europe and five percent in Japan were produced on narrow-web flexo presses.

Older flat-bed or semi-rotary presses are being replaced by rotary printing presses, with flexo taking an increasing share. Where highest quality is demanded, letterpress is often preferred to flexo. Rotary and flatbed screen printing are used in many premier consumer market areas.

Figure 2-16. Labels, wrapper and tag are printed by gravure or lithography.

Labels are also classed as primary and secondary labels. Primary labels are applied by the manufacturer, secondary labels are applied by the retail outlet.

Primary labels are printed by manufacturers who specialize in labels. Secondary labels are usually printed by in-plant shops for anything packaged in the store: meat, deli products, produce, bakery products, even fresh-squeezed orange juice, although many are printed by specialty printer/converters and by commercial printers.

Materials for Manufacturing Labels

Most labels are printed on top-quality, pigment coated paper, but other materials are also used. Pressure sensitive labels are adhered to a specially coated release stock, and bottles are often labeled with plastic labels or wrappers commonly made of oriented polypropylene.

Pressure-sensitive labels are typically printed on a composite consisting of facestock, adhesive, release liner and topcoating. In addition to paper, facestock may be prepared from PVC, PET, PS, PE, PP, acetate and other polymer films. Offering good esthetics and mechanical properties, their use has grown dramatically in the past decade for labeling of plastic containers, especially for health care products.

Security labels are manufactured with an adhesive that is stronger than the paper on which they are printed. Any attempt to remove the label will merely deface the paper.

Pressure-sensitive labels are printed on bleached or unbleached kraft that is usually clay-coated to give an excellent printing surface. Some are printed on plastic. The adhesive layer is an acrylic emulsion or a hot-melt rubberbase formula. The base sheet, again usually bleached kraft, is coated with a silicone release coating.

When pressure-sensitive labels are printed by flexo, the inks are commonly waterbase. They are often overprinted with a clear varnish, frequently UV cured.

There are many variables that affect the converting of pressure-sensitive materials, including heat used during processing, facestock strength, die configuration, width of materials, quality of tooling, performance expectations, and, certainly, the condition of the base material. The most common problems encountered while processing pressure-sensitive material are: breaking of the waste matrix, labels traveling with the matrix (called "predispensing") and labels popping off the release liner after the matrix is removed.

Labels for prestige products such as cosmetics, wines and liquors are often embossed and hot stamped with gold or silver foils.

New products for labels and labeling include ink jet printers: three-dimensional labels, diecut to shape, to magnify the design; in-mold labels; labels recyclable or degradable in alkali; thermal transfer printers and color coded labels that change color on heating.

How Labels are Printed

Every major printing process is used to produce labels: flexo, gravure, offset, letterpress, screen, electrostatic, ink jet and others as well as by combinations of flexographic, gravure, screen and hot stamping, often in a single pass. Ink jet or letterpress equipment often adds serial numbers or batch numbers to printed labels. In the early 1990s, flexo was growing more rapidly than gravure and letterpress for European label printing.

Narrow web presses (less than 24 inches or 609 mm) print by gravure, flexo, letterpress, litho and screen, producing a wide variety of products such as pressure sensitive labels, tags and tickets.

Run lengths differ. Gravure is ideally suited to continuous long runs, those in excess of a million labels and often in the three-to-four million range. Sheetfed offset is chosen for intermediate runs. It is slower than web processes, but because of the large sheet sizes of many label presses (up to 44 x 64 inches) it can generate a higher throughput than flexography or rotary letterpress.

Rotary letterpress tends to be used for longer runs with fewer copy changes than flexo. For the most part, flexo presses can be set up to handle copy changes more quickly. Flexo is used for both primary and secondary labels. Rotary letterpress supplements flexo by being used for high quality primary labels, generally of longer runs than those for secondary labels.

With its ease of setup, inexpensive screens and comparatively slow running speeds, web screen printing is reserved for shorter runs.

Ink jet and electrostatic printing are ideal for printing labels when the run length is one—when every label is different.

Offset Lithography is a major process, accounting for more than half of all label printing. Offset labels are printed by specialist and commercial printers whereas flexo and gravure label printers often print flexible packaging as well, although increased specialization is occurring.

Offset and gravure are the processes best able to reproduce soft halftones and vignettes in four-color process. Such graphics are often required for high-quality food, wine, liquor and other primary labels. Printers of liquor labels, for example, will add gold bronzing, embossing and hot stamping techniques to the printing.

Offset label printers are unique in providing a "combination label" service. The labels on a large (e.g., 44 x 64 inch) press sheet may all go to a single manufacturer, but at certain weekly intervals the labels may be for different manufacturers. Smaller label users are guaranteed spots on a press sheet and benefit by sharing makeready costs. Quality four-color labels by a leading printer thus become affordable. Combination labels have a run length of 30,000 sheets on a large-format press. Users can therefore buy one spot on a press, resulting in an order for 30,000 labels, or additional spots, which result in multiples of 30,000.

The difficulties in producing variable cutoff with web offset keeps sheetfed the major offset printing process for producing labels. The six-color press is standard in sheetfed offset label printing.

Gravure. Three categories of labels are commonly printed by gravure: extremely long-run paper or paper laminate labels and wrappers, spiral-wound labels and in-mold labels. Gravure accounts for less than 10 percent of label and wrapper printing.

The long runs, for which gravure is best suited, are disappearing. Beer labels are a good example. For each brand, there is a light and regular version. Cigarettes are the same. They come in different lengths, with different filters, and in different packages. One leading brand uses seven packages domestically and 27 internationally.

There are reasons for choosing gravure other than cost, which becomes increasingly favorable as runs lengthen. Gravure offers full, rich color and the ability to reproduce metallic and fluorescent ink better than offset or flexo. Its consistency is superior to all other printing processes.

Spiral-wound composite cans, such as those containing orange juice concentrate or biscuit dough, are commonly printed by gravure or flexo.

Gravure is the leading method for producing in-mold labels, While they can be produced by gravure and offset, the gravure press can apply the adhesive inline along with the printing and top coating.

Label printers can conceivably use any size gravure press, depending on the size of their print runs. Most long label runs, however, are efficiently handled by 25 to 44 inch presses. There is interest in smaller 13 to 20 inch presses as printers respond to reduced run length requirements.

Flexography. Flexo has the largest share of the fragmented specialty label market. Flexographic labels are used for 80 percent of specialty secondary, promotional and price-marking labels. Improving flexo technology results in small, but significant inroads into primary labels, an area traditionally served by offset, gravure and rotary letterpress. Pressure-sensitive material is used for the bulk of flexo labels, but flexo is also used on heat-seal, gummed stock or plain paper.

Flexo has the advantage of faster in-register makeready than web offset since plate cylinders can be removed from the press for off-press mounting and proofing. Printing sleeves also are helpful in rapid changeover.

Few narrow-web flexographic printers print more than 10 percent of their label orders in process color. Relatively few primary labels (where process color is apt to be used) are printed by flexo. On the other hand, when the press is well equipped, controlled and operated, flexo produces excellent results. The ability of flexography to print top-quality labels is becoming recognized, and in recent international competitions, first prize has repeatedly been captured by flexo.

Flexo label presses are considered narrow-web, between 6.5 inch and 16 inch width. There has been a trend toward an increasing number of print stations to print more colors. The average flexographic label press can now print five to six colors with an increasing number of eight- to nine- and up to 14-unit machines being sold.

Flexo presses do not usually print screens and solids satisfactorily with the same color unit. The solid is printed on one unit and the halftone screen on another.

Letterpress. Users now pay a premium for rotary-letterpress-printed labels. Nevertheless, the majority of supermarket or pharmacy labels printed by rotary letterpress do not have design characteristics that would require such an expensive process.

The development of UV ink systems has allowed the design of a high-speed rotary letterpress with intercolor drying. Developed in Japan in 1975 and introduced to the world market in 1977, the specially designed narrow-web rotary letterpress is a growing factor in label printing. By the mid-1980s, there were over 1,000 rotary letterpress machines in use worldwide. The hard photopolymer plates, mounted with stickyback, give rotary letterpress more controllable dot gain than flexo.

Primary labels printed by rotary letterpress are gaining popularity in pharmaceuticals, cosmetics and toiletries requiring top-quality, multicolor labels. The pharmaceutical industry often specifies letterpress because of its ability to hold accurate print definition throughout a run. The process is the dominant method of web label printing in Europe; in Japan flatbed letterpress predominates. In Europe, flexo has been growing at twice the rate of rotary letterpress and will soon surpass letterpress as the leading process for label printing. Flexo is the more important process in the United States.

Screen. Screen printing provides brilliant colors, excellent coverage, durability and high coating weights for varnishes. It is especially useful in printing fluorescent and pastel colors where the heavy ink film gives excellent performance. Screen printed labels are especially good for resisting deterioration, and are used where weatherability is important and for printing delicate substrates that are too rough or weak for printing by other methods.

Until the introduction of UV-curing inks and roll-to-roll web screen presses, screen printing was too slow and expensive to compete effectively with other processes for multicolor label work. The advent of UV-curing inks made such web screen presses practical, overcoming the problems of solventbase inks and hot air drying which slowed the process. Top rated speed for a web press is around 3,000 impressions per hour. Web screen presses have advantages over the other narrow-web processes: they do not require highly skilled operators, they generate little scrap, and they are easy to set up and clean up.

Electronic Printing. Several techniques are used to convert electronic information into printed images on labels: ink jet, electrostatic imaging and thermal transfer printing among them.

Computer-generated labels enable the user to print variable information in a multitude of different formats stored in the computer. Companies require a variety of labels for internal and external use: quality control labels, shipping labels, product information labels (model number, technical specifications, bar codes) and material request labels. The capability to print bar codes has become important as their presence on labels is demanded by major customers, industry associations and federal purchasers.

Electrostatic printing equipment that can print and apply labels is being used mainly for secondary packages—pallets and shipping cartons—because speed is limited to roughly 25 packages a minute.

Thermal transfer printing can be used to produce bar codes, and a printer is available to produce four-color labels including bar codes.

Holography. Holography is growing popular as costs of materials come down and equipment becomes more productive. The uniformity of deposition now being achieved on metallizers is also a major advantage. Rolls of holograms up to 60 inches wide are available.

A hologram is produced by first splitting a beam of laser light into two parts. One part is reflected off the object, and the other illuminates a photosensitized, glass plate directly. The reflected and the direct laser light interact at the plate to form the holographic record of the object.

Developing the plate forms the holographic image as tiny micro-corrugations (about 20,000 to an inch) on its surface. A layer of nickel is deposited on the plate surface and is used to make a second generation nickel replica. A special micro-embossing machine uses these second-generation replicas to impress the microcorrugation pattern into the coated films, using heat and pressure.

Different kinds of film are used for various types of holograms: roll leaf for hot stamping, polyester for pressure-sensitive labels, or metallized and sized films suitable for laminating.

Printing and Converting Equipment

The presses used for flexo, offset, gravure and screen-printed labels are not technically different from presses used for producing other types of commercial or package printing. The size of the press can be a distinguishing factor. High-volume producers of cut-and-stack labels use large sheetfed offset presses for the bulk of their production. The main market for sheetfed offset presses greater than 50 inches wide is for label or folding carton printing. The few web offset presses in the U.S. that are used for producing labels have limited print repeats and variable cutoff capability.

RIGID BOXES AND COMPOSITE CANS

Rigid or set-up paperboard boxes are used for candy, returnable beer bottles and cigars. They are distinguished from folding cartons because they are already set up when they are supplied to the company that is packaging the product (figure 2-17). The name "rigid" is preferred by the National Paperbox and Packaging Association; the name "set-up" is still used by the U.S. Department of Commerce in the Census of Manufactures.

Rigid boxes have the advantages of strength and reuse and variety in size and shape, since they are made in short runs on equipment that is relatively simple to operate.

They are usually made from recycled paper with a layer of white paper or pigmented coating applied to improve the graphics. Graphics can be applied during manufacture or on a label which is applied after the box has been shaped. Converting papers, like giftwrap, are printed in rolls and applied to the boxes after they are set up.

Rigid paperboard boxes now consume a half million tons of board annually in the U. S. Rigid boxes, often in combination with other materials, are used for such fragile and varied consumer products as fine chocolates, ceramics, cosmetics, electronic and photographic equipment, tools and other luxury goods.

Cigar boxes are now paperboard boxes, replacing expensive and hard-to-obtain wood. At the most utilitarian level, rigid paperboard boxes were widely used to package shoes, but rigid shoe boxes have been replaced with folding cartons.

Composite containers include small fiber cans and large drums. They are similar to each other in construction. They are made by winding layers of paperboard and other materials such as aluminum foil and plastic films around a mandrel that can be either round or rectangular. Smaller canisters are used for retail products, mostly such foods as frozen orange juice concentrate, oatmeal, cocoa, salt, baking powder and refrigerated biscuit dough. Until plastic bottles replaced them recently, fiber cans were used for motor oil.

The core of the composite can is generally recycled fiber. It can be lined with moisture-resistant materials to create a package with good barrier properties, and resistance to corrosion. Laminations of foil or film can be used as an inner or an outer liner. The outer "label," often applied in the winding operations and generally of bleached paper or foil, is an ideal surface to convey brand identification, printed instructions, and nutritional and contents data.

Some 50 million fiber drums are manufactured annually in the U. S. They are usually decorated with a label. They can be lined with film bags or molded plastic inserts. Many are still used for dry chemicals, food ingredients, pharmaceuticals and plastic resins.

Figure 2-17. Rigid paperboard box and wooden cigar box.

Figure 2-18. Glass and plastic bottles decorated by screen printing.

Rigid and semi-rigid plastic packaging is a rapidly developing packaging field. Plastics are extensively used in bottles, pails and jugs, closures, semi-rigid formed trays, cups and blisters, for household chemicals, foods and beverages, toiletries, automotive and many other uses. They account for 1.6 million tons of plastic annually in the U. S. An early use was for squeeze bottles that dispensed spray deodorants and other personal care products.

These plastic containers are light and have high strength-to-weight ratio and generally good product resistance. The bottle material can be made to suit the needs of the product, including transparency or opacity. Plastics can be formed in a larger number of shapes than glass.

Setup boxes and composite cans are sometimes printed during manufacture using offset lithography, flexography or letterpress. Printing is sometimes followed by coating. But the great majority of setup boxes and composite cans are decorated with labels or wrappers using techniques described above.

BOTTLES

Bottles are rigid or semi-rigid containers with a narrow neck, usually used for liquid materials. Bottles, whether glass or plastic, may be clear, colored or opaque, and they may bear a label or they may be printed directly (figure 2-18). With the exception of flexible packaging, bottles are the oldest form of packaging we have, going back to pre-Biblical times.

Glass. Glass is the material from which bottles were made over the centuries, although use of clay and other ceramics is even older. The major drawbacks of glass are its weight and fragility. Actually, if it is unscratched, glass has great strength, and glass performance has been markedly advanced in recent years by the development of plastic, mineral, silicone and stearate coatings. While the coatings actually add no strength directly, the surface protection they provide increases the strength of glass containers by as much as three times over that of uncoated glass.

The thin foam plastic wraparound label for two-liter beverage bottles protects the bottle and can also be printed. It cushions the bottle from impacts and could be the forerunner of many more composite containers that combine the best qualities of glass with one or more additional materials.

Molded bottles are sometimes formed in a mold that has the bottler's or manufacturer's name in it (figure 2-19). They are often printed by screen process that can easily apply a thick, opaque ink to a curved surface.

Plastic. Plastic bottles are replacing glass bottles for a wide variety of products. Plastic bottles have made the spiral-wound motor oil can obsolete. Plastic is much lighter than glass and less susceptible to breakage or damage.

Decoration. The most common form of decoration of bottles is a label, paper or plastic, that is applied to one side or two sides or wrapped around (figure 2-20). Many bottles, such as wine bottles, typically carry two labels; the 2-liter soft drink bottle typically carries a single plastic wrapper printed by flexo. Wine and liquor labels are usually printed by sheetfed offset, although letterpress and flexo are making inroads into this market.

Labels for transparent bottles are sometimes printed on both sides, a rare exception to the one-sided printing of labels.

Bottles are often printed by screen, a printing process that easily handles

Figure 2-19. Glass bottles with name and decoration molded in glass.

Figure 2-20. Glass bottles with foam wraparound labels printed by flexo.

round or oval shapes. Special presses are built for printing products with such shapes.

COLLAPSIBLE TUBES

These products, originally made of tin, are now mostly made of aluminum and plastic. Plastic tubes are often made from complex laminates, and they are generally less susceptible to puncture than metal tubes.

Although invented by a painter who was looking for a better container for oil paints, collapsible tubes did not become a major form of packaging until toothpaste was introduced in a tube at the end of the 19th Century. The collapsible tube is now used for pharmaceuticals, personal care products, cake decorations and caulking materials. They are produced by about 20 companies in the U. S. They are used to dispense food products in Europe, but find little use for food in North America.

The tube is extruded, the closure is attached, and the tube is filled from the bottom, which is then crimped. The dispensing tip can also be completely sealed with a puncture or break-off feature for products of unusual environmental sensitivity.

Collapsible tubes are usually printed by flexo. They are often coated with a titanium dioxide coating to provide a white base for printing (figure 2-21). The presses are specialty presses, designed as part of the manufacturing line.

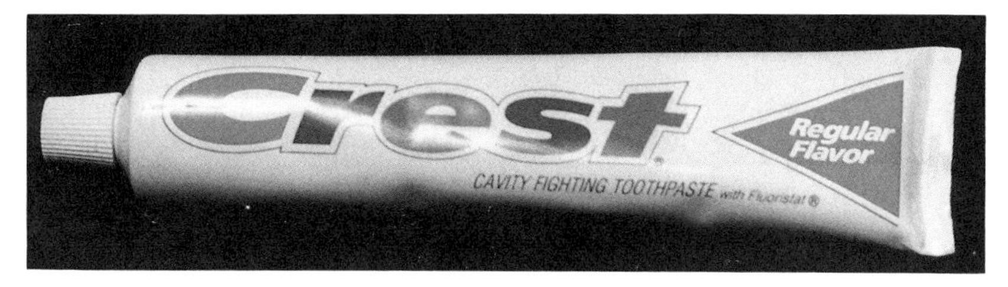

Figure 2-21. Toothpaste tube printed in three colors by flexo.

Figure 2-22. Plastic bottle printed in two colors by screen. Note decorative dispensing closure.

CAPS AND CLOSURES

Molded plastic caps have largely replaced metal screw caps. Such plastic caps come in a wide variety of colors which add interest to the package (figure 2-22).

Metal caps also provide color and interest. Caps may be printed on the metal sheet before forming, but the molded plastics are printed, usually by imprinting, after they have been formed.

ICE CREAM AND DAIRY CUPS

These cups, that contain from eight ounces (200 g) up to a gallon (around four or five kg) are an important packaging item worth around $500 million annually in North America. In spite of the size of the market, it remains something of a specialty with only a couple of companies manufacturing the printing equipment.

The cups, used for ice cream, butter, cheese, yogurt and other products found in the dairy department at the supermarket, are formed from cartonboard or from molding resin before they are printed. Owing to the conical taper of the cups that are nested when they are shipped before filling, special printing equipment is required and the design must be specially distorted to give it an appropriate appearance after printing.

The packages are printed by flexography.

FURTHER READING:

Packaging Encyclopedia. Cahners Publishing Co., Des Plaines, IL. (Annual).

Modern Plastics Encyclopedia. Robert J. Martino, Editor. McGraw-Hill Publications Co., New York, NY. (Annual).

Fibre Box Handbook. Fibre Box Association, Rolling Meadows, IL. (1992).

Corrugated Package Printing and Converting. Techno-Economic Market Profile No 15. J. A. Workman, Editor. Graphic Arts Technical Foundation, Pittsburgh, PA. (1989).

Packaging in Today's Society, Third Edition. Robert J. Kelsey. Technomic Publishing Co., Lancaster, PA. (1989).

Flexible Package Printing and Converting. Techno-Economic Market Profile No 13. N. R. Eldred, Editor. Graphic Arts Technical Foundation, Pittsburgh, PA, (1988).

Folding Carton Printing and Converting. Techno-Economic Market Profile No. 12. N. R. Eldred, Editor. Graphic Arts Technical Foundation, Pittsburgh, PA. (1987).

Polymers as Materials for Packaging. J. Stepek, et al. English translation. Ellis Horwood Ltd.,

Chichester, England. John Wiley & Sons, New York, NY. (1987).
Label Printing. Techno-Economic Market Profile No. 7. N. R. Eldred, Editor. Graphic Arts Technical Foundation, Pittsburgh, PA. (1986).

Chapter

3

The Printing Process

CONTENTS

LIST OF ILLUSTRATIONS

LIST OF TABLES

Chapter

3

The Printing Process

INTRODUCTION

There are many ways to print an image. Relief or typographic processes produce printing from raised type—the process of Gutenberg (figure 3-1). Flexography and letterpress are the typographic processes currently used for printing packages. Planographic processes produce images from flat surfaces such as lithographic plates that distinguish image from non-image areas chemically. Image and non-image areas are wetted differently by ink and by water (figure 3-2). Rotogravure prints from an engraved surface with the image recessed into non-image areas (figure 3-3). It is an example of intaglio printing. Currency and a few other specialties are printed with flat, engraved plates. Gravure package printing is performed using engraved or etched cylinders. Other printing processes include printing by screen or stencil (figure 3-4). The term "serigraphy" is sometimes applied to screen printing. New, non-impact processes that generate images directly from digital data, notably ink jet printing and laser printing, are finding a few applications in package printing.

The ability to print on rough surfaces such as corrugated or uncoated kraft paper varies with the printing process. Flexo compensates for the rough surface with flexible plates, litho with an offset blanket. Gravure and letterpress, which print from metal cylinders or plates, cannot print a sharp image on a rough surface. Screen or stencil printing can form a good image on rough substrates.

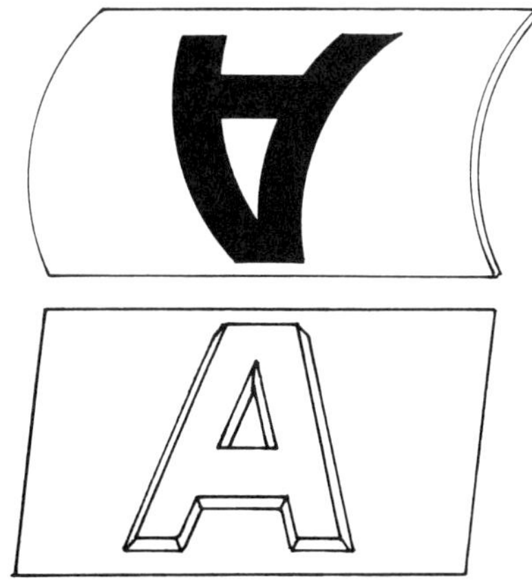

Figure 3-1. Relief printing. Printing from raised images as in letterpress and flexography.

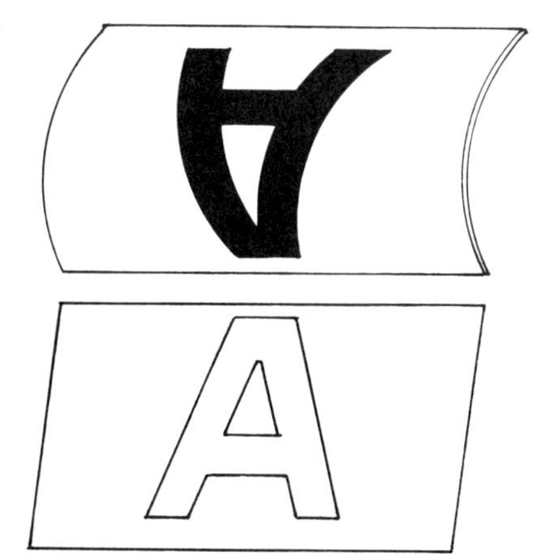

Figure 3-2. Planographic printing. Printing from level images as in offset lithography.

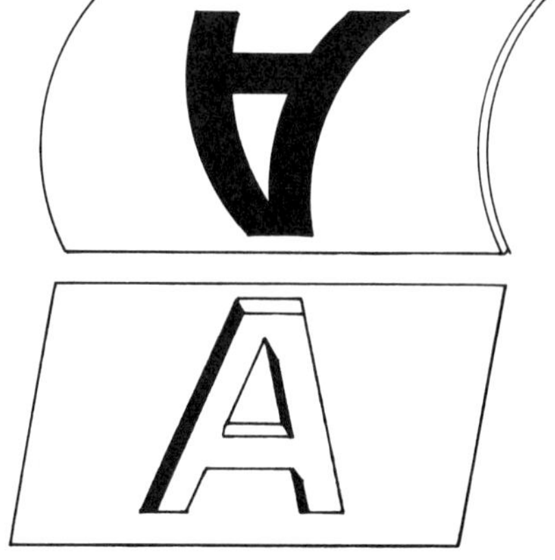

Figure 3-3. Intaglio printing. Printing from recessed images as in rotogravure.

Figure 3-4. Screen printing. Printing through a screen.

TABLE 3-I. TYPICAL INK FILM THICKNESS (WET FILM ON SUBSTRATE)

Process	mils	millimeters	microns
Sheetfed offset	0.2–0.3	0.005–.008	5–8
Web offset	0.3–0.5	0.008–.012	8–12
Sheetfed letterpress	0.3–0.5	0.008–.012	8–12
Flexo	0.4–0.8	0.010–.020	10–20
Gravure	1.0–1.5	0.025–.040	25–40
Screen	1.0–5.0	0.025–.130	25–130

Ink jet does not print from a plate or cylinder, and it can print well on rough and uneven surfaces. Electrostatic and laser printing transfer the image from a metal drum to the paper using an electric charge, but paper smoothness is not critically important to produce an acceptable image.

For printing on materials that must not be crushed, notably corrugated and pressure-sensitive labels, the amount of pressure required to print the image is an important consideration. Flexography prints with very light or "kiss" impression. In lithography, high pressure is required to transfer the paste inks from roller to plate and then to the blanket and substrate. Rotogravure requires high impression to force contact of the substrate with the cylinder. Letterpress requires high pressure both to transfer paste inks and to force contact of the substrate with the metal or hard plastic plate. Screen requires little pressure, only enough to keep the screen in contact with the substrate. Ink jet is a non-contact printing method and requires no pressure. It can print on anything. (Advertising for ink jet printers has shown images formed on raw egg yolks.)

As a general rule, offset prints the thinnest ink film, screen the thickest. Typical ink film thicknesses are presented in table 3-I. The actual thickness can be varied greatly.

This chapter emphasizes flexo,* gravure, litho, letterpress and screen, but other, newer processes are also discussed since they are used to some extent on packaging. Ink jet printing (figure 3-5) is used to print bar codes and for coding and marking cans and other packages, in addition to general commercial printing. Laser printing is used primarily for office duplicating, but it is being used for special labeling such as UPC symbols.

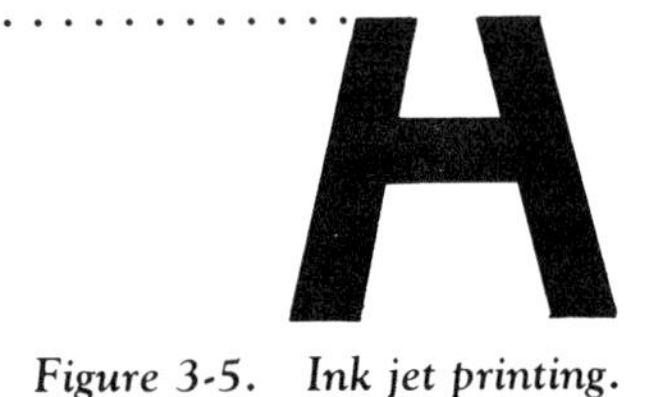

Figure 3-5. Ink jet printing.

Flexography has the broadest markets in packaging because of its ability to print on a broad range of packaging materials, its ability to print any required repeat length and its improving technical capabilities. Gravure gives excellent color and color stability on very smooth substrates such as coated board, film and foil, but the present cost of preparing gravure cylinders limits it to long-run jobs, often in the millions of impressions. Sheetfed offset lithography is popular for folding cartons and labels. The high speed of web offset is gaining it new markets for printing folding cartons. Letterpress is still used for some label printing where the sharpness of the letters and quality of the halftone color are advantageous. Offset letterpress (dry offset) is the chief method of decorating two-piece cans. Screen printing is used on a broad variety of packaging substrates: paper, board, plastic, glass and metal. Ink jet is used for coding or marking packages as well as for producing bar codes, but it is generally done on previously printed packages. Bar codes are also printed by laser printers using electrostatic copying processes.

There are also significant differences in the way that fluid and paste inks transfer to the substrate. Fluid inks used in flexo and gravure readily wet the substrate and are drawn from the printing plate by surface tension. Paste inks used in offset and letterpress are pressed into or onto the substrate. Press operators who have learned to operate offset presses often have difficulty learning to control pressure on a flexo press.

Control of dot gain is a problem of every printing process. It requires that the ink be properly formulated for the substrate to be printed. Dot gain is discussed in Chapter 5, but its control is so important that it must be mentioned here. In process color printing, the image is composed of many tiny, halftone dots. The dots are printed in four colors—magenta, cyan, yellow and black. If the size of the dots in any color varies from the required size, the color of the image will vary undesirably.

Dot gain varies depending on ink, substrate and press. Generally it is greater in flexo than in web offset, greater in web offset than sheetfed offset, and least with screen printing. However, if it is kept under control, color separations and printing plates can be prepared to compensate for the expected dot gain.

Makeready waste is another item to consider. In offset lithography, several impressions must be printed to bring the press "up to color." Waste in flexo and screen printing are extremely low. For gravure, the waste factor is nearly as low as with flexo, but it does take some time for the ink to penetrate the cells.

* The proper names "flexography," "rotogravure," and "offset lithography" are commonly abbreviated to "flexo," "gravure" or "roto," and "litho" or "offset," even in formal writing.

SELECTING THE BEST DESIGN FOR THE PRINTING PROCESS

Any printing process can be made to produce good printing. The choice of printing process is made on the basis of economics and the nature of the substrate, not on the basis of the design.

On the other hand, once the printing process has been selected, the design must be one that will reproduce well and take advantage of the process. Design of artwork for the printing process is discussed in Chapter 4. Some generalizations may be helpful. Offset prints the best vignettes; letterpress prints the clearest small print or reverses. The structure of the gravure halftone and its consistency of color make it excellent for printing pictures in which skin and hair colors are important.

Each process has its own characteristics, and when the printing is less than excellent, these characteristics show up.

Flexo has a halo around each letter or figure if the impression pressure is too high and skips in solids if the pressure is too low. Poor choice or control of ink gives striations through solids.

Snowflaking, skips and missing halftone dots are characteristic of gravure printing on a rough surface and failure to use electrostatic assist.

Litho shows water bubbles or streaks if excess fountain solution is run. The thin film printed by litho is particularly susceptible to dirt of any kind, producing spots and hickeys. Ghosts develop in solid areas, and if the ink has emulsified in the fountain solution, color appears in the non-image areas.

Letterpress, like flexo, shows halos when pressure is too great and uneven, irregular solids if pressure is too low. Excessive pressure embosses the substrate or even punches through the paper, characteristics never seen with flexo.

Screen printing gives the edges of letters and lines a ragged edge if the mesh of the screen is not fine enough.

For good results, it is imperative that the designer understand the printing process that will be used to print the process and create an appropriate design. Usually, flexo or gravure printers or trade houses will have designers on staff to alter or adjust the design for the process to be used. Offset printers usually do not need to have such people available.

The packager must decide which process to use based on production costs and product requirements, then insist on an appropriate quality level. The cost, and perhaps the overall quality of the job, will be better if it is properly designed in the first place.

FLEXOGRAPHY

Flexography is a method of direct rotary printing that uses resilient relief plates of rubber or photopolymer materials. It can print on virtually any substrate. The fluid inks are metered with an engraved metal roll and a doctor blade or resilient metering roll (or both simultaneously in some specialized applications).

The key features of flexography are the flexible, relief (raised-image) plates;

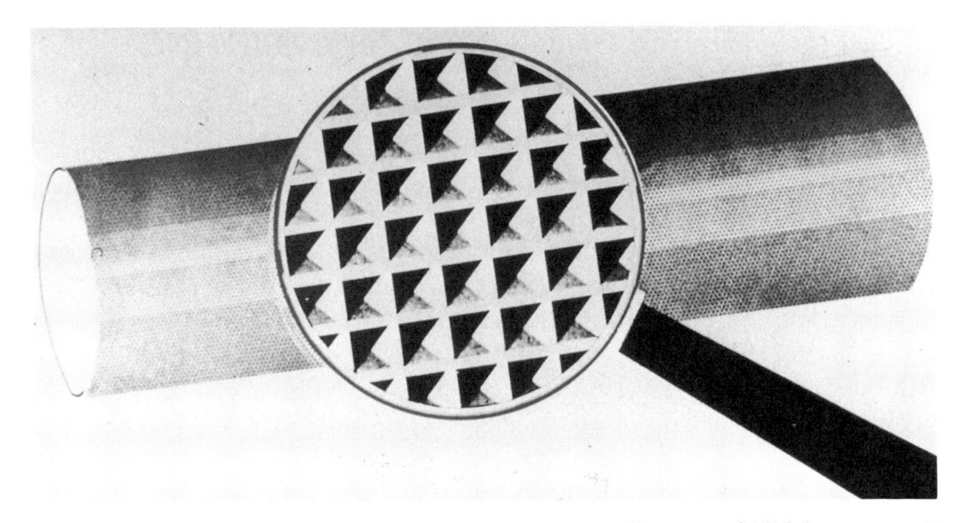

Figure 3-6. Anilox roll, the metering device in flexo printing. (Courtesy of ARC International.)

fluid inks; and an engraved roll, called an anilox roll (figure 3-6), which is the heart of the ink metering system. The soft plates give flexo the ability to print well on a wide variety of substrates.

The term "flexography" was adopted in 1952 after the Packaging Institute polled converters and suppliers. The earlier term, "aniline printing," described the original coal-tar-derivative inks used for products first printed by the process, such as paper bags. The process has shown so many capabilities that it is now the most widely used process for printing packaging materials, notably corrugated, milk and beverage cartons, flexible packages, labels and multiwall bags.

Improvement over the years in the quality of process color printed by flexo is due partly to better color separations, but mostly to better plates, better presses and better training of press crews. Ink color standards are now being adjusted to SWOP (Specifications for Web Offset Printing) standards for most printing on white substrates.

Flexo productivity is increasing as a result of continuing improvements in presses, inks, plates, pin register systems, electronic register control and many other developments. Unlike web offset, the print length of any plate cylinder can easily be adjusted to fit the size of the package or label.

As the newest of the major printing processes, flexography is changing the most rapidly. A few years ago, it was not possible to print the full range of halftone dots, but flexo now approaches offset in that capability. Flexo traditionally employed low viscosity inks, which were very fluid—almost the consistency of water, but package printers can now use "high-solids" inks to achieve increased ink density and improved appearance. High-solids inks give better mileage than the older inks, and they dry faster, facilitating higher press speeds and faster in-line converting operations.

Markets for Flexo

Flexography is the process most widely used for printing packaging materials.

Flexo is used in every major packaging market: flexible packaging (film and foil), corrugated and preprinted linerboard, tags, labels and wrappers, folding cartons and beverage carriers, paper bags, cans and sanitary packaging, milk and beverage cartons and spiral-wrap containers.

Flexo can economically print orders from several hundred packages up to a few million, a range that covers most of the market.

Flexo shares the flexible packaging market with gravure. It dominates the printing of corrugated, although a small percentage of E-flute corrugated is printed by offset and some linerboard is printed by gravure. Folding cartons are usually printed on wide web presses, but more and more are being printed on narrow web presses. Flexo is a major process for printing tags and labels, a market it shares with gravure, offset, letterpress and screen process. Folding cartons have historically been printed by offset or gravure, but with improved quality, flexo is increasing its share of folding cartons and the entire package printing market.

Gravure has produced a large number of beverage containers, but flexo is making major inroads into this market. The average speeds of these flexo printing operations is over 1000 feet per minute.

The low printing pressure required by flexo gives it an advantage for printing pressure sensitive labels.

Flexo has taken most of the letterpress markets for package printing, and it is taking market share from gravure. As with other processes, the press, plates and ink required for high-quality work are expensive, but the overall cost of a flexo job is often less than the cost of a job of equal quality printed by other processes.

Substrates for Flexo

An important property of flexo is its ability to print on a wide variety of materials: rough or smooth, coated or uncoated, paper or board, as well as plastic and metal. Strong, smooth paper is expensive, and flexo does a good job of printing on low-cost paper that is suitable for many printed products.

Characteristics of Flexo Printing

Any process can print on a surface as smooth as extruded plastic or cast-coated paper, but the flexible rubber or photopolymer plate enables flexo to print on relatively rough surfaces. The flexible plate can be used for printing of surfaces as rough as kraft grocery bags, corrugated and uncoated top liner. A foamed stickyback (the adhesive material used to attach the flexo plate to the cylinder) helps the plate conform to rough or uneven substrate. This is accomplished with a rubber blanket in litho, but it is not possible with gravure or letterpress. Even with coated carton stock, variations in the thickness create problems for gravure and letterpress. However, levelness is required for flexo. The amount of variation in sheet thickness that can be tolerated by a flexo press is strictly limited. A thick or heavy area in the paper or board creates high pressure and causes ink to squeeze out. A thin or low spot gives poor or uneven ink transfer.

Flexo is an inherently simple process, but it takes a good press, good training, good ink and good plates to produce top-quality printing. Although top-quality printing is regularly achieved by leading flexo printers, flexo is also capable of producing low cost "pizza boxes." A great deal of low-cost flexo printing is produced, and packagers often accept low-quality flexo printing, or use a more expensive process instead of insisting on high-quality flexo. Flexo is generally lower in cost than printing of equal quality produced by other package printing processes. As with every process, high-quality flexo printing costs more than low-quality output.

Printing Pressure. Flexo prints with a "kiss impression," the least possible squeeze between plate and substrate. Less impression is required to transfer the liquid (fluid) ink to a smooth surface than to a rough one. Tests have shown that surface tension plays a role in flexo printing, and each successive step must exhibit higher surface tension than the previous step.

Control of printing pressure is critically important in producing good quality flexo printing.

Versatility. Flexo's ability to print on a wide variety of substrates is a key characteristic. Another reason for flexo's versatility is its ability to print with either solventbase or waterbase inks. Although there are significant differences in how these inks behave on press, the end result is often indistinguishable, even to the trained eye.

Dot Gain. Dot gain is a problem that requires special attention. The soft flexo plates are easily distorted by excess squeeze or impression pressure, requiring that impression pressure must be carefully controlled. Automatic pressure sensor/control systems are available for flexo presses to help control dot gain.

Technical progress is reducing dot gain in flexo and thereby making a major contribution to improved flexo print quality. Thinner plates produce less dot gain than thick ones. The higher ratios of anilox to line screen also reduce dot gain by reducing the thickness of the ink film required for proper flow out, permitting the use of higher viscosity inks. This also means that the ink dries faster and stops spreading sooner.

Process Color Printing. For printing process color, fine halftone screens give better resolution than coarse screens, and they make the image sharper and more realistic. The limitations of screen ruling depend on both the plate and the substrate. Plastic films have an exceedingly smooth surface and can be printed with very fine screen rulings that depend almost entirely on the properties of the printing plate. On the other hand, paper and board often contain dust that can fill or clog the halftone screen of flexo plates, blurring the pattern. Coarse materials such as corrugated will generally not satisfactorily reproduce halftones that are finer than about 85–100 lines per inch (33–40 lines per cm). By coating the top liner, a much smoother surface is produced and images of 150 or more lines per inch can be produced using the best flexographic plates.

Characteristic Problems. Striations are caused by ink starvation and result from a variety of causes: the amount of ink in the system, the speed at which the ink flows into the pan or the well of the chambered doctor blade system, poor ink pickup on the anilox roll, doctor blade pressures set too high, metering rolls pressures too tight against the anilox roll at low press speeds, poor or uneven ink release from the anilox rolls, and combinations of these and other factors.

Control of squeeze or impression is of great importance in relief printing processes—flexo and letterpress. If the pressure is too low, ink fails to transfer evenly and skips or misses occur in the image (figure 3-7). If the pressure is too great, ink squeezes out around the edge of the image, distorting letters and halftones (figure 3-8). This creates serious problems with flexo and its soft plates and fluid inks. Excess pressure with hard letterpress plates may emboss the substrate, but with flexo, excess pressure will distort or damage the plate. If the ink viscosity is not properly controlled, striations occur which give the print an uneven streakiness (figure 3-9).

Figure 3-7. Spots in solids owing to light impression in flexo. Note spots in solid bar at bottom of figure.

Figure 3-8. Squeeze-out or halo in flexo printing.

Figure 3-9. Striation, a typical defect in flexo printing.

Flexo Presses

There are many types of flexo presses: wide web and narrow web, in-line, common-impression-cylinder (C-I-C or CI), and stack presses, but all use flexible plates, liquid inks and an anilox-roll inking system. Flexo presses are available in a wide variety of design configurations. Each has a particular capability not matched by the other types.

The ink is metered on the anilox roll by a reverse angle doctor blade or with a fountain roll. In some corrugated printing, both systems are used simultaneously.

The anilox roll is an engraved ceramic roll or chrome coated metal roll. The depth and shape of the engraving determine the amount of ink that it delivers to the plate, and the fineness of the engraving (number of lines per inch) affects the resolution of the print (ability of the press to print fine patterns). The chrome rolls are usually engraved mechanically, but lasers are used on the ceramic rolls.

Chrome anilox rolls have historically been used for flexo, but recently, ceramic rolls have been manufactured that produce printing equal in quality to that which can be achieved with chrome rolls. Chrome rolls are more subject to damage and wear than ceramic rolls. The cell configuration must enable the engraving tool to be extracted easily, placing a limit on the shapes and sizes of cells in chrome rolls.

Ceramic rolls resist damage and wear better than chrome rolls, but their major advantage is the ability to shape the cells and, when laser engraved, to increase the packing of the cells, producing cell counts of up to 1000 per inch. This becomes especially important since the ratio of cell count to halftone screen size is generally in the range of 3.5:1 or 4:1, meaning that a cell count of 1000 can be used to produce a halftone screen in the range of 250–285—well into the finest quality printing. Ratios of 4.5:1 and even 5:1 are coming into use as printers find that the higher ratios yield more even ink coverage and reduced ink film thickness to produce even higher quality printing than is possible with the traditional ratios.

Wide web presses are usually defined as 24 inches (60 cm) wide or greater; they are used to print linerboard and flexible packaging. In recent years, wider presses have been introduced to print folding cartons (as well as commercial printing). This makes it possible to eliminate off-line secondary converting operations and lowers the production costs of the package. The super-wide presses used for preprinting corrugated linerboard range in size up to 98 inches wide. These generally are CI presses, although other configurations have been produced and are on the market. Nevertheless, even at these widths, halftone screen counts of 150 lines per inch (60 lines per cm) have been achieved on production jobs.

At the 1990 DRUPA show, Windmoeller & Hoelscher printed a superb image on film with flexo using a 300-line screen. Although it took special equipment and special care, it shows that such work is possible.

Narrow web presses are usually between 4 and 20 inches (10-50 cm) in width, but most commonly 6, 8 and 10 inches. Narrow web presses are usually made in an in-line configuration and are used for labels, tickets and tags—products that are printed on paper or other stiff materials. Materials that are easily stretched cannot be registered from one unit to the next. As many as 14 print stations can be placed in line, along with diecutting, slitting, gluing, folding, windowing and other converting operations. This is true not only for narrow web presses but for all in-line presses.

Stack presses are used to print paper, plastic laminates and other relatively stiff materials. They are also used to coat or to provide an over-all tint to the product. Stack presses are also used in line with other converting equipment such as extruders, laminators and bag machines.

CI presses are used for flexible packaging materials where any stretch between printing stations would result in misregister* of succeeding colors.

Flexo press speeds are usually lower than those of gravure or web offset. There are apparently no published comparisons of the speeds of various printing pro-

* Misregister occurs when the images produced by the four plates used to print process color do not lie exactly on top of each other. The image becomes blurred, and the colors don't fit at the edges.

cesses, partly because there are no standards for evaluating press speed. Speed varies with the substrate and product, the press configuration, and perhaps, most of all, press crew skills and training. Similar presses running products that appear to be identical may vary as much as 100 percent in operating speed from one plant to another.

In flexo, speed is frequently limited by mechanical factors that are inherent in the process, at the present state of technology. For instance, speed can be limited by the recovery rate of the compressible cushioning. Once the press is up to speed, adjustments in pressure have to be made to account for this characteristic. Suppliers are working on cushions that will recover sufficiently rapidly that there need be no difference in impression pressure at high or at low speeds. Nevertheless, when flexo is run at different speeds, consideration has to be given to changing the impression to assure proper ink transfer.

Makeready Time and Waste

Makeready on a flexo press is often considered to be a lengthy procedure, and it is lengthy if the plate is mounted while the press is down. However, it is often possible to get the plate properly mounted on the cylinder before the cylinder is mounted on the press. Then, changing jobs requires merely that the old cylinders be removed and the new ones mounted on the press and brought into register with each other. Makeready is a reasonably quick procedure when plates are mounted off the press. Use of flexo sleeves makes it even faster.

If the press is equipped with automatic register control, the makeready time is reduced, and control of the process is enhanced. Automatic register controls are installed on most new flexo presses that print four or more colors. Details of automatic register controls differ between in-line presses and CI presses (and gravure and web offset presses, for that matter).

On presses for printing paper, which is subject to stretch from the moisture in the inks, automatic register controls are essential. On the large CI presses, special controls balance the stretch, the drying oven temperatures and the drying time to achieve the same results.

The most important factor in controlling makeready time in flexo, litho or gravure is the organization of the process and the training of the crew. This aspect of productivity has been much neglected in the past, but competitive pressures and new business practices such as JIT (just-in-time manufacturing) are requiring close attention to makeready organization and crew training. Under the best of conditions flexo makeready is fully competitive with other printing processes.

The subject of proofing and getting agreement on color at the earliest possible moment is dealt with in Chapter 6. It should never be part of makeready, but when a customer requests a change after the first copies of the job have been printed, the cost of correction becomes exorbitant.

Factors Affecting Flexo Makeready Time. Makeready time is a key test of management's ability to organize and control the operation. As with press speeds, there are no standard comparisons of makeready times for various processes.

Generalizations about makeready time for flexo and for competing processes are difficult. For one thing, there is no generally accepted definition of which steps are included in makeready. The number of steps to be taken before the press produces a saleable product varies from job to job. The lack of standards is partly responsible for the fact that makeready often takes more time than necessary.

There are great variations in the amount of makeready time for any particular process: the time varies according to the difficulty of the job and how much it differs from the previous job, to the crew and crew training, to the equipment available, and to the processing at the delivery end of the press. Most of all, it depends on management skills and the care with which prepress operations are controlled. If everything were planned in advance, kept under control, so that the plate or cylinder and ink printed a package that matched the proof, makeready would be much faster than it usually is.

The length of run is one consideration: on short runs, makeready time may be longer than the press run, and makeready time and procedures are critical. For products like beer cartons or cigarette packages that are put on the press and run for days, makeready time has less impact on the cost of the job.

In flexo, as in other processes, if platemaking is not carefully controlled, small differences in plates (and the resulting prints) will occur. Photopolymer plates get different exposures or development, rubber compounds may vary. Sometimes the press crew alters the ink or changes anilox rolls to adjust the color, further delaying the makeready.

Roll-to-roll printing requires the least time to adjust delivery, sheeting rather more, and in-line cutting and scoring the most. Cutting and scoring are often done off press by feeding the printed roll into the cutter. This speeds press makeready but requires makeready of the cutter/scorer.

In packaging, the product is rewound (sometimes sheeted) and folder adjustments, which take up most of the makeready time for publication printing, are not required.

The mounting of the plate or cylinder and the rest of the makeready procedure are only part of the actual production time that is considered in JIT manufacture. What actually counts is the elapsed time from receipt of the order to delivery of a satisfactory job. Pressroom makeready is a significant part of this time.

Comparison of Makeready Time with other Processes. Makeready time depends on the size of the press and several other factors. Quantitative comparisons are impractical because of disagreements about the definition of makeready. In general, litho makeready is faster than flexo, which is faster than gravure, because it often takes longer to change gravure cylinders or flexo plates than it does to change offset plates. In gravure the crew must usually change the doctor blade. All three processes require adjustment of registration. Washup is faster in flexo and gravure because the cylinders can be washed off press. Ink-water balance slows down offset makeready: ink balance is faster on gravure than flexo, which is faster than web offset. Mounting the plates on the cylinder should not consume makeready time because plates mounted on printing cylinders can be proofed off press and quickly placed into the press. However, mounting plates is an essential part of offset makeready.

The greatest waste of makeready time is poor management and poor organization and training of the crew. Required materials are not at hand, instruments and equipment necessary for a fast makeready are not available, and people stand around waiting for someone else to complete a necessary job.

Comparison of Makeready Waste with other Processes. Low makeready waste is one of the chief advantages of flexo and gravure over litho, but as with other printing processes, high waste often occurs because of poor skills or organization. The first sheet printed verifies that all the plates are properly placed on the press. If makeready is done off press and done correctly, there should be no need for more than one sheet to verify the accuracy of the makeready. With computer controls, impression is controlled and the compressible stickyback can accommodate whatever variations are likely to occur.

Flexo Plates

Flexo plates are discussed in detail in Chapter 7, but because they are important to the characteristics of flexo printing, they must be mentioned here.

Types of Plates. There are three types of flexo plates: molded rubber plates, laser engraved rubber plates and photopolymer plates (figure 3-10). Molded rub-

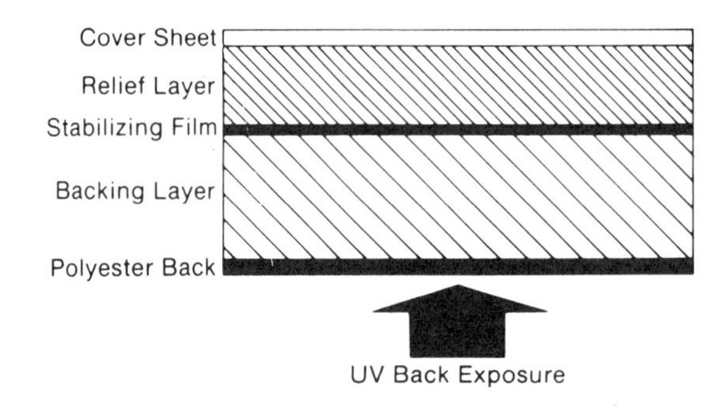

Figure 3-10. Structure and processing of a photopolymer flexo plate. (Source: Flexography: Principles and Practices, Fourth Edition.)

ber plates are produced by curing natural or synthetic rubber against a matrix or mold; laser plates are engraved into cured rubber with a computer-controlled laser, while photopolymer plates are produced by exposing a reactive material to a UV lamp through a photographic negative. The cost of producing an engraving and a mold makes the rubber plates more expensive if only a single plate is to be produced. However, because rubber is cheaper than photopolymer, it is cheaper to produce an engraving and a mold and make duplicate plates if more than four or five plates are required.

Molded rubber plates for printing corrugated remain very popular while photopolymers are moving to the thinner dimensions. One reason is cost; another is that high quality printing is being done more and more with photopolymers. Rubber plates are capable of printing high quality, too. It is often a matter of attitude more than technology.

The best photopolymer plates are considered to be somewhat better than the best rubber plates for producing halftones; rubber is often still preferred for line work. But, the difference in print quality produced by the two methods is usually small enough that economics usually governs the choice of plates. On the other hand, the three types of plates differ enough that color separations should be made for the process to be used. The same set of color separations should not be expected to give satisfactory color prints from molded rubber, laser engraved rubber and photopolymer plates.

Lasers can be used to engrave flexo plates. A laser-engraved, continuous, rubber printing cylinder is truly seamless and highly precise, providing competition for gravure.

Plates with steel backing can be mounted with clamps or they can be mounted on magnetized cylinders. Rubber plates can be vulcanized to the magnetic material while the plate is being molded and then mounted on a metal cylinder.

Printing Sleeves. Flexo plates can be mounted on nickel or plastic sleeves and easily put onto or removed from the press in register. This allows the registration of the plates to be completed before the sleeve is mounted on the press. It also allows plates that will be rerun at frequent intervals to be stored without removing them from the cylinder and mounting and registering before each run (figure 3-11).

Sleeves ("cylindrical plates") are also made with rubber or photopolymer material already on the sleeve so that the image can be developed on the sleeve itself. Although development work is taking place on photopolymers, it is not yet possible to reliably use lasers to etch the photopolymer material. Rubber is more versatile, and with laser etching it can achieve fine halftone screens. With laser engraving of sleeves, it is possible to go directly from artwork to plate, without any intermediate steps. More will be noted in Chapter 4 on generating copy.

Distortion. When a flat, flexo plate is wrapped around a cylinder, the length of the top becomes longer than the length of the base, and some distortion is created (figure 3-12). The printing surface is stretched, and the base of the plate may be compressed somewhat. The distortion is in one direction only. If the

Figure 3-11. Flexo sleeve, a means for rapidly placing mounted plates on a press. (Courtesy of Stork Cellramic, Inc.)

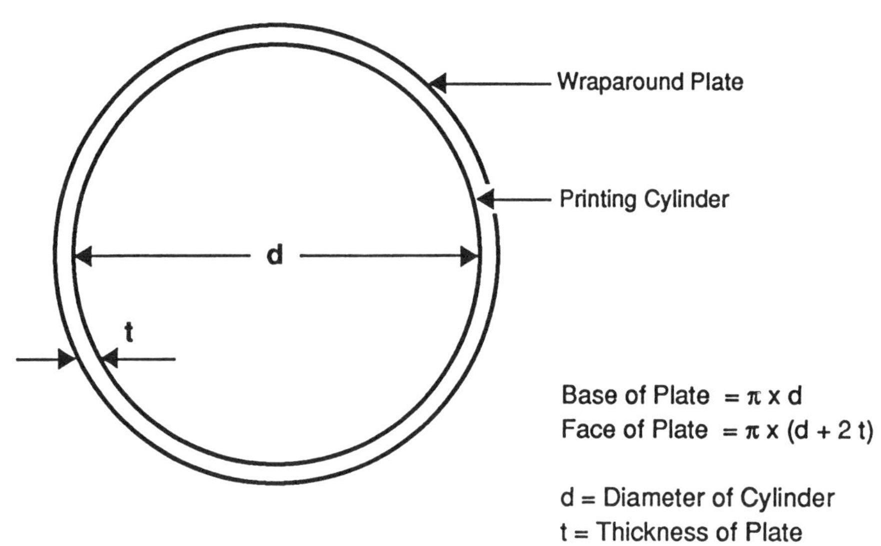

Figure 3-12. Distortion of wraparound plate.

platemaker does not allow for the stretch and if the printer does not keep it under control, circles become ovals and squares become rectangles and vice versa.

This stretching is negligible in offset, non-existent in gravure (where there is no plate to be stretched) and of no significance in screen where the image is formed at the plate-substrate interface. However, with wraparound letterpress and flexo, the distortion can be great enough to affect the length of the printed image.

In addition to the simple geometry of figure 3-12, other distortions occur. The plate tends to stretch more in the middle than at the edges, and the diameter of the plate stretched around the cylinder is different in areas with no image than it is in the solids areas.

The situation is actually more complicated than this. If the plate carries both line work and solid areas or has combinations of heavy line coverage and vignettes in process areas, the undercut or relief will differ, causing an uneven distortion throughout the plate. Furthermore, the plate becomes thinner in the center as a result of the pull around the circumference of the cylinder.

The thinner the plate, the less the distortion, and improvements in materials and technology have made it possible for flexo plates to become thinner and thinner.

Plates have been getting thinner over the decades, and even corrugated is tending to eliminate the thickest plates. Table 3-II lists typical plate thicknesses. Molded rubber plates, which are conventionally mounted on a processed cloth backing, can be made reliably stable because the backing materials resist stretching. Laser engraving machines (figure 3-13) make rubber plates in the round and eliminate the distortion inherent with flat plates. They can work directly from art and eliminate many of the intermediate steps in platemaking.

TABLE 3–II. TYPICAL FLEXO PLATE THICKNESSES

Material	Backing		Thickness		Relief		Total	
	in.	mm	in.	mm	in.	mm	in.	mm
Molded rubber								
thin			.107	2.69	.030	.76	.107	2.71
thick	.060	1.52	.190	4.83	.030	.76	.250	6.37
	.080*	2.00	.250	6.35	.150	3.80	.280	7.11
Laser engraved rubber			.107	2.69	.030	.76	.107	2.72
Photopolymer								
thinnest	.007	.18	.067	1.70	.042	1.07	.074	1.88
typical	.010	.25	.107	2.71	.042	1.07	.117	2.97
thick*	.012	.30	.250	6.35	.150	3.80	.262	6.65

* for printing corrugated

Figure 3-13. Laser engraving machine for making rubber plates (and anilox rolls) in the round. (Photo courtesy of ZED Instruments USA Ltd.)

The color separator works with the printer to take these facts into account and to reduce variability in the image. With modern materials they can achieve prints that compete with the best that is produced by gravure, litho or letterpress.

Stickyback and Compressible Cushions. Flexo plates producing the best work are usually mounted with foam stickyback cushion tape, but some narrow-web printers produce excellent work without using compressible backings. Stickyback is applied to the cylinder to hold the plate in place. The carrier may be a hard vinyl film or a foamed cushion (usually urethane) sandwiched between two adhesive layers of rubber-based adhesive and a protective liner. Proper application of the stickyback is critical to its performance.

Proper selection of foam stickyback improves print quality, with foam density being the most important factor. The advantage of the compressibility is that it "smooths out" the variations of the substrate, so the printing plate remains level throughout, improving print definition and sharpness.

A new photopolymer plate can have a cushioning factor built into it. It sometimes requires the surface of the plate to be hardened and its porosity to be eliminated.

Flexo Inks

Flexo inks contain pigments suspended in a solution of a binder in water or an organic solvent, often an alcohol. The inks have low viscosity, like that of a gravure ink or about that of water. Flexo and gravure inks are called "fluid inks" or sometimes "liquid inks" as opposed to the more viscous "paste inks" used in litho, letterpress and screen printing.

Inks on the fast-moving, printed web dry quickly by evaporation of the water or organic solvent, or, with paper or paperboard, by absorption into the substrate. If the liquid is an organic solvent, it is necessary to recapture and dispose of the evaporated solvent to prevent air pollution; the low boiling point (or high volatility) of flexo ink solvents makes this difficult. Consequently, a great deal of work has been devoted to development of waterbase inks.

While use of water as a solvent reduces air pollution, it creates problems in wetting of water-resistant packaging materials. Air pollution problems are not entirely solved either, because waterbase inks usually contain at least a small amount of solvent to improve the wettability of the substrate and the solubility of the resinous adhesives. There are few packaging flexo inks that are completely free of any alcohol or other organic solvent. Current research and development work is directed to improving waterbase inks as well as improving the wettability of water-resistant packaging materials.

Ultraviolet curing inks are also being used to avoid the problems of solventbase and waterbase inks.

If the solvent or water evaporates from the ink fountain on the press during printing, the ink grows more concentrated and more viscous, and the color density of the print usually increases. However, sometimes the increase in color density is so small that it is not apparent. Thus, much ink is wasted, and costs go up. The viscosity must be monitored during the run. When it exceeds a specified limit, a make-up solvent is added.

Advantages of Flexo

- Flexo gives good print color consistency.
- Press crew levels are often lower than offset or gravure.
- Good waterbase inks are readily available.
- Flexo prints well on rough substrates.
- It prints on a wide variety of substrates, including low strength and lightweight papers.

- If the cylinders are made ready off-press, makeready can be done very quickly.
- Flexo prints faster than sheetfed offset.
- Flexo waste is low.
- Flexo prints large solids evenly and without voids.
- Rapidly evolving technology keeps improving flexo quality and productivity.

DISADVANTAGES OF FLEXO

- Shallow-relief plates can plug easily with dust or dirt.
- Careful control of printing pressure is essential.
- Highlight halftone dots tend to disappear and shadows tend to fill in.
- It is not practical to adjust colors on press.
- It is currently not possible to make a smooth transition of dot size in vignettes, especially at one and two percent dots.
- Flexo speeds are usually less than gravure or web offset.
- Preparation of plates is a lengthy procedure.
- Flexo is often perceived as a cheap printing process, and therefore poor quality flexo is often accepted as satisfactory. As much as anything, this has hampered the growth of flexo.

Comparisons with Other Processes

The ability to print high quality images on a wide variety of substrates at an economical cost has made flexo the most important printing process for package printing.

It is difficult to distinguish top quality flexo, gravure and offset printing without careful examination under a magnifying glass. Each process has its characteristic idiosyncrasies, and it is easy to identify the printing process when the printing is second-rate. For the reproduction of type, flexo is considered to be not quite as good as offset or letterpress but better than gravure. On the other hand, flexo is considered better than offset, but not as good as gravure for the smooth and uniform reproduction of large solids. Flexo plates cost more than offset plates but far less than gravure cylinders. Flexo usually prints faster than sheetfed offset, but not as fast as web offset or rotogravure.

Flexo generates less waste than web offset printing. It is comparable to gravure or sheetfed offset. There are great variations in the amount of waste generated from job to job and from plant to plant so that generalizations are apt to be misleading.

Trends in Flexo Printing

Flexo is gaining market share as quality continues to improve, printing speeds increase, and as packagers realize that excellent quality can be achieved with flexo, often at lower cost. Flexo grows as it replaces sheetfed offset and gravure, especially in printing of folding cartons. Photopolymer plates currently dominate most flexo printing. Waterbase inks are widely used, even on film and foil, and their use is growing. Plates are getting thinner; even corrugated is being printed with plates as thin as 0.067 inch (1.7 mm).

Printing speed continues to increase. Printing speeds of all processes are much higher on publication work than on packaging work. Print speeds for flexo seldom exceed 800 feet per minute (4 meters per second) except on newspapers (where speeds of flexo are soon expected to reach 3000 feet per minute—15 meters per second). In packaging, process color speeds of flexo are generally lower than with gravure. When printing line work, press speeds above 1000 feet per minute are occasionally reached in package printing. Process color speeds are lower.

In 1991, a cooperative research program began to develop test targets and parameters to make flexo more controlled. Control of the volume, pattern and count of the anilox roll proved vital. Stronger inks gave smoother solids, better halftones and improved drying. Capped plates gave better solids and halftones than the non-capped counterpart, and harder plates gave better results than soft ones. Uniformity of impression requires uniformity of plate, mounting tape, press cylinder and cylinder bearings.

ROTOGRAVURE

Rotogravure is a method of direct rotary printing that uses an engraved metal cylinder to print on smooth substrates such as coated paper and plastic films. The fluid inks are applied directly to the engraved cylinder and metered with a doctor blade. Variations of the gravure process include the use of an offset blanket to transfer the image from the cylinder to the substrate.

Gravure printing products are usually grouped into packaging, publication and specialties. This section is directed to packaging, with some mention of the others.

The key features of gravure are low-viscosity inks that are made from a pigment and a binder dispersed in a volatile organic solvent or in water. The inks are printed from a metal cylinder that has the image engraved in it. The cylinder is rotated in an ink bath to fill the cells that form the pattern. The excess ink is removed with a stainless steel doctor blade (figure 3-14).

Figure 3-14. Rotogravure doctor blade. (Courtesy of Allison Systems Corp.)

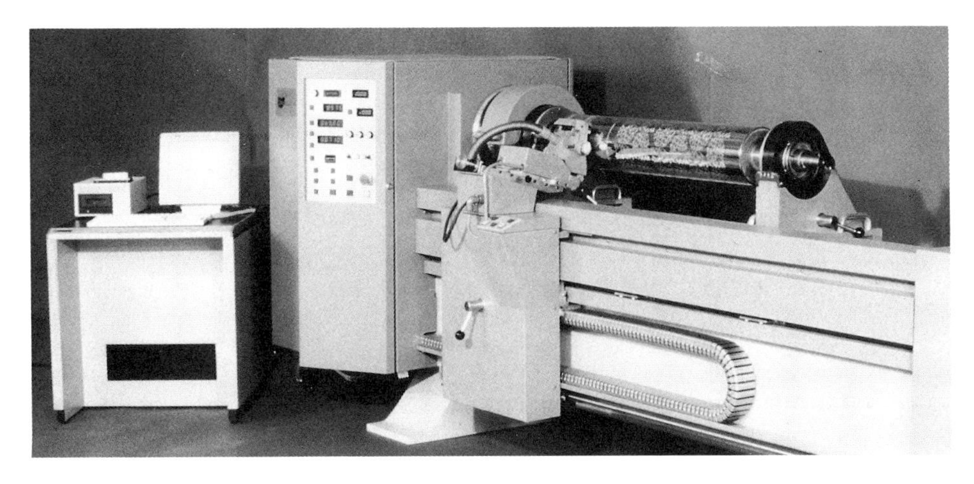

Figure 3-15. Electromechanical engraving machine. (Courtesy of Ohio Electronic Engravers, Inc.)

Offset gravure in which the cylinder prints on an offset blanket is sometimes used to print on paperboard if the substrate is not sufficiently level. Gravure printing is almost always from webs, although a few sheetfed gravure presses are still used to print folding cartons for products such as hair coloring and cosmetics where color consistency is especially important.

With digital color separation, digital proofing and digital engraving, gravure can now be a filmless process, with digital input/output.

At one time, gravure preparation required a intense amount of craft skill. The old "carbon tissue" method of engraving is dead, and supplies are no longer available in the United States. This process has now been replaced by electromechanical engraving machines (figure 3-15) and by direct engraving with chemicals. Most package gravure cylinders are produced on the Ohio Electronic Engraving Machine, while most of the Helioklischographs, made by Linotype-Hell, are used for publication printing.

An impression (backup) cylinder presses the substrate against the engraved cylinder. Cells on the cylinder are engraved to different depths so that gravure is the only printing process that can print halftone dots with planned differences in ink film thicknesses.

To make good contact with the metal cylinder, the substrate must be smooth; paper for gravure must also be somewhat compressible. For most work, the entire image—type and pictures—must be screened, and the screen pattern seen in type matter is a characteristic of gravure. Screening of the entire image introduces a scallop into type and fine reverses, but it also controls the printing of solids. If a 150 line/inch (60 lines/cm) screen is used, the scallop along the letters is invisible except under a magnifying glass. Nevertheless, gravure usually does not reproduce fine reverses as well as flexo or litho.

Because of the high cost of preparing a gravure cylinder, gravure is limited to long run jobs: tags and labels, cigarette cartons and cartons that require a heavy laydown of ink. Shorter run lengths favor flexo and web offset, and the trend toward shorter and shorter run lengths has challenged gravure. Web offset is being used on runs too short for gravure—probably not as short as are profitable with flexo.

In this age of just-in-time (JIT) delivery, gravure is at a serious disadvantage. The customer expects quick delivery of the first job, but frequently there are small changes in the repeat order. For example, on a web of wrap for frozen foods, the original layout had six wrappers for beans, six for corn, and four for peas. The new order specifies six for beans, four for corn, and six for peas. This problem is quickly handled by flexo because appropriate plates can be prepared and mounted, but for gravure, it requires the long and costly preparation of a new cylinder.

Combination presses are available that print process colors by offset and put metallics down with gravure. Others print the main image by gravure and add variable information with flexo.

Markets for Gravure

In addition to printing flexible packaging (plastic film, foil and paper), tags and labels, and carton board (especially cigarette cartons and beverage containers), gravure is used for many non-packaging printed products. It is generally unsuitable for rough materials such as corrugated board or uncoated papers. When gravure is used on corrugated liner, the liner must be coated or filled to improve its smoothness.

The decreasing run length is one societal impact that causes gravure to lose market share: it is losing market share of flexible packaging to flexo and of folding cartons to both web offset and flexo. Gravure printing of packaging materials continues to grow, but less rapidly than flexo and offset.

Substrates for Gravure

To make satisfactory contact with the gravure cylinder, the substrate must be level and very smooth. Coated papers and board, foils and extruded polymer films work very well with rotogravure.

If the paper for gravure printing is not smooth and level, it will not uniformly contact the gravure cylinder, and this creates a problem of "speckle" or "snowflaking" (figure 3-16). There is a difference between smoothness (uniformity of the surface) and levelness (uniformity of paperboard caliper across the web). However, lack of either smoothness or levelness creates ink transfer problems. If a blanket cylinder is added to the press, the ink can be printed on uncoated paper or board, and some offset gravure presses are in use.

Compressibility, also called "cushion," is helpful in overcoming slight roughness. The papermaker improves compressibility or "cushion" in coated paper and board by incorporating a rubber (styrene-butadiene) latex into the coating.

Although the substrate must be smooth, it need not be especially strong or stiff. Gravure is satisfactory for printing on low basis weight papers, even tissue papers.

Figure 3-16. Speckle or snowflaking, a typical defect in gravure printing. The top picture was printed with electrostatic assist. Note scalloped edge of dot patterns. (Courtesy of Gravure Association of America.)

Characteristics of Gravure Printing

The simplicity of the process helps make gravure the fastest of the printing processes—the web runs at the greatest speed. Gravure printing suffers less downtime, and there is less waste than with offset printing.

Principal compensating features of gravure, against the high cost of preparing a cylinder, are the speed and length of run that can be obtained from it. The high cost of preparing a cylinder is less important on long runs. Since gravure presses run at the highest speed of any process, gravure has an important labor cost advantage on long runs when compared with other printing processes.

The halftone structure of gravure gives excellent skin tones. The scalloped edges of line copy are characteristic of the process. Although gravure does not hold fine halftone dots, the ability to print thin films from "highlight" cells gives excellent pictorial color.

The gravure ink film is thicker than flexo or offset. This thicker film promotes gloss when printed on paper or paperboard. The heavy ink film applied by gravure can be particularly advantageous for printing of fluorescent colors and metallic colors—gold and silver.

Another advantage of gravure is its color consistency and stability. Gravure printing shows the best color consistency of any printing process because the cells engraved into the cylinder meter the ink exceptionally well. This also means that the press crew can do little to change or improve the print image after the press is started up. They can make minor changes by adjusting the angle of the doctor blade and the color strength of the ink, but the buyer should plan not to make changes in color or image after preparation of the cylinder has been completed.

Correction of the cylinder is difficult, and price changes or other alterations are more expensive than with any other process. Cylinders can be changed by very meticulous craft procedures.

Storage of original artwork is practical, but the time and expense of manufacturing a new cylinder for a repeat job means that the large, heavy cylinder is stored if the job is apt to be repeated soon—as it often is with package printing.

Gravure Presses

As with flexo and offset, gravure presses now apply many colors. For example, a seven-color press would be used to apply four process colors, two spot colors, and a lacquer or varnish. Gravure presses can have as many as 10 colors.

A packaging press may carry two cyan inks, two yellows, a magenta and a black, in addition to other colors. In theory, packagers probably could use process colors for printing logos, but in practice it is doubtful that satisfactory uniformity of color would be achieved. Corporate colors such as Jolly Green Giant green and Campbell's Soup red must not vary, and they are generally printed with spot colors.

Image Carrier

Gravure Cylinder. The gravure image carrier is normally a steel cylinder with a layer of copper that is engraved or etched to carry the image, then chrome plated. This produces a cylinder that will run for millions of impressions. A gravure cylinder for printing labels may be as small as three inches (7.5 cm) in diameter and an inch wide, while a cylinder for beer cartons or cigarette cartons may be as long as six feet (1830 mm) and 14 inches (355 mm) in diameter. These big cylinders weigh several tons.

The cylinder is engraved electromechanically by special diamond styli or etched by a direct transfer process. Both processes can be controlled to give three different types of cells: variable depth cells, variable area cells, and cells that vary in both area and depth. The nature of these cells helps to create smooth halftones that render excellent appearance to facial tones and hair color.

Engraving the Cylinder. Before engraving the gravure cylinders, a color separation must be produced from the original colored artwork. To do this, the finished artwork is wrapped around the cylinder of a color scanner. As the cylinder turns, a light beam scans the image and the color density is measured and recorded electronically. The computer determines the cell depth and area needed on each of four cylinders to reproduce the colored image.

The color scanner separates the light into its components (red, green and blue), and the intensity of each component required to reproduce the colored picture (cyan, magenta, yellow and black) can be recorded digitally on a disc or optically on a special opaque white plastic. The prints are called "bromides." The electromechanical engraving machine can be driven from these bromides, positives or negatives, or from the digital data.

Equipment is available that makes it is possible to separate the colors and engrave the cylinders in a single step, but it is usually carried out in this two-step process. The single step process eliminates the need for photographic film in the preparatory operations. Saving film significantly reduces the cost of short-run printing but has little impact on the cost of long runs produced by gravure.

Electronic Engraving Machines. The first electronic engraving machine was introduced about 1960, but it was little used in packaging before the 1980s. By 1990, there were at least a hundred such engraving machines in the packaging industry worldwide.

Such a machine can engrave 3600–4200 cells per second, a small packaging cylinder in an hour with a single head. Two or more heads are used on large cylinders. While an electron beam engraving machine has been developed, it is not yet possible to generate data as fast as the machine has the capability of running. With modern electronic engraving, cylinder preparation is much faster than it used to be. It should be unnecessary to retouch the cylinder if it is made with the best modern technology.

Direct Engraving. A common alternative to electronic engraving is direct engraving that involves spraying a photoresist on the copper cylinder, then exposing through a film positive and developing the image. After the resist is removed, the cylinder is etched with acid (or by an alkaline process), creating the cells.

Makeready

Gravure makeready time varies greatly with the particulars of the job, the skill and training of the crew and available technology. New robotic handling of gravure cylinders can change heavy cylinders in less than a minute.

Makeready of a gravure press is as fast and simple as any press. There are few alterations or adjustments that can be made to change the amount of ink that is fed. Adjusting the blade angle makes some difference in the ink film thickness, and consequently on the color, but the stability of the color of gravure printing is one of its greatest strengths. On the other hand, if the print is not satisfactory, the cylinder must be remade or repaired. While such a procedure is to be avoided in any printing process, it is particularly costly in gravure. Changing the color of the print to please the customer after the cylinders have been engraved is prohibitively expensive.

For rotogravure, if inks are not changed, makeready is fast: eight colors per hour, or less if the printer uses a double crew. If the inks must be changed, makeready time is longer.

Color variations in gravure arise from variations in ink pigment level and from viscosity variation. The ink fountain is a shallow pan with an applicator on the cylinder. The applicator keeps the cylinder wet in order to avoid trapping air in the engraved cells. The ink is fed to the fountain by a pump which is often hand operated on packaging presses but runs automatically on most publication gravure presses.

Use of robots is reducing makeready time. Robots are available for removing and replacing gravure cylinders in a matter of seconds. While expensive, they pay off in reduced makeready time (see Chapter 8).

Cylinders are easily transported to the press on trolleys. These trolleys are not limited to gravure, but they can be used to replace gravure cylinders with flexo cylinders.

Register Control

One of the problems of web printing has been keeping all of the cylinders in registration with each other. This has been largely solved with the development of an electronic monitor control system. Reading a control spot with an electric eye, the system accelerates or slows down each cylinder to keep all images in register. Developed for gravure, the system is now available for all web presses, and it is helpful in maintaining print quality and product.

Getting all the cylinders into register is a major part of makeready. With an electronic register control system (sometimes called an "electronic line shaft"), register is readily achieved and maintained.

Electrostatic Assist

A process called electrostatic assist (sometimes ESA or electroassist or "Electrosist") helps attract ink out of the cylinder onto the substrate, and it compensates to some degree for lack of smoothness in the paper or board being printed. The impression cylinder is given a charge opposite to that on the cylinder, and the charged ink is attracted to the substrate. Although it is sometimes used on film, printing of paperboard benefits more from ESA than does film. ESA has found greater use in packaging than in publication gravure printing. It is useful with both petroleum base and waterbase inks.

PROOFING

The subject of proofing is discussed in Chapter 6. It is one of the biggest headaches in gravure printing.

The apparent effort and devotion that go into a long drawn-out job are sometimes perceived to be the marks of a careful production manager. It should not be assumed that a slow, expensive proofing process is necessarily better than a fast, well-controlled process. "Job security" is dying, although it is still around. Because of the unacceptable costs involved in the old gravure press proofs, the packaging industry is moving to digital proofing.

It is a simple, inexpensive task to convert the digitized color separation data into a colored image on a video display terminal, but interpreting the image on the terminal in terms of what it will look like when it is printed on paper is more art than science. Art directors usually like to see a conventional hard proof, not an electronic image. There are many ways of converting the digitized image into hard copy, and while the best prepress proofing gives an excellent idea of what the printed image will look like, none gives an exact replica of the final product. The best replica is obtained from a wet proof (a print) that is obtained from the final cylinder. Correcting the image at this point is excessively expensive.

People complain that prepress proofing methods don't give the same image as a "wet" proof, but they've learned to work with them. The artist, packager and printer must learn to "read" the digital proof and make suitable judgments and adjustments. They are assisted by continuously improving technology.

Leaders in the packaging industry are working towards the day when one off-press proof will be prepared and approved, the cylinder will then be mounted and printed without further proofing. To achieve this requires the latest technology, control of the process, training of crews, and cooperation between customer and printer. The financial returns will be worth the great effort and will help to make gravure more competitive with flexo and offset.

Proofing is an area where skilled, thoughtful leadership can greatly reduce the cost of a printing job. As in most areas, personal skill and judgment coupled with the best technology can produce an improved product at greater speed and lower cost.

Halftone Gravure

Some of the problems involved in proofing a gravure cylinder are avoided by halftone gravure, a process by which gravure cylinders can be prepared from offset separations. The offset separation is proofed, and it can then be used for both offset plates and gravure cylinders. To prepare a gravure cylinder, the separations are mounted in the engraving machine. By selection of proper computer corrections, a cylinder is prepared that will match the halftone print if the gravure inks match the color of the halftone inks.

Halftone gravure is advantageous in publication printing because ads in magazines printed by either offset or gravure can be printed from a single set of color separations. Halftone gravure has been accepted in packaging because it allows offset proofing. On the negative side, the required SWOP/gravure inks (GAA Group VI) are less brilliant than the earlier standard inks and packaging gravure inks. Also, gravure naturally lays down inks of greater intensity and strength, especially in the bright colors, than does offset. Halftone gravure does permit ease of working and proofing. It is easier than proofing from continuous tone separations.

Gravure Inks

Standard rotogravure ink colors have been defined and maintained for many years by the Gravure Association of America and its predecessor, the Gravure Technical Association. These are known as Gravure Standard I, II, etc. Standard colors are not always used in gravure printing. Many package printers believe they can secure brighter, cleaner colors by using their own inks. Currently, with the conversion to halftone gravure that uses lithographic color separations as a basis for engraving the cylinder, the so-called "SWOP Standards" (Specifications for Web Offset Publication) have been adopted for gravure publication and packaging by many printers.

Solventbase gravure inks, like solventbase flexo inks, have low viscosity and high volatility. The inks evaporate readily and give fast, low-cost drying. A greater variety of solvents can be used with gravure than with flexo because the metal cylinder resists softening and swelling in strong solvents. Strong solvents like ketones or acetates "bite" into a plastic substrate, resulting in better adhesion after drying. Organic solvents, however, cause environmental problems, and they must be captured or disposed of at considerable expense. Inks must be formulated to flow rapidly into and out of the fine cells. Foaming must be minimized.

Etching of the cylinders is different for waterbase than for solvent systems. Waterbase inks differ from solventbase inks in surface tension, viscosity, color strength and printability. Waterbase inks carry higher pigment loading (less solvent). The reduced amount of water helps reduce the amount of energy required to dry the ink. Surface tension causes waterbase inks to bead up and resist wetting the lower energy surfaces (such as untreated polyolefin), affecting ink transfer and causing skips, particularly on the third and fourth colors printed. Special care is required in making waterbase inks for electrostatic assist. It is reported that waterbase inks prepared with distilled water work well with ESA.

Advantages of Gravure

- Rotogravure prints at the highest speed of any process.
- Productivity is high, waste is low.
- Consistency of color reproduction is excellent. The press design avoids mechanical ghosting.
- Rotogravure is excellent for reproducing facial tones and skin colors. Different cells can be engraved to different depths to vary the thickness of the printed ink film.
- Long runs into many millions of impressions are practical because the cylinders resist wear.
- After the cylinder is prepared, production costs are modest, so that repeat runs are relatively inexpensive.
- Press makeready is fast and simple.
- Gravure prints well on low strength and lightweight papers.
- The heavy ink films help give bright, glossy prints.

Disadvantages of Gravure

- Preparation of a gravure cylinder is a lengthy and costly procedure, making the process economical only for long runs. Short runs are cost-effective only if they are repeated often.
- Gravure does not print well on rough or unlevel paper and board.

- Small type, especially reverse type, is often ragged.
- Gravure gives poor resolution of fine details.
- It is difficult to make corrections in the cylinder.
- Development of good waterbase inks has lagged.
- Storage of the large, heavy cylinders is costly.
- Gravure technology has not kept up with flexo and offset.

Comparison With other Processes

Intense colors and good color consistency are the strengths of gravure printing. The relatively thick layer favors printing of fluorescent colors. Only screen printing lays down a thicker film.

Gravure presses usually run the fastest of any printing process. The cylinders run for many millions of impressions, so that on long runs, productivity is unequaled. Gravure can produce high quality printing consistently and, on long runs, economically.

Gravure waste is usually much lower than it is with web offset, especially on jobs requiring high color consistency. Lower waste augments the high productivity and lowers the cost.

Because type must be produced from gravure cells, gravure does a poorer job of reproducing small type, fine reverses and bar codes than flexo, offset or letterpress. On the other hand, it prints heavy solids more uniformly than other printing process. The heavy lay of ink produces an attractive glossy appearance that sometimes avoids the need for additional coating.

Because technology is improving flexo quality faster than it is reducing the investment cost of gravure, flexo is gaining market share in competition with gravure.

Trends in Rotogravure

Gravure's problem has been the time and cost of getting the cylinder ready for the press. Electromechanical engraving helps overcome this. Electronic registration control helps to reduce the makeready time and to improve gravure quality. Robots accelerate makeready. But even with the advantages of new technology, shorter run lengths or smaller print orders make gravure impractical for a great many jobs and challenge the industry to make further progress.

Waterbase gravure inks have been used on paper and paperboard packaging since the early 1980s, and they are now being applied to film. Environmental requirements will assure their future growth. Inkmakers are continuing to study and improve the performance of waterbase inks.

The chemistry of copper engraving creates another ecological problem. Plastic coating technology is available and may replace copper when ecological demands require it.

Halftone gravure, the preparation of gravure cylinders using offset color separations, makes possible the use of offset proofing technology to proof gravure separations. The technology is finding applications in packaging as well as in publication gravure.

LITHOGRAPHY

Offset lithography is a method of indirect rotary printing that generates images from a plate surface that is neither raised nor engraved. The plate separates image areas from non-image areas chemically. The non-image area is wet by water; the image area is wet with ink. The image is transferred to a rubber blanket that conforms well to a wide variety of surfaces and prints well on uncoated paper. The stiff or paste inks are metered by a series of rolls and a doctor blade or a series of short blades called "ink keys." When applied to a litho plate dampened with water, the inks adhere only to the image area. Plain water does not work very well, and other materials including phosphoric acid and gum arabic are added to prepare a "fountain solution" or "dampening solution."

The lithographic process includes both sheetfed and web presses, and variations include printing from plates that do not need water for dampening. Although direct lithography (directly from plate to substrate) is possible, it is rarely used in packaging. The name "offset lithography" or "offset" therefore is descriptive of the actual practice of litho in package printing.

The process gives images with excellent sharpness or clarity, and productivity is excellent.

Plates are inexpensive and quickly prepared. They cost a tiny fraction of the cost of a gravure cylinder. Paper plates (used in office duplicating) are even cheaper than the aluminum plates used for most production work—packaging and publications.

The mixing of ink and water on the press can result in emulsification, causing streaks and weak, washed-out colors. Control of ink/water balance requires skilled press operation.

The offset blanket serves the same function that a soft plate plus cushioning serves for flexo: it conforms to rough or uneven surfaces. The blanket also makes possible the printing of metal, so that offset lithography and other offset processes are used for metal decorating.

Like flexo, litho is economically practical from just a few hundred copies to a million or more, covering most package printing runs. Sheetfed presses can profitably produce process color runs under a thousand impressions, and web presses can profitably produce above a million. The crossover between sheetfed offset and web offset lies between 20,000 and 50,000 impressions for process color work, much lower for one-color work.

Unlike the liquid or fluid inks used in flexo and gravure, litho uses paste inks that are much stiffer or higher in viscosity. Litho inks must be worked enough on the press so that they flow readily but still print a sharp image. The work requires a press with many ink rolls. The stiff inks exert more force on the substrate than the inks of flexo or gravure, and this requires stronger paper, which is often more expensive.

Web offset inks are finding some use in packaging. Although they are less viscous or tacky than sheetfed inks, they are still classified as paste inks. They dry by evaporation (like flexo and gravure inks) or by ultraviolet (UV) or electron beam (EB) curing.

Lithography requires water which must be carried together with the ink in the inking system. Accordingly, use of waterbase inks is impractical. Dry offset, printing from a special lithographic plate that does not require water, has attracted attention during the 1990s. (The term "dry offset" is sometimes used for offset letterpress, an unfortunate confusion.)

Drying of heatset web offset inks dries out the substrate and requires use of lots of chill rolls, but it is faster than sheetfed offset, and the increased speed often tips the economics in favor of web. When UV or EB ink curing systems are used, the heat/chill cycle is avoided, but the costs of ink and curing system are greater.

In spite of the fact that offset is a mature process, the economics of the process added to continually improving technology keep it growing in packaging and publication markets.

Offset lithography can print on both sides of the web or sheet at once (called "perfecting") more easily than any other process. This is of great significance in publication printing; of less importance in package printing.

Markets for Offset Lithography

The principal types of packaging printed by offset lithography are labels and wrappers, folding cartons and some metal cans. These have traditionally been printed by sheetfed offset because the image size varies from one job to another. Because of its greater speed, and because technology has made it possible to produce excellent quality printing, web offset is starting to grow as a package printing process. A major disadvantage of most web offset presses is that they have a fixed cutoff—they take only a single-size plate. This limits them to printing products of a single size. For package printing, the press must print the size demanded by the packager, and package sizes are not usually standard.

In packaging, some products such as single-serve cereal boxes, citrus juices or half-gallon ice cream cartons generate enough volume to keep a fixed cutoff web busy.

Variable cutoff web offset presses are opening up new markets for lithographic printing of cartons and labels, where they compete with gravure.

Substrates for Lithography

Because the offset blanket conforms to a rough surface, offset lithography can print on a rougher surface than gravure or letterpress, and paper or board smoothness is not a critical requirement for offset. However, paper must be sufficiently smooth to give a good appearance to any print. No process of printing can produce a really attractive print on a brown kraft bag.

Offset and letterpress inks are stiffer or more viscous than gravure or flexo inks. It takes a great deal of force to split the thin film of viscous ink between the paper substrate and the offset blanket. The force required to peel the paper off the blanket often picks the paper, causing hickeys, pickouts, curled paper and other problems. Offset therefore requires paper and board with high surface strength. The paper must also be stiff in order to avoid stretching and to feed properly in the sheetfed press.

Since the offset blanket contacts the entire surface of the paper or board, litho also requires especially clean paper and board.

Non-absorbent materials such as film and foil cause special problems in litho. The ink contains some emulsified water that tends to wet the substrate, repelling the ink and causing water spots, streaks and other problems.

Dry offset plates have been used for printing plastic films, but the market is not large. Other printing processes are usually more economical. If the process is controlled, lithography prints well on metal surfaces, and it is used for metal decorating.

Most web offset presses have an in-line configuration that is unsuitable for printing thin plastic film, a problem aggravated by the tacky offset inks and the dampening solution. Most flexible films are not stiff enough to feed into a sheetfed press.

Characteristics of Lithographic Printing

Offset lithography produces the thinnest film of ink used in any major printing process. The thin, stiff inks give sharp dots. Highlight dots are faithfully reproduced and shadow dots remain open. Because the ink film is thin, the color density of the ink must be higher than for ink used in any other printing process. Under the best conditions, offset can print a screen of 500 lines per inch or even "continuous-tone" prints.

The sharp edges of fine print are characteristic of lithography. Gravure edges are serrated or scalloped, and letterpress and flexo have squeeze-out marks or halos around the edges of letters and solids, a characteristic that becomes objectionable if the process is not carefully controlled.

Litho inks used for sheetfed printing dry slowly, and they sometimes set off onto the back of the next sheet before they are dry. For most jobs, setoff is intolerable.

Figure 3-17. Water streaks or wash marks, a typical defect in lithographic printing.

Unless it is carefully controlled, the water in the process results in emulsification and streaks. If the ink does not accept water easily or if the press is run with too much water, water may emulsify in the ink. Water streaks (figure 3-17) are often seen when the press carries too much water. Emulsified water causes pale "washed-out" colors and may show up as spots in areas of heavy coverage. Ink emulsified in water appears as small specks of ink in the non-image area. These defects are characteristic problems of lithography.

If the ink is too stiff, it may pull or "pick" bits of paper from the stock. These collect on the printing plate and print as "hickeys," small donut-shaped specks or spots. If the ink pulls fibers from the stock, and these lodge on the blanket, they may print "lint," the pattern of the fibers on the blanket. Although most frequently seen with lithography, letterpress inks can occasionally cause picking. Gravure and flexo inks are usually too fluid to pick or "pluck" the paper.

Picking, linting and piling are problems of offset lithography (figures 3-18 and 3-19) where the substrate collects on the blanket, interfering with the sharpness of the image. When the paper or coating dust works its way into the ink on the blanket, it alters ink transfer and creates unusual patterns. When this "press contamination" occurs, the operator must shut down the press and clean it. Paper, especially paperboard, tends to be dusty, and a vacuum sheet cleaner on the press can greatly improve productivity.

Paper and board must be essentially free of lint and loosely bonded pigments. Cutting and trimming generate a lot of dust if the paper or board is not finished using the best equipment and procedures. Even the best board contains some dust. If it is not manufactured under the cleanest of conditions or if it is poorly cut or trimmed, good productivity may be impossible. Dust and dirt accumulating

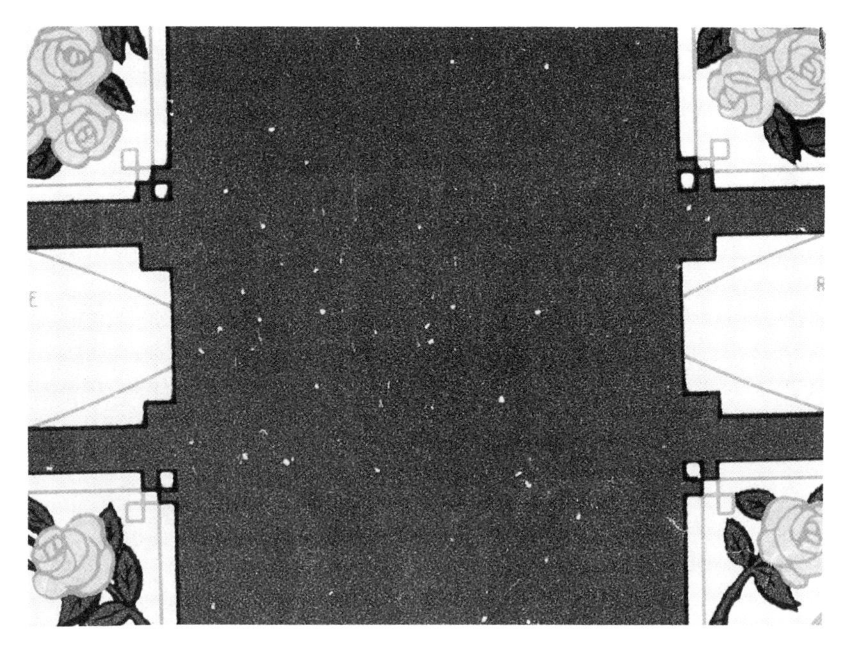

Figure 3-18. Picking, a typical defect in offset lithography.

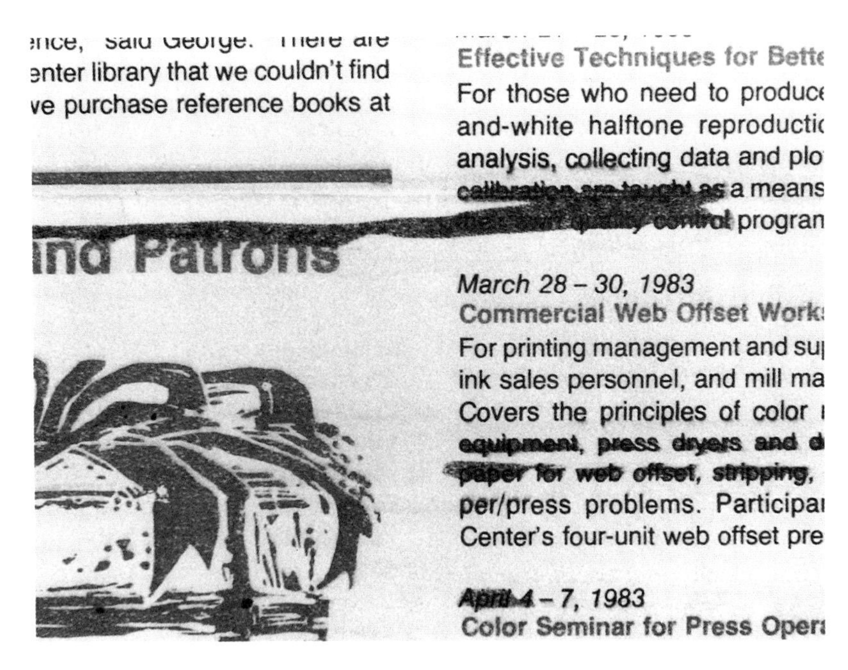
enter library that we couldn't find
ve purchase reference books at

Effective Techniques for Bett
For those who need to produc
and-white halftone reproducti
analysis, collecting data and plo
a means
progran

March 28 – 30, 1983
Commercial Web Offset Work
For printing management and su
ink sales personnel, and mill ma
Covers the principles of color
equipment, press dryers and
paper for web offset, stripping,
per/press problems. Participa
Center's four-unit web offset pre

– 7, 1983
Color Seminar for Press Oper

Figure 3-19. Piling, a typical defect in offset lithography.

on the blanket reduce print quality and require that the press be cleaned, reducing productivity.

The design of the offset press makes mechanical ghosting possible. Unlike gravure, where the image on the cylinder is completely filled with ink each time, or flexo where the plate is completely inked each revolution, the offset plate is inked by transfer from a rubber roll. When the plate prints a heavy image, the rolls may not completely replace the ink on the next revolution, leaving a ghost (figure 3-20).

Sheetfed Offset Presses

Offset presses are immensely more complicated than flexo or gravure presses. They include a plate cylinder, a blanket cylinder and an impression cylinder. They require many rolls to distribute the ink, and they also require a few additional rolls to apply water to the plate (figure 3-21). The mechanism for feeding sheets to the press is complicated. Press speeds for a sheetfed press usually run around 10,000–15,000 sheets per hour, equivalent to 300–400 feet per minute (1.5–2.0 meters per second) depending on sheet size. A press printing 10,000 26-inch-long impressions per hour, runs at 360 feet per minute. Press speeds for both sheetfed and web offset have been increasing in recent years, but web offset presses continue to print at about five times the speed of sheetfed presses.

Sheetfed offset presses can economically print a wide range of sizes of product. This is a valuable asset in packaging where package sizes vary, and where several boxes or labels may be ganged onto a single plate. The size of a sheet can be varied from one job to the next. If the size of the sheet is reduced from 60 x 40

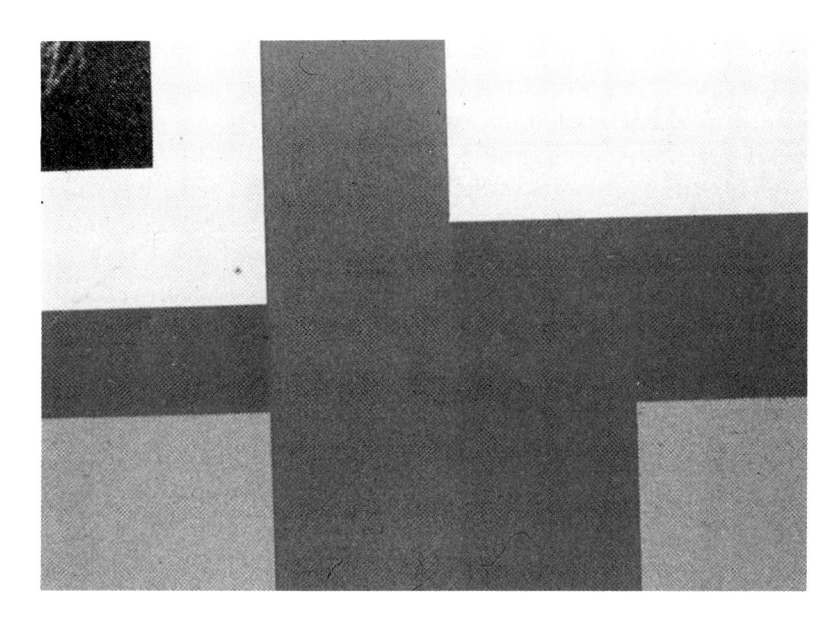

Figure 3-20. Mechanical ghosting in offset lithography.

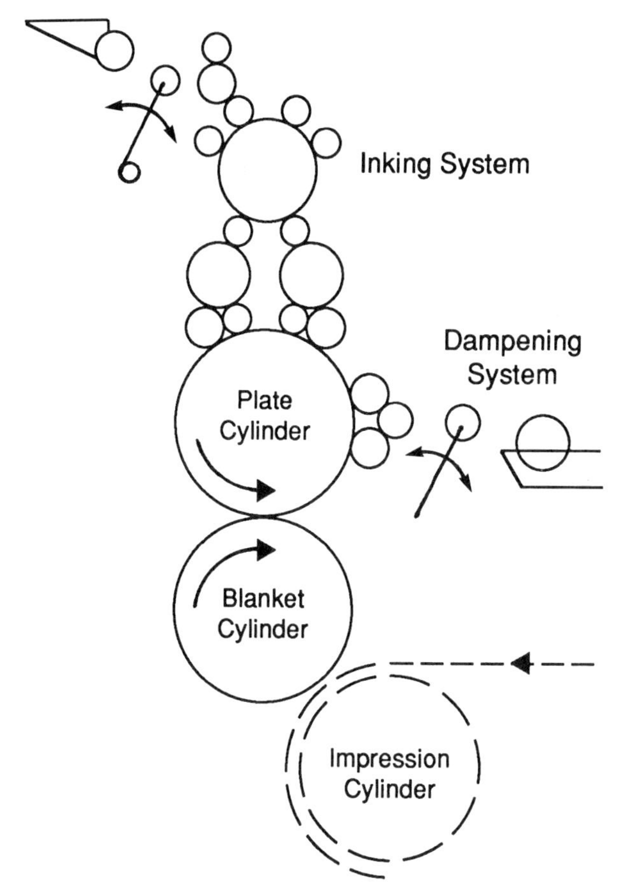

Figure 3-21. Roll diagram of an offset press.

inches (1524 x 1016 mm) to 57 x 36 inches (1448 x 914 mm), the productivity of the press is somewhat reduced, but most web offset presses can run only a single fixed cutoff so that on a press with a 40-inch cutoff, four inches of valuable board or label stock would be wasted for each image produced.

Sheetfed offset presses for packaging tend to be wider than those used for publication work. Sheetfed presses as wide as 78 inches (1981 mm) have been used, and presses 60 inches wide are commonly used to print labels and folding cartons. Very wide presses are hard to control—ink color and water feed vary across the plate, and the paper sheet tends to stretch and expand, creating register problems.

Web Offset Presses

Web offset serves those segments of the industry that use boxes of standard size such as half gallon ice cream boxes, citrus juice or individual-serving cereal boxes.

The inherent productivity and the ease of preparing and mounting offset plates have been gaining markets for web offset package printing. Web offset presses have many characteristics that are disadvantageous for package printing, but their high productivity is so important that press manufacturers and other technologists have developed means of overcoming most of the disadvantages.

Variable cutoff presses, ones that can change the cylinder size, make possible the printing of boxes and labels of different sizes. While these variable cutoff presses are not new, they are gaining more attention in recent years. Plate/blanket cylinders come in cassettes that can be changed when the size of the package changes. Process color requires changing four of them. Changing the cassettes is an arduous task, and makeready of a six-color press can be lengthy.

If cartonboard is to be printed with conventional web offset inks on a web offset press, it must be heated to a high temperature, then chilled in order to dry and then set the ink. The process requires a lot of heat and a lot of refrigeration. Chilling a heavy material such as cartonboard requires huge amounts of refrigeration. Ultraviolet inks that cure by radiation can be used to avoid the heat-chill cycle. Furthermore, because the final steps in makeready for the web offset press must be done while the press is running, they waste large amounts of expensive packaging material.

Makeready

As with other processes, makeready times can vary greatly depending on the requirements of the job, the skill and training of the crew, and the attention that management has devoted to organizing the work. Makeready of an offset press is often considered fast because of the long times required for letterpress that it replaced. The chief problem of letterpress makeready is getting color to match the proof and getting the colors into register. Registration of offset printing resembles registration in other processes, but getting up to color includes achieving ink/water balance that is unique to offset. Bringing the process into ink/water balance consumes time and paper. While sheetfed makeready can be done on waste sheets, makeready on web offset is done on the unprinted web. This contributes to the high waste of paper or board that is typical of web offset.

Lithographic Plates

Lithographic plates print a very sharp image; they show good resolution.

Aluminum is used for most lithographic plates, although plastic, paper, steel and other materials have been used. The original plates were Bavarian limestone (hence the name "lithography" or stone writing). Modern aluminum plates accept water easily, and will run with very little moisture, overcoming many of the old problems of lithography: emulsified ink, washed-out color and poor drying. They are much less costly than flexo plates, and they are quickly prepared from a photographic negative or positive. Long-run lithographic plates now have plate life in excess of one million impressions, under appropriate running conditions.

Offset plates that require no water or fountain solution have been available for many years, but they require special inks and careful control of the temperature of the press. Because of the problems caused by water, waterless plates should be a natural for printing flexible packaging by web offset.

Sheetfed Lithographic Inks

For sheetfed offset, used for printing labels and folding cartons, the inks are dispersions of pigments in drying oils. Sheetfed inks should dry in two to four hours when everything is done right, although printers often allow 24 hours for drying.

Ultraviolet inks do not require high temperatures to cure. They eliminate the need for high heat that reduces the moisture content of board and increases its brittleness.

Sheetfed inks dry by a complicated chemical process that ties together all the molecules in the drying oil, turning them into a solid. Such inks dry in a matter of hours, not seconds (like web inks). UV or EB, that likewise cure in seconds, are also applicable to sheetfed litho.

Because of the thin ink film, lithography is especially susceptible to hickies and other print defects caused by dirt and debris in the ink or paper or pressroom.

Pigments used for lithography must not bleed in water or in alcohol. Pigments used on any package must be selected to withstand the contents of the package, to avoid ruining the print if the contents spill.

It has long been a practice to add a variety of materials to the ink in the pressroom. When a job causes unexpected print problems, printers often doctor the ink with various additives. This is not only time-consuming, but the final color is unpredictable, and the resulting ink usually causes additional printing problems. It is greatly preferred that inks come press-ready out of the can, but this requires planning and good communications between printer and inkmaker.

Web Offset Inks

Web offset inks are less viscous than sheetfed offset inks, but much stiffer or more viscous than flexo or gravure inks. They are grouped with the "paste" inks (sheetfed offset and letterpress inks) rather than with the "fluid" or "liquid" inks (flexo or gravure). Web offset inks can be formulated to cure by UV or EB, but usually they are dried by evaporating a heatset solvent or "heatset oil" that is a petroleum product somewhat like a highly refined kerosene. The print is hot as it exits the dryer, and it must be chilled. The solvent contains volatile organic compounds (VOC) that must not be discharged to the atmosphere. The solvent is usually carried out of the dryer and incinerated.

The dryer is often referred to as an oven, and if it is not carefully controlled, it can dry out the paper or board, causing it to shrink and making it brittle.

Advantages of Lithography

- Litho is economically practical for print runs from less than a thousand to greater than a million.
- Litho uses inexpensive plates that are processed rapidly.
- Offset lithography gives good image resolution, it reproduces fine reverses well and holds highlight and shadow halftone dots well.
- The offset blanket prints well on rough substrates.
- The press is made ready quickly and easily.
- It is simple to remake plates or to make new plates for a rerun that requires any change.
- Web offset prints at high speeds, exceeded only by rotogravure.
- Plates can easily be stored but are often remade for reruns.
- New technology in prepress, printing and finishing keep litho competitive.

Disadvantages of Lithography

- Color tends to vary across and around the sheet and from sheet to sheet.
- Offset requires stiff, clean, strong paper and board, which are expensive.
- Sheetfed offset has difficulties printing on film or lightweight paper.
- Sheetfed inks dry slowly.
- Waterbase inks are impractical.
- Water emulsified in the ink can cause printing problems, especially on non-absorbent substrates.
- Sheetfed offset has low production speeds.
- Web offset wastes a great deal of substrate. Much of the press makeready requires running of clean paper or board.
- Tinting and toning sometimes color the background areas.
- Offset lithography is a complicated process, requiring the highest skill and crew training of any process.

Comparison with Other Processes

Offset litho and flexo share many advantages: high productivity, the ability to print profitably on short and long runs, relatively inexpensive plates that are easily and quickly prepared, the ability to print on rough surfaces and metal surfaces. These have made litho the leading process for publication printing where the substrate is paper. Litho's problems of ink/water balance and the inability to handle plastic films easily have given flexo much broader markets in package printing.

Litho prints letters and reverses better than gravure or flexo, but it does not handle large solid areas as well. Color variability is the greatest of all of litho's problems. Sheetfed offset is slower than most web processes. Web offset is faster that flexo yet still slower than gravure. Offset lithography is the most common printing process; this means that there are more offset printers and more experienced personnel, and it sometimes offers the packager a greater choice of printers. Offset technology is well advanced, giving, on balance, good quality and productivity.

Control of impression pressure on press, while important, is much less critical than it is with flexo, and dot gain is easier to control. Offset inks are stiffer and generally cause less dot gain than is experienced in flexo or gravure. The ability to hold a small highlight dot easily makes litho more versatile than flexo when soft highlights are desired. Prepress factors that affect dot gain are similar for all processes.

Web offset waste is the highest of any printing process, partly because the press is made ready with unprinted paper or board running through it. Sheetfed offset can be made ready on waste sheets, while flexo and gravure generally require only a few makeready impressions before producing salable copy.

Web offset and flexo capabilities limit the gravure market. For a comparable width of press, web offset machinery is more expensive than flexo or gravure. Flexo presses and gravure presses can also be built in widths wider than the widest web offset press.

In spite of important limitations in package printing, web offset keeps gaining market share.

Trends in Lithography

Litho speed, as in all other processes, is increasing year by year.

Although it is by no means unique to lithography, new offset presses often have six, seven or eight units, allowing the printer to print combinations such as four process colors, two spot colors and a varnish. As markets for colored prints continue to grow, so will the number of units per press.

One of the most important trends in lithography is the use of non-volatile materials to replace alcohol in the fountain solution. Isopropyl alcohol has been used as a cure-all for dampening problems, but it causes environmental problems inside and outside the printing plant.

Web offset is being used in packaging, which increases speed some five-fold over sheetfed offset, but it causes a similar increase in waste. The variable cutoff web offset press is allowing offset to capture folding carton markets. It will probably capture some folding carton or label orders that would otherwise have gone to gravure and perhaps to flexo.

Thermography

This is not really a printing process, but it is a technique of producing raised letters, used sometimes on business cards and occasionally on packaging. A varnish is printed onto the substrate, usually by offset lithography, and an expandable powder is dusted onto the wet varnish. The excess powder is removed and the printed sheet is heated to dry the varnish and expand or "blow" the powder. The resulting print looks something like a print from an engraved plate. It is sometimes used on cosmetic and alcoholic beverage packages.

LETTERPRESS PRINTING

Letterpress is a method of direct rotary or flatbed printing that uses rigid relief image plates of metal or photopolymer material. These require a very smooth substrate such as coated paper. The stiff or paste inks are metered by a doctor blade and a series of resilient rolls. Variations of the letterpress process include offset letterpress which can print on rougher surfaces than can a metal plate.

Although letterpress, like flexo, is a relief printing process, there are major differences in the plates, the ink and the inking system of the press. Letterpress prints from a rigid plate, it uses a stiff or paste ink, and the press configuration is very different. The letterpress process requires a complicated array of ink rollers to work and distribute the ink uniformly. Letterpress was once printed with metallic type or plates, but metal has largely been replaced by plastic. Plastic letterpress plates are much harder than flexo plates—Shore A 90 as opposed to Shore A 30 to 60 for flexo. Letterpress uses a stiff ink (a paste ink) while flexo uses a fluid ink. Flexo uses an anilox roll to meter the ink, but letterpress uses ink rolls much like offset. Accordingly, the presses look quite different (figure 3-22).

In package printing as well as in publication printing, letterpress has been replaced by flexo, offset lithography and by gravure. In packaging, flexo has taken a large share of the market, and flexo is sometimes referred to as "the new look of letterpress."

Letterpress can produce very sharp lines and sharp letters. Fonts developed for letterpress printing featured sharp serifs (figure 3-23) while those for offset and gravure printing were often sans-serif, although offset can also produce sharp lines and serifs.

Letterpress lays down a heavier ink film than offset, about equal to flexo. It is used for top-quality printing where it can hold fine reverses and vignettes.

As the quality of flexo improves, it tends to take markets from letterpress, which runs at speeds of only 200–250 feet per minute (1.0–1.25 meters per second) on the narrow web label presses. Unlike litho, letterpress requires no dampening. Flexo is usually significantly less expensive than letterpress.

Offset letterpress (dry offset or "letterset") is used mostly for the printing of two-piece cans. It is possible to print a multi-colored image on the blanket and transfer it to the can in a single stroke or rotation.

Pharmaceutical labels are often printed letterpress. The plates are a lot

Figure 3-22. Letterpress label press. (Courtesy of Gallus Inc.)

cheaper than gravure, and fine print required on the labels is reproduced clearly. Under magnification, the individual halftone dot is usually seen to be a tiny circle (figure 3-24).

Markets for Letterpress Printing

Letterpress is used for labels, blister cards and decals. The major market, labels, includes labels for health care (vitamins, shampoo), cosmetics (perfume, makeup, hair care), indestructible labels for the auto industry, and pharmaceuticals. Pharmaceutical labels are frequently printed letterpress because they often carry a large amount of very small type that must be legible on small labels or packages. Covers for compact discs and audio cassette tapes are often also printed letterpress.

Abt   Abt

Letters with Serifs   Sans-Serif Letters

Figure 3-23. Serifs and sans-serif letters.

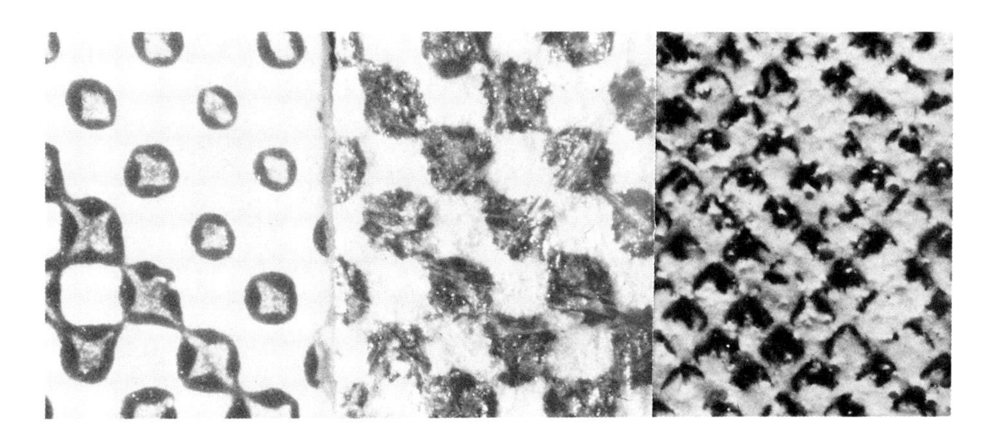

Figure 3-24. Magnified halftone dots produced by letterpress (left), lithography (center) and gravure (right).

The advantage of letterpress for label printing is its ability to hold a fine screen and print serifs and fine lines sharply. Narrow web letterpress is normally done by UV, and the quality and cost combination have given it a firm market.

Substrates for Letterpress Printing

If the printing plate is very hard and incompressible (as is true of letterpress and gravure), the substrate must be very smooth and somewhat compressible. This limits letterpress to printing of paperboard and paper, usually coated paper or board, and to plastic. Glass and metal cans are not printed by direct letterpress although offset letterpress is used to print metal cans.

Characteristics of Letterpress Printing

Printed by letterpress, letters have hard, sharp edges, and process color half-tone screens display typical rosettes. If the substrate is not smooth, the hard plate does not contact it evenly. The plate digs into the high spots while missing the low spots. The resulting embossing is a sure sign that the job was printed letterpress. In addition, ink on the image area is squeezed out, producing a halo around the image, much like poor quality flexography.

Letterpress Presses

New letterpress machines are mostly narrow-web machines used for printing labels. They are rated at speeds as high as 450-500 feet per minute (2.0–2.5 meters per second) while, as with most processes, actual production speeds are apt to run only 50-75 percent of the rated speed. To maintain top productivity on a multicolor job, the press must be equipped with automatic registration; the more colors, the more important electronic control of registration becomes.

Most letterpress machines print directly from the plate to the substrate, but offset presses are also used, producing dry offset or letterset printing. The blanket makes it possible to print on rough or uneven surfaces or onto the surface of a metal can.

The label business wants "letterpress quality." There's a feeling in some quarters that flexo isn't good enough for premium labels or top-quality folding cartons. Rather than find a flexo printer who can handle the job, packagers often turn to letterpress despite the increased cost.

There are still many small plants that are using old platen presses to print specialty work, including die stamping and engraving. Some platen presses are used for diecutting. They do not produce a large volume of package printing or decoration.

Letterpress Plates

The modern letterpress plate is a photopolymer; a polymeric formulation that is hardened by ultraviolet light. The plate is exposed to UV light through a photograph, and the plate is developed, removing the non-image areas. The process resembles exposure and development of flexographic photopolymer plates.

Although metal letterpress plates are disappearing, they are still available. One method of making letterpress plates for flat-bed presses is to mount photopolymer plates on wood.

On cylinder presses (rotary presses), photopolymer letterpress plates are usually mounted magnetically or with stickyback. Typical stickyback is 0.004 to 0.005 inch thick, while the plate is 0.030 to 0.032 inch thick (compared to plate thicknesses of 0.067 to 0.125 inch for flexo). Letterpress plates are often rolled and heated to give them the form of the printing cylinder so that they will adhere better.

The small gap in the cylinder can be held to 1/64 inch (0.4 mm), although some printers get by with 1/32 inch.

If the letterpress plate is not perfectly level, thin spots will fail to print and thick spots will emboss the print. Press crews pack the plate so that its height is increased in the shallow places. This takes a great deal of time—a prime cause for the disappearance of letterpress. The addition of a blanket cylinder that converts letterpress to letterset or dry offset overcomes the need for a perfectly level plate, but this complicates an otherwise simple printing process. Halftone dots often grow by unpredictable amounts, changing the color and appearance of illustrations when they are printed by offset letterpress.

Letterpress Inks

Since the ink film printed by letterpress is much thicker than a film of offset ink, the ink requires lower color strength. Furthermore, it need not work with water because no dampening is required. Like offset, the ink is worked by a complicated ink roll train to reduce its viscosity so that it will more readily flow from roller to plate to substrate.

The bulk of letterpress label printing is now done using UV inks.

Makeready

The great drawback of letterpress used to be slow makeready. Photopolymer plates have greatly improved makeready times so that letterpress is now competitive with other processes. Precision in finishing the impression cylinder has also reduced this problem.

Advantages of Letterpress

- Letterpress gives high print resolution, reproducing fine lines and serifs on small letters and excellent halftones for process color.

Disadvantages of Letterpress

- Letterpress does not print well on rough or uneven substrates.
- Letterpress is usually a costly printing process.
- New technology required to keep letterpress competitive has lagged.

Comparisons with Other Processes

In its label markets, letterpress is sometimes preferred to flexo because it prints a heavier ink film and because flexo has not been able to hold open fine reverses and print small letters sharply.

Color uniformity is better than offset, equal to flexo, but not as good as gravure. The ability of letterpress to print fine detail is usually not sufficiently better than offset or flexo to make up for the additional cost. New letterpress technology has lagged for many years, allowing the other processes to grow faster.

Trends in Letterpress Printing

Letterpress has been around for a long time, and its use will continue for many years. But as flexo continues to improve and print at lower cost, it will push letterpress still further out of its markets. This dreary future discourages the investment in research and development work required to make letterpress printing once more competitive. General technical improvements improve all processes including letterpress. Its ability to produce bright colors and print sharply keeps the letterpress process alive and going.

Many of the presses already installed will remain in production for years, and new rotary letterpress and platen label presses are still being manufactured and sold, mostly in the Far East, some in Europe, and fewer in North America.

Two developments have revived a dwindling letterpress industry: UV inks and photopolymer plates. UV speeds the drying of ink, and it can be used between units to allow dry trapping of inks to give brighter colors and better color control. Photopolymer plates are less costly and faster to produce than the old etchings and electroplates. In addition, press manufacture benefits from improved engineering and technology.

SCREEN PRINTING

Screen printing is a method of direct flatbed or rotary printing that uses a stencil supported on a wire or fabric screen to separate image from non-image areas. It is sometimes called "silk screen" because the earliest screens were made of silk. The light paste ink is moved by a squeegee that wipes and forces the ink through the mesh and stencil. The thickness of the fibers used to make the screen determines the film thickness of the print. The process can print on virtually any substrate.

The stencil is prepared from a screen made from synthetic fibers. The screen is fastened to a frame that holds the ink. The screen can be attached to a flat frame (figure 3-25), or it can be attached to a cylindrical frame required for rotary screen or web printing (figure 3-26).

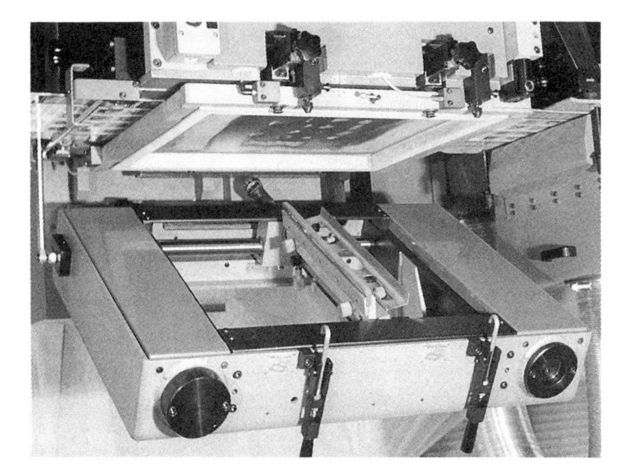

Figure 3-25. Flatbed screen press in operating position (left) and with printing head opened (right). (Courtesy of Gallus Inc.)

Screen printing is an immensely versatile decorating process and is used in applying patterns to all sorts of manufactured products. Screen printing can give especially brilliant colors and produces the finest fluorescent and metallic colors, fade-resistant pastels and opaque whites and blacks. It applies an extremely uniform coating with no ghosting or streaks.

The versatility of screen printing enables it to print a wide variety of inks and coatings of various film thicknesses on almost any substrate, hard or soft, strong or weak, smooth or rough, flat or shaped—paper and board, glass, metal and textiles.

Figure 3-26. Cylindrical screen for rotary screen press. (Courtesy of Gallus Inc.)

Some screen presses can print on unusual shaped packages, including cylindrical and conical packages. The ability of the screen process to handle so many inks and substrates makes it suitable for many short-run, specialty jobs.

Rotary screen presses run up to 500 feet per minute (2.5 meters per second); continuous flat-bed screen presses run around 40 feet per minute. In both cases, production speeds are far less than the rated speed of the press—as is true of most printing presses.

Screen printing is capable of applying the thickest film of any commercial process. It can also apply a fairly thin film. The thick film is useful in applying opaque white backgrounds onto transparent substrates as required to print bar codes. It can achieve 98 percent opacity, whereas flexo usually runs around 60–65 percent. This high opacity makes screen printing advantageous on some flexible packaging and on bottles.

For rotary screen printing, a digitally controlled squeegee is available that delivers a predetermined amount of ink, avoiding any excess that must be discarded.

No tension, friction or rubbing of the substrate is required to transfer ink, as in other processes. The laydown is determined by the mesh of the screen, the pressure on the squeegee, and the angle of the squeegee blade.

Waterbase and solventbase inks are usually dried by hot air; UV inks and coatings are dried (cured) by UV lamps. The word "ink" is usually used for the pigmented liquid that prints type and patterns, "paste" is sometimes used for the clear liquid used for coating. It is a simple matter to convert a screen press from printing to coating by merely changing the screen.

Fluorescent colors are more brilliant and retain their color better if they are printed in a thick ink film that the screen process prints so easily. The thick film that can be applied by screen process is helpful in printing opaque ink on glass and plastics.

The thick film also allows printing of dispersions of coarser particles than can be carried by other processes. Screen printing has been used to apply the glass frit that is sintered onto glass bottles that carry their own insignia.

Markets for Screen Printing

This versatile process can print with a wide variety of inks onto a wide variety of substrates of various shapes and sizes. The development of the rotary screen press has given this oldest of all printing processes new industrial and commercial significance for printing of labels, cans, bottles and other packaging materials.

Tags and labels probably represent the largest packaging market for screen printing, but it is also used to print cans, drums and bottles as well as for specialty products packaged in folding cartons. It is used to print opaque white backgrounds on bottles and even on flexible packaging.

In packaging, screen printing is used on tags and labels where its thick film gives vivid colors, notably for logos. It also gives a three-dimensional effect, and a feel or "hand" to the label, a texture that is favored for cosmetic cartons and bottles.

Screen printing produces very attractive metal decorating. Three-piece cans, metal trays and the like exhibit brilliant pictures when printed with screen.

Application of a transparent label to a polyolefin bottle avoids the need to inventory different bottles for every product. The packager can use a few types of bottles and apply the appropriate label at the end of the bottling line. The transparent label can give the appearance of printing directly on the bottle. Screen printing is preferred for printing such labels because of the high density, or opacity, of the print.

Production speeds, even on rotary presses, are not very high, and screen printing is used mostly for specialty and short-run printing jobs. The relatively low speed limits the length of run for which screen printing is economical.

Substrates for Screen Printing

Since ink is extruded onto the substrate through the screen that carries the pattern, the printing puts no more requirements on the substrate than the ability to carry it past the screen. Screen printing can produce images on the roughest of substrates, including textiles and corrugated boxes. Dusty and linty substrate are always troublesome in quality printing, but the screen press has more tolerance than offset.

Screen can print on a wider variety of substrates than any of the conventional processes. In addition to board, paper, plastic, metal and textiles, screen is used to print on glass. It is commonly used to print bottles (see Chapter 2).

Characteristics of Screen Printing

The heavy ink film that can be produced by screen printing accounts for many of its attractive properties: brilliant colors and fluorescent colors with good permanence, high opacity in white backgrounds, black solids that can be applied with a single coat and do not require a halftone underprint to give good intensity. If the ink film on a package is especially heavy, it was probably applied by screen.

On the other hand, screen does not commonly print as small a halftone dot as other processes. (Specialists can print 150 lines per inch or finer for P.O.P. displays.) Process color requires transparent inks. Transparent screen inks are available, and while screen can produce process color, its strength is in producing opaque colors and thick films. Its resolution is poorer than flexo, offset or letterpress.

Image Carrier

The screen is a fine mesh made from synthetic fibers. In choosing a screen, the printer must consider the desired film thickness, the nature of the substrate, the viscosity of the ink or paste, and the size of any solid particles in the paste.

Images can be hand cut and then applied to the screen, but most commercial work is done with photography. The screen fabric is filled with a photosensitive film, and after exposure through a photographic positive, the areas that form the image are washed out, leaving a stencil through which the ink passes.

The thickness of the ink film is controlled by the thickness or diameter of the fibers used to make the screen, by the angle of the squeegee, and by pressure on the squeegee.

Screen Presses

The press consists of a frame with the screen held firmly in place. Ink placed inside the frame is forced through the screen with a squeegee. There is no standard size press or screen. Flatbed screen printing is often automated to place the screen on the substrate, draw the squeegee and move the substrate. If multicolor printing is required, subsequent colors can easily be registered to the first color.

Fully automated presses printing large formats can print six or more colors in register. Interstation drying by air or UV makes it possible to print good quality process color on corrugated cartons and P.O.P. displays.

Presses for cylindrical or conical objects (cans and bottles) usually use a flat screen. The container to be printed is rotated against the screen, which moves at the same speed as the surface. The squeegee does not move, but it presses the screen against the object.

The flat screen is still used for many small-scale operations and for the production of very short-run products. However, for manufacture of packages, the higher speed rotary press is commonly used. Both the ink and the squeegee are inside the cylinder, which rotates as the material passes beneath first one cylinder and then the next.

Rotary screen presses are available from 15 inches to 15 feet (380 to 4500 mm) wide with repeat lengths of a few inches up to 55 inches (1400 mm).

Screen Inks

Not only does screen print on a great variety of substrates, it can print with almost any kind of ink or coating. All screen inks, however, are paste inks (high viscosity inks). A great variety of inks can be used, including solventbase, waterbase, UV, plastisols, foamed inks and adhesives.

Sheetfed screen can also apply inks that dry by evaporation. The wet prints are dried by placing them on a moving web that carries them through an oven.

UV and EB inks can be applied and cured in line with the press. UV and EB inks solve one of the screen printer's greatest difficulties—slow ink drying.

Rotary presses use solventbase inks that resemble web offset inks and waterbase inks that can be dried in-line with the press in a matter of seconds by heating the web.

Screen printing can also apply inks that cure by baking. These inks are most frequently used on metal, as for the production of metal cans.

Oil base inks for sheetfed screen printing usually require hours to dry and are rarely used in commercial work. The prints are racked or hung for drying.

Advantages of Screen Printing

- The image carrier is inexpensive and easily prepared.
- Screen printing can apply a broad variety of inks to a wide variety of substrates.
- Screen prints brilliant, vivid colors as well as fade resistant pastels, metallic and fluorescent colors.
- Large solid areas (black, white or colored) printed by screen are uniformly dense and opaque.
- Screen is capable of printing the thickest ink film of any process.
- It is suitable for applying heavy films or coatings.
- Screen printing is readily applied to cylindrical and conical objects.

Disadvantages of Screen Printing

- Production speeds are low.
- Great care is required to achieve the high resolution suited for the finest halftones.

Trends in Screen Printing

Screen is a process of the greatest diversity, and it is used in a wide variety of printing and manufacturing operations. New technology evolves slowly, often as a result of developments by the users of the process. The variety of applications may restrain concentrated efforts to develop broad markets for screen printing equipment suitable for long-run packaging work. Screen will continue to play a small part in the printing of packaging materials, notably labels, bottles, plastic tubes and specialties.

ELECTROSTATIC AND LASER PRINTING

Electrostatic and laser printing are methods of direct rotary printing that generate the image by selectively discharging a charged semiconductor drum through exposure to reflected light or a laser beam. The solid toner is applied to the drum where it adheres to the charged area. It is then transferred to the paper substrate. Variations of the process include generating the charge on paper coated with a semiconductor material, then exposing it to the image and applying a liquid toner consisting of toner particles suspended in a hydrocarbon liquid.

Electrostatic copying is office copying of memos and reports. Laser printing produces similar products from digitized information instead of visual copy. Electrostatic and laser printing produce prints from original copy or digital information faster than any conventional printing process. This makes it ideal for producing variable information such as office memos, consecutive numbering, lot codes, dates, etc.

Laser printing can produce excellent process color prints, and it is used to produce proofs of color separations from computerized information. The process is too slow for printing of packages, except for very short special runs.

Electrostatic images can be formed on a semiconductor drum (one that will conduct electricity in the light but remains an insulator in the dark.) The drum is charged in the dark, then discharged with light to form an electrostatic image that is brushed with a powder (toner) to develop the image (figure 3-27). When the light used to discharge the non-image area is a laser beam, the process is referred to as laser printing. Plain-paper copiers produce the image from a visual picture or paste-up.

Laser printing is controlled by an electronic system that produces halftone

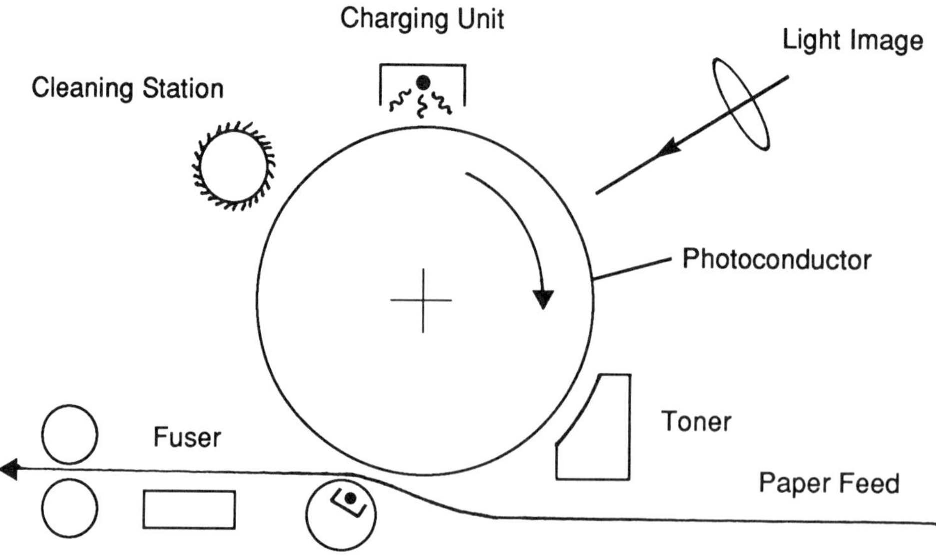

Figure 3-27. Diagram of electrostatic printer showing charging the drum, imaging, toning, transfer of image to paper, fusing the image and cleaning the drum.

dots directly from the computer and generates them on the semiconductor drum. For production of bar codes or alphanumeric symbols, computer information is easily transformed into electronic signals that direct the laser beam to produce an image. For production of a halftone image, a huge amount of digital information is required, and production times are long. It may seem strange or unusual to say that computers are too slow, but it takes a great deal of computer information to make a flexo halftone plate or a gravure cylinder, and all that information is printed out with every rotation of the press. With laser printing, the information must be converted to the image for every print.

Markets for Electrostatic and Laser Printing

Copiers and laser printers are well known in the office and quick-printing shop. Unfortunately, they have not yet combined high resolution with the production speeds required for printing pictorial matter at high speed.

In the packaging industry, this printing technology is largely limited to the production of labels carrying bar codes that are printed in the store to identify the weight, price and identity of the object.

Substrates and Toners

Electrostatic printing is largely limited to paper.

The colorant used to develop and print the image formed on the semiconductor drum is usually referred to as a toner. Dry toners are quite unlike printing ink. These powders consist of a pigment and a resin that softens when it is heated, producing a film that adheres to the paper.

Advantages of Electrostatic Printing

- Copies of printed information are produced quickly and simply.
- Little or no training is required.
- Reproduction equipment is inexpensive.
- Digitized information is printed in a few seconds.
- Bar codes are printed rapidly and inexpensively at the point of sale.

Disadvantages of Electrostatic Printing

- Speed is too low for commercial decoration of packages.
- Prints from these processes cannot be satisfactorily deinked on a commercial scale.
- Most toners produce poor halftone images.
- While proofs of color separations can be produced by laser technology, the process is expensive and a trained crew is required.

Future of Electrostatic Printing

New technology is evolving, and new equipment is introduced every year.

Changes in electrostatic and laser printing are occurring rapidly, and the technology is developing many uses in graphic the reproduction processes. The speed of computers continues to increase.

Although the production of prints from digitized information is an attractive idea, computers are still far too slow to make such laser printing into a commercial production process for multiple images. Application of electrostatic printing to commercial printing so that repetitive images can be produced directly from computer data will require computer speeds far greater than those now known.

The rapidly evolving digital and electronic technologies promise to make these into potentially commercial printing processes for the decoration of packages at some future time.

INK JET PRINTING

Ink jet printing is a method of printing that directs droplets of ink to form an image on any substrate. The image is controlled by charging the droplets of ink, then deflecting them in an electric field. Because the charge is controlled by a computer, ink jet printing provides a simple method for converting digitized information into printed images.

Like laser printing, ink jet printing is output directly from digital information. The two principal uses of ink jet printing on packages are the coding of cans and the production of bar codes. Ink jet is well suited for printing all sorts of variable information. It is also used in the preparation of process color proofs.

Ink jet printers are usually divided into two groups: continuous jet and impulse or drop-on-demand jets (figures 3-28 and 3-29).

Impulse and drop-on-demand printers generate a drop by various methods:

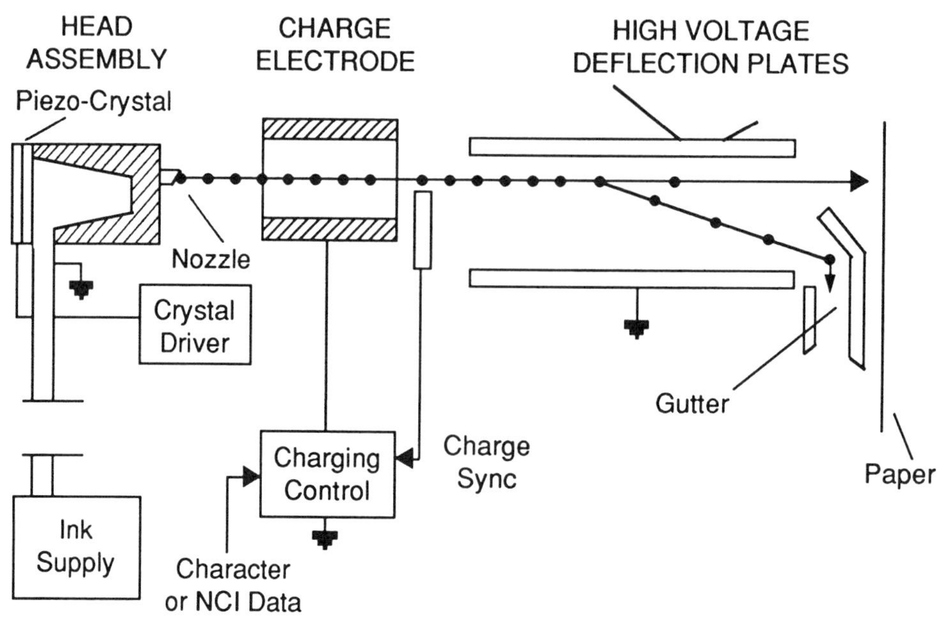

Figure 3-28. Diagram of continuous-jet ink jet printer.

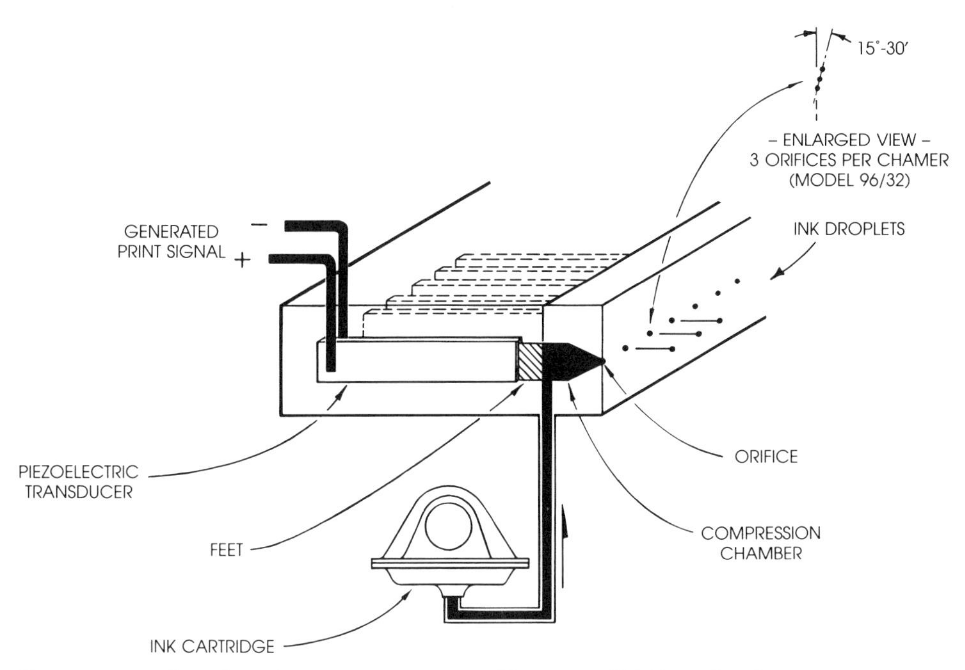

Figure 3-29. Impulse ink jet printing system. (Courtesy of Trident, Inc.)

by heating to generate a bubble, by compressing the container with an impulse generated from the computer, and by other techniques.

Markets for Ink Jet Printing

In package printing, they are used to print dates and lot codes on cans (figure 3-30), labels and other packages and for printing of bar codes. For coding of cans and bottles, the continuous jet is commonly used (figures 3-31 and 3-32). Ink jet printing is also used to prepare proofs of printing.

Ink jet printing is also seen on corrugated cartons. These large-size ink droplets form a crude image, but one that is useful in identifying the contents.

Characteristics of Ink Jet Printing

Ink jet printing is made up of ink droplets that can be seen individually. The individual drops are usually visible, but even when they overlap, the serrated edges are characteristic. The printing process is easily identified.

Substrates for Ink Jet Printing

Ink jets can print on anything, hard or soft, smooth or rough, paper or board, flat or round or irregular shaped. On the other hand, for good quality prints, special coated papers have been developed that will absorb the ink without causing it to spread or feather.

Figure 3-30. Ink jet code printed on can.

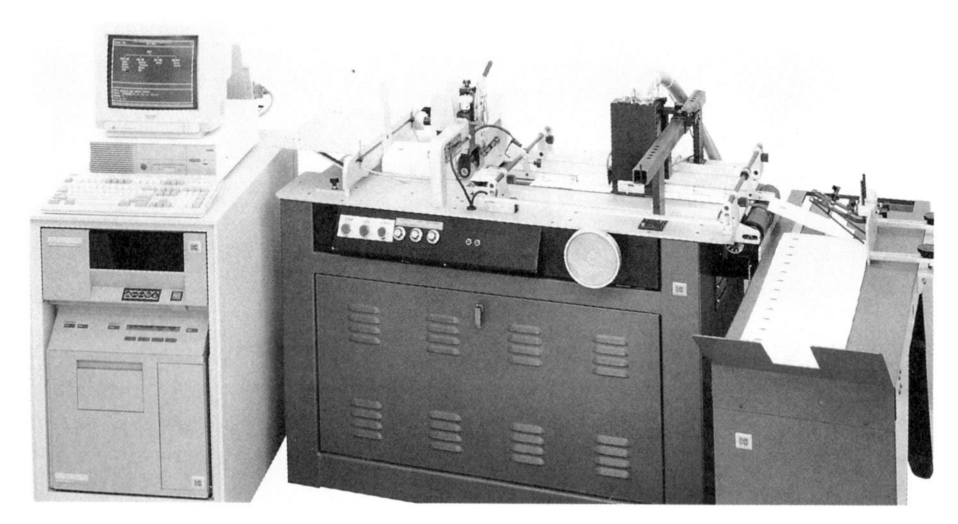

Figure 3-31. Ektajet 5100 Printing System. The system can be set up for on-line or off-line work. (Photo courtesy of Eastman Kodak Co.)

Inks for Ink Jet Printing

Ink jet printers require an ink that strongly resembles writing or fountain pen ink, not printing ink. The ink is a solution of a dye in a solvent, often water. Inks containing pigment particles are apt to clog the nozzle, and they are avoided in printing.

Advantages of Ink Jet Printing

- The process prints directly from digital information without having to go through any optical or photographic steps.
- Ink jet printing is continuously variable on the press; each image can be changed after it is printed. This makes it useful for coding and numbering.

Disadvantages of Ink Jet Printing

- The process lacks the combination of speed and image quality required to print multiple copies economically.

Trends in Ink Jet Printing

This printing method is growing rapidly in addressing and coding, but it is not expected to gain significant markets in the decoration of packaging materials in the near future. The continuing rapid growth of the speed of computers may eventually make it practical for some new applications.

Figure 3-32. Videojet Series 170i ink jet printer. (Courtesy of Videojet Systems International, Inc.)

COMBINATIONS OF PRINTING METHODS

It is often possible to produce a desired pattern or image at lower cost if more than one printing process is used. For example, gravure presses often have a flexo unit for line copy or text. The gravure units produce pictorial matter and the flexo units print variable text such as special offers and prices that are subject to change.

Ink jet printers are easily added to any web press to add variable information to the printer pattern.

A trend in the industry is to use combinations of printing methods such as gravure and flexo or screen and letterpress (figure 3-33). Altering the gravure cylinder is expensive, but it is often possible to print the main body of the package with gravure and to print with flexo the information, such as price, that may vary from one run to the next. Screen and flexo or letterpress are used for printing labels when it is desired to lay down a white opaque coat over a transparent film and to apply printing to the white film. Screen lays down a highly opaque white background film; letterpress prints sharp, fine letters on top of the white background.

Web offset presses often have an anilox roll that applies a coating to the blanket for coating the printed product. Offset presses used for package printing incorporate special coating units.

Figure 3-33. Combination letterpress/screen press with UV drying. (Courtesy of Gallus Inc.)

FURTHER READING

Flexography: Principles and Practices, Fourth Edition. Frank N. Siconolfi, Chairman. Flexographic Technical Association, Ronkonkoma, NY. (1991).

Gravure, Process and Technology. Gravure Education Foundation and Gravure Association of America, Rochester, NY. (1991).

The Lithographers Manual, Eighth Edition. Ray Blair and Thomas M. Destree, Editors. Graphic Arts Technical Foundation, Pittsburgh, PA. (1988).

The Printing Ink Manual, Fourth Edition. Chapter 2. R. H. Leach, Editor. Van Nostrand Reinhold (International), London. (1988).

Printing Fundamentals. Alex Glassman, Editor. TAPPI, Atlanta. (1985).

Chapter

Design for Printing

CONTENTS

LIST OF ILLUSTRATIONS

LIST OF TABLES

Chapter

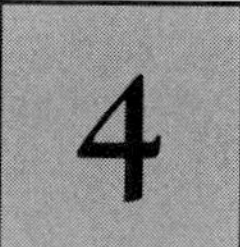

Design for Printing

INTRODUCTION

A package must meet four objectives: It must sell the product, it must inform the potential and actual buyer, it must protect the contents, and it must be convenient to use. The designer is concerned with all four aspects. Graphic design is involved chiefly with the first two.

A designer who understands printing knows how to challenge the printing press without defeating it and how to create a design that takes advantage of the characteristics of the printing process and the substrate to achieve the maximum results.

The design of the package affects product sales and market share. The package is the silent salesperson, whether the package figuratively shouts from the store shelf or quietly taps the consumer's shoulder. Packaging design often yields higher returns than advertising, while accomplishing the same goals of product awareness, image formation and incentive to purchase.

Customer Perceptions

In any supermarket you will find unappetizing or unappealing packages. A pie carton or box of frozen vegetables that is poorly designed or poorly reproduced is worse than a package with no illustration. It can generate a negative product image.

The packaging buyer must make the marketing decision. It is not cost-effective to spend a lot of money on packaging products that will return little profit. On the other hand, an effective package can often increase the profitability of the product. The graphics represent the product, and the higher the perceived quality of the graphics, the more likely the customer is to consider the product a high-quality item. This is one reason for the tremendous growth of preprinted linerboard.

Design Definition and Function

The book, **Flexography: Principles and Practices**, gives a useful definition of design: "Design is the orderly combination of formal elements to produce a composition. Design is the visual plan of line, mass and color, selected and assembled to accomplish a specified goal. That goal may be to convey beauty or simply to provide information by the arrangement of copy on a label. Often, the goal is to sell a product. In that case, the design must have visual impact, and it must provide identification and information about the item. Sometimes, the design goal involves the printed product itself, as with gift wraps, textiles, cups and containers."

The function of the design is to translate the client's ideas into a printed product that will please the consumer. With labeling, packaging and shipping containers, design is often the only thing that identifies the product, the brand and the manufacturer or packer. Many products depend heavily on package design to establish their image for merchandising, advertising and promotion.

The design must be printed on a mass produced package. Doing this economically without sacrificing quality is one key objective. Shipping and distribution requirements, product claims, and UPC and other requirements must not be overlooked.

THE DESIGNER'S FUNCTIONS

We make statements about ourselves by the products we consume. The package design enunciates that statement and is thus responsible for a significant part of the consumer's emotional involvement and ultimate satisfaction with the product. The brand name ties the package displayed on the retail shelf to the advertising that the customer sees on TV or in print.

The graphics, together with the container's style, color and text quickly send the customer a message about the product's category and the targeted market: older or younger people, women or men, high income or low income levels. Different brand names and packages have been used to get a single product into different markets.

The design must both decorate the package and provide information. The printed design must identify the product, the brand and its uses. The purpose may be to convey pleasure, comfort or some other emotional or environmental state. It telegraphs subconscious information about the product's category. The austere design of "generic" brands conveys low cost. The fact that these products do not sell very well suggests that low price is not a primary objective of most buyers.

Because graphic design is aimed at triggering a reaction, visual impact is a major objective. The package must not only attract attention, it must sell the product. In addition to attracting attention on the shelf, the design can sometimes serve a secondary purpose in home decoration and furnishings.

The designer brings the marketer and the printer together. The designer must be aware of the marketing plan, of substrates and their characteristics, of printing facilities available. The design must reflect the marketing plan, and it must make the package desirable to the targeted audience.

The designer must establish a basic fund of experience and knowledge in package design. The staggering investments in today's advertising and sales promotion budgets demand speed and skills from the designer. In one unfortunate instance, a designer who did not understand that it requires a separate printing plate to reproduce each color insisted that the design must include 48 different spot colors. Good designers are also schooled in the fundamentals of package manufacturing.

The designer must be a skilled artist, but he or she must also be a problem solver, a visual communications specialist and client oriented. The designer must add technical knowledge of the requirements of the package to the artist's concepts of design and decoration. The designer must understand graphics and be aware of the client's concerns as well as the mechanics of printing and the ways that images are manipulated in going from art to finished package.

The designer must bring together the artwork, which includes layout and color, with the information that must be carried on the package or label. He or she must design the artwork for printing. The text must be typeset and illustrations may have to be enlarged, reduced or cropped.

Legally mandated labeling requirements must be included although they often pose a major challenge to the designer in maintaining visual continuity.

Designers must work closely with the potential printers to stay within printing limitations and keep costs within budgets. This is necessary if the designer is going to incorporate features such as eyespots, clear channels, glue channels, gold-over areas, nonprintable surfaces, screens and color requirements for the matching of colors into a workable design.

The designer must solve the relationship between design function and production and manufacturing concerns. Designs prepared with knowledge of production art, color separation methods, substrates, engravings, platemaking, inks and press operation will generate greater efficiency and profitability.

Design of the image matches the shape of the package; rectangular boxes and cylindrical cans require different designs than the exotic bottles one often sees. Bottles, both glass and plastic, are made in an amazing array of shapes, challenging both the designer and the printer.

Esthetically, whether a design is bold or delicate, it must reflect good taste and a proper balance of line, mass and color. Each element of design—subject matter, color scheme and typography—is part of the layout, and it must relate to the overall theme.

To maximize the designer's creativity, he/she should not be concerned, in the creative phase, with the physical aspects of production. When the design must be implemented, the rest of the production team gets involved. This is the "value chain" (figure 4-1). Now, production criteria must be considered such as

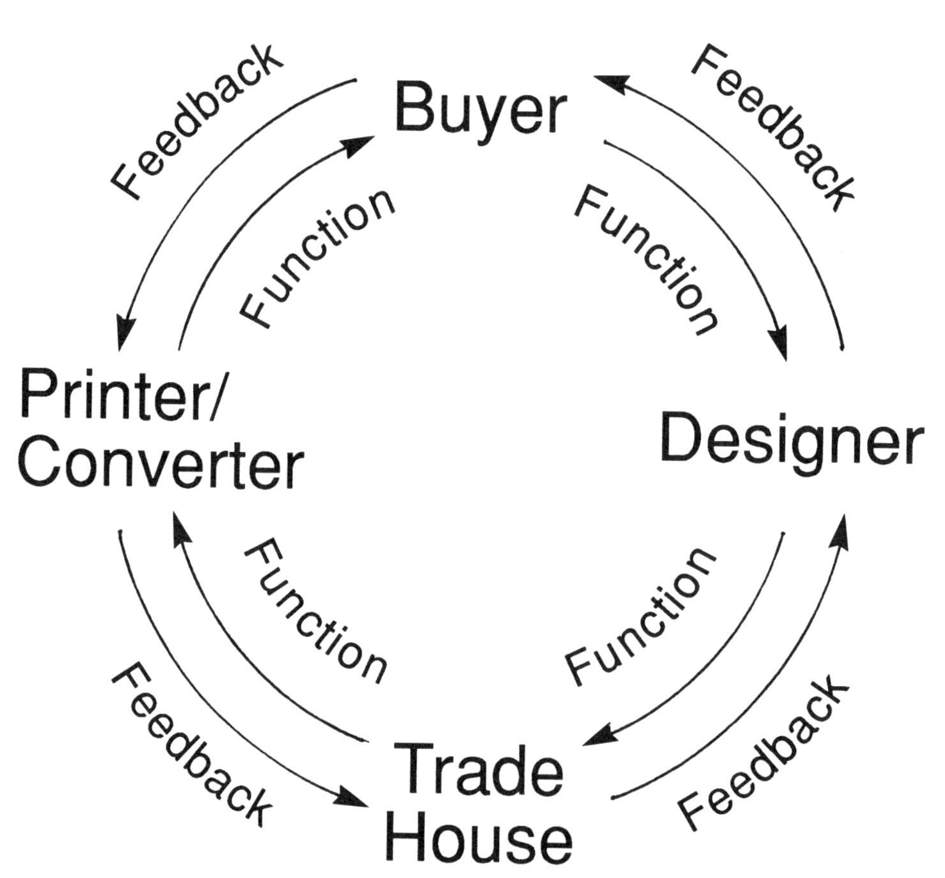

Figure 4-1. The value chain showing relationship of buyer, designer, trade house and printer/ converter.

step-and-repeat, bar codes, mandatory warnings, and die cuts for displays. The graphics must now be presented full scale. This can be cumbersome for a large package, but a proper evaluation cannot be made from less.

THE DESIGN BRIEF

Planning the Package

Prize-winning packages result from more than a stroke of genius. They result from a well-planned approach to solving the problems of the marketplace and of package printing and manufacture. The package must be part of a coordinated plan. Packaging should use simple, everyday language, but should also be designed for high visual impact and coordination with the store environment.

To avoid costly mistakes, all information relating to the package should be written down in a document called a "Design Brief." A design brief is necessary so that all problems are solved before the packager goes out for bids, or worse, into production. The design brief covers all the marketing, functional and graphic aspects of the package. The design brief brings together all the facts the designer needs to carry out the job. It contains information about the marketing of the packaged product, the design of the package to protect the product, and all information required by the packager and government agencies. The data include physical information about the product, the results of research, and management decisions about marketing policy as well as specifications concerning production of the package.

Artwork must fit into machinery specifications, and it must serve during distribution and on the retailer's shelf. The design brief will help the designer make sure that all technicalities have been answered and technical problems have been solved before beginning the printing process. It will also help prevent overlooking important considerations.

Questions tabulated below are suggestive of ones that must be fully answered. A full answer should be set down against each question.

Functions of the Package

Fully as important as the design is the construction of the package to meet its requirements. In fact, the design and construction are so intertwined that the package cannot be designed without a thorough understanding of its structure and mechanics. Study of the structure and design of the package is called "structural design."

To produce an effective package, the graphics must be considered part of the design. Pasting graphics onto a package after it has been designed does not produce the type of impact that is required if the package is to compete with well-planned packages.

Often, structural design is performed first and the graphics are added. Unfortunately, when the structural design has been hardened and little attention has been paid to graphics, packaging costs skyrocket. Therefore, structural designers often work closely with graphic designers, often in the same organization. The synergy of the two is essential if the package is to fulfill all its functions.

The functions of the package can be considered by answering the following typical questions:

- What is the product?
- What are the physical characteristics of the product (solid, liquid, powder, breakable, hygroscopic)?
- What are the end-use requirements?
- What are the functions of the package?
- What size is the package to be? (Individual serving, multipack, family size or large economy size?)
- What are the form and shape of the package?
- Will the package (including its graphics) withstand end-use conditions? (Will a detergent package collapse on a wet drain-board? Will a microwave package withstand heat and moisture? Will the printing on beverage cans withstand jostling in an ice bucket or tub?)
- What problems must the package face during distribution?
- Are dates (such as "sell by") and prices to be shown on the package?
- What special packaging features must be considered?
- If the package is an odd shape, how will it be printed?
- Is it convenient to stock on the retailer's shelf?

- Is it convenient to use?
- How will the package be filled and closed?
- Can the package be sealed adequately?
- Is the type of closure dictated by marketing considerations or by law?
- Does the package conform with all other regulations?
- Is the package part of a family and is it compatible?

Marketing

A major objective of the package design should be a clear, simple statement of what the brand stands for in the consumer's mind. It must support other corporate promotional efforts for recognition. When designing the package, the designer must consider unity of all promotional efforts and the need for the package to reflect images presented elsewhere—appearance, logo and corporate colors. The design and logo ensure continuity and brand identity.

The image shown on the TV screen or in a printed ad should be easily recognizable at point of sale, or the advertising and package design are wasted. The package seen on the retailer's shelf should remind customers of what they have seen on the screen.

Packages are featured in many ads, and the package design should be considered an advertising feature. TV commercials that end by showing the package are known to be more effective in changing brand preference than commercials that don't.

David Ogilvy says: "Over and over again, advertising research has shown that *photographs* sell more than *drawings*. They attract more readers. They deliver more appetite appeal. They are better remembered. They pull more coupons. And they sell more merchandise. Photographs represent reality, whereas drawings represent fantasy, which is less believable. Abstract art does not telegraph its message fast enough for use in advertisements." The designer should consider the relationship between package design and Ogilvy's observations about advertising.

Pictures of the actual product may be uninteresting. A picture of a pie is better than a picture of the packaged flour or sugar. This is known as "selling the sizzle, not the steak."

The cook is always on the lookout for new ways to please the family, and recipes on packages help to sell foods.

Design for export grows increasingly important. Package characteristics important in one locality or nation may be unimportant in another. Every country has its own associations, traditions and prejudices. Consumer conceptions and practices in the U.K. and the U.S. and Canada are close at some times but not so close at others. Packages for sale in Canada must be printed in both French and English. In the United States, Kentucky Fried Chicken containers must remain clean and repel the cooking oils. In England, because of their tradition of oils on their fish 'n' chips, there is no such necessity. The British will accept a product in an unprinted bucket; Americans want brand identification.

The following questions emphasize the marketing considerations that should be clearly answered:

- Does the design identify the product properly, the brand, the manufacturer?
- How can the package best project product image, corporate identity, brand image?
- What selling features (taste, flavor, nourishment, cleansing power) should be reflected in the design?
- What image is the manufacturer trying to project for the product?
- Does it draw attention to the product?
- Does it sell the product?
- Who is the customer (type, age, sex), his/her purchasing habits?
- What type of customer is expected? Chewing tobacco generally appeals to a different type of customer than babies' clothing.
- Is the package design suitable for use in a TV feature?
- How is the product to be advertised?
- How does the package fit into the advertising program?
- Does the design remind customers of the advertising? Does the customer recognize the product as one that is advertised?
- Is the design suitable for all promotional media used for the product?
- Is it for promotion at Christmas or some other holiday?
- What is the nature of the sale, impulse or planned? (Color strongly affects impulse sales.)
- Is the package color important?
- Does the color of the package associate with the product (peas are green and orange juice is orange)?
- Is the product to be illustrated on the package?
- How are related products or varieties to be identified?
- Is the product to be identified by color? (Different flavors may be indicated by changing the color of the label or closure.)
- Should instructions for use or assembly appear on the package?
- Will a spot color illustration or process color package do a better selling job?
- Are a special color or a number of special colors a prerequisite of the package graphics?
- What are the selling conditions, including types of sales outlets, methods of display and lighting?
- What are the needs of the distributor and retailer?
- How will the package be displayed for sale?
- Will the package stack easily on the shelves of a self-service store?
- Does the package help the retailer as well as the buyer? (After all, the retailer is the first or second customer!)
- Will the print on the package withstand the light of the supermarket?
- Does the design color avoid unwanted changes under supermarket lighting and take advantage of retail lighting practices?
- Is the product a low-price or a luxury item?
- Is the price to be shown on the package? "Cents off"?
- Is the design contemporary? The design should not suggest that the product is out of date, nor should radical changes make old stock appear out of date.
- How will the package be displayed? The structure and package design must be coordinated to suit the display when the package stands on the shelf in a

"natural" position.

- Will the package be attractive after purchase? Perfumes and facial tissue will stand open on the bedroom or bathroom shelf and should have an attractive design.
- Is the package to be exported? Will a special package be required for each country, or can a design be found that will appeal in many countries?
- Is there space for two or more different languages?

Machinery Limitations

Although today's sophisticated printing presses are able to maintain tight register, it is still a good policy to avoid hairline or butt register. The best presses can maintain register within a row of halftone dots (0.003 inch or 0.075 mm), but variability of register by older presses and those in poor repair may be many times greater. To be confident and safe, the designer should obtain the fingerprints of the press on which the work is to be run or should obtain specifications of presses that are likely to be used. These types of variations in tolerance are to be found in all sorts of presses—gravure, flexo, offset and others.

The ideal design from a marketing point of view may be limited by constraints imposed by existing machinery that may dictate the shape and size of the package. These have to be identified and assessed when developing the brief.

- Does the design fit conventionally available sizes of rolls of paper, board or film?
- What machinery will be required to produce and print the package?
- Is the machinery common enough to allow competitive bids on the printing and manufacture of the package?
- If the package or family of packages is to be made of more than one substrate, can both or all be printed?
- What effect will the substrates have on the color and the printing? Different substrates take ink differently; a color printed on a white coated board will not match the same ink color printed on brown corrugated or white uncoated board. Furthermore, a white board will permit the printing of a wider range of colors (a wider gamut) than will colored boards.
- How many press units will be required to print the design? Most packages are printed on one side only, but if the package must be printed on two sides, it takes one unit for each color on each side—except for web offset that can print on both sides on one unit. (The same color is usually, but not necessarily, printed on both sides.) Preprinted containers can be a headache if there is a change in the product.
- Can labels be applied at the packaging line?
- Can the package really be produced within the budget, or must the budget be revised?

Substrates and Ink

The substrate is chosen to protect the product, to make its use convenient and to produce attractive printing.

However, environmental considerations and pressures now are as important as graphics and structural properties. Recycling has become an environmental necessity, and recycled papers are often less bright than papers made from virgin pulp. Furthermore, the bleach required to produce high brightness generates environmental problems of its own. Recycled metals, glass and plastic may not be as strong as virgin materials due to the presence of post-consumer impurities.

Cosmetics and pharmaceuticals are packaged in solid bleached sulfate while most other folding cartons that need a white surface use a gray recycled board that carries a white top layer of bleached white fibers. Sometimes a solid unbleached kraft is coated with bleached white fiber. To obtain good color reproduction, the surface of the board must be white. If environmental forces require use of a board with lower brightness, the design must be changed to take this into account.

Smooth, glossy, white surfaces give the best graphics reproduction. Coated paper and paperboard are widely chosen for this reason. Plastic films give excellent gloss and smoothness, and when they are white or underprinted with a white coat, they meet all the requirements of a superb printing substrate.

Aluminum and other foils, although not strictly white, have a high gloss or reflectivity and give startlingly attractive graphics. Aluminum cans can be given a white undercoat for even better looking graphics.

However, if the design is to be printed on a paper bag or an uncoated corrugated box, the designer must be up to the challenge. These rough surfaces will not reproduce halftone dots or fine print, but attractive, colorful designs can be reproduced even on such materials if the substrate is taken into account. The dark brown color of the unbleached, uncoated bag or box will produce oranges, browns and yellows with poor contrast. Blues are turned gray or black, but they will give excellent print density or masking power. Strong colors mask the background better than pastels. An opaque ink will usually do better than a transparent one on such a material.

Colored substrates have limited ability to reflect color. The eye can see only those colors that are reflected, and a yellow package substrate (the surface is yellow because it absorbs blue light) cannot print brilliant blues.

Printing inks absorb light that falls on the surface of the package (see Chapters 5 and 9), and a good quality ink gives the best colors.

The designer must also consider other properties of the ink to be used. It is wise to check with the ink supplier concerning ink toxicity and ability to withstand processing such as freezing or cooling. A package that is likely to remain on the shelf for long periods must have inks with good resistance to fading. The popular fluorescent colors are particularly likely to fade. The ink should be matched to the stock. An ink of a given color on one type of material may appear different on another type of material. To match the color of a printed carton with the color of the ink on a metal box or a plastic container requires different ink formulations.

- Is the ink properly formulated for the substrate?
- Is the ink properly formulated for the printing press and the end use?
- Will the printed ink meet the end-use requirements of the package?

Information and Typesetting

Good typography helps people read the copy, while bad typography prevents them from doing so. Many shoppers have poor or improperly corrected vision. Design that favors simplicity of layout, large lettering and logotype sizes, and maximum contrasts will be seen by these shoppers. This is particularly important as the population continues to age.

An instructional tag or label should provide information with accuracy, brevity and clarity. On shipping cases, the graphics should instantly identify a product, along with coded data, to assist shippers, handlers and warehouse people from the packing line to the final display area.

Federal, state and even local regulations require that certain information appear on packages. Cigarette warnings, ingredients and nutritional data must be suitably printed. The size of type is sometimes set by law. There have been many circumstances where printed packages had to be destroyed, then redesigned and reprinted because they failed to meet legal requirements.

- Is the text visible at a suitable distance? Product and brand name should be clear at 10 feet (3 meters) in the supermarket, but the list of ingredients need be legible at only 10 inches (25 cm).
- Does the package or label design follow government requirements concerning what must be printed: contents of the package, warning labels, FDA requirements, nutritional information, etc.? The requirements grow more complex year by year. Federal requirements are published daily in the Federal Register. The designer must seek well-qualified advice to be sure that all legal requirements are met before completing the design. It is the responsibility of the packager to provide the legally required data, but the designer should be aware of the need to have the mandated material available.
- What other information is required—weight, hazards, price, name and address of manufacturer or distributor, instructions for use, storage information (especially on frozen foods), product ingredients, cooking instructions, recipes, bar codes, special notices "FOR RESTAURANT USE ONLY," "NOT FOR RESALE," instructions for use or assembly of the product?

The Printing Press

The designer usually finds a press that will produce the design, but if the design is well planned, a larger number of printers or converters can be asked for bids. On web presses there must not be more colors than there are printing stations. Most web presses now have five or six colors, some more. An advantage of sheetfed presses is that the product can be put through the press twice or even three or four times. This is, of course, more costly than printing all colors in one pass. In any case, the addition of more colors to the design usually increases the cost of printing it.

The design must not be wider than the press. Cutoff length must also be considered. If the design has to be an unusual size, the packager may be unable to get competitive bids for the printing. Most processes have considerable latitude in cutoff length, but web offset is usually limited to a single size. It costs about the same to run a press for an hour on a full web as it does on a narrower web, and the design should be as wide as practical. On the other hand, it may be less expensive to run fewer across and to run the job on a faster narrow press.

These questions should be taken up with the printer and with the trade house that prepares the color separations. There are few designers who know as much about such design factors as the printer and the people at the prep house, and experienced designers frequently call on them for help.

In printing labels, and even larger packages, it is usually possible to gang several different products on a single layout and print them all at once. If the job is suitably planned, they can be ganged to print different quantities of different products.

Furthermore, the designer must take downstream functions into account: slitting, diecutting, gluing, cutting of windows, etc. Will the job be produced more economically if these functions are carried out in line or with off-press operations? How will the printer handle the work?

Thought must be given to considerations even further downstream. Will the carton open on the filling machine? Will the glue hold varnished surfaces together, or should glue lines be included in the design?

If the design must wrap around the package with lines to match, special press and design considerations are required. If the match is on a bias, the problem will be even more difficult.

Every designer, user, buyer and specifier must be aware of the opportunities to save money or create problems. Printers all too often tell their customers that if they had been called in for the discussion during design of the job, they could have saved them thousands of dollars of unneeded expense. The printers themselves decry this situation because of the frustrations of producing a poorly planned job, even if they get paid for it. A well-designed job that proceeds easily is just as profitable and much less expensive, without all the aggravation.

Labels and Wrappers

Simple gummed labels are still used on a great variety of cans and bottles. A label is part of the whole package and must be considered in the original planning. Wrappers are grouped with labels because it is often difficult to distinguish between them. The wrapper around chewing gum or a highly decorated corrugated box is simply a large label.

Food cans are almost always labeled. It is cheaper than printing a different message on every different can: 25 different varieties of soup can be packaged in the same kind and size of can. Conventional, ungummed printed labels account for a very large proportion of the packaging market; and there are also pressure sensitive labels and heat seal labels.

An awkwardly shaped package is difficult to print, so a label may be the recourse. Preprinted containers can be a headache if there is a product change or a change in marketing policy. Design of a label that can be readily changed protects against this problem.

Sometimes, a label is used to promote sales of natural products. The "SUNKIST" stamp on oranges was one of the earliest attempts to label individual pieces of fruit in support of an advertising campaign. Now many fruits, such as bananas and kiwi fruit, and even tomatoes carry individual labels, but these are separate, printed labels. The information is not stamped directly onto the product.

- Should the design be printed on the package or on a label?

Product Protection

Although protection of the contents is a major purpose of packaging, and proper package design is essential to this purpose, graphics play only a minor role. Warnings, such as "This End Up," "Protect from Freezing" and "Do Not Drop" are printed prominently when necessary.

Convenience

Except for instructions on its use, graphics have little to do with the convenience factor of graphic design.

- If special instructions on opening the package or using it are required, are they shown clearly on the package?
- Should they be printed in an inserted booklet?

Approval of the Design

Now that the designer has a design that will meet all legal and marketing requirements, there is one major step to take before going to the customer. The designer must bring together the separation (prep) house and the proposed printer. If they cannot produce the design within budget, it is unacceptable. Usually, the printer and prep house can suggest changes that will reduce the cost of the project. A job that is difficult to execute is going to be more expensive than an easy job—even for the same design.

Now that all of the problems have been laid out and discussed, the designer is ready to bring everything together into a design that will meet all requirements. The design is then presented for final approval to the customer, the package buyer, who may approve it or find that some unanticipated problems have come up. The customer should have been involved in all the prior planning and execution phases. A good planning job will improve the chances that there are few problems and that they are not serious.

The design is ready to be given to the trade house for color separation and platemaking. Other problems may be discovered at this point, but again, if the planning was done carefully by an experienced designer, the chances that the design will prove workable are greatly improved.

THE PACKAGER'S CHOICES

The printing process is usually chosen by someone else, seldom by the designer. Nevertheless, the designer must consider the printing process in the plans for the work.

There is a widely held misbelief that some printing processes give better reproduction than others. The best process gives the best results for the money, with quality appropriate for the package. Gravure is sometimes considered expensive and flexo cheap, but this is no more true than with any other well-considered or poorly considered choice. Every printing process can and does give outstanding printing as well as poor printing.

Creativity can do wonders with any printing process. Award-winning packages are not limited to any particular printing process. The key is how well the designer makes use of the printing process, and how well the design projects and sells the package and the product.

The packager's choices are broadened by technological change that has improved the quality of reproduction that can be obtained from any of the conventional package printing processes. Most package printing is done from "relief" plates (flexo or letterpress plates with raised type) or with engraved cylinders (gravure), but offset lithography is used in several package printing markets, as summarized in table 4-I.

TABLE 4-I. PRINTING PROCESSES COMMONLY USED TO PRINT PACKAGING

Package	Printing Process
Labels	Flexo, gravure, offset, letterpress, screen
Tags and wrappers	Flexo, gravure, offset, letterpress, screen
Corrugated	Flexo, letterpress, offset
Top liner	Flexo, gravure (especially in Japan)
Folding cartons	Offset, gravure, flexo
Flexible packaging	
Foil	Flexo, gravure
Plastic	Flexo, gravure
Paper bags	Flexo, gravure
Grocery bags (Paper or plastic)	Flexo
Beverage cans	"Dry litho" (offset letterpress), offset
Metal boxes and 3-piece cans	Offset, screen
Plastic bottles	Flexo, screen
Caps and closures	Offset, flexo
Plastic (butter) tubs	Flexo
Squeeze tubes (toothpaste tubes)	Flexo
Metal caps	Offset
Plastic caps	Flexo
Blister packs	Offset, flexo

The designer must be aware of the limitations inherent in the printing characteristics of the chosen substrate—film or foil, corrugated, carton board, coated and uncoated paper, metal, glass or plastic. A packaging material has been changed more than once because the chosen one could not be suitably printed in a cost-effective manner. The result is a costly compromise that maintains the most important characteristics needed for protection and the graphics needed to enhance sales.

If the designer understands the process, he or she can take advantage of its characteristics to design an image that will achieve maximum value for the least cost. The packager can always choose the artwork, but the printing process that will be used for any job is usually dictated by the printing and converting economics and the requirements of the package.

Nevertheless, characteristics of the printing processes make them suitable for some jobs and unsuitable for others.

For example, rotogravure gives excellent reproduction of flesh tones and excellent uniformity of color. It is sometimes chosen for cartons containing cosmetics, but cosmetics cartons are usually printed offset because of economics on short to medium run lengths of up to one million impressions. Gravure does not do as well in the reproduction of type, especially small type, and it is expensive on short runs.

Gravure cylinders and letterpress plates do not conform to hard surfaces, and they cannot be used to print metal, rigid plastic or rough paper and board. Gravure is avoided for printing kraft bags or uncoated paperboard.

Offset uses stiff inks and is very difficult to control on thin plastic films. The water from the offset plate/blanket is not absorbed by plastic, making the process difficult to control. With careful control, however, offset can be used to print on plastic, and it is commonly used to print metal for three-piece cans. Recent gains in waterless offset may result in major changes toward use of offset for printing on non-absorbent substrates.

Developing technology has made flexography into a versatile process that is suitable for printing almost all packaging materials. When process colors and intricate designs are to be printed, top-notch press control is required. Nevertheless, the smoothness and uniformity of the substrate limit the final print quality.

Screen printing, although it prints brilliant colors and metallics, is still slow, and it is at a disadvantage with long runs.

The packager usually has a limited selection of cost-effective printing processes for any package.

Flexography, rotogravure, lithography, screen and letterpress, embossing and foil stamping, ink jet and thermography all have their places in packaging, and while they are not interchangeable there is always a multiple choice. However, if the job is properly planned, designed and printed, exciting illustrations can be printed by any printing process. (These printing methods are discussed in Chapter 3). Understanding of the methods is a prerequisite for any designer.

When they are well controlled, all processes give excellent linework or halftones. On the best work, even an expert cannot tell which process is used without using a magnifying glass—and sometimes not even then! It is only when the printing is faulty that the process is readily apparent. The best modern flexography will print mouthwatering ice cream cartons or snack packs, but if the process is out of control, halftones plug and halos form around solids and large letters and striations run through the solids.

Misregister is a problem with any process, but especially with gravure where the distance between each unit allows the paper or film much room to stretch. Skips and missing dots are particularly troublesome in gravure, especially if the printer does not use electrostatic assist.

Offset lithography has more color variability than other processes, and ghosting, tinting (ink dots in the non-image area) and hickeys are characteristic of offset.

The heavier the ink film, the greater the gloss and color intensity and the ease of handling metallics and fluorescent colors. On that basis, the designer would choose screen over gravure, and gravure over flexo and letterpress, with offset the poorest. But well printed offset can give excellent gloss, color depth and metallic colors.

A major consideration in choosing the process is length of run. Gravure is uneconomical on runs of less than several hundred thousand impressions. Flexo and offset are practical with runs of a few thousand impressions, and screen is economical on runs as short as a few hundred.

Although it is possible to print a design by any process, if the design is printed by two different processes, the prints will seldom match. Different processes have different ink film thickness and produce halftones differently. Furthermore, inks for each process are made of different ingredients, even for the "same" colors.

Table 4-II is an attempt to summarize the characteristics of the five major printing processes. Like all generalizations, there are exceptions. Some of the ratings are more subject to quantification than others. Nevertheless, the table points out characteristics of five leading processes. Some of these characteristics can be more easily changed by skill and careful planning than can others.

TABLE 4-II. COMPARISONS OF FIVE MAJOR PRINTING PROCESSES
(A Ranking of "1" indicates the preferred process.)

	Flexo	Gravure	Offset	Letterpress	Screen
Reproduction of Type	3	5	2	1	4
Reproduction of Solids	3	2	5	4	1
Reproduction of Highlights	3	1	1	3	5
Reproduction of Shadows	5	3	1*	2	3
Resolution	3	4	1	2	5
Register Control	1	3*	3	1	5
Color Consistency	2	1	5	4	2
Plate Cost	3*	5	1	3	1
Speed of Platemaking from Original	2*	5	1	2	2
Ease of Plate Correction	4	5	2	1	2
Plate Length of Run	2	1	3	3	5
Paper Cost	1*	1	5	2	2
Tolerance for Paper Roughness	3	5	2	4	1
Paper Strength Requirements	1	1	5	4	1
Tolerance for Low Basis Weight	1	2	5	4	3
Ink Cost	1	1	4	3	5
Thickest Ink Film	4	2	5	3	1
Operator Skill Required	2	3	3	5	1
Press Make-ready Time	4	3	2	5	1
Economy on Long Runs	2	1	3	3	5
Economy on Short Runs	3	5	2	4	1

* See text

The numbers in the table rank each process with "1" being the process that does best and a "5" being the least suitable process. Where two processes are essentially equal, they are given the same number. Asterisks are placed next to numbers that require further explanation.

1. Web offset is more apt than sheetfed to fill in 95–98% dots (shadow dots) because of the thicker ink film and lower viscosity of the ink. Web offset might rate a 2, being the equal of letterpress. For the same reasons, its resolution is apt to be lower, a rank of 2.

2. Sheetfed offset ranks better than web offset on register control. Register control depends more on the type of press than on the process. CI presses give the best register control. In-line presses with a long lead (as are used in much gravure printing) cause the most problems.

3. Plate cost of flexo varies depending on the type of run. Rubber plates require a matrix that makes them costly for a single plate but less costly when multiple plates must be prepared.

4. Rubber plates used in flexo take more time to prepare than photopolymer plates and would be ranked 4, ahead of gravure. However, if multiple plates are to be prepared, rubber is faster.

5. Costs of plastic or metal substrates do not vary with the process, but gravure and flexo papers are less costly than offset paper.

KEEPING COSTS DOWN

Holding to the Budget

While keeping costs within budget is important, the packager must remember that poor quality printing suggests a poor quality product to the consumer. Skillful design and planning are necessary if the packager is to get good value for the money paid for the package.

The designer needs to know whether the budget can be adjusted or whether it is firm. In any case, the designer must keep the cost of the package as low as possible.

The cost of printing is part of the cost of the package. The preparatory work and plates or cylinders are usually a one-time charge while ink, substrate and press time depend on length of run. Keeping the number of colors down without sacrificing the graphics will decrease the upfront costs of preparation.

Any additional work such as cutting of windows or stamping or embossing will increase the cost of the package. Diecutting and folding are equipment common with most printers of labels and folding cartons, but foil stamping and embossing often require special equipment which may mean shipping the package to a separate company or operation.

Good photographs cost more than bad ones, but clean, fresh art is important. The packager needs good artwork in order to capture an increasing market share for the product. At least one packager pulled a wrinkled print from his pocket for the designer to use as an illustration. Poor original artwork can never be brought up to the level of good artwork, even with electronic touchup and the wonders of modern computers and technology. The cost of bringing it up to mediocrity is exorbitant.

Quality control is important; bad printing and faded inks suggest a bad product. Bad printing or bad color can damage a brand image. Furthermore, Total Quality Management (TQM) keeps costs down by preventing the production of faulty packaging or printing.

The designer can sometimes reduce printing costs by letting the product itself do its appetite appeal—a window carton or a transparent film or lamination that shows the pie or the quality of the strawberries in the package.

The cost of a short-run flexo job can be increased more than 20 percent if the designer does not understand the requirements of the flexographic printing method. In fact, in extreme cases, redesigning the job has been reported to cost as much as the original design. Every flexo printing house requires an in-house design staff, and the cost of redesigning a flexo job is higher than the corresponding cost of a job printed by some other method. Even under such adverse conditions, flexo often has an economic advantage over other printing processes. Eliminating the cost of correcting the design can significantly lower the cost of printing the packaging.

Standard Procedures and Effective Communications

Smooth, well-established standard operating procedures and effective communications are required if a job is to be produced economically. A skilled designer can achieve the required objective more economically than a novice.

A communications gap between designers and artists and printing technologists can cause misunderstandings and misinterpretation of procedures, and costs can skyrocket. Designers and paste-up artists must understand the production problems and technical limitations faced by graphic arts photographers, film assemblers, platemakers, press operators and finishers. These people, in turn, often do not understand the means and techniques that artists employ to achieve desired esthetic effects. Information, careful planning, carefully written specifications and experience help to avoid high production costs, unnecessary meetings and high tempers.

Holding frequent meetings at which people raise problems and agree on solutions is good insurance against expensive, insoluble problems and even more recriminatory meetings after the job is in production. It is a good idea to have a standard viewing box available because even if 5000K lights are used, different viewing locations give different appearances. If one or more people on the team have different color perception, this problem should be ironed out before final approval. (See Chapter 6.)

One of the most common problems discovered after the package has been designed is that consideration has not been given to printing requirements, e.g. the package design requires just one more color than is available on the printing press; the size of the package does not fit available prepress, printing or finishing equipment. The more colors the design calls for, the fewer printers will be able to perform the job, and fewer competitors generally yields less competitive pricing.

Press Limitations

It is possible to run paperboard through a sheetfed press twice, doubling the number of colors and increasing the printing cost. There are ways to split the ink fountain on an offset press, making it possible to print more than one color per unit, but this is a complicated procedure; nevertheless, some sheetfed folding carton printers specialize in this. Accordingly, there will seldom be more colors on the final package than there are printing units on the press.

Package printing presses can print only one color at a time on each printing unit. The printing plate or cylinder transfers ink to the substrate which is backed up with an impression cylinder. A 10-unit flexo or gravure press can put 10 colors on one side or eight on one side and a white ink and a coating on the other. The same 10-unit press can also put five colors on each of two webs at the same time.

SPECIFYING THE QUALITY

Specifications of the job, detailed for each step from preparation of specs for the artwork through the delivery of the printed piece, are of greatest importance in assuring a top-notch package within budget. These specifications are the product of a well-conceived design brief. Few people understand properly each stage of production and the attitudes, requirements and customs of the people who will carry out their art. Printed jobs are often incorrectly prepared, poorly scheduled and subject to many last-minute corrections. It is a credit to the skill and ingenuity of the printer/converter that many jobs are successfully produced.

Specification of quality is part of the TQM process. If the product quality is to be specified, then all suppliers must also have certified TQM programs. The subject is mentioned in Chapter 12 and dealt with extensively in numerous books.

PROOFREADING

Proofreading is part of the quality control process. Proofreading should be assigned to a single, skilled individual. It has been demonstrated that when many people are responsible, many more errors are overlooked than when only one person has responsibility.

Checking the bar code symbol for accuracy and machine readability is the responsibility of the proofreader.

PACKAGE RESEARCH AND PACKAGE DESIGN RESEARCH

Skilled package design is essential to the success of the packaged product, but corporate packaging decisions are increasingly based on research rather than hunch or opinion. The designer is not only an artist, but he/she is a scientist, keeping an eye out for new developments and measuring the results of the design of the package.

Advertising research is well established with an impressive array of published literature and reference material. Package design research is building this type of background. Package design research started in the 1950s, long after the development of package research and package engineering. It is not yet as well developed as package research or engineering or advertising research.

Research can find out which of several package designs will sell best. It is also useful in finding out whether the package will remain sealed until ready for use and whether it can be easily opened by the customer when it is ready for use. We've all had experiences with "leakers" and with packages that are frustratingly difficult to open. Market researchers in the United States have selected Terre Haute, IN, Syracuse, NY, Phoenix, AZ, and Columbus, OH, among other cities, for key testing

People react to a product's perceived reality rather than to its actual, factual attributes. Research measures people's reactions to the package. Package design research examines the success of a package design in communicating the marketing objectives. It is still more of an art than a science.

Research is used to test a package's performance characteristics (protect the contents, promote the product and inform the customer) and especially how well it performs its functional requirements during filling, storage, shipping, on the merchant's shelf and in the home or factory.

COMPUTER DESIGN

Computer graphic systems now available greatly increase the designer's productivity. Computer-generated graphics have touched virtually every aspect of printing production, and they have touched the design function as deeply as any. A glance at the computer-oriented environment reveals many of the changes.

The following quotation from *Flexography: Principles and Practices* neatly illustrates the use of computer graphics: "Let's suppose that a designer has the title, copy matter and pertinent legal description of a new wine about to be introduced. The client wants flowers on the label, and market research agrees. The designer pores over the file full of photos, gleaned from many magazines and chooses some. Then the designer sends an assistant out for a dozen roses.

"With roses and pictures in hand, the designer moves to a copy stand with lights and a small video camera. The flowers are arranged nicely and some colored paper is fixed up as a contrasting background. The computer scans the three-dimensional arrangement using the camera and captures the scene.

"The designer's next step is to put the title and copy into the system. There are several alternatives: A modem can capture copy from any place that the designer has access. Text can be transferred by floppy disks, or copy can be entered with a keyboard.

"The computer can change the color of the roses and the color of the background. The computer can add a mountain scene in the background or make the roses float among the clouds."

The designer can easily check out a broad range of ideas or can "spin the dials," looking for inspiration. Last-minute changes can be incorporated, and small changes suggested by the buyer can quickly be accommodated. Rather than finding the computer a fearsome competitor, artists have learned to make it their friend.

FURTHER READING

Flexography: Principles and Practices, Fourth Edition. Frank N. Siconolfi, Chairman. Flexographic Technical Association, Ronkonkoma, NY. (1991).

Confessions of an Advertising Man. David Ogilvy. Athenium Press, New York, NY. (1988).

Ogilvy on Advertising. David Ogilvy. Crown Publishers, Inc., New York, NY. (1983).

The Lithographers Manual, Eighth Edition. Ray Blair and Thomas M. Destree, Editors. Graphic Arts Technical Foundation, Pittsburgh, PA. (1988).

Selecting Colour for Packaging. E. P. Danger. Gower Technical Press, Aldershot, Hants., England. (1987). The book contains a bibliography on use of color in design and lists periodicals that carry information related to packaging.

Packaging Design 2: The Best of American Packaging and International Award-Winning Designs. Paul Schmitt et al. PBC International, Inc., New York, NY. (1985). The book illustrates more than 250 outstanding package designs and comments on the features that made them successful. Names of designers and the design firms that produced each package are included. Nine categories: foods, beverages, pharmaceuticals, cosmetics, housewares and hardware, sports, electronics, industrial and specialty.

Handbook of Package Design Research. Walter Stern, Editor. John Wiley & Sons, New York, NY. (1981).

Chapter

Color and Color Printing

CONTENTS

LIST OF ILLUSTRATIONS

LIST OF TABLES

Chapter 5

Color and Color Printing

INTRODUCTION

Although the earliest packages were printed in a single color, usually black, packagers soon recognized the value of using a variety of colors. Use of color has continued to increase to the present day.

Color sells. In the supermarket and in the retail warehouse, the package is the point of contact for the sale, and this usually calls for colorful packages. While almost all packages on the supermarket shelves are highly colored, containers in the retail warehouse or "purchasing club" still offer many opportunities for improved presentation of the product (figure 5-1. See color insert). Packages can be more colorful than the product itself. Although the item itself is often displayed, packages on the shelves that are dull and undecorated reveal opportunities to motivate shoppers.

Color identifies the brand and the product. The brown of Hershey's chocolate is associated not only with a product, but with a brand of that product. Likewise, French's mustard has always had the same yellow label or package. The color of hair rinse, lipstick and other cosmetics identifies the color of the product in the package. Red distinguishes strawberry ice cream from vanilla and chocolate.

Reproduction of color is notably important in package printing, and the printing and viewing of color require close attention from the package printer and the package buyer.

Color is a complicated and confusing topic: it must be considered as both a psychological and a physical phenomenon. We must understand both additive and subtractive color. Color on a package is affected not only by the ink but by the color and texture of the substrate and the viewing conditions. Although most packages are printed with "spot color" (a different ink is chosen for each color to be printed), many packages are printed with process color where a wide variety of colors is produced from tiny dots of four "process-color" inks. There are major differences in the techniques used to produce the two types of printed colors and in the types of packages that are decorated by them.

Control and communication of color require knowledge and proper instrumentation. The great waste and expense associated with so much color printing are caused by lack of knowledge of its control and inadequate use of instruments. An understanding of color printing and color measurement helps the printer keep variability and costs under control.

WHAT IS COLOR?

To understand color, one must consider both the physics and the psychology of color. The physicist talks of color as part of the electromagnetic spectrum or as light waves with varying wavelengths. The psychologist talks of color in terms of what people (and many animals) see and how they respond.

It is not possible to completely reproduce a real, three-dimensional image with ink on a flat (two-dimensional) surface. The artist, print staff and packaging buyer must agree on what is an adequate representation of the real object, or more important, perhaps, whether the reproduction conveys the message that the packager desires.

Color and the Packager

The packager must consider that feelings and memories are associated with color. For example, red and green are the STOP and GO signals on traffic lights. Cherries are red and plums are blue. Red and orange attract attention on store shelves. Package designers must use color to produce a package that will both attract the consumer's eye and create a favorable attitude toward the product. The printer and converter must faithfully reproduce that color. Printer and packager must understand the ideas behind color, color matching and color viewing. Understanding these ideas helps improve the efficiency and performance of the printing operation. It also improves the relationship between the package designer, the package printer and the package buyer.

Physical Properties

Light is one form of radiant energy. The spectrum of radiant energy is shown in figure 5-2. Light is radiant energy that is visible to the normal human eye. Color is a combination of physiological and psychological reactions to light.

Color is determined by the object to be viewed, the illumination and the viewer. The wavelength of light can be accurately measured, and light of different wavelengths produces different sensations of color in the human eye. Humans

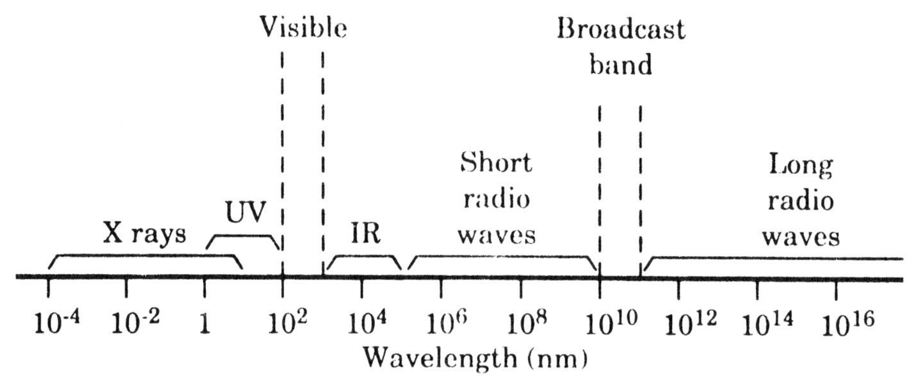

Figure 5-2. Spectrum of electromagnetic radiation.

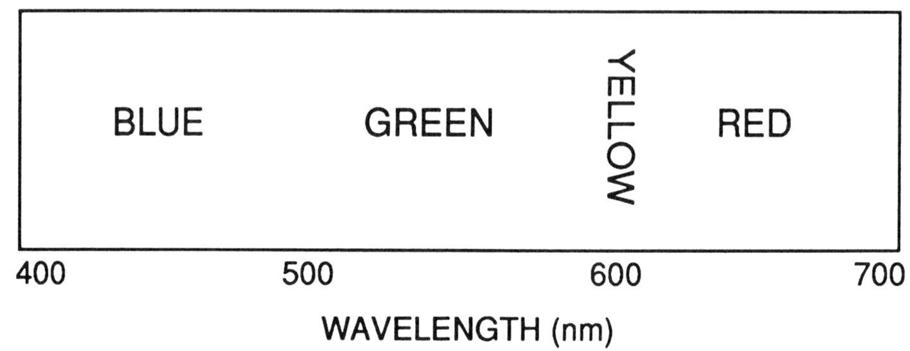

Figure 5-3. Spectrum of visible light.

can generally see colors with wavelengths from around 400 nm (blue) to 700 nm (red) (figure 5-3). (A nanometer abbreviated—nm, also called a "millimicron," is one-billionth of a meter or roughly 0.000,000,4 inch.) Radiation shorter than 400 nm (ultraviolet) or longer than 700 nm (infrared) is invisible to the human eye, as are x-rays and radio waves. Some birds, insects and animals can see wavelengths invisible to humans).

Light has the properties of both waves and particles. It can be refracted and reflected like waves, but it transmits its energy to a colored surface or to a photographic film in particles of energy called "quanta." Physicists characterize visible light by its wavelength (the crest-to-crest distance between waves) and by the amount of light present (its intensity).

A glass prism can be used to demonstrate how white light can be separated into a rainbow or a spectrum of all the colors. Red light (with longer wavelength) is bent more than blue, and by bending the longer wavelengths more than the shorter ones, the prism separates white light into a rainbow or spectrum. Similarly, the spectrum can be recombined into white light (figure 5-4).

The eye receives many colors—although only green, blue and red are required to produce a good picture or a realistic reproduction of a colored subject. Radiation from around 400 to 500 nm is blue, from around 500 to 600 nm is green, and 600 to 700 nm is red. Yellow is a strong band around 520 nm, but the eye also combines green and red to produce the appearance of yellow.

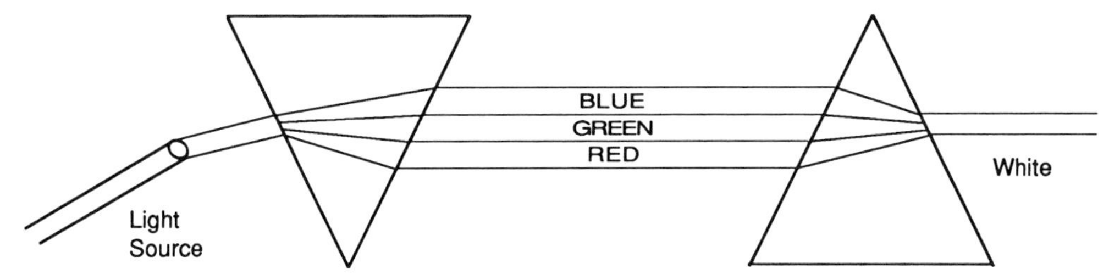

Figure 5-4. Separation of white light with a glass prism.

Additive and Subtractive Colors

There are two ways to combine colors to make new colors: the additive system and the subtractive system. The additive system is the one used in television tubes where blue light and red light combine to make lavender, or red and green combine to make yellow. In printing, we use a subtractive system where magenta ink subtracts the green from a white paper and cyan subtracts the red leaving blue. On the TV screen, if all three colors are present, the color is white: on the printed carton, where all three colors are present the color is black (or a very dark brown).

A colored material reflects and absorbs the various wavelengths that shine on it. For example, a red package seen in white light absorbs all wavelengths except red. The blue and green are absorbed and only the red is reflected. This is why we see red.

A filter or other transparent colored material absorbs light in the same way. Broad-band filters are made to absorb light with a wide breadth of wavelengths, for example 400 to 500 nm, while narrow-band filters absorb light over a narrower range. This difference is important in building color measuring instruments, and those using densitometers or colorimeters should be aware of these differences.

It is essential that original artwork, proofs and prints be viewed using the same kind of illumination. This will ensure the same interpretation. Colors of objects and prints change when they are viewed under different conditions. Even different "white" lights, like fluorescent lamps, incandescent lamps and skylight are composed of different wavelengths. The same sample appears slightly different under these different lights. That's why the same steak sometimes looks bright red under the "warm" lights of the meat department at the supermarket and a drab red under the fluorescent lights of a home kitchen.

Additive Colors. The colors on the TV screen are additive. Blue, green and red light combine to produce white light. Red and green produce yellow, green and blue produce cyan, and blue and red produce magenta. Red, green and blue together produce white. You can run a simple experiment with three slide projectors and a white screen. Place a red filter over one, a green filter over the second and a blue filter over the third, and point the projectors toward the screen. Where the red and green lights overlap, the color is yellow; where the red and blue lights overlap, the color is magenta; and where the blue and green lights overlap, the color is cyan. Where all three lights overlap, the color is white, as shown in table 5-I.

TABLE 5-I. ADDITIVE COLORS

Additives	Result
Red & Green	Yellow
Red & Blue	Magenta
Green & Blue	Cyan
Red & Green & Blue	White

TABLE 5-II. SUBTRACTIVE COLORS

	Transmits	Absorbs
Yellow	Red & green	Blue
Magenta	Red & blue	Green
Cyan	Blue & green	Red
Black	Nothing	Red, Green & Blue

Subtractive Colors. Films of transparent printed ink do not emit light, they absorb it. A cyan film transmits blue and green but absorbs red, a yellow film absorbs blue, and a magenta film absorbs green. If all three colors are printed over each other, they will absorb all color and the film will appear black (or almost black, depending on the precise colors of each ink pigment).

The same colors can be produced by placing cyan, magenta and yellow films over a light table. The cyan and magenta yield blue by removing red and green, yellow and cyan yield green, and yellow and magenta yield red. All three filters on top of each other make a dirty black. (See table 5-II.)

It must be emphasized that red ink absorbs both blue and green, magenta absorbs only green. Similarly, blue absorbs both red and green, cyan absorbs only red. The error of confusing magenta with red and cyan with blue causes many problems.

When sunlight or white light falls on white paper, the reflected light looks white because the sheet reflects all colors uniformly. This is why it is important to print on a white background if the printed image is to contain all colors. If the substrate is not naturally white—say a sheet of transparent film—it can be made white by preprinting it with a white ink or coating.

To obtain good color reproduction, each of the three colors must absorb the same amount of light. When cyan, magenta and yellow are printed in equal amounts (say 50%), the color should be gray. Gray balance patches are included in many print control bars, and if one of the three printers goes out of control, the color of the gray bar changes (and, in fact, the entire print color shifts). For this reason, among others, gray balance is required in process color printing.

The rule of thirds may help one understand additive and subtractive colors. If one adds green, red, and blue one obtains white.

$\frac{1}{3}(g) + \frac{1}{3}(r) + \frac{1}{3}(b) = \frac{3}{3}(\text{white})$

If one subtracts red with a cyan filter one obtains blue green, and subtracting blue with a yellow filter yields green.

$\frac{3}{3}(\text{white}) - \frac{1}{3}(r) = \frac{2}{3}(\text{blue green})$

$\frac{2}{3}(\text{blue green}) - \frac{1}{3}(b) = \frac{1}{3}(\text{green})$

Definitions of Color

To define a color requires three properties or three parameters or dimensions. Different instruments use different properties, but three are always required.

It is useful to consider hue, brightness and saturation of a color. Hue is the shade or color (red or green or blue), brightness or lightness refers to the absence of grayness in the color (brick red as opposed to the bright red in the U.S.,

Canadian, British, French or Japanese flags), and saturation or color strength refers to the intensity of color (a pink sweater has less color strength than the red of the flag).

The color attributes of hue, brightness and saturation can be measured instrumentally or they can be located in a collection of color samples or described verbally.

Hue. The first dimension of color is hue. Red, green and blue, like brown and orange, are hues. Skim milk is sold in a blue carton while chocolate milk is sold in a brown carton and orange juice is sold in an orange carton.

Brightness. The second dimension is brightness, which refers to lack of grayness or to color purity. A shirt may be described as bright blue or gray blue. "Clean" and "dirty" are sometimes used to refer to the degrees of brightness.

In describing color, the designer must specify whether the color is to be a bright (clean) blue or red or a grayish one. If you compare the two, you will note that the Sun Maid color is brighter or cleaner than the Campbell's red. The Campbell's red is grayer.

The word "brightness" is also used to describe the intensity of illumination. A bright light is more intense than a dim one. A piece of black velvet has no reflectance, that is, it has zero brightness. A piece of barium sulfate pigment that the physicist uses as a standard for white has 100 percent brightness. We can arrange grays, then, in a line ranging from 0 to 100 percent brightness. We can also specify the grayness of colors. The STOP sign is a bright red, while a red brick fireplace has low brightness.

Saturation. The third dimension of color is its intensity, strength or vividness—technically called saturation. The pink sweater of the model whose picture appears on the package of cosmetics has low saturation, but the reds on Campbell's soup or Sun Maid raisins have high saturation.

Other Systems. The Hunter L, a, b is a different color scale (figure 5-5), but it also has three dimensions as do the CIE* systems that bear dimensions L*, a*, B* and L*, U*, V*. These are sometimes used in printing, but more frequently in the scientific laboratory.

THE APPEARANCE OF COLOR

Several visual phenomena affect the way we see color. The designer can use these to achieve desirable results, or the design may cause them to detract from the desired affect.

Adaptation. The eye adapts to various levels and colors of light. The black bark of a tree in bright sunlight reflects more light than a white candy box in

* Commission Internationale de l'Éclairage, established in 1931.

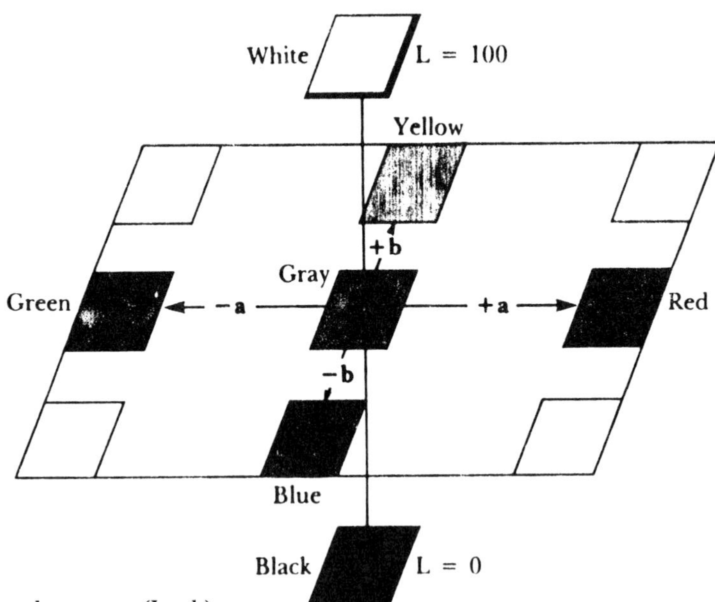

Figure 5-5. Hunter color system (L,a,b).

subdued light, but the eye adapts to the illumination level and reports that the tree trunk is black and the box is white.

Association. We tend to see the expected color. The color of a TV screen may be seriously out of adjustment, but the eye will still see the blue of the sky and the green grass of the football field.

Adjacency effect. Colors tend to blend with one another. In halftone printing, this makes the colors of the dots blend into a single color. The same effect occurs to a lesser extent in fine patterns.

After-image. The visual process of the eye tires and a design can be spoiled because the after-image on one part of the design alters the appearance of color on another part of the design.

Color-blindness. This important problem must be considered in designing warning labels.

Color variability. Some printing processes and some printers suffer less variability than others.

Metamerism. Metamerism is the change in appearance of a color that occurs when a print is illuminated under different lights. Because colors are composed of reflections at many wavelengths and different lighting systems emit different wavelengths, colors that match under one light may not match under another.

The packager must not only avoid unwanted color effects but must take advantage of retail lighting practices. The lighting at the supermarket meat counter is different than the lighting at the soap gondola. The lighting is chosen to enhance the graphic qualities of the package. In the meat area, it is the meat itself; in the soap area it is the printing on the packages, which are usually red, orange and yellow.

Usually we don't pay much attention to TV color fidelity. We've grown used to the color on our favorite set. But when we go into an electronics store, we're shocked to see the variability among different sets. Such variability would be intolerable in cereal boxes or perfume labels that stand side by side on the shelves at the retail store.

Standard lighting conditions for viewing and comparing print with original call for 5000 K, a light that is a compromise between northern light and home incandescent lights. It is approximately the color of fluorescent lights in a supermarket, but it is much bluer than incandescent home lighting. Packagers may sometimes specify that another illumination be used, and the package printer must conform to the customer's request. Both the design and the color separations made by the printer must be viewed under the same light.

Figure 5-6 (see color insert) shows the difference in appearance of a photograph under different illumination such as indoor incandescent lights, ordinary fluorescent lights and outdoor illumination.

It is possible to achieve a broad range of colors using the four process colors in flexo, gravure, litho, screen or letterpress, and the hue, brightness and cleanliness of any color can be varied over a wide range. If a particular shade or background color is required to match the logotype or to be part of the overall design, it should be printed as a separate color called a "spot" color. A good printer will control variability within stated limits that differ with each of the printing processes. Spot colors or PANTONE colors should give the least variability.

The color of the stock or substrate is also important—even whites vary in shade. A blue-white gives greater contrast with black and with blue, but a blue-white will produce muddy yellows while a warm white reduces the iciness of snow or the clarity of the ocean, both of which depend on blue reflectance from the paper. Where color variations in the design are to be minimized, samples must always be tested under different types of lighting in addition to standard lighting. An ink formulation that gives the desired color under standard illumination may not be suitable for lighting in the home or the gift shop. The designer should frequently consult with the inkmaker. Colors are usually specified visually, that is, they are to match a sample.

In addition to printing, color can be added to the package by coloring the plastic or glass container. The molded plastic cap, used for bottles, is replacing the still-familiar metal cap. The plastic cap can easily be colored to add interest and identification to the package and the product.

Rules for color usage are based on physiology, psychology and careful research, but mostly on observation. In the U.S. the most acceptable colors are bright gold for canned corn, darker gold for peaches and bright red for cherries. Sales of baked beans improved when the color of the label was changed from tan to a rich, dark brown. This color is wholly unsuited to laundry bleach or detergent. Blue is very suitable for health products because it conveys cleanliness and coolness.

As David Ogilvy observed, a milk carton printed in brown shouts "chocolate milk" and a blue carton says "skim milk." Green says buttermilk and an amber widemouthed jar communicates only confusion regardless of the label.

OBSERVING COLOR

The ultimate criterion by which color printing is judged is the human observer. If the print or proof is satisfactory, someone approves it, the presses roll, the job is delivered and an invoice is issued. Disagreements concerning the success or failure of the job to meet agreed-upon specifications often arise from variations in the way the colors are viewed.

Color Vision

There are several theories of color vision. The retina of the human eye is made up of many levels of nerve cells, called rods and cones, that are sensitive to light. There are an estimated 120 million of them in each eye. The central area of the retina contains only cone cells and is primarily involved with daylight vision. The rods, which are involved with night vision, are found in the periphery of the retina. There are apparently different kinds of photopigments in the cone cells, each of which is sensitive to different colors of light. Field's text contains an excellent discussion of the current theory of color vision.

Variations in Color Perception. Different people see color differently and even the same observer sees color differently depending on the illumination, background, viewing conditions and time of day. Surround effects, after-image and the tendency of the eye to blend colors all affect the way we see color. Other problems, such as metamerism, are problems of illumination, not of vision.

Psychological Effects. Psychological effects include the ability of the observer to see grass as green and sky as blue, even when the color of the print is far from green or blue. The eye records the color of the bark of a tree in the bright sunlight as black, and the color of the page of a book in subdued light as white, even though there is more light reflected from the black tree bark than from the white page of the book. The other side of the coin is that when we see certain colors, we associate certain results with them: white and red for pharmaceuticals, earth tones for natural foods, and green for mint-flavored products.

Surround effects are the characteristic of the eye to see the same color as a different color when the surrounding color is changed. The eye also tends to blend colors over a small area, and this makes possible the effective use of process color or halftone printing. The eye blends the cyan and magenta halftone dots, sending to the brain the color of blue. When we stare at one color, our eyes change their perception. The familiar effect of after-vision is illustrated in figure 5-7. (See color insert.) If you stare at the black spot for 20 seconds, then shift your eyes to the blank square, a white spot will appear in the center. A similar experiment can be performed with an American flag printed in black, yellow and green. After staring at it for about 20 seconds, shifting your vision to a white space next to it will produce the appearance of a flag in its usual colors. Staring at a Canadian flag for 20 seconds and shifting the vision to a white background will produce a green maple leaf.

Color Blindness. There are several forms of color blindness, but the most common is a form in which the eye confuses red and green. In addition to specific color blindness, there are differences in people's ability to detect or discriminate between small differences in color. There are numerous tests that measure color perception.

The normal human eye can see colors with wavelengths from about 400 nm to around 700 nm. The vision of most people corresponds rather closely, but about five to eight percent of all men and a much smaller percentage of women do not perceive color in the same way as the majority. In addition, both the cornea and the lens become yellower as people grow older, and this means that older viewers are less sensitive to blue. As there are tests for color blindness, there are tests for yellowness of vision.

Management should test people for color perception if they are to have responsibilities for judging color. If color match is critical, it may be wise to have more than one person view it. With the instruments available on modern presses, color blindness should be only a minor hindrance for press operators.

There are various tests for color blindness, the best known of which is the Ishihara test where numbers composed of dots of varying sizes and colors are superimposed on backgrounds made up of similar dots.

Illumination

It seems obvious that a sample cannot reflect colors that are not present in the illumination; we cannot see good blues under red light. Under red light, a well-printed blue appears black. Similarly, if the light has a yellowish cast to it (incandescent light), the colored picture or sample will appear different than if the light has a bluish cast (outdoor or fluorescent light).

Light is often described in terms of its "color temperature." When a black object, such as a blackened steel rod is heated, it begins to emit light. We have all seen pictures of hot iron in the steel mill or the blacksmith shop. As the temperature of the steel is raised, the amount of light emitted increases and its color changes. At first, the light is a dull red, but it becomes increasingly white as the temperature is raised and the color shifts towards the blue end of the spectrum.

It has become customary to describe illumination by noting the temperature of a black radiator that gives off the light. If we take a perfect radiator—cast iron or a tungsten wire approach this—and heat it to 2000 degrees Kelvin it will glow with a dull red color. (The temperature in the Kelvin or absolute temperature scale is the Celsius reading plus 273 degrees, and the color of light given off at 2000 degrees Kelvin is often described simply as 2000 Kelvins or 2000 K.) If we heat the steel to 5000 degrees it will be much whiter. At 7500 degrees, it has a bluish tint like light from a northern window on a clear summer's day. This is one of the reasons why it is important to examine proofs and prints under standard viewing conditions, which includes illumination at a color temperature of 5000 K. Figure 5-8 shows the distribution of colors at different temperatures.

A tungsten filament used in incandescent lamps is around 2800 K. This light is very yellow, and packages viewed at 2800 K do not show good blue colors. A

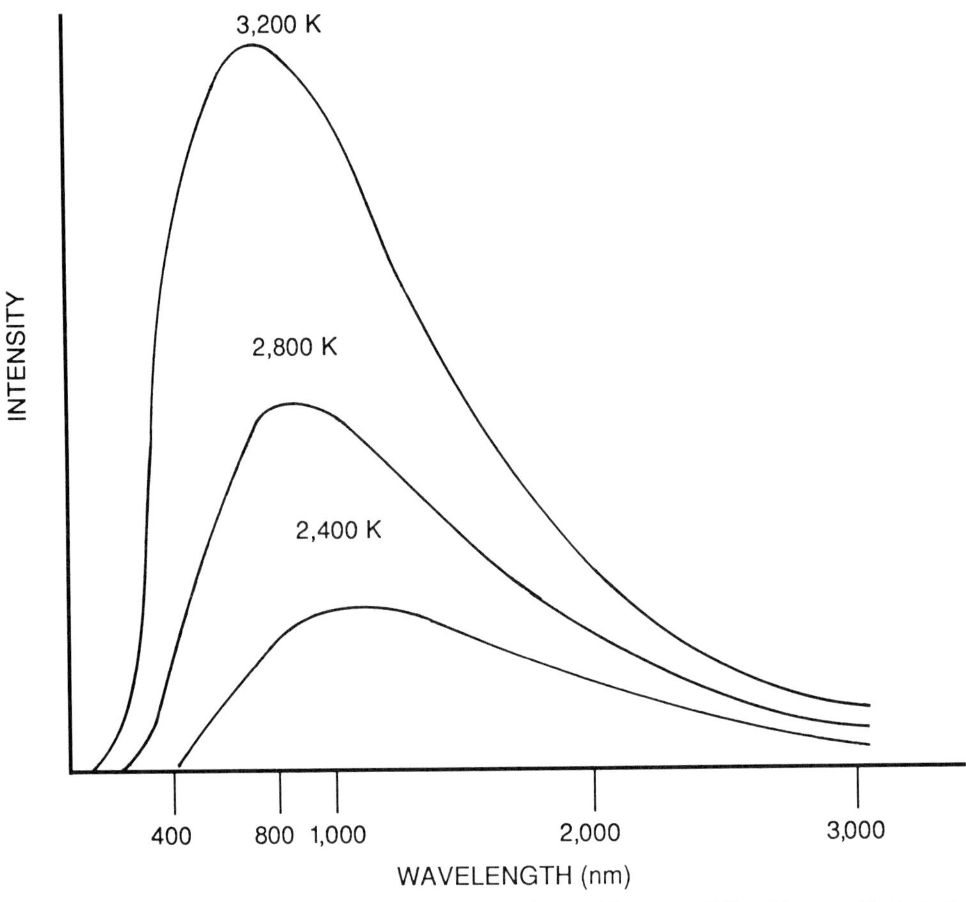

Figure 5-8. Distribution of light from a heated radiator. (Courtesy of Graphic Arts Technical Foundation.)

light temperature of 5000 K is closer to outdoor light or fluorescent light, and a light temperature of 7500 K is the blue, northern light that artists wish for in their studio. It gives much better viewing of the blue colors.

Light Sources and Viewing Conditions

We receive light from many sources: fluorescent lights, incandescent lights, and sunlight either direct or reflected. All of these have different color distributions or spectra. The package designer and buyer must consider the probable viewing conditions under which the customer will make the decision to buy.

Since the viewing conditions most important to the packager are often those in the supermarket, the color of lights in the supermarket may be chosen for viewing the design. Illumination in the meat department at the supermarket will be skewed towards the red to make the meat appear fresh. General merchandise is usually displayed under fluorescent illumination which is rich in blue.

However, because changing viewing conditions changes the apparent color of the package, standard viewing conditions should be used after the design colors have been chosen. When design, proofs and final package are to be compared, it is vital that standard viewing conditions be used by all viewers at all stages of production. If the package is designed to be viewed under non-standard conditions, these must be considered and planned from the start.

Standard viewing specifies not only the color of the illumination but the intensity of the light and the color of the surrounding space. Walls should be painted the standard neutral gray. If the sample is viewed where the walls are painted green or pink it will appear different than it will under standard conditions. A color transparency or print that is held against the window will not appear the same color as the same transparency or print when it is viewed under a tungsten light or a fluorescent light. Proofs that match the original under one type of illumination will not match under another because the pigments used in the film are different than those used in printing ink.

What is satisfactory color under one set of conditions is unacceptable under another. The artist who designs a print under "northern light" will be disappointed with the print that is viewed under standard lighting.

Viewing of design, proofs or print under yellow or green fluorescent lights is such an obvious error that it is amazing that anyone ever does this. Yet such blunders are reported frequently.

To avoid these errors, comparisons should be made with a standard transparency viewer (figure 5-9) or in a standard viewing booth (figure 5-10). Standard viewing booths (built according to international standards) should be available in every design studio, every color separation house and every plant producing colored printing. In addition, the marketing department needs access to standard viewing.

A control tool to determine whether the lighting meets standards is the GATF/RHEM Light Indicator, a small adhesive-backed patch printed with magenta ink. Under standard lighting conditions no stripes are visible, but under

Figure 5-9. Standard transparency viewer. (Courtesy of Macbeth Division of Kollmorgen Instruments Corp.)

Figure 5-10. Standard viewing booth. (Courtesy of Macbeth Division of Kollmorgen Instruments Corp.)

non-standard conditions, stripes appear across through the patch. The Light Indicator (figure 5-11) takes advantage of metamerism in such a way that the inks used to make up the patch match only under standard lighting.

Color Viewing Standards

Color reproduction Specifications for Web Offset Publications (SWOP) are gradually being adapted for all kinds of color reproduction. According to SWOP "Recommended Specifications for Web Offset Publications;"* "Accurate perception of color and tonal values requires that subjects be illuminated by light of the proper color quality, and viewed in a chromatically neutral, controlled viewing environment. There is worldwide agreement that light with a correlated color temperature of 5000 K which has essentially equal amounts of energy in

* This report is available from American Business Press, 205 East 42nd St., New York, NY 10017 or American Association of Advertising Agencies, 666 Third Avenue, New York NY 10017

the blue, green and red portions of the visible spectrum (400 to 700 nm), permits the viewer to most easily detect minor color differences. Accordingly, it is best for use where critical color and tone comparisons are to be made."

Color viewing standards are specified by ANSI and ISO.* They specify the following conditions for viewing copy, artwork, proofs and printed products. The color rendering index is a measure of how effectively the light source conforms to uniform light distribution throughout the visual spectrum without peaks or differences at any wavelength.

The standard color viewing conditions issued by ANSI in specification PH2.320 1972 are:

Color temperature of lighting: 5000 K
Color rendering index to 90–100
Level of print illumination 204 ± 43.6 footcandles
Geometry of illumination to minimize reflected glare
Level of transparency illumination: 409 ± 88 footlamberts
Surround: matte, neutral gray (Munsell notation N8/)

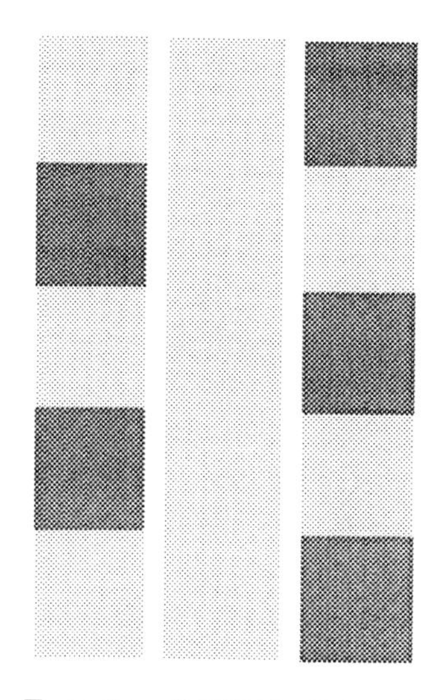

Figure 5-11. GATF/RHEM light indicator to indicate proper illumination. (Courtesy of Graphic Arts Technical Foundation.)

Metamerism

In music, if three different notes (that is, notes of three different wavelengths) are played at one time, the ear hears all three, and the chord is known as a triad. Color perception is quite different. If two or three different wavelengths are present in a color, the eye combines them and sees only one color. For example, the presence of red and green wavelengths together produces the sensation of yellow. In fact, a given color can be produced by several different mixes so that it is possible to produce a desired color by mixing different combinations of pigments in ink. However, since the inks absorb rather than transmit color, the color that is perceived will depend on the illumination. If a certain lavender or brown or cream color is produced by blending the pigments in an ink, the color seen with the eyes will depend on the illumination, and colors that match under one type of light will not match under another light.

This phenomenon is called "metamerism" (met-am'-er-ism). It is another reason that judgments of color must be made under standard conditions. Two different inkmakers matching a given color will probably produce inks that will not match at all wavelengths unless the customer specifies the match must be nonmetameric.

COLOR MEASUREMENT AND SPECIFICATION

The brain interprets the visual impact of light to give the sensation of color. Because psychology is involved, objective measurement of color requires physical instruments. It cannot be done by the eye alone.

* American National Standards Institute, 11 West 42 Street, New York, NY 10036. Phone (212) 642-4900. ANSI is the American section of ISO, the International Standards Organization, which is affiliated with the United Nations.

There are three principal instruments for measuring colored objects: densitometers, colorimeters and spectrophotometers.

Transmission and Reflection Densitometers

In the printing industry, the most common instrument for measuring color is the densitometer. The transmission densitometer (figure 5-12) measures the amount of light absorbed by a photographic film. The reflection densitometer measures the amount of light absorbed by a printed image. The transmission densitometer is usually used in the camera department to measure the density of an image on photographic film. Strictly speaking, these instruments do not measure color but they measure the amount of light absorbed by a transparency or a print over a rather broad range of the spectrum. The range is determined by a color filter. The amount of light absorbed is reported as the optical density.

Densitometers range from fairly simple hand-held models that report only the density, to sophisticated, on-press models that calculate several factors related to optical density (figures 5-13 and 5-14). They are sometimes connected to feedback control systems on press.

The reflection densitometer is supplied with three color filters so that it can measure the absorbency of a magenta, cyan or yellow print. The black density is usually measured without a filter. Since magenta reflects red and blue and absorbs green, a green filter is used when measuring the absorbency or intensity of a magenta print. Similarly, a blue filter is used to measure the absorbency of the yellow and a red filter to measure cyan. Because no ink completely absorbs one color, nor completely reflects the other two, the reflectivities of all three colors are measured on each color of the color bar.

In the pressroom, the densitometer is essential in controlling the amount of

Figure 5-12. Transmission densitometer. (Courtesy of Macbeth Division of Kollmorgen Instruments Corp.)

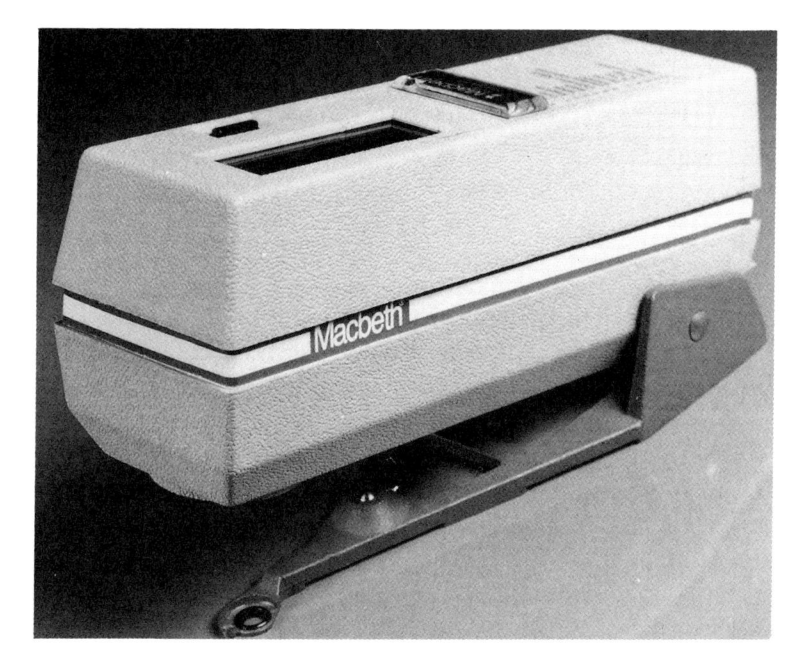

Figure 5-13. Macbeth Series 1200 portable pressroom reflection densitometer. (Courtesy of Macbeth Division of Kollmorgen Instruments Corp.)

ink printed, because the thicker the ink film, the greater the optical density or absorption of light. Printers who practice Total Quality Management will have data relating color strength of the ink to density so that they can reduce makeready time, waste and spoilage while keeping the color variability within specs.

Figure 5-14. Scanning densitometer. (Courtesy of Tobias Associates, Inc.)

TABLE 5-III. TYPICAL OPTICAL DENSITIES OF PRINTED INK FILMS

Ink	Blue Filter	Green Filter	Red Filter
Yellow	1.04	0.06	0.02
Magenta	0.45	1.14	0.08
Cyan	0.08	0.30	1.02

Table 5-III shows typical density numbers for printed ink films on the color control bar. These will vary with each press, ink and substrate combination.

These numbers show typical problems with modern ink pigments. An ideal yellow absorbs strongly in the blue while transmitting all the green and red. An ideal magenta absorbs strongly in the green while transmitting all the blue and red, and an ideal cyan absorbs strongly in the red while transmitting all the green and blue. In reality, magentas absorb a great deal of blue and a little red as well, and real cyan pigments absorb too much green as well as some blue, creating a color that is more gray than we would like. Figure 5-15 shows typical absorption curves for process color inks. The yellow pigments are the best, absorbing the blue and reflecting most of the green and red light.

We can measure the reflected wavelengths with a densitometer and plot them on a GATF Color Circle. Figure 5-16 shows the actual absorptions of several yellow, magenta and cyan inks. The yellows fall close to the ideal mark, the magentas fall too close to the red, and the cyans are too gray—they lie too far from the 100 percent reflectance circle which is the outside circle.

Densitometers must be cross-calibrated against one another if the numbers are to be used for communication between plants or between locations within a single plant.

Densitometers are important pressroom tools. Their numerical density readings of the printed package and the standard (including light and dark tolerances) provide an objective method for measuring the conformance of the printed color to the approved color.

Colorimeters

Perhaps the most familiar colorimetric measurement of color is the Hunter L,a,b system (figure 5-5). In the Hunter system, "L" stands for lightness, "a" for redness and "b" for yellowness. (Green has a negative "a" value and blue has a negative "b" value.) Thus the bright red in the flag would have a low "L" value but a very high "a" value; the sweater would have a high "a" value and a high "L" value, and the bricks would have a low "a" value and low "L" value. The instrument used to develop these numbers is called a colorimeter (figure 5-17).

Spectrophotometers

A spectrophotometer measures the light reflected at each wavelength in the visible spectrum (or UV or IR in the research laboratory). The reflectance (or

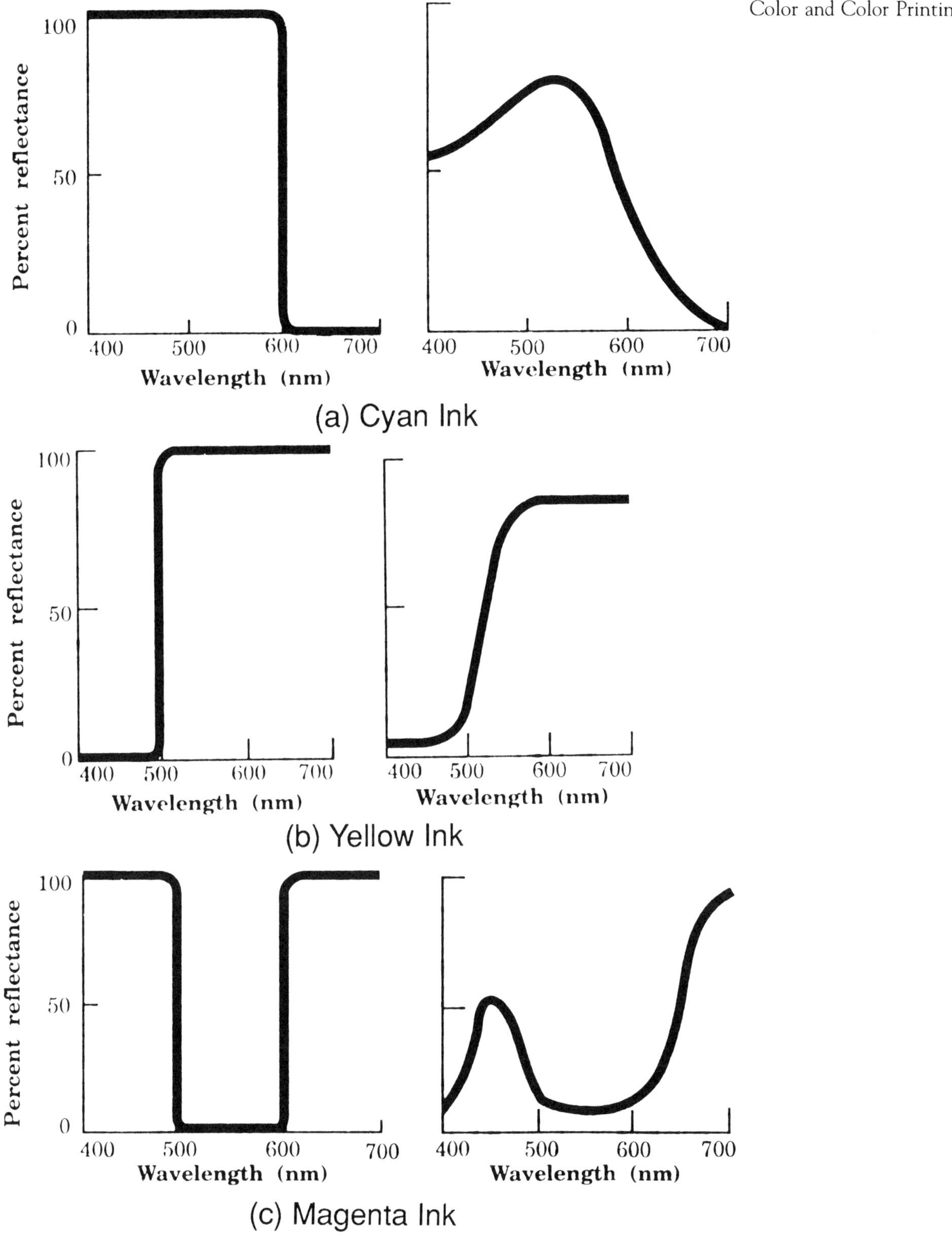

Figure 5-15. Ideal and typical color absorption curves for process color inks; (a) shows ideal and typical absorption curves for cyan, (b) shows ideal and typical absorption curves for yellow and (c) shows ideal and typical curves for magenta. (Courtesy of Graphic Arts Technical Foundation.)

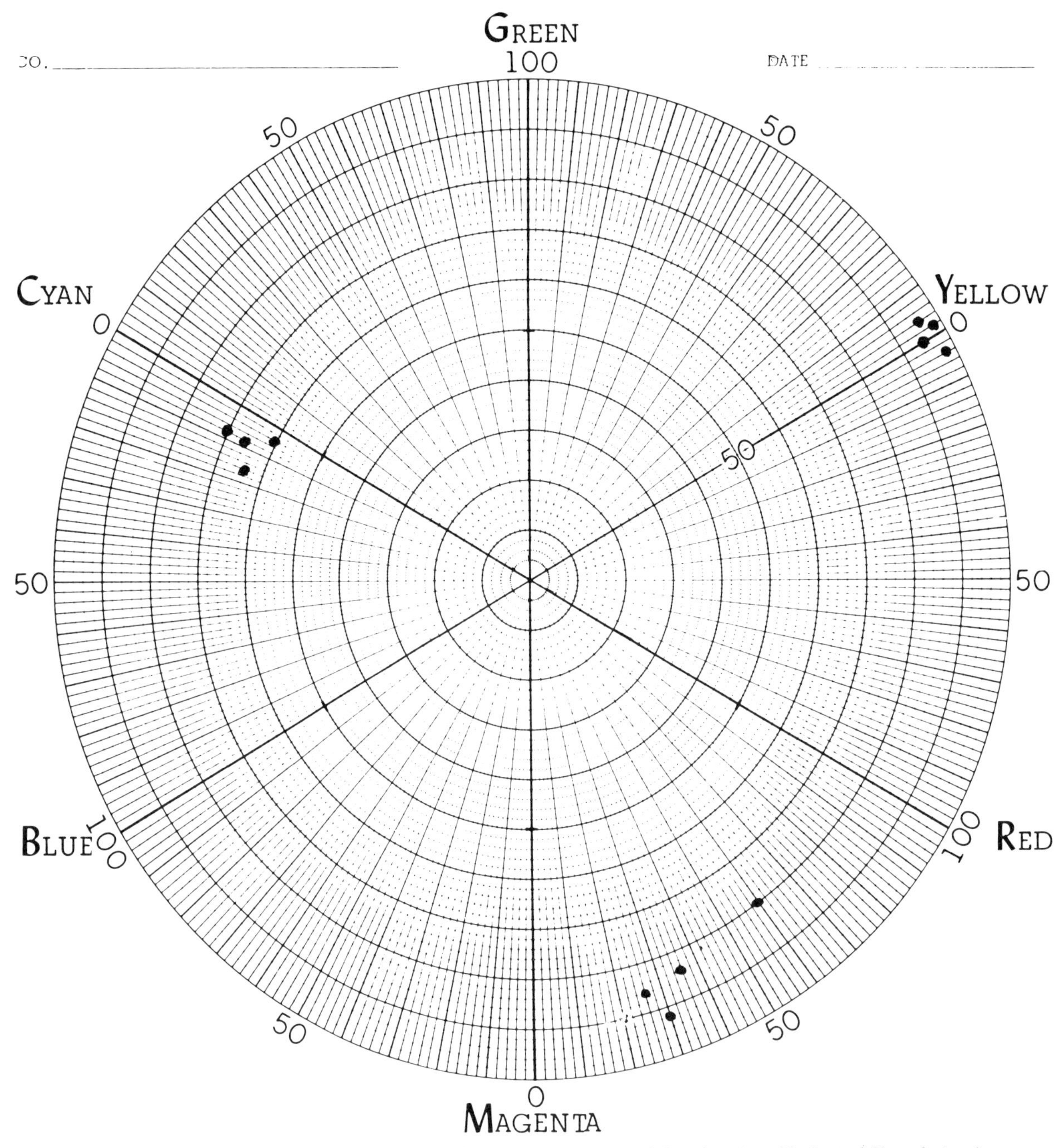

Figure 5-16. GATF color circle. (Courtesy of Graphic Arts Technical Foundation.)

Figure 1-3. Label produced for Budweiser beer, used from 1883 to 1891. (Photo courtesy of Anheuser-Busch Companies Inc.)

Figure 5-1 C ... *erages, appliances, pharmaceuticals and furniture.*

Figure 5-6. Colored picture viewed under three different lights. The photo in the middle is viewed under standard illumination. The photo on the right is too blue, that on the left is too red. (Photo courtesy of Graphic Arts Technical Foundation.)

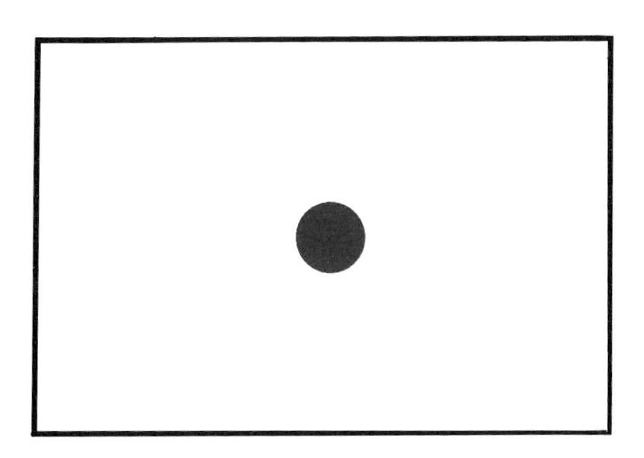

Figure 5-7. Effect of aftervision. Look at the red dot in the left hand box for 20 seconds, then look at the center of the right hand box. The color in the right hand box will be cyan, the complementary color to red. If the dot were magenta, the complementary color would be green.

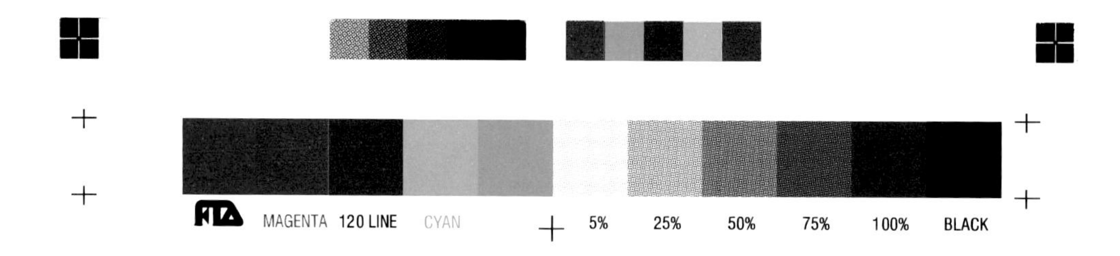

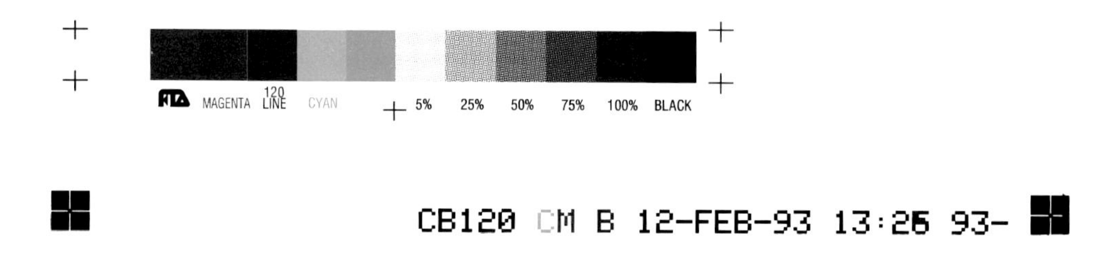

Figure 5-26. FTA Color Bars. (Courtesy of Wilson Engraving.)

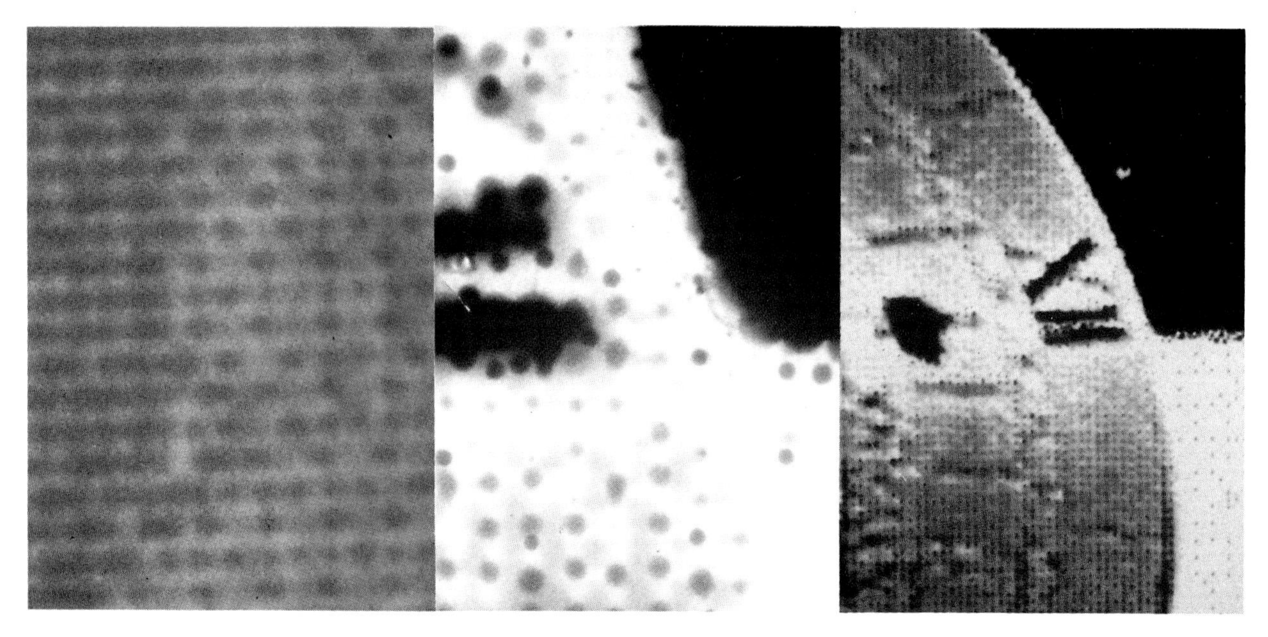

Figure 6-11. Ink jet proofing. The pictures show three degrees of magnification: the figure on the right is the lowest magnification; the center picture shows a small portion of the right-hand picture, and the left-hand picture shows the actual dot structure of the ink droplets. (Courtesy of Iris Graphics, Inc.)

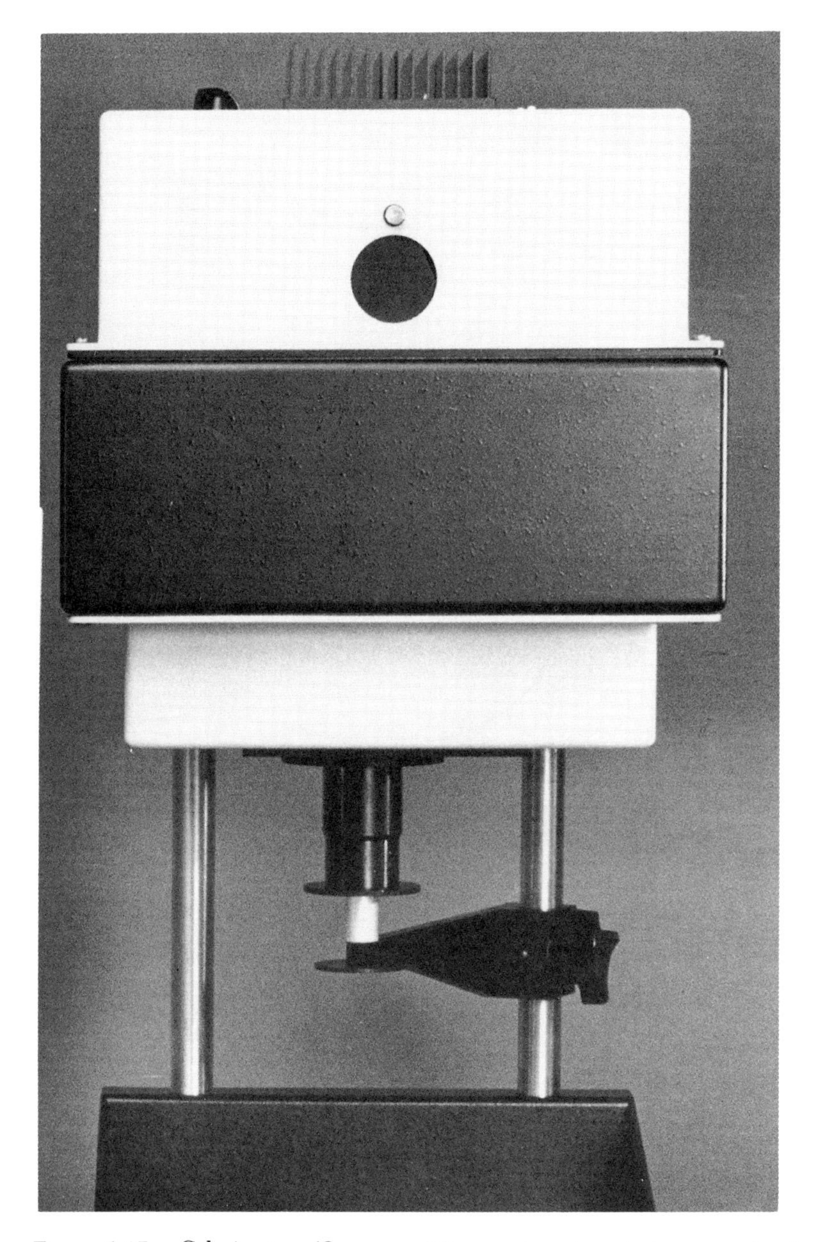

Figure 5-17. Colorimeter. (Courtesy of Hunter Associates Laboratory, Inc.)

absorbency) of each wavelength is plotted to give a spectrophotometric curve or spectrogram. These are expensive instruments, used primarily in the research laboratory or in the quality control lab in ink manufacturing operations. Some large packaging companies use spectrophotometers to control the color of printing inks. A typical spectrogram is shown in figure 5-18 and a spectrophotometer is shown in figure 5-19.

Unfortunately, numbers do not convey any sense of what color looks like. Designers and artists like to use the Pantone Matching System,* which has two

* Trademark of Pantone, Inc., 55 Knickerbocker Road, Moonachie, NJ 07074.

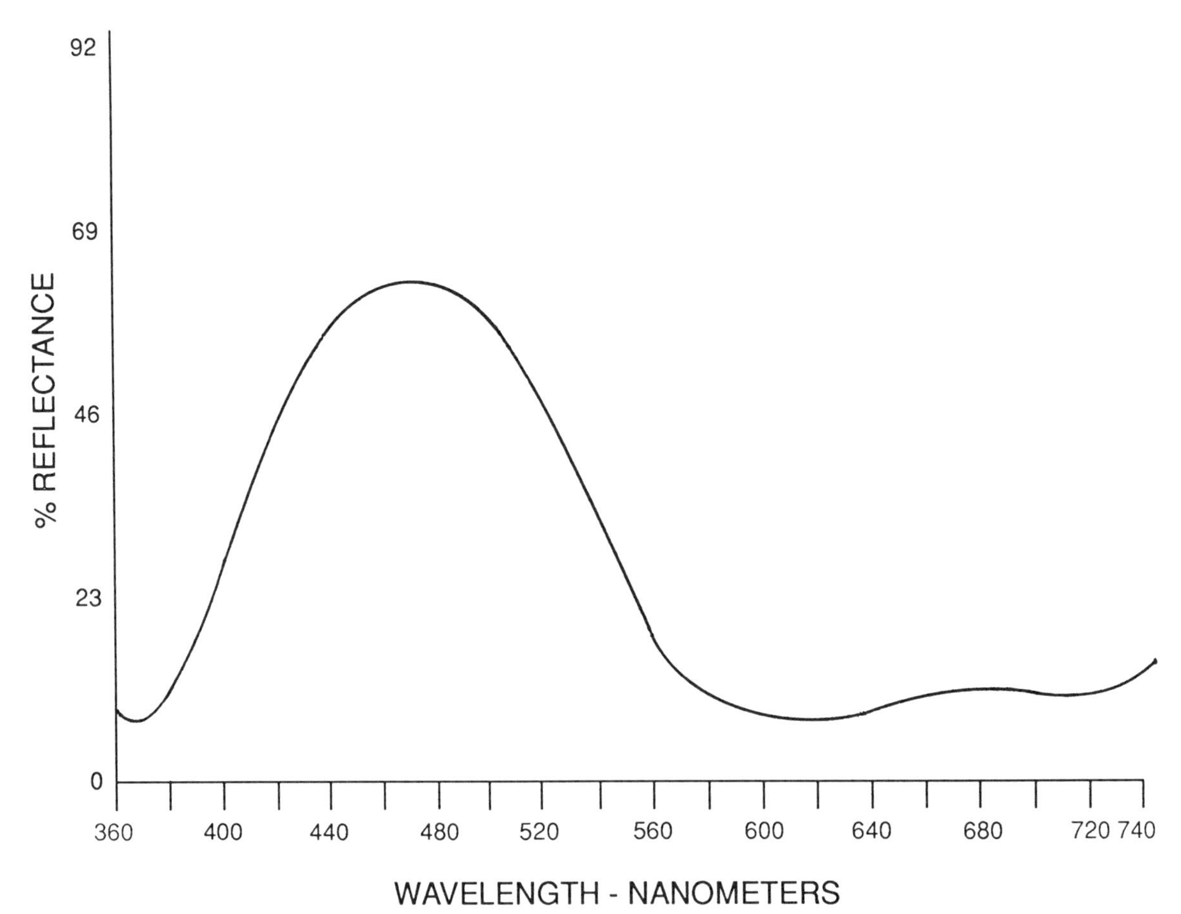

Figure 5-18. Spectrogram of a cyan ink. (Courtesy of Graphic Arts Technical Foundation.)

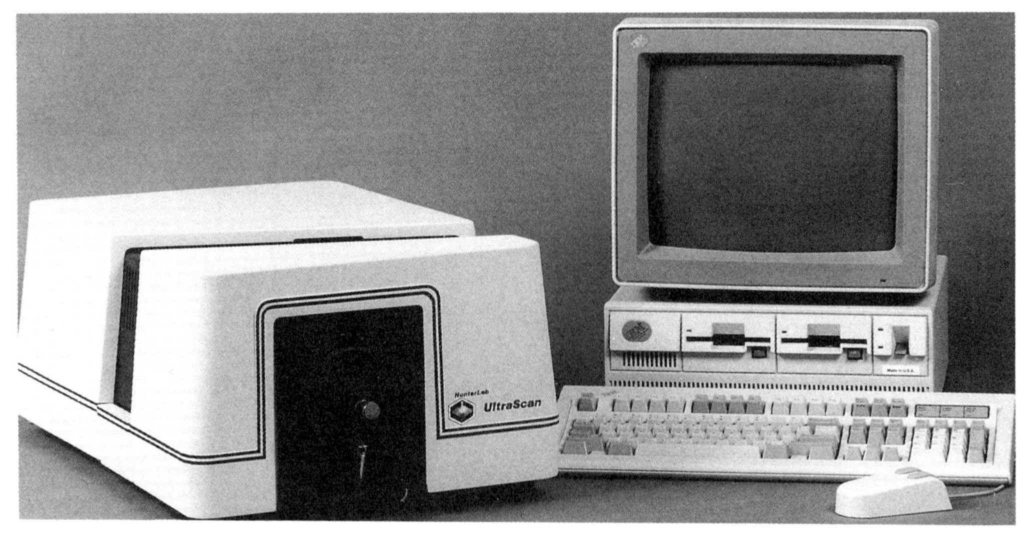

Figure 5-19. Spectrophotometer. (Courtesy of Hunter Associates Laboratory, Inc.)

advantages. It not only gives a visual representation of the color, but it gives an ink formulation that will produce that color. Pantone also produces the Pantone Process Color System for use in estimating blends of process color inks to achieve a given color.

The following instrumental usage is convenient to control color in package printing: densitometers for controlling ink film thickness on press; colorimeters for paper color, and spectrophotometers for ink color quality control. In addition, the Munsell Color System is helpful in explaining the attributes of color; and the Pantone Matching System for determining ink mixing formulas for matching special colors.

"PMS" is a widely used contraction of Pantone's trademarked Pantone Matching System. One hears the term "PMS" used to refer to spot or line colors.

THE APPROPRIATE USE OF EYES AND INSTRUMENTS

Ultimately, judgment of the package must be made with human eyes. Eyes can do things that instruments cannot do. Eyes can look at the entire package and judge the overall effect, and they can be used to decide whether or not the package is satisfactory. Eyes are very good for deciding whether two colors, or even two prints, match. A careful observer can detect color differences more precisely than can a densitometer.

However, there are many things that eyes cannot do. They cannot put a number on a colored patch—a number that can specify a color and be transmitted by telephone or fax to a supplier or customer. Eyes have a very poor memory of color. Although an observer can compare two prints side by side, the observer cannot accurately compare two prints presented one after the other if there has been as much as one second between the viewings. Eyes are influenced by surroundings and by memory of what color is expected.

The best control, then, is obtained by the appropriate use of both eyes and scientific instruments together with an understanding of the uses and limitations of each.

COMMUNICATING COLOR

Communication of color is a major problem for the designer, the printer and the packager. The most common method is verbal, and this is the least accurate of all. But the fact that communication of color is difficult does not mean that it should be ignored nor that it cannot be remedied. By a combination of visual matching techniques and instrumental controls, color control can be much less of a problem than it usually is.

Although differences among color prints produced by flexo, gravure and offset are subtle and minor when they are well printed, the experienced buyer or artist learns what differences are to be expected when a job is printed. They may not be great when judged by an amateur, but they are often significant to the packager, because even subtle differences in package color have been shown to affect buyers' choices.

We have emphasized the technical difficulties of observing and measuring color, but communicating color involves an additional problem, that of psychology. Numbers, samples and words are helpful but not sufficient in every case. An education program directed toward the subject of color communication can be very helpful for buyers and sales representatives involved in the commerce of colored packages.

Communication Channels

Two communication channels are required: the technical and the esthetic. Errors in color communication that result from technical faults, such as the use of nonstandard viewing conditions, are fairly easy to identify and remedy.

Reasons for differences in esthetic values have been argued for centuries and are still discussed. Failure to communicate color is believed to depend not only on differences in individual perceptions, but upon the fact that different people do not always evaluate the same image in the same way. Evaluation of the color of the image or print may even be influenced by the observer's bias for or against the subject. Such failure to communicate is difficult to identify and almost impossible to correct.

Marking Color Proofs

There is no standard way to indicate the changes that are desired on a proof. Use of instrumental measurements, physical samples and verbal description all have their strengths and weaknesses. This is source of much frustration and wasted effort and materials. A lengthy discussion in a viewing booth is not always possible, and discussion over the telephone may be fruitless. Numbers and color swatches, for all of their limitations, are almost the only means of communicating color corrections desired in proof or print. Use of a PMS number or the indication that a yellow halftone area should be increased 10 percent on the proof are at least more helpful than many of the remarks found on some markups.

The GATF* Color Communicator (figure 5-20) can be helpful in communicating color. Available from GATF*, the device permits the operator to judge the color that will be achieved by blending fixed percentages of cyan, magenta and yellow, or the amounts of each process color required to give the desired color. Colored films that represent varying amounts of cyan, magenta and yellow can be moved to achieve the desired color. The device includes numbers that describe the amount of cyan, magenta and yellow required.

LINEWORK AND SPOT COLOR

Some 90 percent of packages are printed with spot color. In linework and spot color, each color is produced by one ink. If two colors are called for, the press applies two colors of ink. If the artist selects two colors plus black, they take up three units on the printing press, leaving the fourth unit on a four-color

* Graphic Arts Technical Foundation, 4615 Forbes Ave., Pittsburgh, PA 15213. Phone (412) 621-6941.

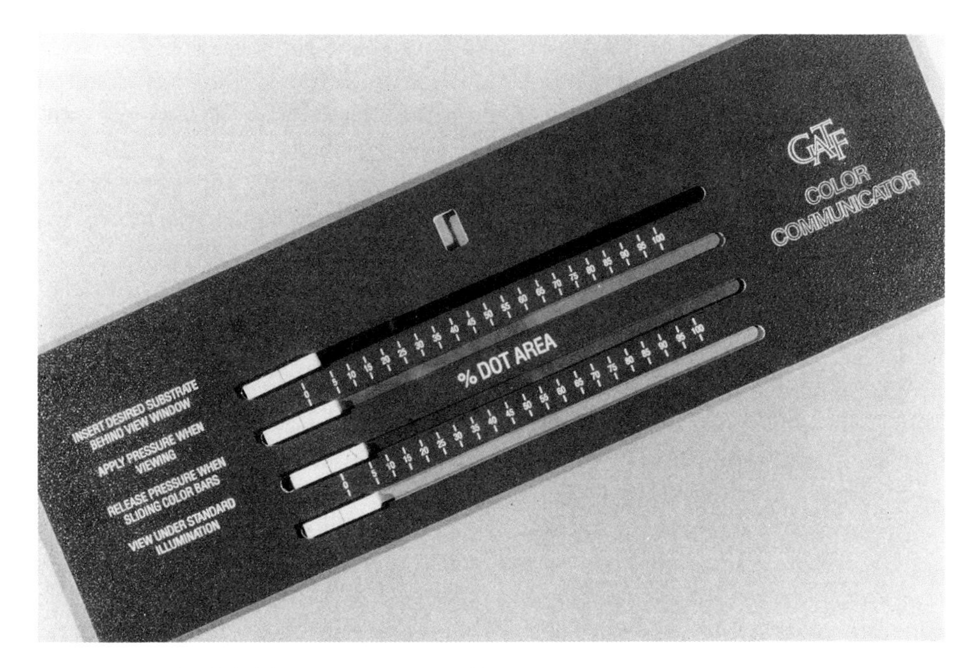

Figure 5-20. GATF Color Communicator. (Courtesy of Graphic Arts Technical Foundation.)

press available to apply varnish or coating. If process color is chosen, four units are required for the color.

It is usually less costly to reproduce linework and spot color than it is to reproduce process color because process color requires color separations and a minimum of four plates and press units. Spot colors can reproduce only those colors for which inks have been selected, whereas when process color is used, a wide gamut of colors can be achieved from four inks. With spot colors, no color separations are necessary, and variability of color and color balance on press are not usually major problems. Except for the fact that well-reproduced color photography often sells products better than drawings, packagers would use even more spot color. Spot color cannot match the color range or gamut of a color photograph.

Choosing Spot Colors

When designing and printing linework and spot color, the artist chooses the desired color and selects an ink that matches it from a book of samples. Color matching systems give a wide variety of colors and shades that can be obtained by mixing two or sometimes three inks with basic colors. The most popular of these is the Pantone Matching System. Each color in the sample book is given a number, and communication between artist, packager and printer can be maintained by using that number.

Spot colors on shipping containers are commonly matched to the Glass Packaging Institute (GCMI) kraft color guides. Wide color variation can no longer be accepted. If ink properties are correctly specified, only the viscosity need be altered by the printer.

Nevertheless, even printed color books show slight differences. For instance, the base color of the paper may differ slightly with different printings, resulting in some change in the appearance of the print. Papers and inks change color with age, and the color in the books also changes. Different lots of pigments used in inks often vary slightly. Handling and fingerprints affect the printed color. And, most likely of all, two people viewing in separate locations will seldom have identical viewing conditions.

Reverse Printing

Transparent films are not always printed on the top. The reverse side of a plastic web is often printed, either by flexo or gravure. The film protects the ink and adds gloss to the print. This calls for a different ink sequence than used in the customary printing on the outside of the package. It is usually necessary to provide a white background, and this is accomplished by printing a white coat after the design is printed. This is especially true when the printing is to be on the inside of a laminated sandwich. Different layers have various properties needed to protect the product, such as barrier to moisture, odor, oxygen and nitrogen. If the film is slightly tinted, color adjustments are made in the ink or coating. An opaque white coating can mask tints or yellowness.

PROCESS COLOR PRINTING

The flatness of color obtained from printing spot colors is unattractive for many designs or subjects. The print of a woman's face or hair for use on a cosmetics box requires printing a broad range of colors. This flatness is overcome by process color printing in which a wide range of colors are produced from only four inks. On the other hand, the ability to print many hues and tints makes process color printing variable and hard to control.

Theory of process color printing is very simple, but in practice it has taken three hundred years of experimenting and development to bring process color printing to its present state of development. Since the eye blends colors that lie close to each other, it is possible to produce a wide gamut of colors by printing bits of color next to each other. Impressionist painters did this by putting small dots of ink on a canvas. The modern printer does it by breaking the image up into small dots, called halftone dots, and printing them beside or on top of each other.

Theory and Practice of Process Color

Process color printing is the technique of separating a full-color original picture into its basic three or four colors, converting the separated colors into halftone dots and then printing the colors sequentially. This reproduces the original in full color. Originals such as color transparencies or full-color photographic prints can be reproduced in print. In process color, three primary colors of cyan, magenta and yellow are blended to produce the illusion of a wide variety of colors. Each ink absorbs part of the white light reflected from the white substrate, producing the desired colors.

This "three-color process" is almost always converted to a "four-color process" by adding black. Actually, it is possible to print process color with only three inks: yellow, cyan ("process blue") and magenta ("process red"), but because the commonly available pigments do not perfectly absorb the proper colors, a black printer must be added to improve the contrast and/or tone balance and to deepen the shadows.

Printers often used the term "red" for magenta and "blue" for cyan. However, red is very different from magenta and blue is very different from cyan. The packager who is working with process color must watch out for this common source of confusion.

The shade of an ink printed on a substrate can be changed by varying its thickness which, in turn, changes the amount of light reflected or absorbed. However, gravure is the only printing process that can produce different colors on a sheet by accurate control of ink thickness from one point to another, but even in gravure the variation in optical density is severely limited. Therefore, we control the amount of light reflected from the substrate by controlling the size of tiny halftone dots.

If we print yellow and cyan dots next to each other we produce green. If we print small yellow halftone dots next to larger halftone dots of cyan, the color will be a bluish green. If we print a little cyan next to a great deal of yellow, the color will be a yellow green. Except in some gravure processes, the actual thicknesses of both the yellow and cyan ink films or dots remain constant. If three or four different hues or color are controlled in this way, we can produce a broad range of colors.

"Color separation" is the name given to the process of making four negatives or positive photographic films from a continuous-tone original. The end result is four photographic films, each containing the color and tone values of one basic color from the original. The continuous-tone image of the original can be made into a halftone image either electronically with a color scanner or photographically with a camera and a screen which is a piece of film with lines. In photographic separation, the original is photographed through a color filter that cuts out the undesired colors and a screen that breaks the image up into halftone dots. The process is repeated three times to give four different images.

A halftone print appears continuous because at a normal viewing distance, the eye cannot detect the tiny dots. It requires a magnifying glass to see them.

The colors are recombined by making a plate to print each of the colors and printing each color sequentially in register. The screening process not only breaks the light up into tiny dots (circles, squares, ovals, etc.) but it controls the area covered by the dots. Where the color is heavy, most of the area is covered by the dot. Where the color is light, most of the area is uncovered (figure 5-21).

When using an electronic color scanner, the electronics control the size of the halftone dot, either with a laser beam that exposes halftone film or by generating an electronic image. Lasers may also expose photochemical printing plates, drive an engraving machine that engraves a gravure cylinder, or laser engrave flexo printing plates with moderately fine resolution.

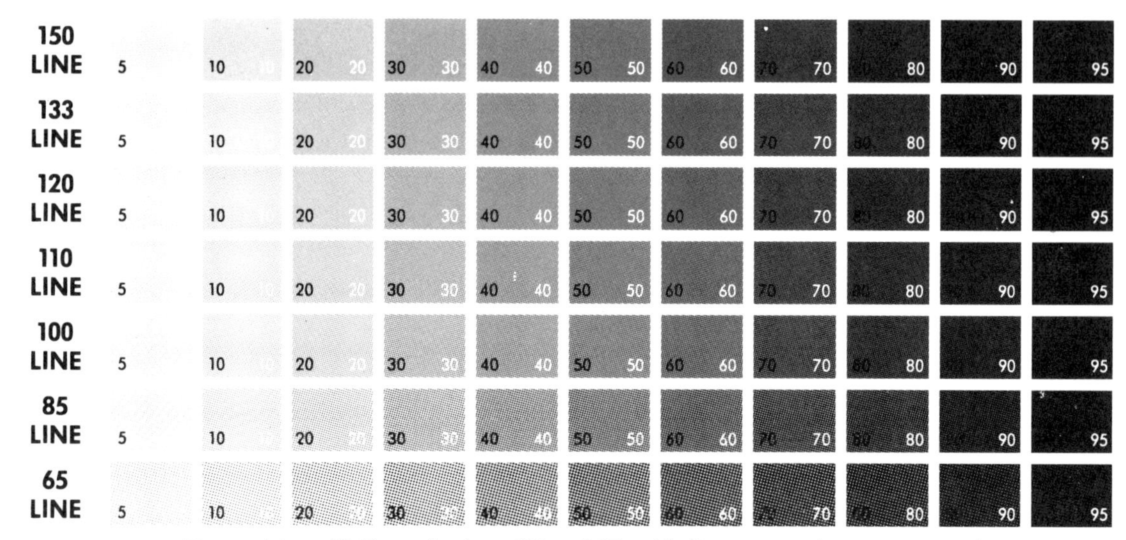

Figure 5-21. Halftone dot from 5% to 95% with line screens from 65 to 150 lines per inch. (Courtesy of E. I. duPont de Nemours & Co.)

Process color inks must be transparent, because many of the halftone dots fall on top of each other, and the top dot must not mask the color of the dot underneath.

Advantages of Process Color. When using process colors, a press can print hundreds of colors with just four inks. As a general rule, spot colors or PMS colors can print only as many colors as there are color units on the press. A four-color press can print only four spot or line colors (including solids); a six-unit press can print only six. Color dividers on an offset press can sometimes be used to print one color on one side of the form and another color on the other side.

Process color is required if the package is to carry the wide gamut of colors seen in a color photographic print or transparency. As photography becomes more popular, so does process color.

Disadvantages of Process Color. As mentioned above, it is usually more costly to print a design with process color than with spot color.

Process colors can drift so that the print often does not match the proof, the colors at the end of the run do not match those at the middle or the beginning, and a rerun does not match the original. The problem is particularly troublesome with offset, but it can occur with any printing process. Improving technology and improving controls help the press crew hold color variability to a minimum, and they have also made process color printing less costly. The best way to reduce variability is to apply the techniques of statistical process control.

Nevertheless, no color variability whatsoever is acceptable in a Campbell's Soup red or a Kodak yellow or a Jolly Green Giant green, and these colors are customarily printed with an ink mixed carefully to match the required color with minimum tolerance.

If the substrate is not white, then the colors absorbed by the substrate are not reflected. A brown corrugated box cannot produce brilliant blues with process color because the brown color absorbs much of the blue light. The unbleached kraft can be also be coated with a white pigment, or a white top can be applied to the paper to provide a substrate suitable for process color.

Growth of Process Color. The replacement of spot color with process color continues to grow as packaging still finds increased opportunities to use photography as a merchandising tool. Growth of process color is related to the improving quality of flexographic, gravure and offset printing equipment and materials. Process color presents manufacturers and suppliers with an unending challenge to improve presses, proofing, printing ink and the surface of packaging materials. Since most packages are printed in one or two colors, increasing use of process colors increases markets for printing inks and for printing plates as it demands the highest performance. At least four different inks and four plates are required. Adding spot colors and varnish further increases the requirements for inks and plates.

Printing four colors is a great deal more difficult than printing one color. Not only must all of the colors be kept in balance and in register, but there are four times as many printing plates to handle and prepare.

Successful Process Color Printing. Successful process color printing requires a team approach that includes the color separator, ink supplier, platemaker and printer. The customer is a key team member because his/her color judgment will determine whether the job is successful. Clear communication among team members is essential. A major factor for success or failure is the skill of the color separator. The best color separations are prepared with knowledge concerning the printing press, substrate and inks that will print the package. Presses and press equipment set limitations on the quality of the color reproduction. The highest quality of color printing requires top-notch printing equipment, a highly skilled crew and a management that understands how to avoid problems.

Because color judgment is subjective, it is mandatory that the desired results be agreed on before the production run. Usually, this is achieved through the use of a prepress or press proof that the customer approves before starting the press. Failure to invest in correct equipment and failure to establish a sophisticated quality control program will lead to very expensive production problems. The cost of printing without an agreed standard can be very high—delays in production, reruns and even lawsuits.

Anyone producing or buying process color printing should standardize the operation as much as possible. Setting up procedures with this in mind will help make it happen. The process begins with good recordkeeping so that when a job is repeated, the operating procedures can be duplicated. The records are an essential part of building a quality control program.

Converting Color Separations. If the same set of films is used to make plates for offset, flexo and gravure, the printed result will not be the same. Even within the same process, changing the press, inks, paper or substrate will change the results. If a color separation is made for a job that is to be printed, for example, by both offset and gravure, adjustments are made so that the prints obtained by each process strongly resemble each other. The process of halftone gravure in which offset color separations are used to produce gravure cylinders depends on the computer to adjust the offset separations.

Implications for Package Printers and Their Suppliers

Printing of colored packages is a challenge to the printer, but use of process color is especially challenging. When everything is printed in one color, registration of the images on separate press units is no problem. Variations in paper and ink have less effect on spot color than in the printing of process work. When the package buyer orders process color work, all of these variables must be kept within specs. The fact that leading printers are able to do so is a challenge to the entire industry.

The packager's ideal, of course, is to have a wide variety of shades, tints and hues available with minimum variability once the job is approved. From this standpoint, gravure is the optimum choice. Flexo also gives a good stability of color when it is operated skillfully. The process that gives the most variability is offset lithography. One finds that cosmetic packaging is sometimes produced both by gravure and by offset, gravure for color stability and offset for good economics. It is common practice to make minor adjustments to the offset press to overcome color deviation from the proof or original. The stability of gravure and flexo makes such adjustment very difficult.

Press manufacturers are selling presses with more and more printing units. A one- or two-color sheetfed press cannot produce process color competitively with a four-color press, and a process color job that also requires one spot color and a varnish can be done more efficiently on a six-color press than on a press with fewer stations. It is impractical to run webs through a second time, and the number of colors required to print the job and the number of units on the press must be reconciled while planning the job.

Quality Control

Good quality control is essential to the economical reproduction of process color. Anything that goes wrong with the color separation, proofing or platemaking, poor selection of paper or ink, and almost anything that goes wrong in makeready or during the running of the press is going to affect the color of the print.

The old way of printing in the offset job shop was to make the color separation, make plates, put them on the press, start up the press, run a few copies and find out what must be changed to achieve the desired results. Because of the inherent variability of the offset press, changes could often be made on press without remaking the plates. This procedure, obviously, is very time consuming and expensive.

The old concept of starting the job, correcting the most obvious mistakes and then discarding printed pieces that are judged to be too far out of line can no longer be tolerated. To print process color profitably requires a management dedicated to Total Quality Management—dedicated to the proposition that everything will be done correctly the first time and that color variability will be maintained within established limits.

Press proofing is more expensive than prepress proofing, but it is widely practiced in gravure and flexo. The cylinder or plates are put onto a proof press and printed. Because of variations between press, substrate and ink, even press proofs do not exactly predict what the production run will look like, and some judgment is required. New proofing techniques are reducing costs and improving results, but just as there are variations between the prints during the run, exact matches between proofs and production prints will not be achieved.

Improved technology, and even more important, improved management of technology, has made the printing of process color faster and less expensive, which is the major reason driving the growth of process color printing in both packaging and publication printing. Electronic processing of colored copy is reducing the cost and time required to go from original to satisfactory print. Modern electronic technology also reduces variability in color.

Process Color Inks

There is a great deal of confusion concerning process color inks, partly due to carelessness and partly due to technical limitations.

Magenta may be called "process red," which like "process blue" for cyan is a cumbersome name. Red ink absorbs green and blue light; magenta should absorb only green. The fact that magenta inks tend to absorb too much blue light adds to both the confusion and the difficulty of producing good blues by process color printing. There seems to be no excuse for terms like "process pink" or "process purple" that have no technical significance and further confuse things. Ink colors are discussed further at the end of this chapter.

Dot Gain

The chief cause of color variation in process color printing has been shown to be variability in dot gain. Accordingly, control of dot gain is extremely important in process color printing. Dot gain is caused by many factors in prepress operations as well as in platemaking and on the press. A certain amount of dot gain is associated with each step in the printing process.

Dot gain is the change in size of the printing dot from the photographic film that made the plate to the size of the dot on the substrate. There are two types of dot gain: physical or mechanical, and optical.

Halftone dots change size as they are transferred from film to plate to substrate. It is not possible to avoid this change in size, but it is possible to keep it under control so that the final job is the desired color. Growth in size of halftone dots from film to print is called "physical or mechanical dot gain." There is another phenomenon, called optical dot gain, that causes the apparent color of the print to change. It is caused by absorption of light between the halftone dots (see figure 5-22). Like physical dot gain, it can be controlled and compensated for.

Dot gain variability starts with the color separation where under- or overexposure cause dots to become too large or too small. Exposure and development of the film affect the size of the halftone dots, as do exposure and development of printing plates. On electronic scanners, the input from the press should be

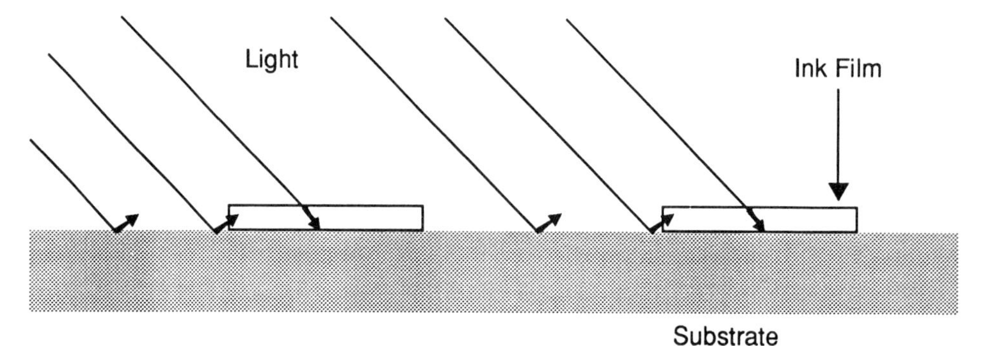

Figure 5-22. Optical dot gain. Reflectance of the print is reduced because light is absorbed by the underside of the ink.

used to adjust the computer controls so that the separations and plates or cylinders will be appropriate for the ink-substrate combination to be run, but errors do occur. Once the color separation films have been made, errors in platemaking may produce improper color tones (halftone dots that are too small or too large).

In most situations, the amount of dot gain is not the problem, it is the variability that causes problems. In a well-controlled pressroom, the amount of dot gain to be expected from a given prepress-press-paper-ink combination is well known and controlled so that color separations can be made to produce the desired color after the expected dot gain has occurred.

Coated papers produce sharper dots than uncoated papers. Plastic films must be properly treated or coated to accept ink, or halftone printing becomes impossible.

Different inks show varying amounts of dot gain. One brand of inks will have different dot gain than another brand. Web offset inks usually show more dot gain than sheetfed offset inks, and the dot gain of sheetfed inks is changed if the crew decides it is necessary to alter the ink. Flexo inks usually have a higher dot gain than web offset inks, but the speed with which flexo inks are dried on the press helps to reduce variability in dot gain during the run. Obviously, characteristics of the inks to be used must be thoroughly known, and the inks must not be altered after they are purchased.

Different presses also show different dot gain. Not only do flexo presses differ from gravure and offset presses, but each flexo press differs as does each sheetfed offset press or each web offset press. This is the reason for fingerprinting the press (see Chapter 6). The same press will change its printing characteristics as it wears and as repairs or alterations are made. For optimum control, the press should be fingerprinted at least once a year.

Flexo dot gain is caused by: plate backing—type and caliper, ink viscosity, ink metering systems, cylinder TIR (total indicated runout), impression, age and condition of the press, anilox cell count and wear, substrate and humidity. Press speed and impression are on-press variables.

Every printing process has a different technique for altering dot gain, and these techniques are used in order to change the color of the print. In flexo and letterpress, dot gain is altered when pressure or printing impression is changed. Adjusting the doctor blade affects dot gain in gravure. In offset, dot gain is affected by the ink viscosity, ink film thickness and printing impression. Pressure on the squeegee and the squeegee angle affect dot gain in screen printing. Altering the inks at press side, an old-fashioned practice in the litho shop that is still sometimes performed, changes dot gain and many other characteristics of the ink, usually unpredictably. All of these techniques have been used to increase or decrease color after the first print has been made, but this wasteful process is being replaced by control processes that keep variables within tolerances so that adjustments on press are minimized.

In practice, things are not usually kept under control, and the press crew must attempt to make adjustments to get the color to match. For best control, printing pressure should be kept close to the ideal level. These adjustments have much less effect on color than does dot gain, so it is critical that the color separations be made properly in the first place, based on good quality control theory and practice.

Unless they are carefully trained and monitored, different crews will achieve differing amounts of dot gain when running the same job on the same press with the same inks. Change of color when the shift changes is a common problem.

Dot gain is usually most severe in the range of the 50 percent halftone dot, because dot gain at two or three percent coverage has little effect on color, and at 95 percent coverage, an increase of only five percent gives 100 percent coverage. Dot gains of 20 to 30 percent in the midtone range (50 percent coverage) are not unusual (figure 5-23). As a result, unless a correction is made in the color separation, the press will apply more color than is called for by the original, and the color will be out of the desired range.

Dot Loss

When the printer is working with negative films, overexposure makes the halftone dots grow. Some printers, especially in Europe, use positive films. Overexposure here results in making the halftone dots smaller. One of the advantages of positive film is that it tends to sharpen during processing, yielding a printing plate whose dots are too small. When printing by litho and flexo, dot growth during the printing process tends to overcome this dot loss. The advantage does not, however, lessen the need for good quality management.

Color Correction

After the color separations have been made, proofs are prepared (see Chapter 6). The customer views them and approves or requests color corrections. Color correction is the altering of the cyan, magenta, yellow and black printers to achieve the desired color. Such alterations are especially important in flesh tones and hair colors in cosmetic packaging, but they are carried out in almost all process color printing.

Good skill on the part of color experts keeps the amount of color correction to a minimum. The idea that a good buyer proves his worth by ordering a large number of color corrections has been replaced by good quality management and by skills that keep the amount of correction to a minimum by getting things right the first time.

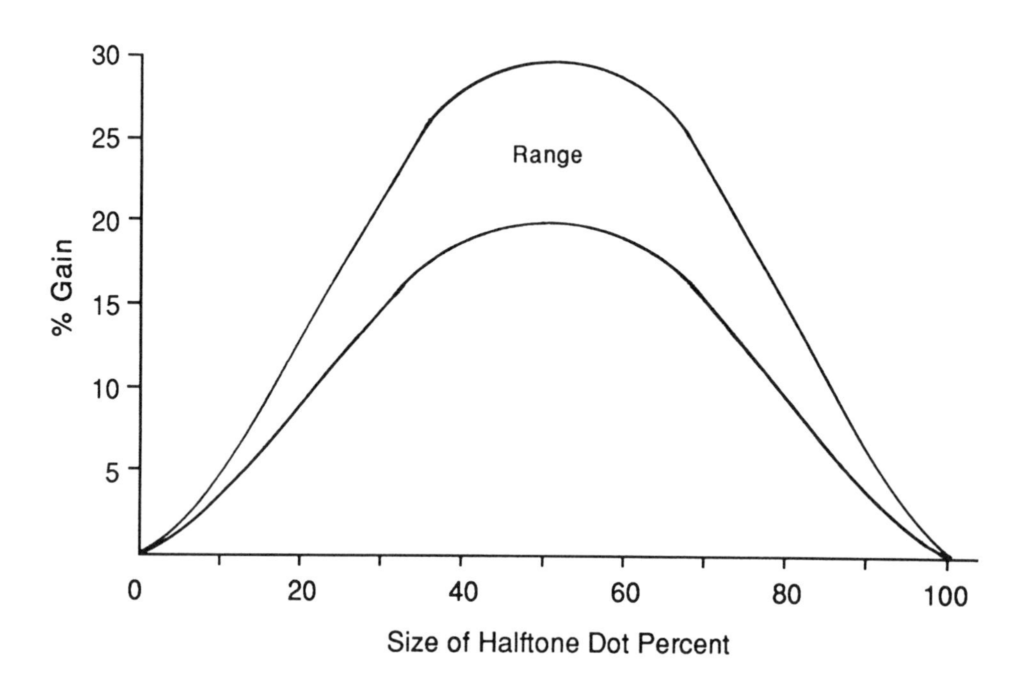

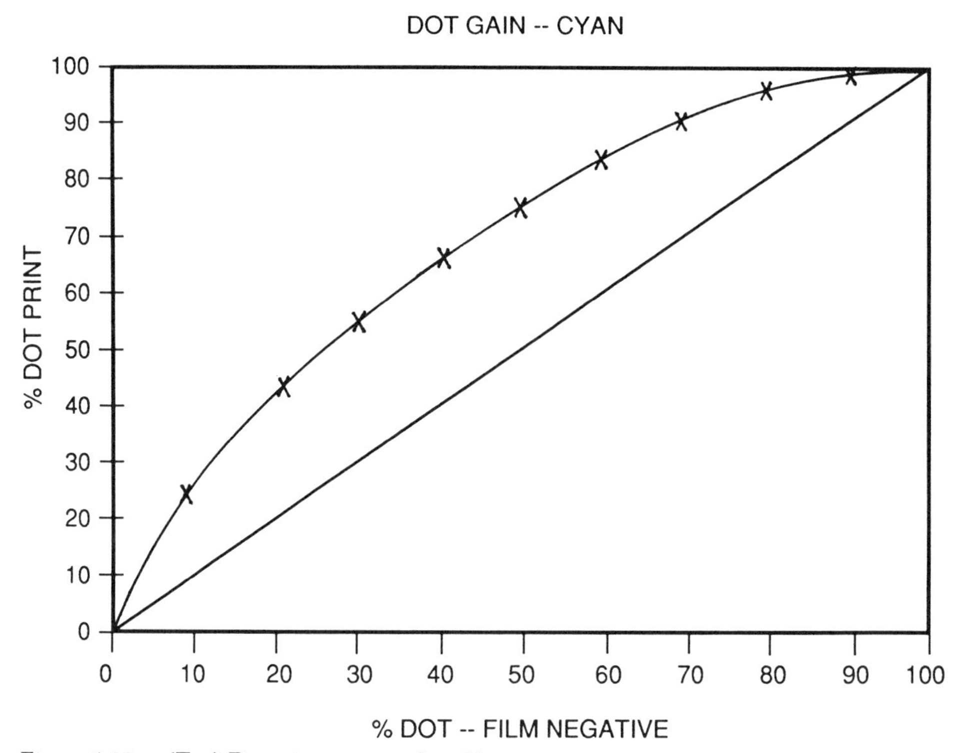

Figure 5-23. (Top) Dot gain curve produced by measuring dot gain at various dot sizes. Gain is 0 at 0% dot and 100% dot and maximum around 40 to 60%. (Bottom) Dot gain can also be represented by plotting the actual size of the printed dot against the size of the dot in the film.

If proofs are improperly produced and approved, improper color corrections may be ordered. Ideally, the process should be under such good control that color corrections are never required, but as long as human judgment of color is involved, this goal may never be reached.

After viewing the proof, the customer often makes numerous suggestions about increasing or decreasing the amount of each of the process colors in various areas. Color correction of separations is carried out with electronic scanners that alter the size of the halftone dots in parts or all of the image. This is an expensive part of the color separation process. Every effort must be made to standardize the process and to arrive at a satisfactory proof as quickly as possible.

It is still common practice to make corrections after the first copies have been turned out on the press, an unacceptably expensive place to decide that previous work was unsatisfactory or not under control. With modern instruments, coupled with a practiced eye, sound judgment and modern quality control practices, color correction after starting the press should be vanishingly small.

Trapping

To produce process color by most printing processes, the first-down ink film on the substrate must pull the second-down ink from the plate or cylinder. This is referred to as "trapping." When all the second-down ink transfers or sticks to the first color printed, we say there is 100 percent trap. Ink seldom gives 100 percent trap. This difference can be measured with a densitometer and is referred to as the percent trap. As with dot gain, the amount of trapping is not very important; it is the variability in trapping that causes unmanageable problems on the press. As with dot gain, the measured results are fed into the computer that controls the color scanner, and the separations are prepared so that the plates are suitable for the amount of trap in the press-ink-substrate system.

There are differences between wet trap and dry trap. Wet trap is the trapping of ink onto a wet film of ink. Dry trapping occurs when the first-down ink film has dried before printing the second. Dry trapping is often more predictable (less variable) than wet trapping, but it is not always practical to dry the first-down ink before applying the next color. For example, it is not always practical to dry flexo inks between colors on a CI press nor offset inks between colors on a web press. One of the advantages of UV inks is that they can be dried between units on a multi-color press.

The percent trap is calculated from the density of the top layer of ink less the density of the first ink and converted to a percent according to the following formula:

$$\text{Percent trap} = \frac{D_{OP} - D_1}{D_2} \times 100$$

where: D_{OP} is the density of the overprint film.

D_1 is the optical density of the first ink printed on the substrate.

D_2 is the density of the second-down ink printed on the substrate (figure 5-24).

If the printer is able to control the printing process to keep this number fairly

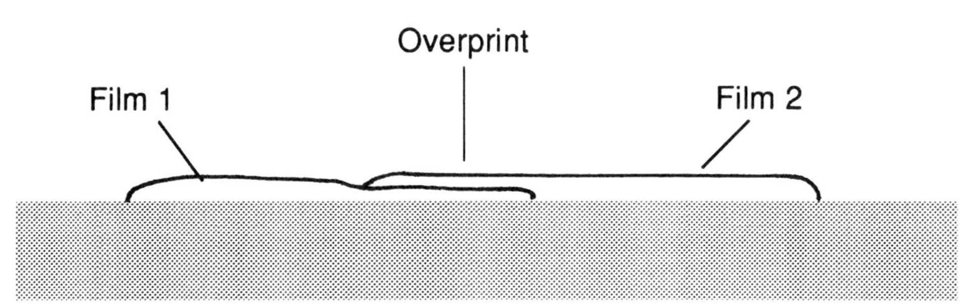

Figure 5-24. Diagram of ink trapping. Sketch shows film 1 on paper, film 2 on paper and film 2 trapped over film 1 (the overprint of film 2 over film 1).

constant, corrections in the color separation can be made to give the desired color. When the number shifts during the print run, the color will vary uncontrollably.

Undercolor Removal (UCR)

If the printing plates were to call for 100 percent each of the cyan, magenta, yellow and black at any one spot this would give 400 percent coverage. Such thick ink films cause problems with drying, trap, fill-in and dot gain.

It is difficult to trap a thin ink film over a thick film, and the printer obtains better color reproducibility by printing thin films of ink. Accordingly, the color separator or color scanner removes enough of each color so that the total ink film thickness is no greater than around 240 to 300 percent. Reducing the amount of the process colors to limit the total thickness of the printed ink is called undercolor removal (sometimes abbreviated UCR).

Gray Component Replacement (GCR)

Process colors are more expensive than black, so where it is desired to print black, combining three process colors is an expensive way to achieve it. Modern color scanners make it possible to produce color separations that reduce the cyan, magenta and yellow ink where they overprint to produce dark colors, and the colored inks are replaced with black ink. This is called "gray component replacement" (often abbreviated GCR). It not only saves on expensive colored pigment, but being a type of UCR, it permits the process color picture to be printed with less ink. This improves trap and reduces cost, dot gain and the tendency of the ink to fill in highlight dots. More important, reducing the amount of ink reduces color variability and makes the press easier to operate.

On-Press Color Control

Adjusting the press to print a color that matches the proof varies from impractical to impossible. If the proof is not truly representative of the design or if the color separations have not been made for the press-ink-substrate combination to be used, trying to match the color will, at best, be costly and unsatisfactory. Waste is increased and it takes longer to complete the job.

There is no better example of the application of the principles of Total Quality Management to optimize quality while minimizing cost than in the control of color and especially process color printing. It is the improvements in technology, procedures and management that have brought the reproduction of process color under control, thereby improving quality and lowering costs to create a market that is growing rapidly.

EFFECT OF SUBSTRATE AND INK ON COLOR PRINTING

Most packages are printed on paper, paperboard, plastic or glass, or on a metallic substrate—foil or metal. The most important properties, from the viewpoint of printing, are the color, ink absorption, smoothness and compressibility of the substrate.

A good substrate for printing must be smooth and nonabsorbent. The color in colored substrates can be hidden by opaque inks, but process colors require transparent inks. If transparent inks are being used, the substrate should be white. Color separations should adjust for the color of the substrate, but no color separation can replace light that is absorbed by the substrate. If the surface is rough, halftone dots will not be sharp, and the print will tend to be grainy. If the surface is absorbent, ink will tend to soak into the sheet or feather out ("wick"), again causing dot gain and ragged halftone dots.

Substrates that are not white do not reflect all colors, and it is not possible to print a full gamut of colors. Brown kraft bags and uncoated corrugated liner have poor smoothness and color, and they are excessively absorbent, tending to soak up the ink. Sizing the surface to give it good holdout is helpful, but to print good images on unbleached kraft top liner or board, the surface should be coated with a smooth, white material with good ink holdout. Corrugated board with white clay or paper coatings can reproduce artwork excellently.

Aluminum metal or foil and plastic films have excellent smoothness and resistance to absorbency of inks. To get a full gamut of process ink colors, however, they must be coated with a white ink. Halftone printing of waterbase inks on plastic film requires careful attention to ink formulation and surface treatment of the film. Ink that does not uniformly wet the surface of a film cannot produce good printing.

Color of the Substrate

If transparent inks are being used, the substrate should be as white as possible. The more light it reflects, the better the substrate will reproduce color. In addition, colored substrates produce their own psychological effects on customers, and, for many reasons, white is the color most commonly chosen for packages. Even opaque inks transmit some light and are directly affected by the unwanted light absorption of colored substrates.

Whiteness has two aspects: the absence of hue and the absence of grayness.

Not only must the substrate reflect all colors equally, but it must reflect high amounts of all colors. Grays or black reflect all colors equally, but at a low level. Similarly, fluorescence of the paper or board can change its color and affect the color of the print. Fluorescence can be troublesome if one is trying to achieve exact color matches, especially of light or pastel colors.

Transparent materials such as plastics and glass do not reflect light. To print process colors on these materials, they must first be coated with a highly opaque white coating. They may also be printed with opaque inks that will hide the substrate or packaged product.

Metals, likewise, do not have a white color, and like plastic and glass they are usually given a white coating or printed with opaque inks. A new technique is to leave metallized areas blank or to omit ink from some areas and employ the metallic color as a design element. Attractive metallics result from printing transparent inks over foil or metal and designers make excellent use of this effect. One often sees such packages in the store. Another technique is called demetallizing. Demetallizing the film makes the otherwise metallized plastic film transparent at that location and makes it possible to view the product through the window.

Absorbency of the Substrate

In general, high ink holdout is preferred because it gives stronger colors. Ink that soaks into the surface of the substrate loses much of its ability to color the package. Most packaging substrates have excellent holdout for ink, but paper and paperboard are often coated to improve ink holdout. Inks that depend on absorbency of the substrate for drying are not commonly used in package printing. Even corrugated containers and kraft grocery bags are printed with volatile inks.

Ink color is affected by paper absorbency, and it has been shown that papers of different ink absorbency yield different colors with the same ink. Therefore, a color separation or proof printed on one paper or board may not match the same print produced with the same plates on the same press but printed on a different paper or board. Differences in absorbency of the substrates are one reason that proofs do not always exactly match the final print.

High holdout does offer one disadvantage when printing with offset lithography or waterbase inks. Water on the surface of the substrate interferes with printing of ink. Printing of nonabsorbent substrates with litho waterbase inks requires great care.

Wettability of the Substrate

Good color printing will not be achieved if the ink does not wet the substrate. Like water on the hood of a waxed automobile, ink and water tend to fall off the surface of many plastic films. In extreme cases, ink will "crawl," forming small beads that are almost invisible; the ink seems to have disappeared. In more usual cases, films that are not readily wetted by ink give a mottled pattern or uneven color.

To overcome these problems, the films are treated with a corona (electric discharge) that alters their wettability or they are coated with a wettable material. Corona treatment sometimes disappears over a period of months, and it is a good idea for the package printer to check films before printing.

Smoothness and Compressibility

One of the advantages of the flexible plate used in flexo is that it can apply a uniform ink film to rough paper surfaces such as linerboard or combined corrugated. Rough substrates, such as corrugated board, do not smoothly contact gravure cylinders or the hard plates used for letterpress, and those processes produce a blotchy image. The offset blanket produces a good image on uncoated paper, and screen and ink jet printing can produce good images on the roughest surfaces.

Roughness also affects the appearance of the print. Textiles or uncoated kraft papers do not give sharp definition of fine lines or halftone dots nor do they produce a smooth, glossy image with any process. Similarly, it is not possible to print a fine halftone screen on unbleached kraft paper or board. A glance at the quality of halftones in the daily newspaper gives an excellent example of the problem. If rough materials such as corrugated board or unbleached kraft are to be printed by process color, they should first be coated to give a smooth, white surface.

Gravure and letterpress printing give better process color on paper or board if the stock or its coating has some compressibility to allow it to make better contact with the plate or cylinder. Special coatings are applied to paper or board for gravure printing.

Ink Color Gamut

The gamut or color range of an ink depends on its composition, primarily on its pigments. From the point of view of color alone, the best magentas print the best blues and reds, and the best cyans print the best blues and greens. In making the choice of an ink, however, other factors such as price, toxicity and resistance to fading or bleeding must also be considered. In practice, even the best ink pigments will not reproduce every color that the eye can see, but they produce an image that is attractive to the package buyer and the consumer. Sometimes, small differences in the ink or paper may make unacceptably large differences in the color of the final printed package.

Color of Ink Pigments

Figure 5-15 shows typical absorption curves of yellow, magenta and cyan pigments. The diarylide yellows have absorption curves that approach those theoretically desired. Magentas typically absorb too much blue (they are too red), while cyans are too gray—they absorb both too much blue and too much green. The more expensive magentas have better transmittance of blue light than the less costly ones. On the other hand, they tend to bleed and have other poor end-use qualities which limit their use in package printing.

Since colored prints are produced by absorbing color from the white substrate, unwanted absorption of the process color inks limits the colors that can be printed. Color correction of the color separation can overcome hue errors in the pigment, but there is no way to reduce the "grayness" of a pigment.

Because the process inks do not have ideal reflectance curves, combining all three of them in process color does not give a satisfactory black, therefore, the printer adds a fourth color, black, to make up for the rather muddy, brown colors that would otherwise be obtained.

Color Strength of the Ink

There are also inherent differences in the color strengths of different pigments. The earliest inorganic pigments, such as ocher and iron oxide, were both weak in strength and grayish in hue. Over the centuries, better pigments have been discovered and developed. Those commonly used today in process colors have both good strength and good brightness.

Because the pigment is the most expensive component in the ink, the price per pound of ink can be reduced by reducing the amount of pigment and thereby reducing color strength of the ink. However, color control of the printed package requires inks of good color strength. When inks of low color strength are used, a thick ink film is required to produce the required intensity or saturation of the color, and thick ink films are hard to control and dry. They have a tendency to soak into the paper or board or to spread, causing excessive dot gain. Finally, inks with low color strength do not save much money because more ink is required to run the job. If the printer practices Total Quality Management, ink color strength will be carefully specified and measured.

Standard Ink Colors

Standard inks are of great importance in publication printing where an advertisement may be printed in hundreds of newspapers or magazines. They are of much less significance in package printing where one printer usually runs the entire job. Nevertheless, some packages are produced by more than one printer, and standard inks will be helpful if the packages are to match under all lighting conditions.

Package printers often prefer to use their own colors. The standard colors often lack brilliance or some other desired property. If the packages are to be printed by different printers, however, ink colors must be made to match if the printing on the packages is to match.

Gravure has used standard inks for years, but the standardization of offset inks has occurred only since the 1970s. Gravure standards are referred to as GAA* Standards. In offset, there are two different groups of specifications, referred to as SWOP (Specifications for Web Offset Publications) and SNAP (Specifications for Nonheatset Advertising Printing).

---

* Gravure Association of America. These inks were formerly called GTA inks for its predecessor, the Gravure Technical Association.

For use with halftone gravure, the SWOP lithographic standards have been adopted as GAA Standard Group VI. SWOP standards are now being adapted for some flexographic printing, but for the most part there are no general standards for flexographic inks.

GCMI color standards for corrugated and kraft paperboard are issued by the Glass Container Institute (GCI—formerly the Glass Container Manufacturers Institute).

Opacity of the Ink

The color of overprints shifts toward the color of the last-down color. Even transparent ink pigments have some opacity and changing the color sequence will affect the color of the print. When printing solid colors this is not a disadvantage, but in process color work, opaque inks strongly affect the color of the print. The printed package will match the proof better if both are printed using the same sequence of colors. If opaque pigments are used to print process colors, the effect may be startling.

Gloss and Reflectance

We must distinguish between specular reflectance and diffuse reflectance. An aluminum film has high specular reflectance; it reflects most of the light, but it reflects images like a mirror. A film coated with white ink also has high reflectance, but because it is diffuse you cannot see images in the reflected light. Brilliant colors are obtained by printing transparent inks over aluminum films, but the images may be obscured by unwanted reflections. A white film that reflects light diffusely provides a more uniform color.

COLOR CONTROL BARS AND PRESS CONTROL SYSTEMS

Over the years, many organizations and suppliers have developed control bars and devices that help the press operator determine whether the press is operating under standard conditions. They indicate all sorts of variations in press operation: dot gain or loss, variations in trapping and color balance, misregister, slur, doubling and other press problems.

Print targets are commonly used in publication printing and have improved quality and reduced cost where they have been used in package printing. They are produced by many commercial and noncommercial organizations.

One of the best known of these is the GATF Star Target (figure 5-25). The Star Target is designed in such a way that dot gain on press causes fill-in around the center of the target. If printing problems such as slur or doubling develop on the offset press, the target develops specific patterns that tell the pressman of the problem more easily than can be judged from looking at the print itself.

The Color Control Bar (figure 5-26; see color insert) helps the crew judge color shifts. Color control targets are easily printed in trim areas on label sheets or on flaps on folding cartons (see Chapter 13). In flexo, it is recommended that color bars be placed in the margin area that will be trimmed or cut out, as in a

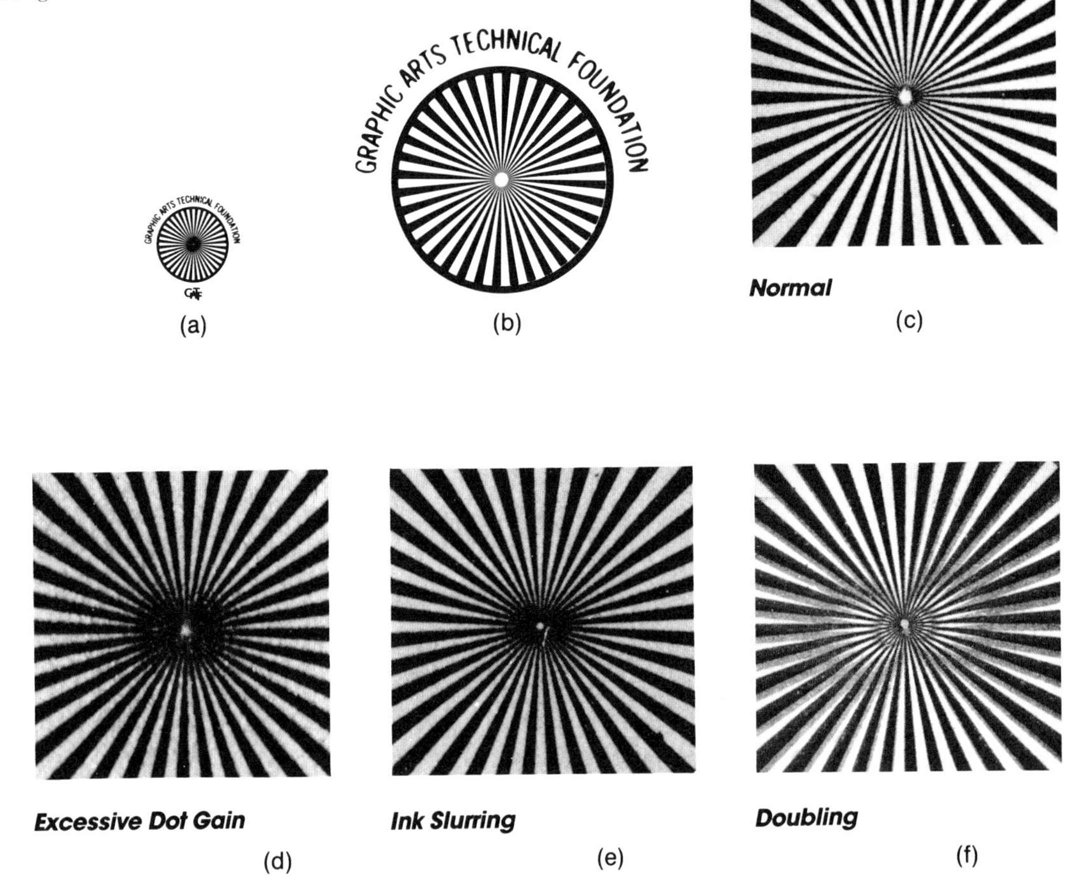

Figure 5-25. GATF Star Targets. (a) Full scale. (b) Enlarged. Printing targets showing (c) Normal printing. (d) Excessive dot gain. (e) Ink slurring. (f) Doubling. (Courtesy of Graphic Arts Technical Foundation.)

diecut window. When color bars are used, they aid in significantly reducing color variation that occurs on press.

Most printing presses are now sold with ink control systems that assist in achieving and maintaining the correct colors. Controls on flexo and gravure presses are available that can add concentrated ink pigment to the ink if the color becomes too light or add solvent if the color gets too dense. Viscosity is the key to color control on flexo presses, and maintaining proper viscosity usually means maintaining proper color. With offset presses, ink flow is usually controlled by setting the ink keys or the ink fountain. Color control is semi-automated on offset presses with densitometers reading control targets to adjust ink feed. An increasing number of modern flexo presses also include such controls.

Press settings are usually based on densitometer readings from color bars, but modern systems can use spots in the design itself as a basis for control. For many types of packages where color bars are impractical, these devices can improve color uniformity.

COLOR TOLERANCES AND SPECIFICATIONS

There are no standard national or international tolerances or specifications for color variability. These must be worked out between packaging buyer and printer as the product demands. Some products require tighter tolerances than others, and as a general rule, the tighter the tolerance, the higher the price.

In most package printing, color variation must be held to a minimum. Color variability is more acceptable for some packaging than for others. Cosmetic packages, for example, require very tight tolerances. Even color variation between jobs may be unacceptable if the new package is to stand on the shelf next to packages printed earlier. If the colors on one package are duller than on another, consumers may decide that one batch is fresher, or they may decide that the packager is not a careful manufacturer and buy a different brand.

Because of all the variations in the reproduction process, it is difficult to establish quantitative control limits on color variability. When a dozen or so randomly selected packages are spread out and viewed under standard light, variability should be barely noticeable under careful scrutiny.

Some variability between the proof and the printed product is inevitable because they are not produced by the same process. Even press proofs are usually prepared one color at a time—that is, each ink is dried before the next is printed. Any variation in the ink or substrate makes it unlikely that two prints will match.

Comparisons should be made under standard viewing conditions. If statistical quality control is to be used, instrumental measurements are required. Recent developments of on-press densitometers and controls help to reduce variability of color and other print factors.

One further problem that complicates the matching of prints with proofs is called "dry back." With offset and letterpress inks, the color of the print changes significantly as it dries. Board or labels printed by offset or letterpress will not match the proof until they have dried.

Since some color variation is bound to occur, it may be wise to prepare samples that bracket tolerable variations. As stated above, numbers alone are helpful but insufficient to decide whether a given package is within acceptable tolerance limits. Visual evaluations are used together with numerical standards.

The standard should be flanked by a light and dark variation, within which falls the color the customer wants (figure 5-27). The color approved at startup will be maintained throughout the press run. Provided it is carefully stored, this standard also will narrow the variation between succeeding later orders for the same design.

The packaging manufacturer or buyer should obtain ink drawdowns of the desired color in its pure form as well as overprinting the colors required for the job. They should be prepared on the substrate to be used. These drawdowns will provide a good idea of the color as it will look on the package (figure 5-28). Ink drawdowns made on a swatch pad will not match those on a folding carton or label because the color of the print is affected by the color and absorbency of the substrate. Opaque inks will match better than transparent or process colors. Manual drawdowns are notoriously unreliable, especially when prepared by different people.

Figure 5-27. Drawdowns of ink on swatch pad.

CONTROLLING THE COST OF COLOR PRINTING

Process color requires color separations and printing of four colors in close register on a high grade of white paperboard, paper or film. Process color always requires a minimum of three and usually four plates and four inks, while most line work is done with fewer plates and inks. Complex line or spot-color work is sometimes called for, with a dozen or more different colors, each with its own plate and ink.

Proofing of a process color job can be a long, drawn out and expensive operation. New proofing materials and processes are being introduced every year. Efficient production of process color also demands that the printer get the proofing process under control to keep costs from getting out of hand (Chapter 6).

Understanding and control of proofing is one of the most important functions in keeping the cost of color printing under control. The later in the preparatory stage or in the printing process that color changes are requested, the more expensive the changes become.

Intricate designs and use of many colors also increase the cost of spot color. The marketing team must decide whether that cost is an acceptable trade-off for the results.

A design that requires a large number of spot colors increases the cost of the printing job, and (except for sheetfed printing) it eliminates from the bidding those printers whose presses have fewer units. Offset, flexo and gravure presses commonly have five or six colors, permitting the printing of four or five colors plus a coating. Presses with more colors are becoming increasingly common.

The packager and designer must carefully evaluate the number of printing units required to print the job and the converting equipment required to finish it. This is a place where creativity can often produce great savings. Nothing is more effective in controlling the cost of color printing than a team that works closely to keep the entire process under control.

FURTHER READING

Color and Its Reproduction. Gary G. Field. Graphic Arts Technical Foundation, Pittsburgh, PA. (1988).

The Lithographers Manual, Eighth Edition. Ray Blair and Thomas M. Destree, Editors. Graphic Arts Technical Foundation, Pittsburgh, PA. (1988).

The Printing Ink Manual, Fourth Edition. R. H. Leach, Editor. Van Nostrand Reinhold (International), London. (1988).

Chapter

Prepress Operations

CONTENTS

LIST OF FIGURES

LIST OF TABLES

Chapter

Prepress Operations

INTRODUCTION

This chapter emphasizes the importance of communication among client, designer, advertising agency, prepress and press operations concerning image quality and color.

Prepress operations include setting the type and preparing the pictures for printing, assembling them into the form to be printed, proofing the form and preparing the plates. By the early 1990s, most preparatory work was done by computer. It has achieved the sobriquets of "Desktop Publishing" (DTP) for most operations and "Color Electronic Prepress Systems" (CEPS) for more sophisticated ones.

This area of graphic reproduction has changed and continues to change more rapidly than any other aspect of the graphic arts processes. Even with modern electronics that can do wonders with words and pictures, the principle of "Garbage In—Garbage Out" still applies. Poor work can be greatly improved with the latest technology, but the cost of obtaining even a fair reproduction from bad art is excessive, and quality is compromised.

There is a mystique that a careful art director will take a great deal of time to do a good job. Gravure cylinders being sent back for changes six or eight times are, unfortunately, not entirely a thing of the past. By insisting that everything be done right the first time, a good manager can reduce time and costs while improving quality.

Although the industry is moving rapidly to DTP and CEPS, a great deal of

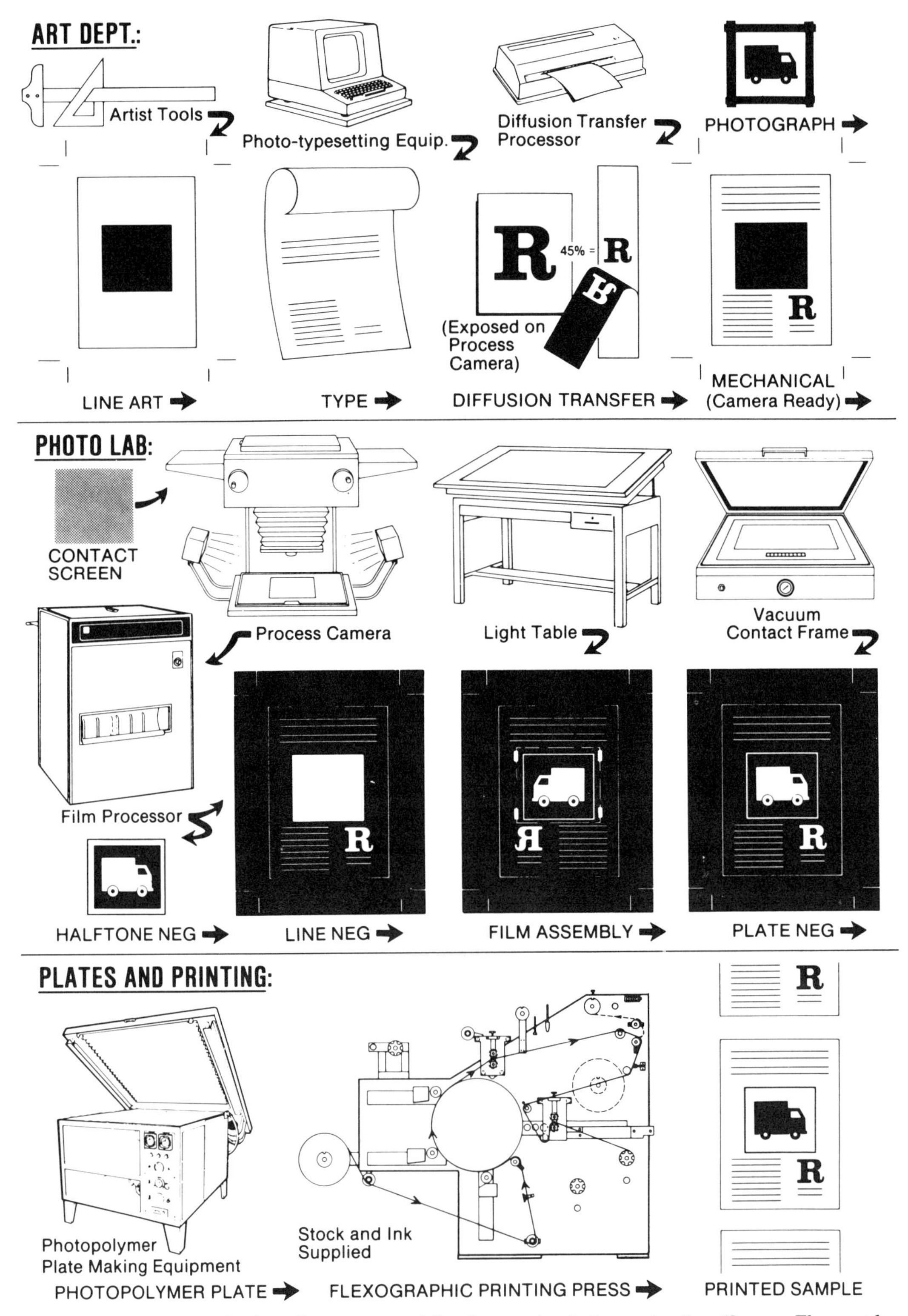

Figure 6-1. Outline of processes used for photomechanical reproduction. (Source: Flexography: Principles and Practices, Fourth Edition.)

film and old technology is still used (figure 6-1). Many of the principles that applied to manual operations are incorporated into electronic operations (figure 6-2) and are discussed in this chapter.

Costs of getting everything prepared with camera, film and manual assembly have held back the advance of color printing, especially four-color process, but modern electronic processes are making process color more economical and practical in package printing as in all other printing.

It is not unusual, and growing more common, to use more than one printing process on a package. Flexo printers are commonly used to put down a coating on webs or sheets printed by gravure or offset. Flexo printers are often used to put variable information (like price) onto a gravure-printed package. Ink jet printers add unique information on all sorts of packages.

The same design may be used for the label on a soft drink bottle, for a two-piece can, for paper cups, and for the folding carton box or carryout basket. Requirements vary, and the design must be adapted for each application.

Computer technology enables an artist to create a design and to adapt that design to print satisfactorily on any package by any printing process. But even with modern digital technology, the team must discuss the design and work with it so that it meets the requirements of the printing press, the package and the customer.

For example, halftone gravure permits color separations designed for litho to be used for rotogravure. Some compromise in quality occurs when originals are not prepared for the printing process that is to be used, but modern electronic imaging and alteration give an acceptable print, and new technology continues to reduce the compromise in quality.

STANDARDS

Pressroom Standards

It is possible, in the 1990s, to control the prepress and press process so that press results are predictable within narrow limits. It requires controlling each step of the process, learning how much variability is in the process, and working to reduce that variability.

Systematic color reproduction is possible if the printing conditions are stable and if standards are established and maintained. If conditions are unstable, the trade house or prepress department is trying to hit a moving target. Printing standards should include the use of plant standard inks, paper or board, plates and other materials. The press must be operated under standard conditions of printing pressure, ink feed, etc. These standards are used consistently in order to achieve standard solid densities, overprint colors, gray scale and gray balance, dot reproduction, gloss, register and other print quality factors.

It has been said that if you cannot measure, you cannot control. Many instruments are available including densitometers, sensitometers, exposure meters, light integrators, packing gauges, ink film thickness gauges, color control bars, standard lighting, registration devices, star targets and pin register systems. If properly used these will help the management and the crew reduce and control

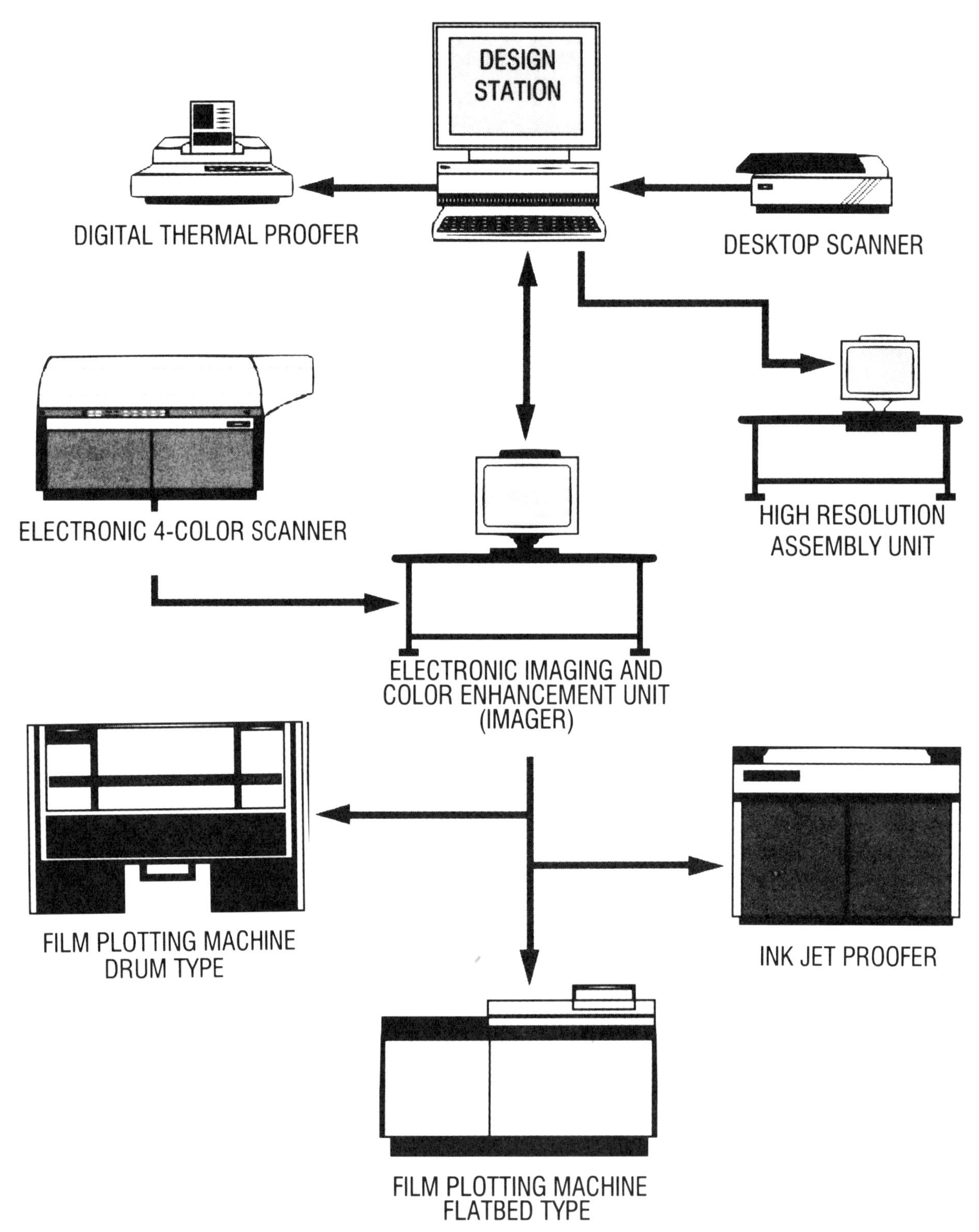

Figure 6-2. Electronic prepress network. (Courtesy of Wilson Engraving.)

variability and produce product to the satisfaction of the packager and everyone else.

Fingerprinting the Press

Printing specifications depend on the substrate-ink-press combination that is being used. To make the color separations and printing plate or cylinder properly, a fingerprint of the press is required. Every press is different. Not only do flexo presses print differently from gravure presses, but no two flexo presses print exactly the same, and the same press changes its printing characteristics over the months.

If the trade house knows the characteristics of the press, plate, substrate and ink to be used, they can prepare color separations and plates that will easily print the desired design. When that information is not available, the prepress trade house must prepare plates to their own specifications, and these may or may not closely match the requirements of the job. Variability in color is introduced when the press crew is constantly attempting to adjust the press because the plates were not made to match printing conditions and materials.

Fingerprinting is done by printing a test sheet containing print targets using the desired substrate and inks. Color, trap, dot gain and other fundamentals are measured, and the color separator and platemaker learn what adjustments must be made in the color separation to produce a plate that will correctly print the desired image fit and color.

One of the most important elements is dot gain. It cannot be eliminated, but it can be measured and controlled. If the color separator and platemaker know how much dot gain to allow for, the dots on the plate or the bars on a bar code symbol can be reduced so that the print will match specifications.

Fingerprinting the press takes time, but like most quality control procedures it pays off in reduced waste and costs, coupled with improved productivity and quality. To keep the process under optimum control, the fingerprinting should be repeated about once a year, because presses slowly change their printing characteristics. Many characteristics of the press can and should be measured on every press run with the use of color and print control devices.

Fingerprinting the press gives the color separator the information required to make color separations that will produce the color to match the proof: dot gain curves, trap, resolution, register, etc. Fingerprinting the press allows the bar code symbol to be designed so that it will read properly after it is printed on the package. Fingerprinting the press allows the company or division responsible for prepress operations to make a plate that will print properly the first time, avoiding wasted time and substrate while inks are changed and the press is adjusted in an attempt to make the print match the proof.

Fingerprinting the press is not new, but it is only in the early 1990s that it has become widely practiced. How did printers get along without it? Because presses were under poor control, fingerprinting was of limited use. The press could not be expected to produce the same results two weeks in a row. When a new job went on press, either there was a great deal of time and material wasted while a satisfactory match of print and proof was achieved, or the customer accepted a poor match. These two conditions are becoming increasingly unacceptable, and fingerprinting is the method used to reduce the problems.

TEXT AND TYPESETTING

Text is usually handled differently than pictures. Text is usually composed at a keyboard, art is prepared with a camera or pen and ink. If large display type is drawn by the artist or incorporated into the design, text and design are handled in the same manner. To be handled by computer, text and design must be digitized, and differences in the way they are handled are growing less distinct. Computers can now "read" and incorporate hand-drawn material. It becomes increasingly attractive to handle the prepress operations in this manner as computers grow faster and cheaper.

The first method of setting type was manual. The desired letter was selected from a case, placed in a composing stick and then locked into a chase. The case with the capital letters was placed above the case with the small letters, and capital letters are still called "upper case" and small letters "lower case." Linecasting machines that set type mechanically were invented in the late 1800s. The best known of these was the Mergenthaler Linotype machine.

Mechanical typesetting machines were replaced by phototypesetters which placed the type photographically onto a light-sensitive film or paper. Electronic typesetting is now done digitally with laser beams and photosensitive film or plates. Type can be set at phenomenal speeds, and corrections are made electronically without requiring manual manipulation of the copy.

Typewriters and many computer printers space all type evenly, but typesetters use proportional spacing that gives more space to wide letters like "M" or "W" than to narrow ones like "i" and "l." Typesetters also carry out a function called "kerning" that reduces the space between letters by fitting them together so that the space between a "WA" is less than the space between "WS." Kerning saves space, but more importantly, it improves the appearance of the printing. Letters are sometimes tied together into characters called ligatures (figure 6-3).

The machines now used are imagesetters that output both text and images—line art and photographs. Imagesetters create a bit map in which the entire image, text and graphics, is represented as tiny dots. Each dot (or pel or pixel) is assigned

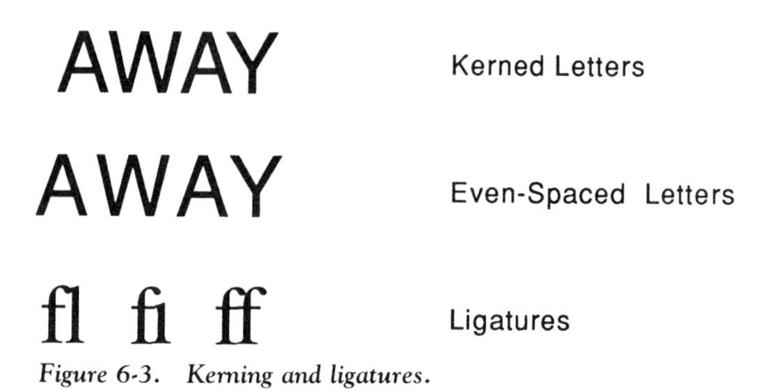

Figure 6-3. Kerning and ligatures.

a specific location, and the computer recreates the image by exposing film or other photosensitive material to a laser beam or other light source that is controlled by the bit map.

Input to the imagesetter is through a keyboard, a mouse, a digital camera or an optical scanner. Optical character recognition (OCR) refers to the types of scanners designed especially to convert alphanumeric characters (letters and numbers) to digital form. Drum scanners and flatbed scanners are usually used to convert pictorial input to digital form, although digital cameras are becoming common. Modern equipment not only inputs the text and graphics but allows the editor or artist to arrange it (page makeup) and to alter shades and colors (color correction and undercolor removal).

Much of the electronic technology has grown out of CAD/CAM operations, Computer Assisted Design (or Drafting) and Computer Assisted Makeup (or Machining).

PICTURES AND DESIGNS

Art Preparation

Art and copy include all materials supplied for reproduction: type, diagrams, drawings, photographs and color transparencies. The artist or designer starts with a layout—a blueprint of a printed job. The designer must be familiar with the printing process and the product. Flexo, gravure and offset have different requirements, just as do folding cartons, labels and butter tubs.

Line Copy and Continuous Tone. Original images are either line copy as in type matter, diagrams, and pen and ink drawings, or they are continuous tone as in a photograph. These are further broken down into art to be reproduced in one color, multicolor or process color.

A colored picture consists of different shades, hues and gradations of those hues. In a photographic film or print, the gradations are created by varying amounts of dyes and silver in the print: the more silver, the darker the tone. This type of picture is known as continuous tone. These tone gradations cannot be represented by varying amounts of ink with a single impression on the press. (Some variation in ink thickness is possible in gravure printing.)

To produce the tones necessary to reproduce a picture, a printing plate must use the halftone principle. This is an optical illusion in which the gradations in color are represented by solid dots that all have equal spacing and ink density but vary in area. The eye cannot see the dots at the usual reading distance, but they are readily visible under a magnifying glass or linen tester.

Linework requires no color separation or preparation of halftones and is less costly than process or halftone color. It allows no subtle gradation of tone and no shadows or highlights.

Halftone work permits reproduction of shadows and highlights. If process colors are used, it is possible to match subtle changes such as those in facial and hair tones. However, such subtle changes make the process difficult to control, and color variation occurs when halftone dots do not print to match those in the artwork. It is the ability to reproduce small variations in color that is both the strength and weakness of process color.

Prints and Transparencies. Artwork ("copy") for color reproduction falls into two classes: reflection and transmission. Reflection copy, such as an oil painting or photographic color print, is viewed and photographed by reflected light. Transmission copy, such as a color transparency or color negative, is viewed and photographed by transmitted light.

Color transparencies are usually preferred over other originals for the reproduction of color. Copies of the original are never as sharp or clear as the original, and in halftone color, sharpness is of greatest importance. Transparencies are flexible and can be wrapped around the drum of a color scanner. Prints on board or other stiff materials must either be photographed so that the film can be scanned on a drum or they must be scanned with a flat bed scanner.

Assembling Art and Copy. A printed piece is often assembled from many different elements. Some may have to be scaled (reduced or enlarged) and unwanted parts may be cropped (cut away). Although modern imagesetters do this work quickly and easily with electronic processes, much of it is still done manually with cameras, film and a sharp knife on a light table.

A major contributor to high costs is the practice of getting the job together piecemeal and from different sources. With good planning and coordination, good artwork can be delivered in one package to the prepress service, making it possible to produce good plates quickly at minimum cost. Electronic systems are capable of receiving portions of the final artwork and can assemble and output the final artwork onto film. Attempts to output the final work directly onto plates are achieveing limited success.

Handling Artwork. Many problems in package printing can be avoided by the proper preparation and handling of artwork and by close cooperation with trade houses or service bureaus and printers before final art is prepared.

Illustrations suffer all sorts of abuses that render them unfit or require extensive additional work to restore them. Writing on the back of an unmounted photograph with a hard pencil embosses the face of the picture, and the embossing will show in the print. A photograph should never be mailed in a tube if it is small enough to be sent flat. Rolling the picture tends to crack or check the emulsion. If it must be mailed in a tube, the photograph should be rolled with the photographic emulsion on the outside. Paper clips also emboss illustrations, and it is never safe to fasten instructions directly to a photograph with a metal clip. Instructions can be taped or otherwise secured without a clip. Original artwork should be marked in a manner that will not interfere with its future use.

Cameras and Photography

With modern processing, halftone dots are generated and reproduced electronically. The vertical camera is still part of the prepress team, but the horizontal camera is almost gone, replaced by electronics. The vertical camera is useful for some fast jobs where quality is not of chief concern. It is also useful for enlarging and reducing copy to the desired size.

The flat bed scanner produces pictures in electronic form, ready for electronic manipulation. It is sometimes practical to convert prints or slides to electronic form with an electronic camera or flat bed scanner.

Film and Film Processors

Whether the copy is reproduced by a scanner or a camera, film and film processors are used. Graphic arts film is almost always based on silver salts that are sensitive to light. Silver salts are coated on a film or onto paper.

There are several types of graphic arts photographic films. Rapid access films can be processed in 60 to 90 seconds. Lith films give a greater contrast but must be processed very carefully, and the special dual developer rapidly deteriorates. In the 1980s, a hybrid system was introduced that combines the high contrast of the lith film with the ease of processing of the rapid access film.

After the film or print is exposed, the exposed salts are converted to silver, the unexposed silver salts are washed away with a developer and the remaining silver is fixed with hypo, a chemical that prevents further reaction. This processing was once done by hand in trays containing the necessary chemicals, but most print shops now use mechanical processors that control the process so that it is repeated precisely for each film. These machines process all types of films, and not only save considerable time but produce more consistent results. Film is developed, fixed, washed and dried in less time than it used to take to develop by hand.

After the films are processed and dried, they usually have pinholes caused by dust and dirt or other defects. These holes are covered with an opaque material rather like the opaque material used to cover errors on typewritten copy.

Color Separations and Halftones

Until 1970, color separations were done by means of a laborious and lengthy series of steps with the process color camera which consumed great amounts of film. The use of halftone dots to create process color is discussed in Chapter 5. Halftone dots are now created electronically.

The camera uses a contact screen to convert continuous-tone images to halftone images. The screen ruling is the number of lines per inch. A 133-line screen creates halftone dots that are just beyond the resolving power of the eye at normal reading distance. Screens with finer rulings are often used for printing color on coated papers. As a general rule, the finer the screen ruling, the sharper the rendition of the detail in reproduction. In color reproduction, fine screens produce purer colors; more of the image is covered with ink and the graying effect of a white substrate is reduced.

In four-color process printing from halftones, there is a problem called moiré, in which the halftone screens create unwanted patterns. With the screen rulings commonly used, a minimum pattern is formed when the axes of the screens are separated by 30°, but since there is room for only three screens in 90°, a compromise is made. For example, in four-color offset printing, the usual screen angles are yellow 90°, magenta 75°, cyan 15° (or 105°) and black 45°. The further two colors are apart, the less the moiré pattern appears. The angles themselves are also very critical. Errors as small as 0.1° can cause objectionable moiré patterns in grays or areas where three and four colors print together.

Moiré is also caused by other patterns. Reshooting a halftone image usually creates unwanted patterns, a reason that printers insist on continuous tone copy. The screen rulings can also create patterns with images in the copy such as textile patterns, roofing shingles, and even the grooves in phonograph records.

In addition to patterns created by copy, moiré can be created by the press itself. Rulings in the anilox roll of a flexo press can cause unwanted patterns as can the screen used in screen printing.

Scanners

Color Scanners. During the 1970s, color scanners were introduced that make it possible to convert a picture to dots of cyan, magenta, yellow and black ink that can be manipulated to produce the desired color. A previewer helps the scanner operator work with the halftone dots.

The original is scanned with a light beam that is split into three beams after passing through, or being reflected from the original. Each beam goes to a photocell covered with a filter that corresponds to one of the additive primaries and this separates each area of the copy into three color components.

Digital color can be captured with an electronic or digital camera that creates an image that is already "scanned." It is a simple matter to convert color prints or slides to digital pictures that can be manipulated electronically (figure 6-4).

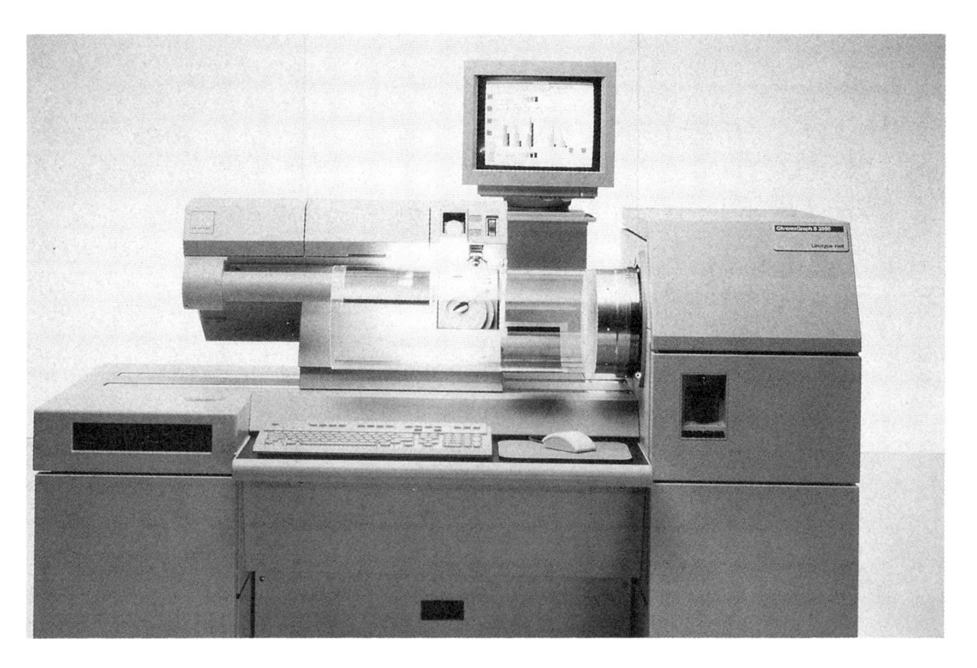

Figure 6-4. Color scanner. (Courtesy of Linotype-Hell Co.)

Figure 6-5. Flatbed scanner (Courtesy of Linotype-Hell Co.)

Flatbed Scanners. With a flatbed scanner (figure 6-5) an original illustration is scanned by a light source. The information is transmitted to a computer which digitizes and manipulates it electronically. The image can be reproduced at any size in either or both directions (anamorphic enlargement); it can be slanted, outlined, with or without borders, reversed, either laterally or from positive to negative; it can be sharpened by techniques known as unsharp masking and edge enhancement; and screen patterns can be altered. Flatbed scanners are also made for color separation.

The flatbed scanner, of course, can scan rigid artwork. Disk or rotary scanners require that the work to be scanned be wrapped around the cylinder.

It is possible to go directly from computer to plate without using any film, and this process continuously grows more common. In flexo, rubber plates or sleeves are laser engraved, and in offset, plates are exposed with computer-controlled laser beams. In gravure, the engraving machine can be driven from digital data.

The cost of color separations keeps dropping as electronic manipulation of the images makes color separation faster and cheaper.

Color Correction

Since each paper/ink/press combination prints differently, it is apparent that artwork must be treated differently depending on its final use. Changes were once made by hand using a process called dot etching in which a skilled craftsman treated photographic film to decrease the size of the halftone dots.

This process is now done electronically by programming the color scanner. If the paper/ink/press combination has been studied, it is possible to program the color scanner to produce a color separation that will often print properly the first time.

Patterns for Coating

The packager frequently uses spot varnish to produce gloss (or matte) finishes over a part of the design. This is easily accomplished using a printing plate that prints a varnish, just as it would an ink, over the desired part of the design or pattern. This is called "pattern coating." It becomes especially effective when a package has to be sealed, and the coating would repel the adhesive in the areas to be sealed.

Suitable patterns can be prepared for printing by any process. Patterns for coating are prepared exactly as are designs for printing ink.

IMAGE ASSEMBLY

When all of the films for a job are completed, they must be cut apart, cut to size, and assembled into a specific layout for exposing onto the printing plate. This is done by cutting holes in a plastic sheet, usually called "goldenrod," and precisely taping or fastening the film onto it.

The hand assembly of four images in register for process color is a tedious, time-consuming and labor-intensive job. It is the most serious bottleneck in production, and it has kept four-color printing expensive. Automating the prepress has lowered the cost and increased the use of process color for package printing.

Electronic Design Makeup

When using DTP or CEPS techniques, the various parts of the image are assembled electronically. The effect of electronic image assembly is apparent in the rapid growth of process color printing. The old techniques of registration, spreads and chokes, undercolor removal and color correction are now done in a fraction of the time required for camera exposure, film development and manual placement of films onto flats or cabs.

Distortion of images to correct for the bending of flexo or letterpress plates is also done electronically. When a plate is wrapped around a cylinder, its face becomes longer than its base (figure 6-6). The figure shows that the amount of distortion is related to the thickness of the plate and the diameter of the cylinder. Flexo and letterpress images for plates made in the flat must be distorted so that after the plates are bent around the cylinder, they will still print correctly on a flat surface. No such distortion occurs in gravure or rotary screen or for flexo

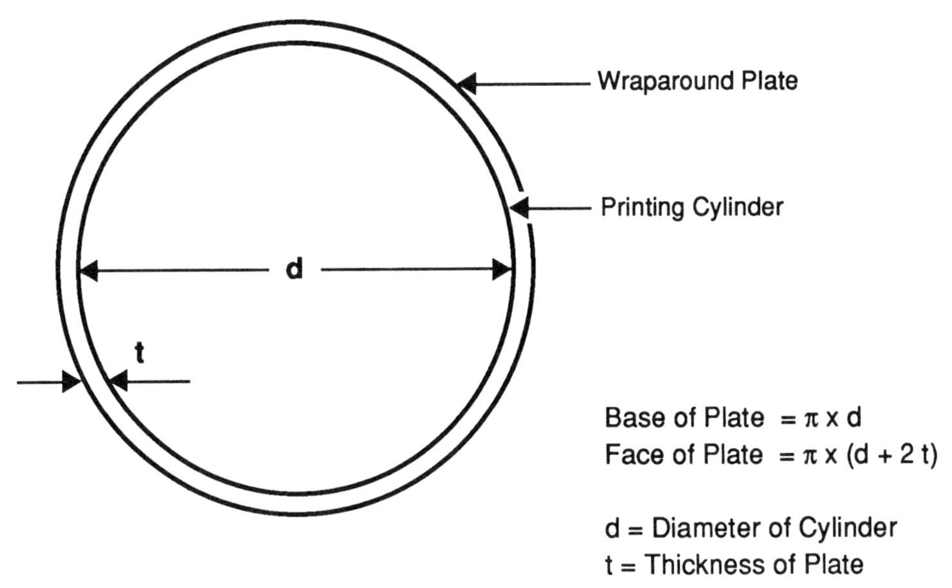

Figure 6-6. Distortion of wraparound plates.

plates made in the round, and the distortion in offset is negligibly small. The amount of distortion depends on the circumference of the plate cylinder and on the overall plate thickness. The distortion correction factor is shown in figure 6-7.

Step and Repeat

When several copies of the image are to printed from a single plate, each image is placed precisely by a computer-operated machine called a step-and-repeat machine. If 50 four-color labels are to be printed from each of four plates, the time required to strip each of two hundred images, in register, by hand is prohibitive (figure 6-8). Mounting the images of six or a dozen cartons on each of four plates is both more economical and more precise with a step-and-repeat machine.

The machine consists of a bed for mounting the plate, a chase for mounting the film, rails for traversing the chase accurately in the X and Y directions, and a high intensity lamp to expose each image.

This is another example of low cost going hand in hand with high quality.

Automated Imposition of Images

Closely related to the step-and-repeat machine is the automated imposition machine or photocomposing machine. If 20 different labels are to be ganged on a plate, the problem of placing each of the labels onto two, three or four plates so that the images fit in each color is a formidable task. The automated imposition machine positions each film correctly for exposure of the plate.

In flexo, the accuracy of color-to-color registration is enhanced by automated imposition instead of mounting several individual plates on a cylinder. If the

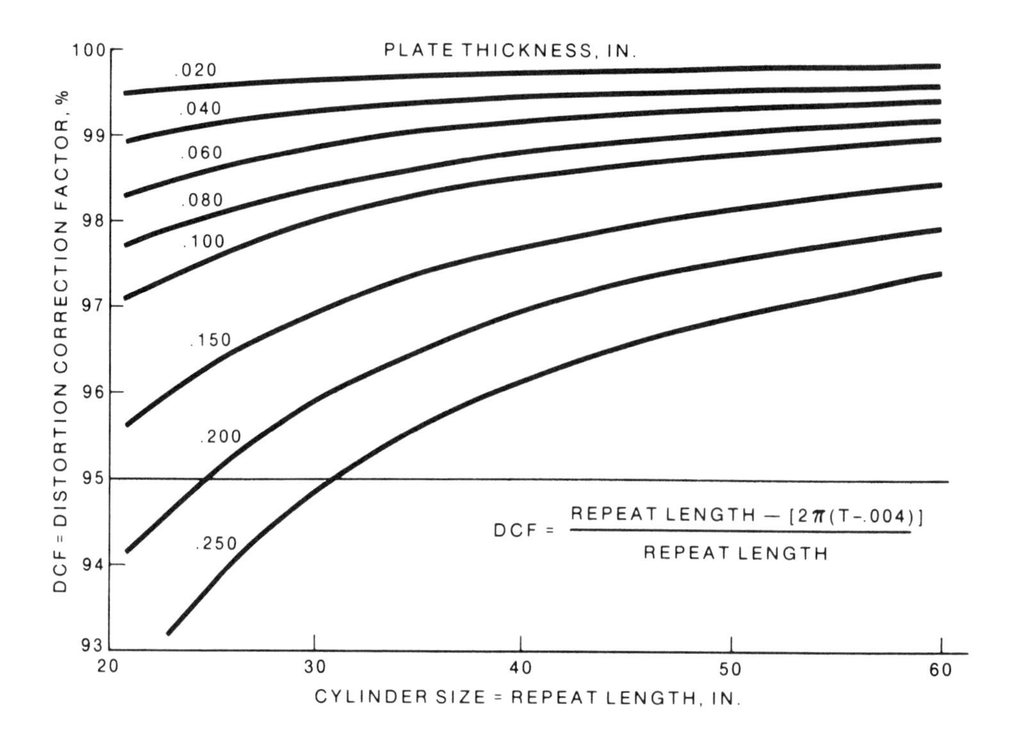

The formula to use to determine how much a film negative must be reduced in order to compensate for image distortion is:

$$\% \text{ reduction} = K/R \times 100\%$$

where "K" is a constant given in the table below and "R" is the repeat length of the printing cylinder or its printing circumference.

The repeat length is usually specified with the job; if it is not given it can be calculated by adding together the bare cylinder radius, stickyback thickness and plate thickness then multiplying the sum by 2 × (3.1416). Circumference = π (3.1416) × Diameter = (2 × Radius).

The "K" factor is dependent upon the plate thickness. Following are K factors for various plate thicknesses in inches and centimeters. K factors are supplied by the manufacturer.

Plate Thickness (inches)	K Factor (inches)	Plate Thickness (centimeters)	K Factor (centimeter)
0.030	0.157	0.076	0.399
0.067	0.389	0.170	0.989
0.080	0.471	0.203	1.197
0.090	0.534	0.229	1.356
0.100	0.597	0.254	1.516
0.107	0.641	0.272	1.628
0.112	0.672	0.284	1.708
0.125	0.754	0.318	1.915
0.155	0.942	0.394	2.394
0.187	1.143	0.475	2.905
0.250	1.539	0.635	3.910

Figure 6-7. Distortion correction factor based on the circumference or repeat length of the cylinder. (Source: Flexography: Principles and Practices, Fourth Edition.)

Figure 6-8. Step-and-repeat machine. (Courtesy of Misomex North America, Inc.)

press can accommodate four across and three around, the 12 images require 48 individual plates for a four-color process job. The photocomposing machine will quickly pay for itself where the printer does this sort of work.

Fit and Register

Fit and register are different but related phenomena or occurrences. If two images are not the same size, or if one is out of square, they cannot be made to fit. Even if they fit perfectly, two or more images may print out of register, but they can be brought into register on press. Misregister of images is one of the most common, troublesome and expensive problems encountered in the printing industry.

The dimensions of most materials vary in response to changes in environmental conditions. Materials expand when heated and contract when cooled; photo emulsion swells with an increase in humidity and shrinks as the air becomes drier. Even the most dimensionally stable film will show some size variation, albeit very little. Significant changes in film size cause fit and register problems during the contacting process. Even in shops with good temperature and humidity control, fit and register problems often become more pronounced with changes in the seasons.

In preparing art where two or more colors are to be printed, registration of the different color images is usually an important factor. Hairline register is difficult to maintain and designs that avoid butt register make production of attractive packages easier.

Registration Systems

Pin Registration Systems. A pin register system consists of a punch for cutting holes precisely and pins that fit snugly through that hole. When properly located, these pins and slots assure that copy and printing plates will be in register.

Pin registration begins with the mechanical artwork for the job. If the system is used in preparing copy, film and printing plates, register is greatly improved, and a great deal of time is saved in prepress preparation and press makeready. Such equipment is available for flexo plates, letterpress, and offset plates and for prepress work for any process.

Optical Registration Systems. Several devices or systems are available to mount plates precisely and in register. One electronically controlled device locks the flexo printing cylinder into position while stickyback is applied to the cylinder. A plate on a hold-down table is held in place while a TV camera is placed over register points that are entered into the computer. The table brings the plate square to the cylinder face as it positions the plate. A computer command rotates the plate cylinder, and the printing plate is mounted in register. The sequence significantly reduces mounting time and improves precision, especially with three-, and four-color work (figure 6-9).

Another device for mounting and proofing flexo plates for flexo printing of corrugated is shown in (figure 6-10). These are only representative of several systems in use.

PROOFS AND PROOFING

There are several types of proofs because they serve different purposes. The first, a "position proof," shows the layout of the job, helps to verify copy quality and reveals errors in typesetting. The proof is an important part of quality control even though good editing will have eliminated errors in spelling and grammar. Good management will avoid changes in text and layout after the job has been photographed.

Figure 6-9. Electronically controlled mounting and proofing machine for flexo plates. (Courtesy of Anderson & Vreeland, Inc.)

Figure 6-10. ***Mounting and proofing machine for flexo plates for corrugated printing. (Courtesy E. L. Harley Inc.)***

Proofing is also required to check color. This is the most difficult and expensive aspect of proofing.

Reasons for Proofing

The farther the job goes before errors are caught, the more expensive it becomes to correct them. It is the purpose of the proof to eliminate all errors and changes before the job goes to press. Changes after the printing press has been started up are dramatically expensive.

Proofing serves many purposes. The first is to be sure that the required information (package contents, bar codes, warning statements, etc.) is present on the job, that the information is properly located and that words are properly spelled.

Proofreading is a highly specialized skill. Not only does the proofreader require a sharp eye for misspelled words, a skill many people lack, but he/she must be aware of legal requirements. A knowledge of chemical nomenclature is also helpful.

It has been observed that if several people proofread a job, more errors will get through than if only one is responsible. Proofreading responsibility cannot be shared.

Because proofing serves many purposes, they may sometimes conflict and become a great source of delay and waste. The primary purpose of color proofing is to determine whether the plate or color separation will produce a satisfactory print. Technically, the proof is supposed to show what the printing plates will produce from the color separations and preparatory material that have been processed.

However, since the customer often "signs off" on the proof, there is a temptation for the color separator or even the designer to doctor the proof and make it as attractive as possible. When these changes are made in the proof, it becomes very expensive or impossible to produce a print that matches the proof. Another problem is that the customer often changes his/her mind when looking at the proof, and part of the process must be repeated, causing delays and increasing the cost.

Proofs are also used by the press crew to determine what the finished job is expected to look like. If proofs do not match the color separations and plates, operating the press becomes a creative art in itself.

Prepress or off-press color proofing systems are used for all printing processes as quality control checks for camera operations and for color separations and corrections. Customers sometimes prefer press proofs, mainly because the etching involved in making plates and cylinders affects tone reproduction and correction. Offpress proofing works well for lithography, because the films, plates and proofs have essentially the same tone reproduction characteristics.

Methods of Proofing

There are many different types of proofs. Type and layout are easily checked with a silverprint, blueline, brownline or Dylux. Color proofing requires more sophisticated and more expensive procedures and materials.

Many different processes are used to prepare proofs. Press proofs are produced on a proof (proofing or proving) press, and prepress or off-press proofs are produced without a press.

The oldest, and most expensive, method of proofing is press proofing, in which plates or cylinders are prepared and mounted on a press, usually one built especially for preparing proofs. Sometimes the customer demands that a proof be taken from the actual press on which the job will be run, aware of the fact that all presses differ in their characteristics and that a proofing press will produce a different result than a production press.

Where high quality is demanded for flexo, gravure and screen printing, a proof can be made directly onto the substrate to be printed, using production inks and plates.

While press proofing has many advantages, this costly process is avoided whenever possible. Even press proofing does not give an exact duplication of the final print because it is run on a different press, prints are usually allowed to dry between colors, and proofing inks usually differ somewhat from printing inks.

This is the reason for the "Why-won't-the-print-match-the-proof?" problem.

The simplest of the prepress color proofs is the overlay proof, of which Color Key, Cromacheck and NAPS are well-known examples. A transparent sheet carries each of the process colors, and when the four sheets are overlayed, the proof gives a representation of the color on the final print. Several brands of overlay proofs are available, each with somewhat different characteristics. Reflections between the transparent layers of the overlay proof cause grayness or yellowness. These proofs are usually used only by production personnel and not for color approval by customers.

Integral or single-sheet proofs, such as Cromalin or Matchprint, are prepared by exposing a special sheet to each of the four colors and developing that color on the sheet before adding the next. This process is more expensive than the overlay proof but also gives a better representation of the final color print. Nevertheless, even here the sample may differ in many respects from the production job because of limitations inherent in the proofing process itself.

Overlay proofs are simple and economical to make, but because they consist of four separate films, the internal light reflections when viewing them cause graying of whites and reduce the light tones. Integral systems come closer to matching the press print in appearance, but most of them use special bases or substrates that do not look or feel like the printing paper. Although they do not perfectly match the final product, the proofs are accepted by many customers as final proofs.

Direct digital color proofs are produced directly from digital data without separation films. They are used with CEPS and DTP and accepted by customers as contract proofs of what the job in the system will look like when printed.

Digital proofs are becoming more common. Digital proofs are made by several techniques such as ink jet printing, thermal dye sublimation and electrostatic printing. Enlargements of a proof produced by the Iris Graphics ink jet system is shown in figure 6-11 (see color insert).

Proofing of digital color separations with an ink jet printer makes it possible to produce a hard-copy proof without using any photographic film. A proof is produced on glossy paper, matte coated paper or transparency film.

One proofing system uses a thin stream of ink that is modulated to produce 625,000 droplets per second. The droplets are electrically charged according to the data stream from the image file in the computer. Charged droplets are deflected, uncharged droplets print the image. A halftone dot can be composed of anywhere from one to 16 ink droplets (figure 6-12), which controls the size of the dot. The electronic controls also simulate the effect of a halftone screen. The proof is printed at a rate of about one inch per minute.

Soft proofs, that is images of the artwork on a VDT screen, are sometimes used as proofs. These are quick and corrections can be made and viewed immediately. However, a hard copy of the proof is needed for many reasons. The problem with soft proofing has been that VDT screens produce red, green and blue instead of yellow, magenta and cyan, and it is difficult to visualize what

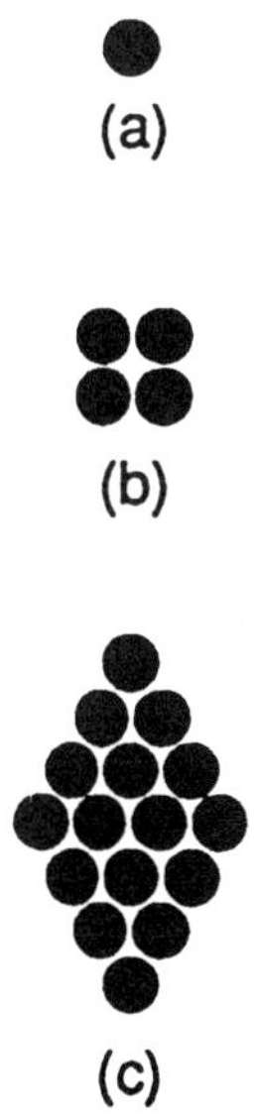

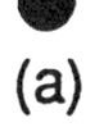

Figure 6-12. Structure of a halftone dot made from ink jet droplets of any circular pixel: (a) dot from one pixel, (b) dot from four pixels, (c) dot from 16 pixels.

the final product will look like. New calibration tools have largely solved this problem for CEPS, but DTP still needs some help.

Dangers and Pitfalls of Proofing

The proof is the means of communication among the printer, the designer, the separator and the customer. A signed proof is part of an agreement to pay for printing the package that matches the proof. While the proof is supposed to show how the press will reproduce the artwork that has been processed and put onto plates, various abuses creep in. One common abuse for the separator is to work with the proof to produce a picture that pleases the customer. Special proofing inks that differ in hue or color strength from production inks may be used, or the yellow ink may be given a "double bump" (printed twice) to increase its intensity. Such a proof is a thing of beauty, but it becomes impossible to match the proof with the press print. The cost of printing to match such unrealistic proofs is far greater than it is to prepare the artwork properly in the first place and to order inks and paper required to produced the desired colors.

Another common abuse is for the customer to make extensive changes in the proof, to order "increased red" in one area or "blue" in another. The practice probably goes back to the early days of color printing 50 to 100 years ago when there was no good way to judge the color that a given set of plates, ink and paper would produce. Now, it is often a matter of pride to make multiple changes on the color proof; the proof is returned to the trade separator repeatedly for corrections—and costs skyrocket.

The goal of good management should be to use the capabilities of the modern prepress technology to produce an acceptable proof the first time. This is possible

TABLE 6-I. COLOR PROOFING BUZZWORDS

Add contrast	N.G.	Greek out
Add snap	Try again	Grubby
Air out	O.K.!!!	Harsh
Beef up color	O.K.???	It's got to jump
Burn down	Lacks snap	Too piny
Pushy	Livelier	Whew!
Raw	Too flat	Whisper more
Redder reds	Make it pop	Doesn't have zatch
Reduce one step	More guts	More shape
Reduce 2%	More metallic	Make it sing
Seems cloudy	More neutral	Color too skinny
Soften back	You went too far	Delete a little red
Subdue	Dull down	Grainy
Tad less	Exaggerate condition	
Too dull	Fleshier face tones	
Flick less	Give pleasing color	
Oops! Too much	Needs more dimension	
Make colors warmer		
Make colors colder		

From Research and Engineering Council Roundtable, Chicago, October 1976.

only some of the time, and it requires unusual skill, experience and cooperation among the designer, printer, trade house and customer. Savings on the cost of preparation can be tremendous. Communication in numerical terms is required if corrections are to be made in a predictable fashion. It is still common to receive incomprehensible comments about changes desired in the proof.

DESKTOP PUBLISHING AND COLOR ELECTRONIC PREPRESS SYSTEMS

An early application of computers, called CAD, is called variously Computer Assisted Design, Drafting or Drawing. The programs are useful in all of those applications. When CAD/CAM programs are used, camera work and film consumption are greatly reduced. With the growth of computer technology, some of these programs became incorporated into a series of programs that make up desktop publishing or DTP.

It is useful to distinguish the fast, inexpensive color and design work turned out with a Macintosh computer and the more sophisticated, top-quality available with a high-end color scanner and imagesetter. The differences in cost of the hardware are commensurate with the differences in speed and capacity. The manager must be aware of both quality demands and costs. Packages that require top-quality four-color separations still require CEPS.

"Desktop publishing" is now able to handle all composition, color separation and image assembly from the artist's desk. It brings the designer and art director closer to the prepress tradeshop and the packager. DTP can avoid use of a color scanner by manipulating original pictures captured with an electronic camera or converted to a digital form from film.

Since newspapers have led prepress technology for decades, it is useful to consider how they handle copy since it can be used to predict what prepress technologies are and will likely be available for package printing.

While filmless prepress has been imagined since the mid 1970s, it has become reality in publication and package printing only since 1990. The camcorder and the digital camera supply pictures for electronic manipulation. Since the early 1980s, TV programs set in the newsroom have featured the computers used by reporters to keyboard their stories. Before that, the stories were typed and edited, then sent to composition for typesetting with a linecasting machine (e.g. Linotype). This was proofed and corrected. The type was locked into a chase and a stereotype was produced, a technique not unlike the production of flexo package printing in the early 1990s.

Now the reporter's story goes directly to rewrite and editing after the spelling program checks for keyboarding errors. Without rekeyboarding, the digital copy goes to page makeup and platemaking. Pictures, often in process color from a digital camera, are added and handled by the system. Direct exposure of plates by laser beam is still under development. The digital copy is still taken to film from which the plates are developed.

PLATEMAKING

The final step in prepress operations is preparation of the image carrier. The most common method is to expose a photosensitive material to light and to develop the image. Some variation of this is used in all processes except gravure. Gravure cylinders are prepared by electronically controlled mechanical engraving.

Use of laser beams to produce an image carrier permits plates to be produced from digitized information without producing an intermediate film. All of the major printing processes have systems for preparing plates or cylinders from laser beams, but none of these have yet become dominant technologies.

A laser system currently used in flexography engraves rubber plates in the round (both on plates and on sleeves). The artwork can be any copy that can be scanned in the round. Color filters separate the artwork into its process components, each of which becomes a separate printing plate. In this process there is no negative stage. Rubber flexo plates made this way will last for many hundreds of thousands of impressions.

Finally, there are electrostatic and ink jet processes that print directly from digitized information without generating a printing plate. These technologies are widely used for printing variable information such as serial codes or batch numbers, but they do not yet offer the combination of speed and quality of the older processes.

Printing plates are discussed in Chapter 7.

PRESS MAKEREADY

Now that all of the prepress work is complete and the designs have been converted into printing plates or cylinders, the plates, cylinders or ink jets must be installed and adjusted so that the press can produce the designed product.

Press makeready can be a time-consuming operation depending on the nature of the job and the organization and training of the crew. Removing the previous job from the press is usually included in makeready.

The operations included in makeready are not well defined, but makeready includes inking the press, placing the paper or substrate into position, putting the plates or plate cylinders into position and preparing the first salable sheet or package.

The time required for the job is reduced if jobs are scheduled so that the same inks can be used and if other changes are minimized.

For in-line operations, converting equipment must be set up to give the right cutting and creasing, folding, etc. This usually takes more time than making the press ready.

TRAINING

The complications involved in preparing art and text for production with a computer are not well handled by a neophyte. Stories in the trade press report that the new equipment was far more helpful than management had expected and that training was far more difficult than anyone had predicted (least of all, the vendor of the equipment). Efficient equipment, procedures and the training to handle them well do wonders for the morale of plant personnel.

Most prepress managers seem to agree that it is easier to train someone with art and production experience to use a computer than to train a computer specialist in all the requirements of converting art and text into a finished plate or cylinder.

In either case, continual training is essential as new techniques and equipment cascade into the prepress operation. Managers usually find it more profitable to spend money and time keeping staff up to date rather than to let staff grope around, making mistakes or doing things the hard way, learning by experience while the competition gains the upper hand because they have learned to improve quality and productivity while reducing costs and turnaround time.

FURTHER READING

The Pocket Pal, Fifteenth Edition. Michael H. Bruno. International Paper Co., Memphis, TN. (1992).

Flexography: Principles & Practices, Fourth Edition. Frank N. Siconolfi, Chairman. Flexographic Technical Association. Ronkonkoma, NY. (1991).

Gravure: Process and Technology. E. Kendall Gillett, III, Chairman. Gravure Education Foundation and Gravure Association of America, Rochester, NY. (1991).

Color and Its Reproduction. Gary G. Field. Graphic Arts Technical Foundation, Pittsburgh, PA. (1988).

Graphic Arts Photography: Black and White, Second Edition. John E. Cogoli. Graphic Arts Technical Foundation, Pittsburgh, PA. (1988).

The Lithographers Manual, Eighth Edition. Ray Blair and Thomas M. Destree, Editors. Graphic Arts Technical Foundation, Pittsburgh, PA. (1988).

Electronic Color Separation. R. K. Molla. Rafiqul K. Molla, Montgomery, WV. (1988).

Printing Technology, Third Edition. J. Michael Adams, David D. Faux and Lloyd J. Rieber. Delmar Publishers Inc., Albany, NY. (1988).

Principles of Color Proofing. Michael H. Bruno. GAMA Communication, Salem, NH. (1986).

Chapter

Plates and Other Image Carriers

CONTENTS

LIST OF FIGURES

LIST OF TABLES

Chapter

Plates and Other Image Carriers

INTRODUCTION

This chapter covers the preparation of plates and cylinders and the use of plateless printing processes. The different printing processes use different types of plates. The first step in platemaking and printing is to have well-prepared artwork. For most processes the artwork must be converted to properly exposed and developed high-contrast photographic films. Although direct-to-plate technology is now being used in flexo, gravure and offset, film is usually used in preparation of plates, and bromides are still used in preparing gravure cylinders.

- Flexo plates are made by molding rubber into a matrix or by removing vulcanized rubber from the non-image area with a laser beam. Solid-sheet photopolymer plates are prepared by exposing a soluble or partially soluble photopolymer to UV through a film negative and washing away the unexposed part. Plates from liquid photopolymers are prepared by exposing the liquid to UV through a photographic film and recovering the unexposed liquid. Estimates of the share of each type of flexo plate in various countries are reported in table 7-I.

Flexo plates may be mounted directly onto the press cylinder, or they may be mounted on a carrier or blanket which is then mounted onto the cylinder. A plate can also be produced by engraving a layer of rubber coated on the cylinder.

TABLE 7-I. MARKET SHARE (%) OF VARIOUS FLEXO PLATES IN THE UNITED STATES, UNITED KINGDOM, NETHERLANDS, AND SCANDINAVIA

	U.S.	U.K.	Neth.	Scand.
Photopolymer	33	65	70	55
Rubber	65	30	25	40
Laser engraved rubber	2	5	5	5

Source: Euro Flexo Magazine. December, 1992.

- Gravure cylinders are prepared by digitally controlled engraving or by chemical etching of a copper-plated cylinder.
- Offset plates are made by exposing a photosensitive layer on a base that is usually aluminum but sometimes plastic, then developing the plate to remove the ink-accepting material from non-image areas.

In gravure and in some flexo operations, the images are placed on the cylinder before mounting on the press. These prepared cylinders can be quickly mounted, saving valuable time on the press. Offset plates must be mounted while the press is standing idle.

- Letterpress plates used for packaging are usually photochemically produced, much like flexo plates but harder. Some metal plates are still used.

Unlike flexo, letterpress and gravure do poorly on rough surfaces because the metal or very hard plastic plates cannot conform to the printing surface. Offset does much better because the blanket, like a flexo plate, has some resiliency or conformity.

- Screens for package printing are prepared by exposing a photosensitive material held in the screen, then washing away the unexposed material in the image area.
- Plateless systems are often digitally controlled: ink jet, xerography, laser printing and others.
- Platemaking processes can create environmental problems, including disposal of solvents and heavy metals. Developmental work on new plates puts heavy emphasis on "environmental friendly" processes and materials.

FLEXOGRAPHIC PLATES

General Characteristics

There are two broad groups of flexographic plates: rubber plates and photopolymer plates. Both types are commonly used for package printing. A color separation made for one type is generally unsuitable for the other. For best results, inks should be matched for the plate—rubber or photopolymer—because their ink release properties differ. The plates must resist swelling by ink or wash solvents. Flexo plates are significantly softer than photopolymer letterpress plates. Flexo plates may be as soft as Shore A 20, or they may be as hard as Shore A 70 for labels. Photopolymer letterpress plates run around Shore A 90 to accommodate the viscous, tacky inks and rough substrates. Metal letterpress plates are much harder.

Photopolymers produce somewhat better halftone reproductions than rubber plates, especially with fine halftone screens. High quality rubber plates can be produced through the detailed process of acid etching of copper to produce the engraving or by laser engraving in the round. The difference in image quality is so small that the decision to use rubber or photopolymer is usually based on other considerations, including the number of plates that are required, the frequency of reruns, solvent resistance required for the ink system and crew experience. The printer may choose photopolymer plates if only one is needed, but if more than three or four are needed, the expense of preparing the matrix may be compensated for by the lower cost of the rubber plate.

In order to avoid moiré patterns, screen angles for flexographic plates must be adjusted so they do not conflict with the screen angles of the anilox rolls to be used on the job. Some anilox rolls are designed with flow-through characteristics that minimize the moiré.

Reproportioning of flexo plates is required because the printing surface of the plate is stretched when it is wrapped around the cylinder. Electronically imaged negatives used for halftone plates should be computer-compensated for dot gain. A formula for the distortion correction factor is given in Chapter 3. In simplest terms, the plate surface becomes longer than the backing when it is wrapped around the cylinder. The lengthening of the surface is directly related to the thickness of the plate, as shown in Chapter 6.

Table 7-II shows the print length distortion and thickness loss when various plates are mounted on a 40-inch (1016 mm) cylinder.

Figure 7-1 compares the time required to prepare different types of flexo plates for a five-color job.

Rubber Plates

Rubber plates are made by pressing or molding uncured rubber into a mold or matrix that has been made from an engraving that bears the image. Rubber plates can also be made by etching the image into rubber with a laser beam. Molded rubber plates 0.250 inch (6.35 mm) thick have long been the standard of the corrugated industry, but new materials and printing presses make possible use of plates that are much thinner and give superior results. Molded rubber

TABLE 7-II. DISTORTION OF FLEXO PLATES MOUNTED ON A 40-INCH CYLINDER

	Plate Thickness (inches)		
	0.250	0.125	0.085
Print length distortion (%)	96.2	98.2	99.2
Thickness loss distortion (inch)	0.004	0.002	0.001

Source: Hercules, Inc., *Flexo* (Sept. 1992)

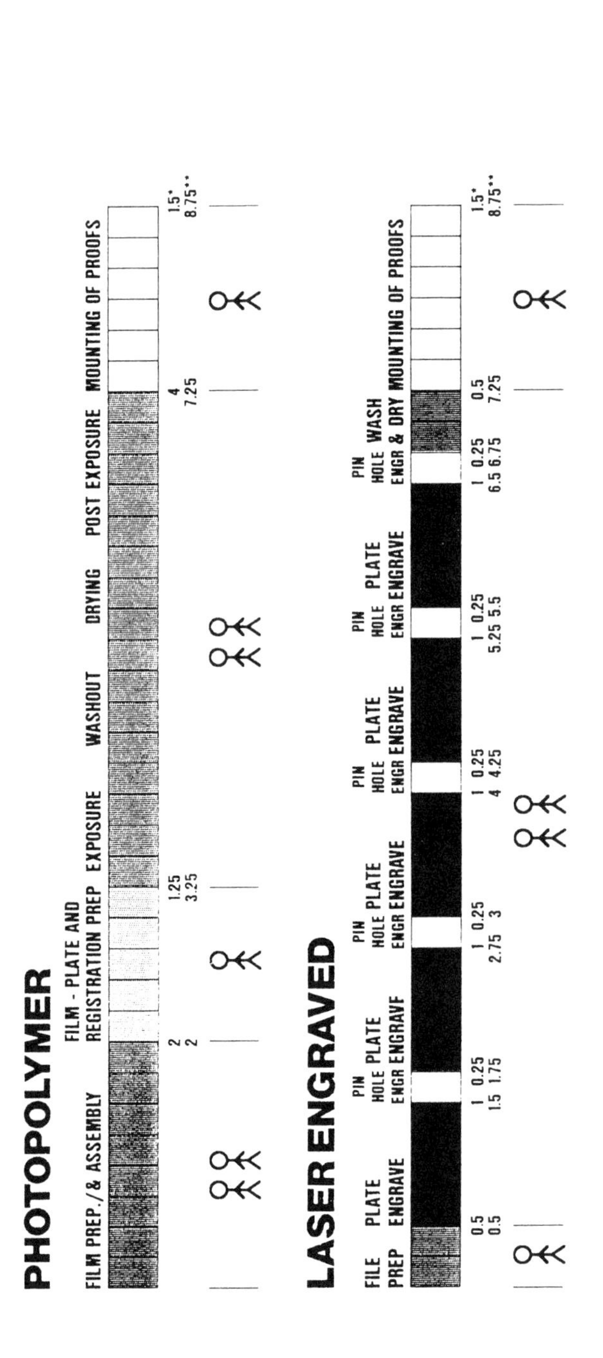

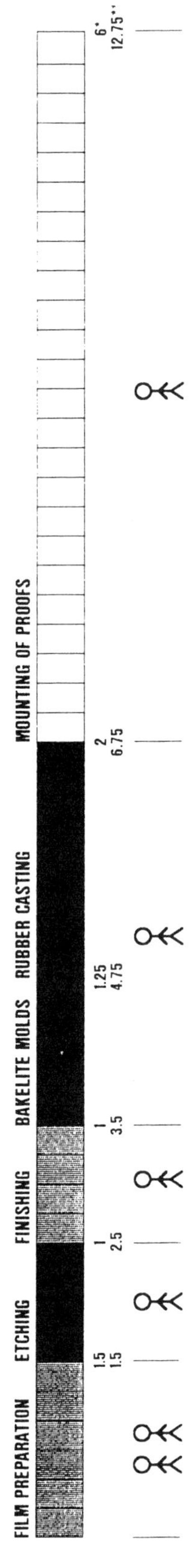

Figure 7-1. *Time required to prepare different types of flexo plates for a five-color job. (Courtesy of ZED Instruments USA Ltd.)*

plates, which are conventionally mounted on a rigid PVC backing, are dimensionally stable since the backing material does not stretch.

Rubber plates are flexible and resilient with excellent affinity for a wide variety of inks and the ability to release ink onto many different substrates. Rubber plates are less susceptible than photopolymers to hardening with UV radiation, but they do grow harder with age and oxidation.

Rubber plates are still used for large market segments, and they will be around for a long time. The raw material cost of the rubber alone is about half that of photopolymers. After the matrix has been prepared, rubber plates can be made in minutes. Sheet photopolymers require longer times for processing and drying, although processing time is being reduced by new developments. Liquid photopolymers can be produced in 30 minutes from photographic film to plate.

Rubber plates offer some advantages over the photopolymer plates. Once the matrix is prepared, duplicate plates can be produced quickly and easily. If repeat jobs are expected, the matrix (or the engravings) can be stored and used to prepare plates for later jobs. It is possible to make some alterations in engravings (changes in photopolymers are made on the negative). Large rubber plates are very heavy. Due to limits in the size of available molding presses, individual rubber plates are rarely larger than 30 by 40 inches (760 x 1016 mm). Photopolymer plates can be made thinner, they are more dimensionally stable, and larger sizes do not become unwieldy. Photopolymer flexo plates are available in sizes up to 52 by 80 inches (1320 x 2030 mm).

Molded Rubber Plates. The engraving is made from a metal block that has been coated with a photographic (or photosensitive) emulsion. The block is usually made of magnesium and sometimes called a "mag." Some zinc is also used. Copper is sometimes used for fine screen reproduction and whenever fine tone or process color jobs are involved. The coated, metal block is exposed to UV light through a photographic negative of the design to be printed (figure 7-2).

The metal is usually 0.064 inch thick, with a line relief etched between 0.030 and 0.035 inch. The corrugated industry still uses 0.250 inch originals etched between 0.140 and 0.150 inch, but thinner plates are being accepted rapidly. Relief as low as 0.030 or 0.045 inch (0.75 or 1.15 mm) is now used in preprinting of corrugated, and around 0.060 inch (1.5 mm) is used for combined board.

Careful preparation of the negatives is crucial. Small defects in the film will

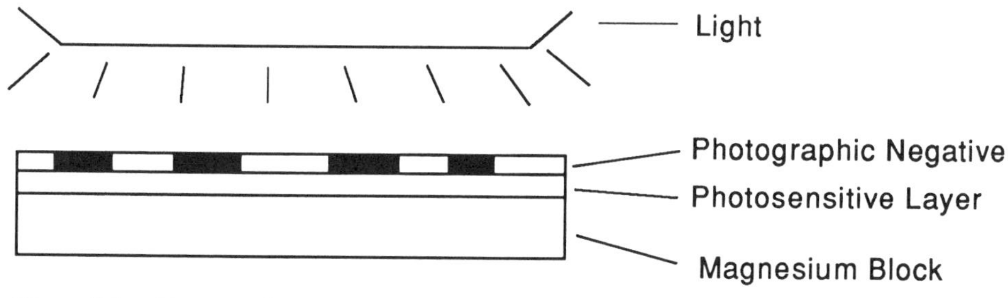

Figure 7-2. Exposing the engraving.

be picked up by the engraving. The film is developed to achieve maximum density and is inspected for defects. Pinholes are covered with opaquing material. The exposed coating is developed, uncovering the non-image area on the metal. The metal is etched in an acid bath, removing the non-image area and forming a relief image in metal.

To print properly, the engraving's shoulders should be almost vertical, with a uniformly smooth taper. A broad, stepped shoulder tends to catch and build up ink, causing a smeared and dirty print. Powderless etching is performed with nitric acid that contains a special filming agent to protect the sidewall or shoulder of the image from being undercut during etching. Master plates are also made from sheet or liquid photopolymers exposed to UV light and developed. Nylon masters are capable of molding up to 150 lines per inch for process color printing. Proper etching and several defects are shown in figure 7-3.

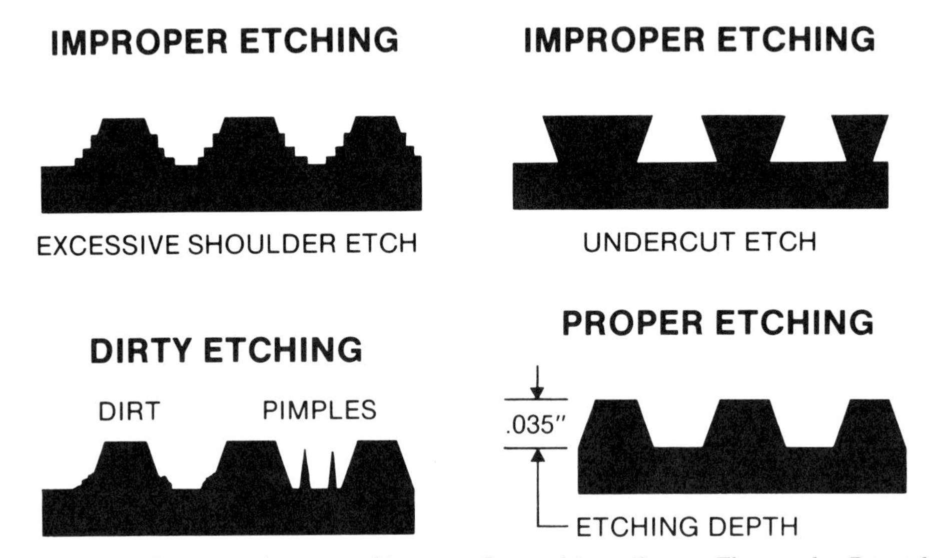

Figure 7-3. Diagrams of proper and improper flexo etchings. (Source: Flexography: Principles and Practices, Fourth Edition.)

After the engraving is produced, a phenolic molding compound (powder and or board) is molded against it and cured at 300°F (150°C) for 15 minutes to make the mold or matrix. Many printing plates can be made from this matrix.

The plate is then produced from uncured rubber compositions (such as natural rubber, Buna N, vinyl rubber and others) by pressing the plate materials against the matrix and curing (vulcanizing) at 300°F for 10 minutes. Rubber shrinks about 2% during vulcanization.

The plate is peeled off the matrix, trimmed and ground to improve the uniformity of thickness. The tolerance is usually ±0.0015 inch (±0.38 mm). If the thickness of the rubber plate is not uniform, the printing impression pressure will vary from one part of the plate to another, and ink squeeze-out will distort the type, lines and halftones as the pressman increases the pressure to print the low spots. A good plate can be improved by grinding, but grinding is insufficient to make a bad plate good.

The entire process must be closely controlled: exposure, developing and etching of the engraving, preparing the mold, curing the plate and grinding. Skill and care are required to make a good plate.

Laser-Engraved Rubber Plates. Often called "design rolls," these printing cylinders are made by adhering a natural or synthetic rubber polymer such as a polyurethane to the printing cylinder and carefully curing it. The rubber-covered cylinder is ground to produce a stress-free roll that is level, concentric, smooth and precise to within ± 0.001 inch TIR.

The laser beam atomizes or ablates the rubber, leaving a clearly defined image. Engraving depth can be varied to meet the needs of the job. Because the printing surface is already curved when it is engraved, stretching and distortion do not have to be considered in providing artwork for laser engraving. The engraver can meet the customer's (printer's) requirements regarding trap, bleed, registration marks, eyespots and other devices.

Artwork can be input directly to the computer. Laser beams can be controlled by computers, making it possible to generate printing cylinders directly from computerized artwork without the use of photographic film (figure 7-4). The system can be programmed to scan and engrave at the same time. Resolution is currently limited to lower line screens and productivity is low, but newer equipment and technology are rapidly improving.

Photopolymer Plates

There are two types of photopolymer plates: solid and liquid plates. Solid photopolymers are made from a soft, solid material. Liquid plates are prepared

Figure 7-4. Machine for preparing laser-engraved flexo plates. (Courtesy ZED Instruments USA Ltd.)

from a viscous liquid. Whether they are produced from liquid or solid, photopolymer plates are generally adhered to a dimensionally stable backing material, usually a stable polyester film.

There are numerous brands of photopolymer plates. The plate material is hardened or solidified (polymerized) under UV radiation. Plates are prepared by first exposing the base to UV to create a uniformly cured floor, then exposing the face through a negative. Developing removes the unexposed, non-image polymer to form a relief image.

Dimensionally stable photopolymer plates make accurate register possible and generally give better halftone reproduction than rubber plates. Going from engraving to matrix to flat rubber printing plate is less reproducible and predictable than a photopolymer exposure and development. The thickness of photopolymer plates is more uniform than that of rubber plates. Solid sheet photopolymers have a stable polyester backing. Run life of photopolymers outpaces rubber plates by two or three to one.

Pin register systems do not work well with rubber plates unless the rubber plates are premounted onto blankets. Pin register provides photopolymer plates an advantage for process color printing, small type and jobs requiring close register.

Photopolymer plates offer many advantages over rubber plates: better print

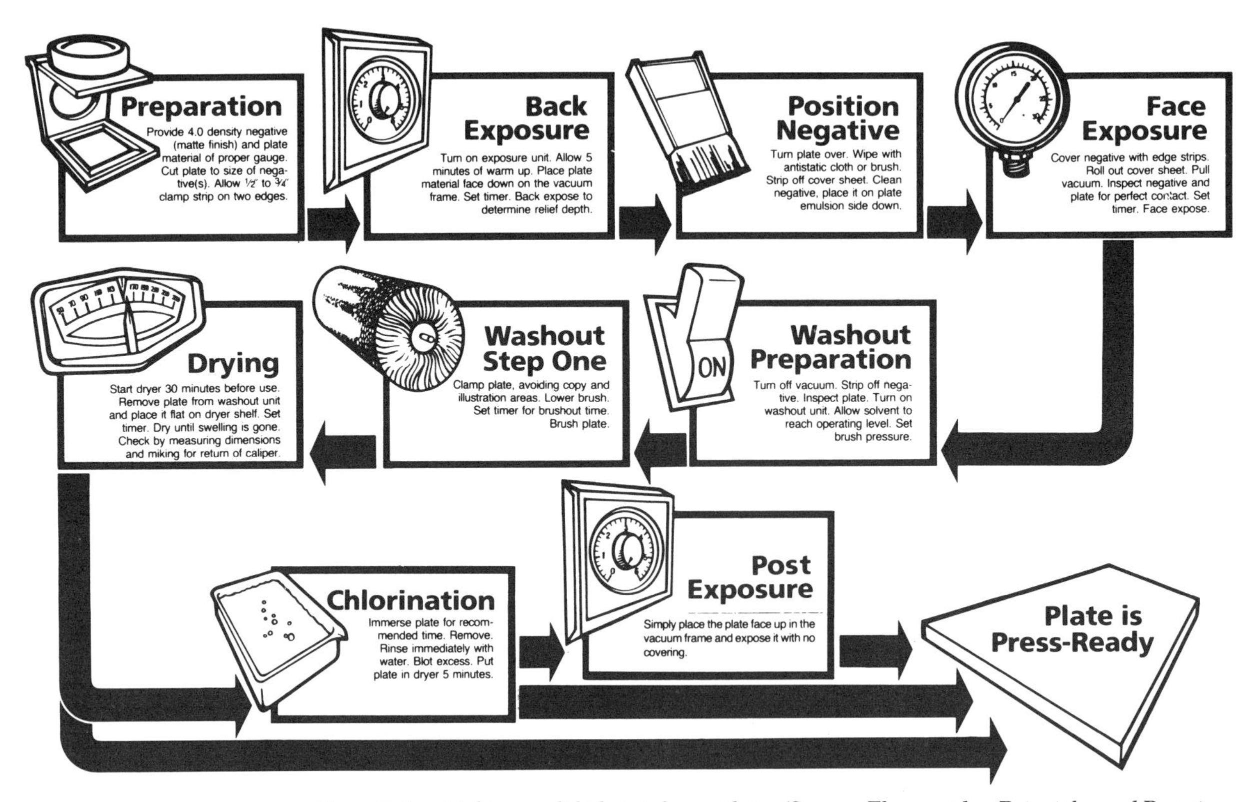

Figure 7-5. Making a solid photopolymer plate. (Source: Flexography: Principles and Practices, Fourth Edition.)

quality, longer plate life, reduced and predictable dot gain, improved color registration and pin registration.

There is a trend to one-piece plates with the repeat pattern produced by a step-and-repeat machine. This greatly reduces the amount of time required to mount individual plates and prevents errors in their mounting.

Solid Photopolymer Plates. Solid plates comprise an elastomer or rubbery material that is hardened under ultraviolet light. The plate is exposed through a photographic negative that hardens the image area, and the plate is developed using a strong organic solvent that must be captured and recycled (figure 7-5).

There are several solvent blends used today; some require special, expensive handling and disposal. Most printers reduce costs with an in-house vacuum distillation system.

For this and other environmental reasons, plate suppliers are working to perfect satisfactory, water-developable plates. Some plates can be developed with aqueous caustic containing some organic solvent; others use detergent solutions for plate processing.

After the plate is exposed to UV, it is developed. It must be re-exposed to completely harden the plate. It must then be further treated, with actinic light or chlorine, to eliminate any tackiness in the plate.

The most common size for photopolymer plates is 30 by 40 inches (762 by 1016 mm), although sheets as large as 52 by 80 inches (1320 by 2032 mm) are available. These large sizes require special processing equipment. They are used for preprinted corrugated liner and for stepped images for many types of packages.

Liquid Photopolymer Plates. To prepare a plate from liquid photopolymer, the negative is placed on the platemaking machine and covered with a thin barrier film (figure 7-6). The photopolymer resin is cast to the desired thickness and the film backing is simultaneously applied (figure 7-7). The photopolymer resin has the consistency of honey at room temperature. It is safe to handle in

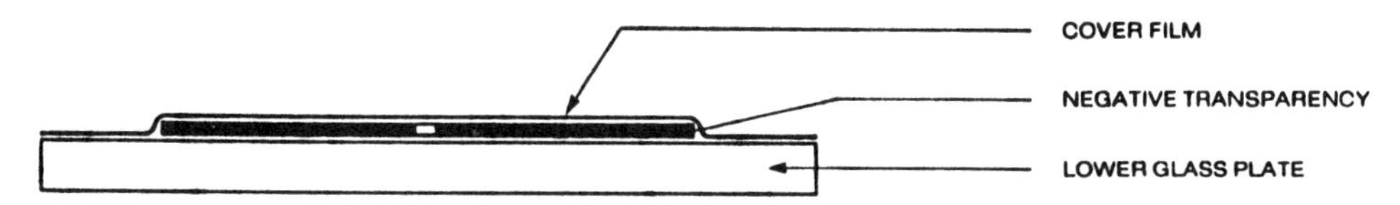

Figure 7-6. Making a liquid plate: placing the negative in the exposure unit. (Courtesy of Hercules, Inc.)

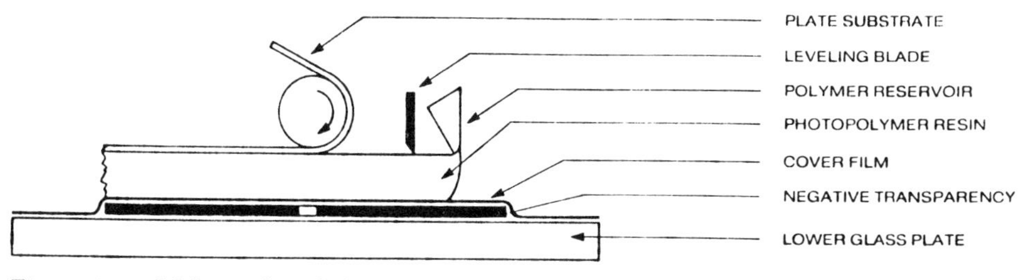

Figure 7-7. Making a liquid plate: casting the photopolymer and applying the plate substrate. (Courtesy of Hercules, Inc.)

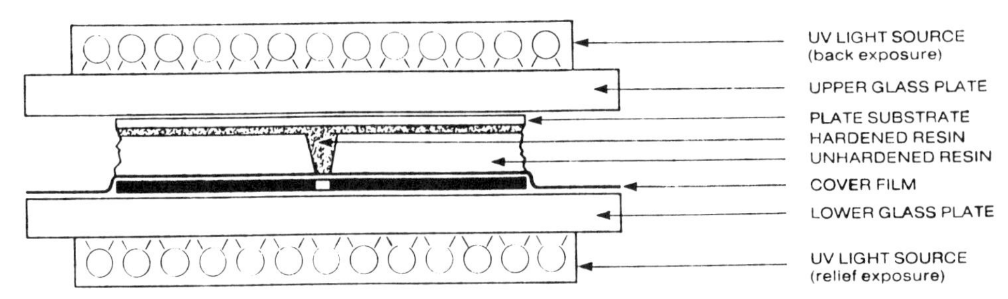

Figure 7-8. *Making a liquid plate: exposing the plate and base. (Courtesy of Hercules, Inc.)*

ordinary light, but it is sensitive to UV and sunlight (which has a high component of UV).

The sandwich is then exposed from both sides (figure 7-8). Exposure through the negative develops the image; exposure through the film backing produces a uniform support for the image. In the unexposed areas, corresponding to the non-printing areas in the plate, the resin remains liquid. The unexposed resin can be reclaimed manually with a squeegee or by washing away with soap and water. An automatic resin recovery unit uses an airknife to recover the unexposed liquid photopolymer. The reclaimed polymer can sometimes be reused for future platemaking. Figure 7-9 shows the machinery.

High-contrast negatives are used to image the liquid resin. The resin is hardened in the clear areas where the UV strikes the photopolymer. After exposure, the cover film is removed and a post-exposure step with UV further hardens and strengthens the plate.

Capped Plates. The liquid photopolymer plate provides sharper lines and halftones if the printing surface of the plate is hard, with a softer layer underneath. Such plates are called "capped plates."

Figure 7-9. *Exposure unit for making a liquid plate. (Courtesy of Hercules, Inc.)*

The cap is a four- to ten-mil (0.10 to 0.25 mm) layer of cap resin on the printing surface, supported by a base layer. The layers are chemically similar except for the photoinitiator components that cause the cap layer to form its unique sidewall structure during plate imaging.

The hard layer or cap helps to reduce dot gain, squeeze out, plugging and halftone dots with holes in the center, typical problems in flexo printing. Figure 7-10 shows deformation of the base resin under pressure, and figure 7-11 shows deformation of the cap. The softer layer gives some cushion or compressibility to the plate, enabling it to print well on surfaces that are not very smooth. It also provides a wider impression latitude to the plate.

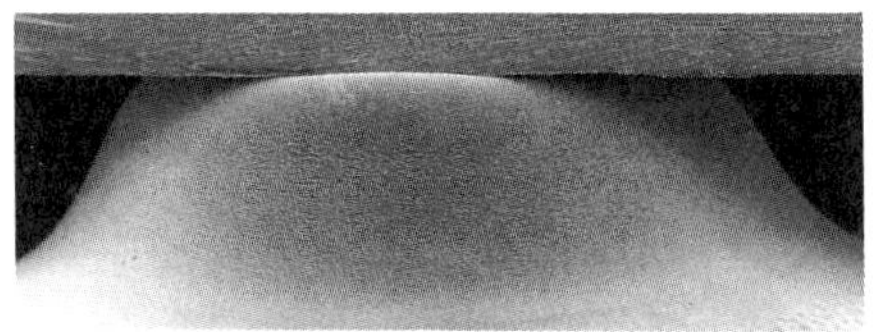

Figure 7-10. Distortion of highlight dot on conventional plate. (Courtesy of Hercules, Inc.)

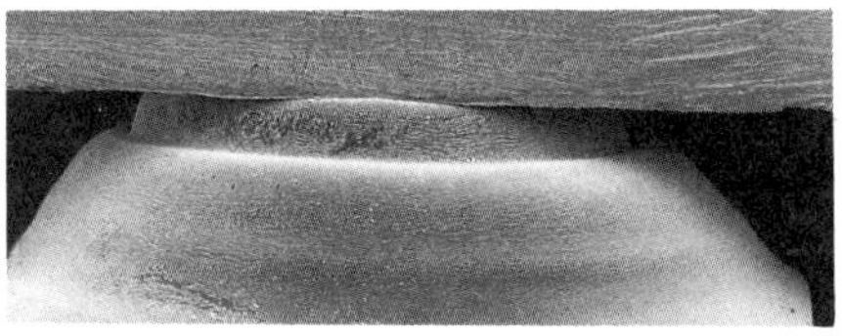

Figure 7-11. Distortion of highlight dot on capped plate. (Courtesy of Hercules, Inc.)

The side wall supporting the cap must also be strong, because if it is too pliable, it will bend under pressure and ink will be deposited on the sidewall instead of only on the face of the plate.

Water-developable Plates for Waterbase Inks. Flexo plates that are designed for use with waterbase inks and that can be developed with water have appeared on the market. After the back of the plate is exposed and hardened, the face is exposed through a negative. The plate is then washed with water to generate the relief areas, dried in a plate drier and post-exposed to harden the image.

Thickness of Plates

Plates are generally classified in three thicknesses: thin plates (0.045 to 0.067 inch or 1.15 to 1.70 mm) are used for tags, labels and preprinted liner; medium thick plates (0.067-0.112 inch or 1.70 to 2.84 mm) for flexible packaging and wide web; and thick plates (0.125, 0.155 and 0.250 inch or 3.25, 3.94 and 6.35 mm) for corrugated. Laser-engraved rubber plates fall in the second and third groups. The 0.250-inch plate includes rubber-impregnated canvas backing with a thickness of 0.060–0.070 inch. Rubber molded thinner than 0.100 inch won't come out of the mold. These thin rubber plates are unbacked.* An example of the trend to thinner plates is shown in table 7-III.

* Some writers refer to plate thickness as the thickness of the plate material plus the cushion and backing or carrier. In this book, plate thickness refers to the thickness of the plate material only.

TABLE 7-III. THICKNESS CHANGES IN MOLDED RUBBER PLATES

	Old Value	New Value	Old Value	New Value
Total thickness	0.250 inch	0.155 inch	6.35 mm	3.93 mm
Relief	.150	.075	3.81	1.90
Floor or base	.100	.080	2.54	2.03

Source: W. R. Grace & Co.

Thin plate technology has enhanced flexo print quality. Use of thin plates was once almost exclusively confined to the narrow web, flexible packaging and preprinted linerboard markets. Thin plates are now making inroads into the corrugated industry as boxmakers realize the significant production, print quality and economic benefits that can be achieved. Thin plates are not just for multicolor process work but are useful for all types of flexo printing. (Thin, of course, is a relative term meaning around 0.155 inch for printing corrugated board and 0.067 inch for labels.)

Corrugated, among other segments of the flexo printing industry, is making increasing use of thin plates because they improve print quality but also because they are light and easy to handle. Such plates are precisely mounted on carrier sheets ("blankets") off the press. The mounted plates are wrapped around the plate cylinder and attached in register with pins or other registration devices. This means the first image printed is in register, something that was not possible when registration was adjusted after the plates were mounted.

The carrier sheets are made of dimensionally stable material and prepared for precise mounting. Carrier sheets can be used with pin register systems. Plates are premounted using an accurate device, usually optical, with accurate mounting tapes or compressible stickyback.

Historically, photopolymer plates for corrugated have ranged in thicknesses from 0.155–0.250 inch with an accepted 0.280 inch cylinder pitch or undercut. The thicker the plate, the greater the chances for cupping, voids and inherent manufacturing defects in the plate.

A compressible plate mounting foam cushion can reduce plate pressure by as much as one sixth. At typical impression settings reduced pressure due to compressibility results in longer plate life. A thin, solid plate with a foam backing has been shown to give 300 to 500 thousand impressions, compared with 150 to 200 thousand with traditional backing material. Plates as thin as 0.067 inch supported by a foam backing are used for critical high-quality work on corrugated. Thin plates are less expensive and last longer than the thicker plates. They offer many other advantages:

- Improved print quality
- Reduced dot gain
- Improved thickness control
- Improved productivity
- Faster plate processing
- Finer detail because of less relief

- Easier, quicker mounting since they weigh less
- Reduced wear of anilox rolls and bearings from the reduced print pressure.

The thinnest large flexo plates usually have about 0.067 inch thickness with 0.030 inch relief and a polyester backing. With compressible backing (0.060 inch) these thin plates produce the highest quality printing.

Mounting the Plates

Flexo plates are usually mounted with a two-sided adhesive tape called stickyback. Adhesives are rubber or acrylic. The tapes may have a foam cushion that helps compensate for small differences in plate and substrate thickness. This is called "cushion stickyback." Hard films based on vinyl resins have been used, but cushion mounts that include a foam layer are gaining popularity because they improve print quality. Gauge control of the tape is extremely important.

Printers who win top prizes in printing competitions generally use some sort of cushioning material although some printers, particularly narrow web printers, do not use cushioning material; they achieve excellent results by careful control of print pressure.

Plates for printing corrugated are often mounted onto carrier sheets. The carrier sheets wrap completely around the plate cylinder, usually over a compressible urethane blanket. Use of carrier sheets plus the compressible blankets permits the use of thinner plates, with lower relief, to be employed. The benefits include off-press registration of color to color, less dot gain, closer register of colors on press, reduction of slurring and other defects that occur with deep relief plates.

A liner, normally polyethylene-coated paper, protects the stickyback adhesive from dirt. The liner must release easily from the tape after mounting. The support layer, PET or PVC, lies between the foam and the adhesive on the plate side. Figure 7-12 shows the structure of the mounted plate.

Plates backed with steel can be mounted with clamps like letterpress plates, or they can be held in place with magnetic cylinders.

Demountable Sleeves

Sleeves are mounts for flexo plates that are slipped on or off the printing cylinder or mandrel. Most systems use compressed air to stretch the sleeve enough to allow it to be moved easily on or off the cylinder. When the air is released, the sleeve fits tight and will not slip.

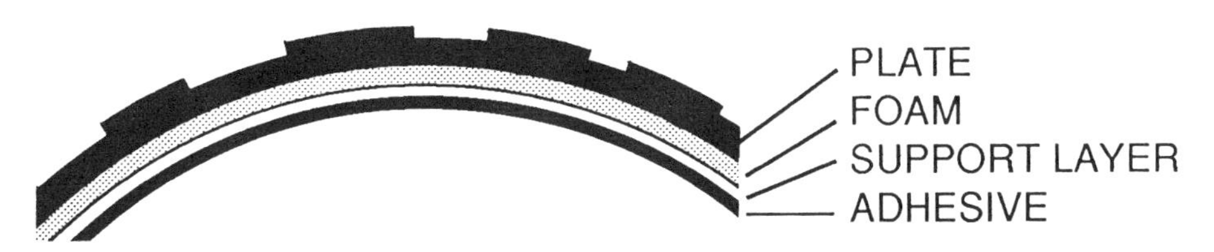

Figure 7-12. Structure of mounted plate.

Figure 7-13. Sleeve with mounted flexo plates. (Courtesy of Stork Cellramic, Inc.)

Sleeves make it possible to store mounted flexo plates for repeat jobs at a cost much lower than removing and remounting the plates on the printing cylinder. It is not practical to store a fully mounted integral cylinder. Sleeves are easily sent by air freight from engraver to printer (figure 7-13).

A demountable sleeve system comprises two elements: a steel cylinder onto which is mounted a plastic or metal sleeve. Today, there are a dozen or more sleeve systems in use. Sleeves may be made of thin nickel or polyester, fiberglass or other polymers. Sleeves carry predeveloped printing plates or rubber covering to be used for laser engraving. Their size must be carefully controlled so that they do not slip when they are mounted on the press.

Sleeves can be mounted onto mandrels off press, in a similar fashion to that of mounting plates on integral cylinders. It takes only a few seconds to remove a used sleeve and replace it with a new one.

Sleeves are used in all flexo printing markets except direct printing on corrugated. Sleeves give consistent, repeatable register and process color, and they reduce startup time. On preprinted liner, sleeves are used for all-over tinting and varnishing for large repeats and width. Reduced run length and an increase in the number of colors together with demand for improved quality have led the bag and sack industry to turn to sleeves.

It is possible, by using a rubber covering on the cylinder, to achieve different repeat lengths while using the same sleeve. However, if much difference exists, different mandrels and sleeves are required so that it is often necessary to have a wide variety of sizes on hand.

GRAVURE CYLINDERS

Gravure is the most stable printing process, and it produces the most consistent work. On the other hand, it is more difficult to make printing changes once the image carrier is finished than it is with any other process. Despite automation and computerization of the process, preparation of a gravure cylinder is still more expensive than preparation of any other printing image carrier, but once it is prepared, the gravure cylinder can produce printed product at a higher rate than any other process.

Structure

Gravure cylinders are prepared by plating copper onto a steel cylinder. An image is etched or engraved into the copper coating, and the gravure cylinder is chrome plated. To reclaim a gravure cylinder for reuse, the chromium is removed chemically and the copper is removed with a lathe or automated machine tool.

Aluminum cylinders are sometimes used. They are much lighter and cheaper to ship, but difficult to electroplate. On small cylinders, aluminum sleeves can be used. A new plastic tube or sleeve that can be electroplated has recently been introduced.

A special technique produces a copper layer called a Ballard shell. Copper is plated, nonadhesively, onto a specially prepared cylinder; it can be easily stripped off by hand when the image on the cylinder is to be replaced.

Packaging often involves seven-, eight- and nine-color work. Computerized, mechanical engravers produce all of the cylinders. Use of film and bromides is slowly being reduced.

Preparation

Mechanical Engraving. Most of the gravure cylinders (perhaps 75%) used for package printing are engraved mechanically with a machine that uses a diamond stylus to cut the individual cells into the gravure cylinder. The plated cylinder is mounted in a device that resembles a lathe, and the stylus cuts cells out of the cylinder as it rotates. The frequency of the stylus is constant (3600 to 4200 cells per second). The speed of rotation of the cylinder controls the number of dots per inch.

Unlike flexo, offset or chemical engraving, mechanically engraved screen angles cannot simply be rotated. However, by changing the horizontal and vertical spacing of the cells (see figure 7-14), the apparent screen angle can be changed. Depth, width and length of the cell can be controlled electronically.

Two companies make electronic engravers, Linotype-Hell and Ohio Electronic Engravers, Inc. The Hell machine is usually used for publication printing and the Ohio for packaging. Although there are many technical differences in the two machines, both are capable of serving both markets.

All the cells are diamond shaped, cut by a diamond stylus. The chisel edge creates a diamond shaped cell. As a chemical etch goes deeper, the cell wall becomes weakened. The electronic engraver generates stronger cell walls.

The scanner can read bromides or screened separations from the color scanner. It is neither possible nor desirable to engrave directly from the color copy, but the engraver can be driven from digital information generated by a color scanner.

Direct Digital Engraving. Computerized handling of the image throughout

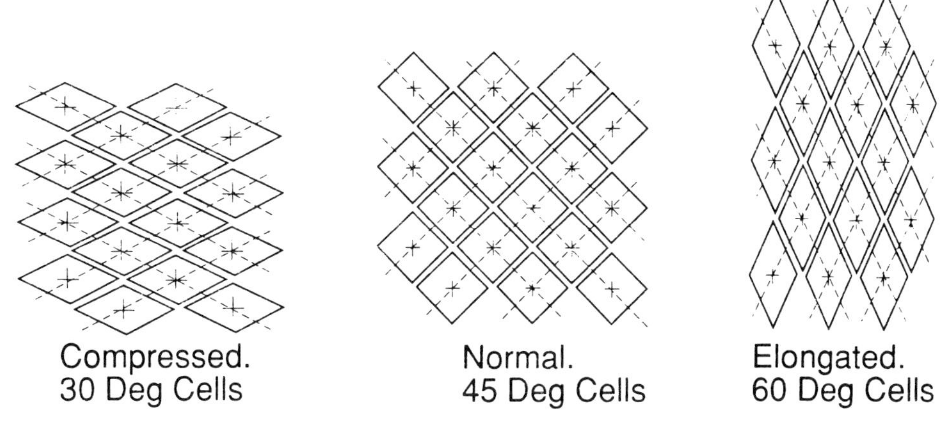

Figure 7-14. Screen angles created with electronic engraving. (Courtesy of Ohio Electronic Engravers, Inc. and Gravure Association of America.)

the process, including the digital engraving of the cylinder, is called direct digital engraving. All, or most, color separation films are eliminated by avoiding the steps of creating copying and rescanning films. This makes the process faster and reduces loss of image quality from repeated copying of the image.

Direct digital engraving speeds the preparatory process. An entire job can be saved on magnetic tape. It has become widely used as the technology has become commercially available.

Chemical Etching. The old "carbon tissue" method of chemical etching is no longer used in the U. S. and very little in the rest of the world. Chemically etched gravure cylinders are prepared by direct transfer, also called direct engraving. In direct transfer, a photographic resist is coated directly onto the cylinder and dried. It is exposed to UV radiation through a positive photographic film. A 150-line screen is usually used, but screens up to 250 lines per inch can be used. The unexposed resist material is rinsed away with water, and the cylinder is etched. The process is complete in a few minutes. If the photoresist is sensitive to laser light, it can be exposed directly from computerized output.

Direct transfer does not produce the long tone scale required for illustrations, and it used mostly for line work in package printing.

Electron Beam Engraving. It is also possible to use an electron beam to machine away the cylinder surface to produce dots. Still being developed, the process is expected to be very fast but also expensive.

Laser Beam Engraving. Techniques have been worked out that enable the engraving of gravure cylinders with laser beams. Copper cylinders with a uniform screen, filled with plastic, can be imaged with a computer-controlled laser beam.

Many variants have been developed. A plastic coated metal sleeve has been engraved on an electro-mechanical engraving machine for short run gravure and label and package printing. Cylinder cost is claimed to be comparable with flexo plates. Although the process is not in industrial use, it is available if and when ecological developments require replacement of copper.

Several other image carriers have been developed, but they find little use.

Wraparound Plates. These permit the assembly of a cylinder from several pieces and make it possible to replace part of a cylinder. The application enhances gravure economy on runs where part of the image must be changed during the run.

Chrome Plating. The steel base cylinder is plated with copper to a thickness of 0.02 to 0.04 inch (0.5 to 1.0 mm.). Copper provides engravability, stability in the press and reproducibility.

After engraving or etching, the imaged copper is chrome plated to make the surface harder and wear-resistant. Chrome has a low coefficient of friction, and it helps to lubricate the doctor blade, allowing the cylinder to print millions of impressions. The chrome film is about six microns (0.00023 inch) thick.

Cylinder Recovery

After the cylinder has served its purpose, it is dechromed with hydrochloric acid, which does not affect the copper underneath. Chemical etching of the copper cylinder, chrome plating and dechroming all create environmental problems that the gravure printer must cope with.

To recover the cylinder, the copper is removed with a lathe or, if the coating is a Ballard shell, it is removed mechanically.

If the job is to be rerun, the cylinder is stored. Storing and handling of large, heavy cylinders is expensive, but far less expensive than remaking the cylinder. It is possible to get the job up and running quickly if a satisfactory cylinder is on hand. If the job is not to be run again, the chrome and copper must be removed from the steel shaft. The entire process of plating with copper, etching and chrome plating is repeated to make a new gravure cylinder.

Gravure is usually economical only for long runs, single long runs or repeated short runs. Gravure is at a disadvantage when repeat orders require any change in layout, text or price information.

LITHOGRAPHIC PLATES

In lithography, the image is generated by the differential wetting of a flat surface. The image is wet with ink but not with water. Water or a "dampening solution" is applied to the plate, putting a thin film on water-receptive, non-image areas and leaving the image dry. When ink is applied to the dampened plate, the ink adheres only to the dry, image area. This method is capable of producing extremely fine resolution. Ordinary offset plates will easily hold a two percent halftone dot, and if they are operated properly, a 98 percent dot will remain open.

Screens for process color halftone work normally run 100 to 200 lines per inch (50 to 80 lines per cm), with 133–150 being the most popular. Special projects have been run with screens as fine as 300 and 500 lines per inch. In fact, it is possible to use the grain in the plate, itself, as a halftone screen, producing what is called "screenless lithography." While the ability of screenless lithography to produce fine detail is amazing, it is very difficult to control on press, and, for most purposes, screenless litho offers little advantage over a screen pattern of 200 lines per inch (80 lines per centimeter).

Offset or litho plates can be exposed through either negatives (negative-working plates) or through positives (positive-working plates) depending on the nature of the plate coating. There are many reasons for preferring either kind, but in the Western Hemisphere and many places in Europe and Asia, negative-working plates are preferred. Positive-working plates are preferred in many Western European countries.

On sheetfed presses, the paper can be cut to the size of the image itself, but with web offset presses, paper or board that remains unprinted at the gap is usually trimmed away and wasted.

Plates at one time were made by printers. The aluminum was grained and a photosensitive emulsion was applied. Printers now buy presensitized plates that are ready for exposure to the photographic film.

Offset plates are inexpensive and are easily and quickly prepared, accounting for their popularity in publication printing where paper is the principal substrate. The need for dampening makes them difficult to use in many packaging applications.

Structure

Lithographic plates are made of paper, plastic, brass, steel and, at one time, stone; but most plates used in commercial production of labels, folding cartons and other packaging materials are made of aluminum. Paper plates are often used for offset duplicators, but not for commercial packaging or labeling. Some plastic plates are used for commercial or publication printing.

The first step in manufacturing an aluminum offset plate is graining. It was once done by hitting the surface with metal balls, but it is now done either chemically or electrochemically. Graining roughens the surface of the plate, giving it a matte finish consisting of microscopic hills and valleys. The process increases the surface area between two and four times. It may be done mechanically with a brush and pumice, or electrochemically with acid under an electric current.

The plate is anodized next, which increases the hardness of the plate and increases its press life. The anodized aluminum surface has the hardness of hardened steel. It accepts water more readily than the untreated plates formerly used.

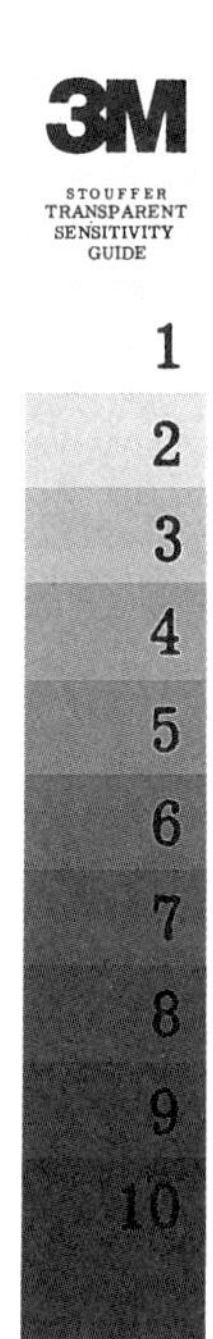

Figure 7-15. Stouffer transparent sensitivity guide. (Courtesy of 3M Co.)

After graining and anodizing, the photosensitive coating is applied, making the plate ready for the platemaker. Such plates are referred to as presensitized plates. Wipe-on plates, those coated by the printer, find little use in package printing.

Plastic plates are made on polyester film. Since the polyester is naturally ink-receptive and water-repellent, it must be coated with a water-receptive coating. The photosensitive layer is applied over this coating, exposed and developed to make the printing plate. If the water-receptive layer is damaged in press, the plates will scum or take ink in the damaged area.

Preparation

Presensitized offset plates are made pressready by exposing them to UV light through a photographic film and developing them with a chemical that will dissolve the unexposed part (in negative-working systems) or the exposed part (in positive-working systems), then hardening the exposed plate. Positive-working plates can be baked to give longer life. The best offset plates will run for hundreds of thousands impressions.

A plate exposure guide or sensitivity guide should be stripped into the film (figure 7-15) to help assure that the plates are properly exposed.

Offset plates are usually developed with an automatic platemaker that controls processing time and temperature, reducing the variability of the old hand-processing method.

To avoid the problems of disposing of solvents used for developing plates, plates that can be developed with aqueous solutions are now being made. Even some of these developers contain hazardous solvents, and attempts are continuing to find coatings that can be processed in pure water after exposing the image. These are referred to as "zero-solvent" systems.

Positive and Negative Plates. Offset plates are made by coating a photosensitive coating onto the base aluminum. They are exposed to UV light through a photographic film. Depending on the nature of the coating, either a negative or a positive film can be used. In negative-working plates, the film on the plate is hardened by light, generating the image area, With positive-working plates, exposure to light makes the exposed areas soluble in the developing solution.

Driographic Plates. The term "dry offset" applies to two different printing processes: to offset letterpress and to waterless lithography. Driography applies solely to waterless lithography. In waterless lithography, the background area of the plate is coated with an ink-repellent material such as silicone, so that the ink will not adhere, and no dampening solution is required. The process makes possible the use of extremely fine screens. Ink-water balance problems are avoided, and strong colors are easily printed. The process requires very careful temperature control because if the ink becomes too warm, it does not release cleanly from the background and "scumming" occurs. Although first developed in the United States, the process has been more popular in Japan.

Mounting Offset Plates

Offset plates are mounted by clamping them into the press cylinders while the press is stopped. This requires a gap in the cylinder that creates a non-printing area and wastes paper or substrate (figure 7-16). Unlike flexo plates, offset plates cannot be premounted because there is no fast and simple way to change plate cylinders in an offset press.

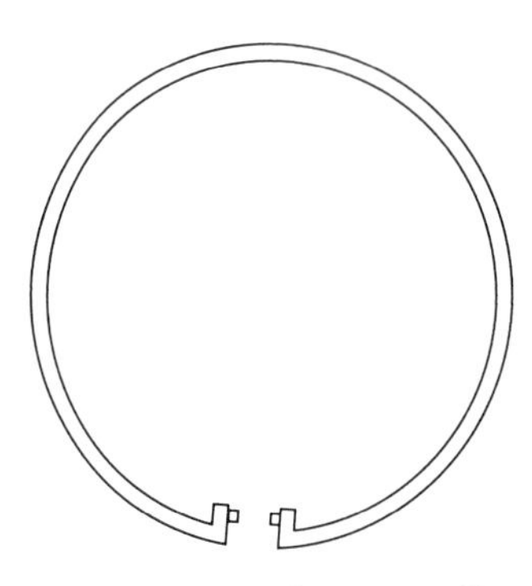

Figure 7-16. Gap in offset plate or blanket cylinder in which the plate or blanket is mounted.

LETTERPRESS PLATES

Letterpress plates find their most important packaging applications in narrow-web labels, printing of two-piece cans and printing of corrugated displays. Letterpress plates print labels and displays directly, but they print metal cans from an offset blanket.

Although letterpress and flexo both print from raised images or relief plates, they are very different processes. They use different kinds of inks, different kinds of inking systems, and letterpress plates are much harder than flexo plates. Letterpress plates are made of photopolymer or metal.

Photosensitive plastic plates used for the production of letterpress relief plates are called wraparound plates. They are made in one piece to be wrapped around the plate cylinder, helping to reduce the time for makeready. They range from 0.017 to 0.030 inch (0.43 to 0.76 mm) in thickness so that they can be bent to fit into the cylinder clamps. Flexo plates are usually attached with stickyback.

Liquid photopolymers, as in flexo platemaking, are used for letterpress. The process is the same as for flexo, but the resulting plate is harder (about 40–60 Shore D) due to the requirements of the printing process.

Metal letterpress plates are still used in dry offset (offset letterpress) printing of two-piece cans, where they print against a blanket that prints the can. The conventional metal has been zinc, but a chrome-plated magnesium plate with superior performance at high speeds is becoming popular. It prints fine-line details accurately together with large smooth solids.

SCREEN PRINTING

There are three groups of screen stencils: hand-cut stencils, tusche-and-glue stencils and photographic stencils. Handcut stencils are prepared by cutting the printing image area from a base or support material. Tusche-and-glue is an art process involving drawing the art directly on the screen fabric with a lithographic tusche (an oil-based pigment) and then blocking out the non-image areas with a water-based glue material.

The introduction of photographic screen printing allowed the screen printer to enter package printing markets. A color image can be screen printed with any ink on any surface shape (flat, cylindrical or irregular).

Package printing by screen is mostly limited to photographic imaging in which a photosensitive material is applied to the screen, exposed to the image and then developed to permit ink to pass through the image areas.

Structure

Monofilament fabrics, such as single strand polyester, nylon or wire cloth (copper or stainless steel) have a uniform weave and allow ink pigments to pass readily through the mesh openings. It is generally necessary to treat or roughen the filament surface to obtain good stencil adhesion. Each fabric type has its own characteristics. Nylon tends to absorb moisture and will react to changes in room humidity. Metal screens absorb no moisture, and they pass most abrasive pigments with little difficulty, but they react to temperature changes by growing or shrinking. Monofilament polyesters have low moisture-absorption, good stability and strength. They are also less expensive than most other screen materials so they are rapidly becoming the main material for commercial work.

Selection of screen material depends on the requirements of the job, especially the ink and the fineness of the line detail of the image to be printed. A coarse screen mesh will pass a heavy layer of ink, but it will not hold a fine line stencil.

Different fabrics stretch different amounts, and proper stretching of the fabric is essential in controlling the printed image. A stretch frame is used in most industrial screen printing operations to increase the uniformity of the stretch (see Chapter 8).

Preparation

There are three photographic stencil processes: indirect or transfer image method, direct image method, and film emulsion or direct/indirect image method.

For the indirect process, a dry photosensitive emulsion on a plastic support sheet (available in rolls or sheets) is exposed through a transparent right-reading positive and then developed. The exposed areas (the non-image areas) are hardened, and the image areas are washed away with a warm-water spray to form the image or printing areas. The stencil is adhered to a clean screen while it is still wet, and the support sheet is removed after the stencil dries.

The direct process uses a wet emulsion that is coated directly onto a clean screen. The emulsion is exposed through a transparent positive to harden the non-image areas. The image areas are washed away with warm water. When the emulsion is dry, the stencil is ready to print.

For the direct/indirect process, an unsensitized film is attached to the stencil side of the screen. The two-part stencil emulsion is mixed to make a light-sensitive material that is coated through the screen to the film support. When the emulsion is dry, the backing sheet is removed, and the film supported in the screen is exposed and developed. The chief advantage of the process is the uniformity of the emulsion thickness.

The diazo sensitizers and ferric salt sensitizers used in preparing stencils should be handled with appropriate care, but they are not highly toxic and can normally be discharged to the sewer.

Rotary screens are prepared by plating a coating electrolytically onto a steel cylinder, then removing the cylinder and applying a photochemical coating to the metal coating. The photochemical coating is exposed through a photographic film and the image areas are etched out to form pores through which the ink flows.

A step-wedge exposure control is used, much like those used in other processes.

The ink film thickness applied to the substrate depends on the thickness of the screen and the shape and pressure on the squeegee.

PLATELESS PRINTING

Ink-Jet Printing

The computer is a binary device (it prints "on/off" or "1/0"), and ink jet printers are also binary. This analogy with the computer has some long-term advantages for ink-jet printing and for laser printing, which also operates on binary principles. The two types of ink jet printers are continuous jet and impulse or drop-on-demand. The continuous-jet printer charges the droplets digitally, and they print or are deflected by an electric field. In impulse or drop-on-demand printing, an ink droplet is generated by an electrical impulse when called for by the computer.

In continuous jet printers, a continuous jet is broken into drops, all of the same size, and each droplet is given an identical charge. They are then deflected according to the electric charge. One type of continuous jet printer deflects the droplets by varying amounts to print a matrix; the other directs as many as 120 jets of ink per inch in a continuous stream of droplets. Droplets that are not to print are deflected into a trough for recovery (see figures in Chapter 3).

Drop-on-demand printers use one of several different methods of generating a droplet when signaled by the computer. They are capable of generating a very fine pattern.

Electrostatic Printers

These can be placed into two groups: xerography and laser printers. Both processes form an image on a semi-conductor cylinder that is electrostatically charged. A semiconductor retains its charge in the dark but loses it when illuminated.

Xerography. In xerography, the image is often an item to be copied, or it may be a paste-up of artwork. The charge on the semiconductor drum is dissipated by light reflected onto the drum from the non-image area. The drum is then brushed with a toner that is attracted electrostatically to those areas retaining a charge. The toner is transferred to a sheet of paper. When it is heated, the toner fuses to form an image on the sheet.

Laser Printers. Like ink-jet printers, laser printers, guided by the computer, generate their images from digital information. The laser beam falls on a semiconductor drum that carries a static charge. Where the light hits, the charge is dissipated. After exposure, the remaining electrostatic image is developed with a toner. The toner is then transferred to a paper and fused.

FURTHER READING

Chemistry for the Graphic Arts, Second Edition. N. R. Eldred. Graphic Arts Technical Foundation, Pittsburgh, PA. (1992).

Gravure: Process and Technology. E. Kendall Gillett, III, Chairman. Gravure Association of America, Rochester, NY. (1991).

Flexography: Principles and Practices, Fourth Edition. Frank N. Siconolfi, Chairman. Flexographic Technical Association, Ronkonkoma, NY. (1991).

Printing Technology, Third Edition. J. Michael Adams, David D. Faux, and Lloyd J. Rieber. Delmar Publishers Inc., Albany, NY. (1988).

Chapter

Printing Presses and Auxiliary Equipment

CONTENTS

LIST OF ILLUSTRATIONS

LIST OF TABLES

Chapter

Printing Presses and Auxiliary Equipment

INTRODUCTION

In every printing job it is important to control the artwork, the ink, the substrate and the press. This chapter centers on the press, but we must never overlook the importance of other factors.

The most common way of distinguishing printing presses is by the type of plate that is used: relief plates for flexo and letterpress, intaglio or engraved cylinders for gravure, planographic plates for offset and stencils for screen printing.

In general, each process has a unique inking system although letterpress and offset use very similar systems. Keyless offset uses an anilox roll to meter the ink to an offset press. The process is limited to publication printing, so far.

Another way of distinguishing printing processes is the system of feeding paper or other substrate: sheetfed or web. Sheetfed systems can be used for any printing process, but in practice they are largely limited to offset and screen printing although some corrugated boxes are printed by sheetfed flexo and letterpress. Web has largely displaced sheetfed printing in other processes, especially in package printing.

A major reason for replacing sheetfed printing with web is the ease of placing converting operations in line with the press. With the exception of flexo folder-gluers in the corrugated boxmaking industry, sheetfed presses seldom have converting in line. The flexo folder-gluer, however, is primarily a converting machine to which printing capabilities have been added. In-line converting operations are of paramount consideration—creasing, slotting, diecutting, jointing, counting, stacking and bundling.

Presses may also be distinguished by the configuration of their printing units. All processes use in-line presses, but flexo presses are often common (or central) impression (CI) presses or stack presses. Gravure presses are largely limited to in-line configurations owing to the great weight of the cylinders. Offset presses may be one-side-only or perfecting presses (presses that print both sides at the same time). Screen presses may be flatbed, flatbed-cylinder and rotary styles. Except for some platen presses and narrow web label presses, flatbed letterpresses are mostly a matter of history.

The terms "unit" and "station" are largely interchangeable. Actually, a printing unit is modular, containing all printing functions. A press station is found on a CI press and does not carry out all of the printing functions.

Press size refers to the width of the press. A 54-inch press is 54 inches wide; however, this may refer to the maximum width of the substrate that can be printed or to the maximum width of the printed image. This difference may amount to as much as one inch.

Presses may print direct or offset. Lithographic presses are almost always printed with an offset blanket, accounting for the fact that "offset" has become synonymous with "lithography." However, the lithographic plate can print directly onto the substrate, a configuration limited mostly to newspaper printing. Offset letterpress ("dry offset" or "Letterset") is commonly used for two-piece cans as well as other packaging materials. Offset gravure is sometimes used to compensate for rough substrates that are poorly printed by direct gravure. Because the soft plate helps flexo print on a surface that is not perfectly smooth or flat, an offset blanket is not useful with a flexo press.

Combinations of press types occur frequently. Flexo units are often found on gravure presses to print price information or other information that may be changed during a long run. Anilox systems are used to meter the flow of varnish on offset presses. A gravure cylinder sometimes replaces the anilox roll on a flexo press, and the flexo plate is replaced with a smooth blanket in a process called flexo-gravure. Ink-jet numbering is often used on cans and bottles printed flexo or offset letterpress.

The type of printing process makes little difference as to whether a web press feeds roll-to-roll or delivers to a sheeter, but it involves the converting equipment at the end of the press. Sheetfed presses are seldom equipped with in-line converting capabilities following the printing. Converting and finishing are therefore discussed independently of the types of printing presses in this chapter.

It is impossible to cover all the many types of presses in a brief chapter. This chapter is directed primarily to the types of presses most important in package printing, omitting specialty presses such as flexographic cup printing/converting presses and envelope presses that do not directly apply to packaging. Neither is it possible to cover every printing process. Pad printing, collotype, DiLitho, heat transfer printing and other processes little used in packaging are either omitted or mentioned only briefly.

Several trends are apparent: the trends to greater speed, quality and automation are universal, but three trends apply specifically to presses—the trend to narrower presses, the trend to web operations and the trend to more printing units or stations—that is, more color. Increased use of color and quality management affects the way presses are operated. One obvious change is the growing use of fingerprinting (discussed in Chapter 6) to determine and control press operating characteristics.

FLEXOGRAPHIC PRESSES

Characteristics of The Flexo Press

Flexo is a simple and versatile process which accounts for its popularity in package printing. Flexo has the greatest variety of press configurations: central impression (CI), stack, in-line, belt and combinations of these. Widths also have the greatest variation, ranging from about two inches for narrow web tags and labels up to 200+ inches for sheet corrugated. The simplicity of the press makes it easy to change the cylinder diameter, thereby changing cutoff length. The ability to print a variety of sizes is a major consideration in package printing.

The soft plate, which is often mounted on a compressible backing, helps flexo conform to rough surfaces such as uncoated corrugated, but it also creates squeeze-out or halo effects around letters and images if the pressure is not closely controlled. To produce high quality printing, the mounted plate must contact the substrate with a "kiss" impression—a very light impression.

If printing pressure is insufficient, ink does not transfer properly; excess pressure produces halos and squeeze-out. The continuing trend to thin photopolymer plates and compressible backing help to reduce halos and dot gain, but careful impression control is still required for top-quality flexo.

The anilox roll turns in a bath that delivers a uniform ink film to the entire plate, and ghosting and ink starvation do not occur as they may in litho or

TABLE 8-I. TYPICAL MARKETS FOR FLEXO PRESSES AND TYPES OF PRESSES COMMONLY USED

Labels: Narrow web in-line, stack and CI
Flexible Packaging: Wide web CI, stack and in-line
Folding Cartons: Narrow and wide web in-line or stack
Sanitary Food Containers: Wide web in-line
Beverage Containers: Wide web in-line or CI
Corrugated Liners: Wide web CI-stack combinations
Paper Sacks: Wide web in-line, stack or CI
Corrugated Boxes: Sheet-fed printer slotters
Fiber Cans and Tubes: Narrow and wide web in-line or CI
Laminations: Wide web CI

letterpress where only part of the ink is replaced on each revolution. (In litho and letterpress, the plate is inked by a roller train, and heavy images leave a ghost pattern.) Flexo, accordingly, produces prints with good color uniformity. It is not possible to change the amount of ink to meet requirements of heavy or light images on the plate, and illustrations requiring both are usually improved if one station is used for halftones and another for heavy solids.

The anilox roll can produce a moire pattern with a halftone image if it is not properly etched, or if the screen angles of the tints and halftones don't accommodate the screen angle of the anilox roll. Other inking problems arise if the ink drying rate is not correct or if the ink dries on the anilox roll.

In flexo inking configurations where the anilox roll runs in the pan, and ink is metered with a reverse angle doctor blade. This configuration permits use of inks with higher viscosity and greater color strength. Use of the chambered doctor blade system dispenses with the ink pan entirely and uses the lower blade of the system as a reservoir of ink.

Its versatility makes flexo suitable for most packaging markets. Many of these are listed in table 8-I.

The Printing System

The purpose of any inking system is to apply a precisely controlled film of ink to the plate. The flexo press prints from a soft or flexible relief image. The image is inked with an engraved roll, called the anilox roll that turns in an ink bath or is inked by a fountain roll that turns in the ink bath. Excess ink is removed by a metering roll or with a doctor blade. The flexo process is described in Chapter 3.

Inking Configurations. There are at least six inking configurations in use on flexo presses used for packaging. These are illustrated in figure 8-1. Figure 8-1(A) is probably the oldest configuration, comprising a rubber fountain roll in the ink pan. that applies ink to the anilox roll. A metering roll (sometimes called a "rubber" roll) levels the ink on the anilox roll mostly by pressure. The metering roll may turn at the same speed or somewhat slower than the anilox roll. When it turns slower, it provides a wiping action as well.

Figure 8-1(B) is a modification of 1(A), using a doctor blade for metering the ink. Figure 8-1(C) is a combination of figures 8-1(A) and 1(B). Figure 8-1(D) shows a configuration in which the anilox roll turns in the ink fountain, and a metering roll levels the ink. Figure 8-1(E) is similar but uses a doctor blade to level the ink.

Figure 8-1(F) is the most recent configuration introduced around 1988. It uses a chambered doctor blade system in which the bottom blade presses very lightly against the anilox roll, but the top blade does the actual metering. The ink is contained in a well formed by the bottom blade. Because the anilox roll rotates in a motion that picks up ink in the cells, virtually no ink slips down. It is necessary to have end seals so the ink does not leak out around the ends of the blades and the anilox.

Anilox Roll. The anilox roll is an engraved roll of ceramic or chrome-

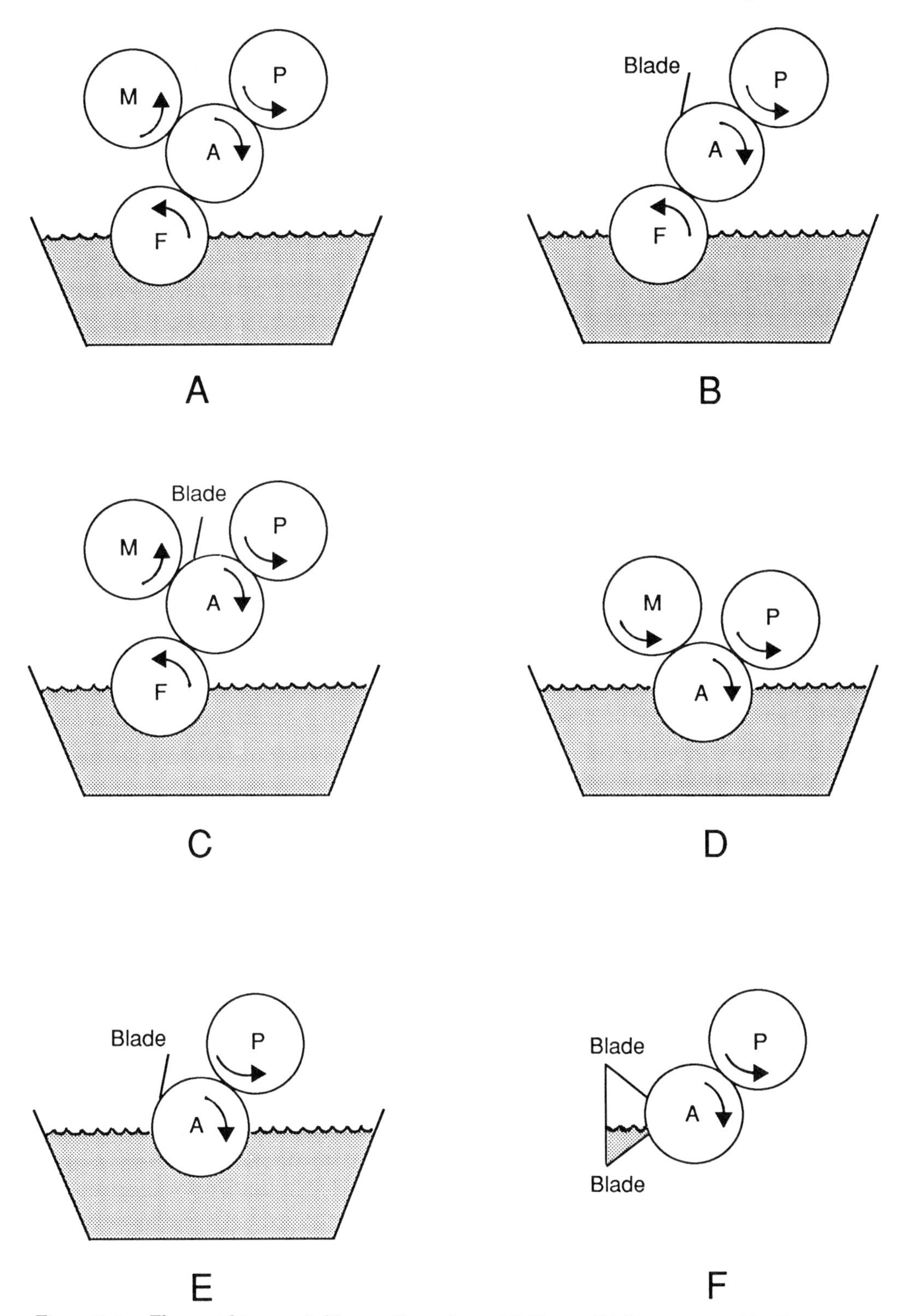

Figure 8-1. Flexographic press inking configurations. (A) Four-roll inking system. (B) Three-roll inking system with doctor blade. (C) Four-roll inking system with doctor blade. (D) Three-roll inking system. (E) Two-roll inking system with doctor blade. (F) Two-roll inking system with chambered doctor blade.

plated steel that carries a precise amount of ink. The ink on the surface of the anilox roll is wiped away with a doctor blade or with a metering roll. The engraved anilox roll is characteristic of the flexo press. The roll is engraved in a variety of ways. It may be a chrome-plated mechanically engraved steel, ceramic coated mechanically engraved steel, a smooth or textured ceramic or a laser-engraved ceramic roll.

Wear of the anilox roll has been a major cause of color variation over long flexo runs. Wear makes it difficult to match colors on repeat orders. A small amount of wear generates a large loss of ink carrying capacity (figure 8-2).

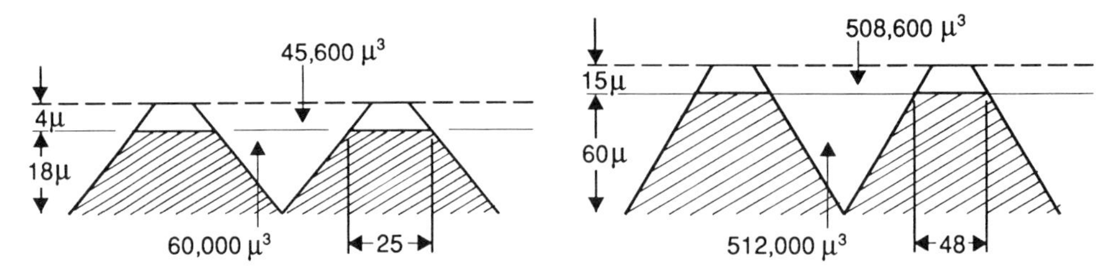

Figure 8-2. Effect of wear on ink-carrying capability of an anilox roll. The left illustration shows a 200-pyramid cell structure with 20-percent wear. Its volumetric capacity has been reduced by 43 percent. In the right illustration, a 120-pyramid structure loses 51 percent of its ink-carrying capacity with 20 percent wear. (Source: Flexography: Principles and Practices, Fourth Edition.)

Laser engraving of ceramic-coated rolls has lengthened roll life and reduced cell volumes while increasing cell count, thereby enabling an improvement in print quality. These laser-engraved ceramic rolls are becoming very popular in spite of the fact that they are priced higher than chrome-plated steel anilox rolls. They last considerably longer with minimum wear and are less subject to impact damage. Figure 8-3 shows the patterns of engraving: Figure 8-3(A) is engraved at a 30° angle, figure 8-3(B) at 45° and figure 8-3(C) at 60°. Laser engraving gives a deep, cylindrical shape to the cell (figure 8-4).

It is generally recommended that the number of cells per inch in the anilox roll be 3.5 to 4.5 times greater than the halftone screen to print halftones without moire patterns. This means that to print a 150-line-per-inch halftone requires an anilox roll with 525 to 675 lines per inch. There is a trend toward higher ratios or finer anilox engraving. Laser engraving of ceramic rolls can produce 1,000 cells per inch. This is a significant development, because with the use of high viscosity inks, these rolls print with low dot gain, giving better control in the reproduction of process color. Ratios of 4:1, 4½:1 and even 5:1 are coming into use as printers find that the higher ratios yield more even ink coverage and finer precision. The higher ratios give more even coverage with less ink and produce higher quality printing than is possible with traditional ratios.

By varying the depth of each cell, laser engravings can be made to provide variations of three- or four-fold in the capacity of the anilox roll to carry ink with a given screen count. This amount of variation cannot be made with mechanically engraved rolls.

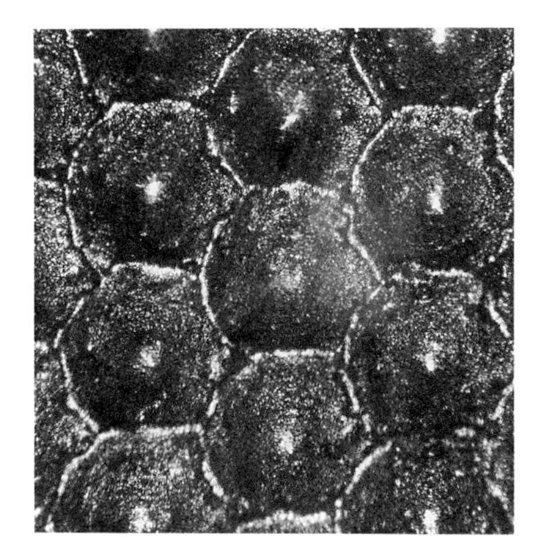

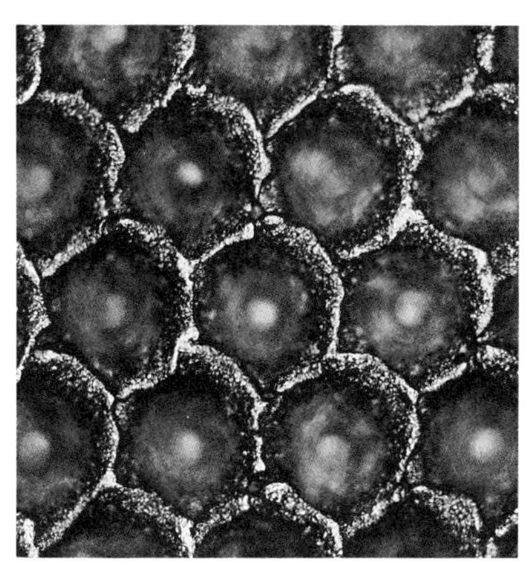

Figure 8-3. Photomicrographs of laser-engraved ceramic anilox roll engraving patterns, all at 200 X magnification. Top left: A 30-degree angle engraving with a screen count of 180 cells per inch and cell depth of 21 microns carries approximately 5.5 billion cubic microns (bcm) per square inch. Top right:. A 45-degree angle screen with a screen count of 200 and cell depth of 13 microns carries approximately 4.4 bcm. Bottom. A 60-degree angle screen with a count of 250 and cell depth of 32 microns carries approximately 7.0 bcm. (Photos courtesy of Praxair Surface Technologies, Inc.)

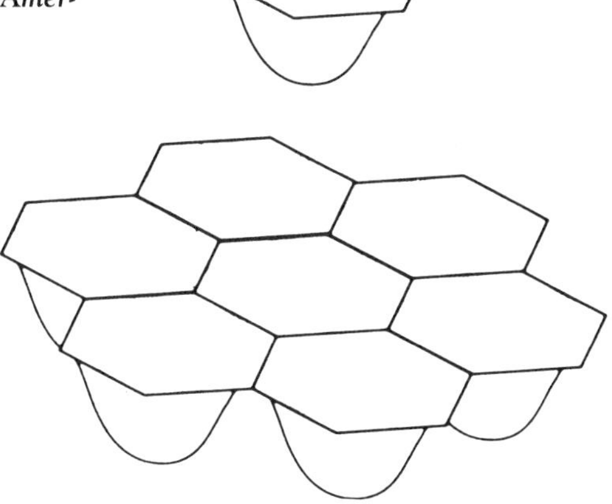

Figure 8-4. Shape of typical laser-engraved ceramic anilox cells. This is a 60-degree pattern. (Illustration courtesy of Harper Corporation of America).

Not only the depth, but the relationship of the screen angle of the engraving to the screen angle of the halftone is critical. The job specification is related to the press used on a job. All package printers, designers and purchasers should know the importance of this relationship. Improper screen angles not only cause moire, but they cause slurring and "dot bridging" (fill in) owing to improper inking in some spots.

Selecting an anilox roll that will achieve the best printing, the longest wear and the lowest over-all cost grows more complex as the number of screens and engraving methods increases.

Ink Feed. Ink feed to the plate is controlled by the cell volume of the anilox roll but also by the uniformity with which the surface of the roll is wiped. The anilox roll must be round, concentric and properly balanced; the greater the speed of the press, the less eccentricity or TIR can be tolerated. (TIR or total indicated runout is a measure of the eccentricity and deviation from roundness.)

A rubber-covered metering roll is a metal shaft with a rubber or elastomeric coating of neoprene, nitrile or Buna N, ethylene propylene or polyurethane elastomer. Natural rubber is sometimes used, but its lack of resistance to oil, ozone and heat have limited its use.

Although wear of the anilox roll and variations in ink viscosity affect the color of printed ink films, there is little the press crew can do to make the print match the proof if the ink is not properly selected and the press is in poor condition. However, it is a simple matter to keep color uniform during the run.

Doctor Blade. The reverse angle doctor blade "shaves," or "doctors," ink from the anilox roll that turns in the ink fountain. This arrangement permits use of an ink with higher viscosity and higher color strength than does the metering roll arrangement, improving the quality of color produced by flexo.

Doctor blades are made of Swedish blue steel, stainless steel, plastic or composites such as plastic, fiber and carbon. Most blades come prehoned or presharpened, but the press operator should check the blade carefully before installation (figure 8-5).

The change to waterbase inks and laser engraved ceramic anilox rolls has required changes in doctor blade technology. Ink and doctor blades, even plastic blades, wear out chrome anilox rolls. The future appears to belong to the ceramic anilox roll in many—but not all—applications. On the negative side, ceramic rolls wear out steel doctor blades more quickly, and efforts have been made to find plastic blades that will give the performance quality of steel blades or steel blades that will give the wear life of plastic blades.

The chambered doctor blade (figure 8-6) applies and meters the ink while eliminating evaporation of solvent from the ink fountain. The chambered doctor blade unit reduces the variation in ink density when printing at high speeds. The ink starvation once seen at high speeds is minimized. Reducing evaporation reduces the need for frequent additions of solvents to the ink and sometimes has been sufficient to bring printing plants into compliance with environmental regulations.

Quality and JIT demand repeatability. The reverse-angle, enclosed chambered doctor blade system and the laser-engraved ceramic anilox roll improve repeatability, as do thinner plates and improved cushioning.

Ink Fountain. Sometimes called an ink trough, this is the pan that holds the ink. For waterbase inks, which are now preferred for most flexo printing on paper and board, ink fountain components are made of stainless steel.

Types of Flexographic Presses and Their Markets

There is a great variety of flexo press configurations, and the variety is further

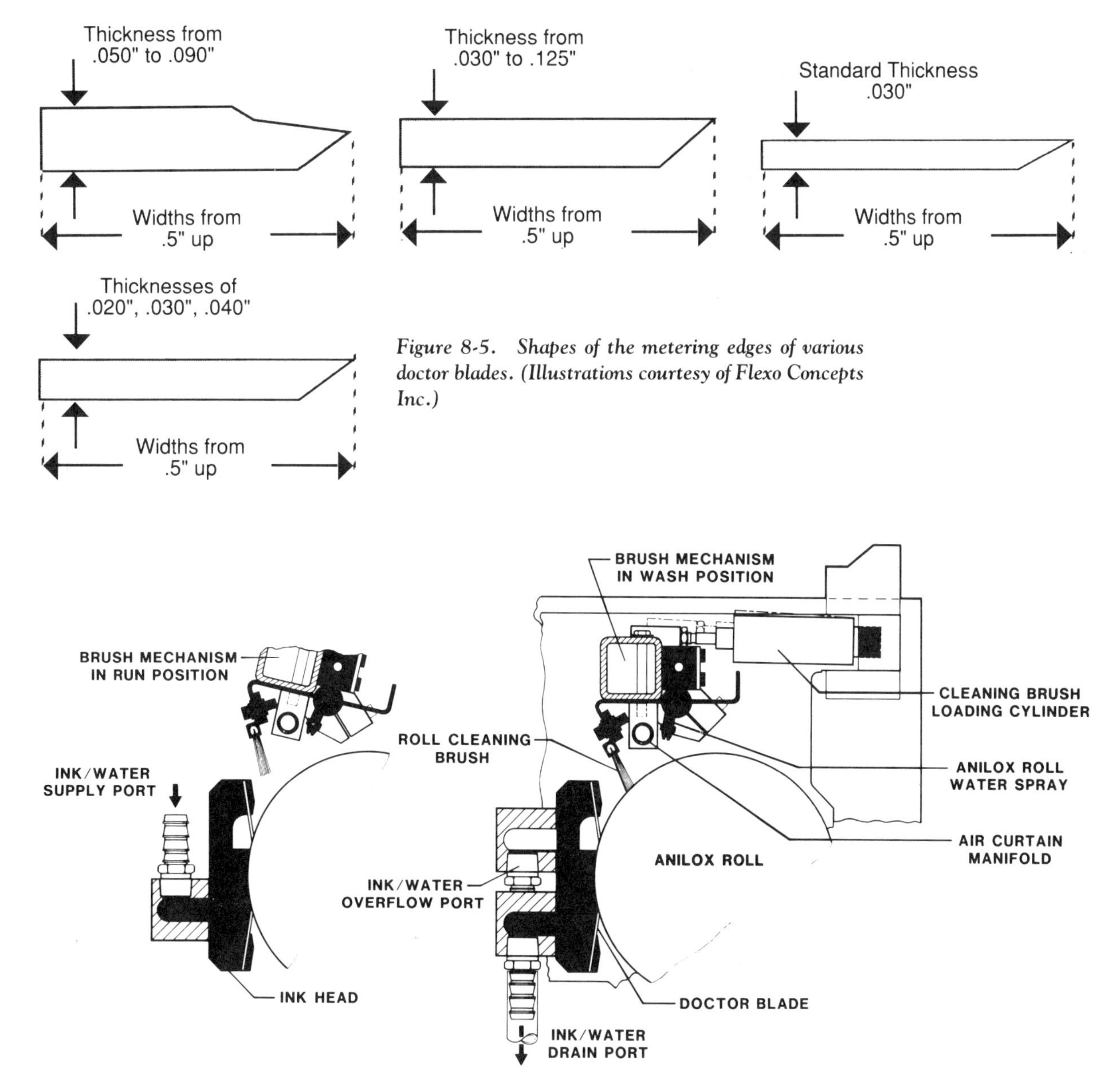

Figure 8-5. Shapes of the metering edges of various doctor blades. (Illustrations courtesy of Flexo Concepts Inc.)

Figure 8-6. Diagram of a chambered doctor blade ink metering system. This system is used on a corrugated flexo folder-gluer. (Illustration courtesy of the Langston Corp).

complicated by configurations that use combinations of press styles. The major types are mentioned in this section.

There has been a trend toward an increasing number of print stations to print more colors. The average flexographic label press can now print five to six colors, with an increasing number of eight to nine unit machines being sold. There are presses with as many as 14 (or more) print stations in line, plus converting.

Flexo presses are usually divided into two groups; narrow web, running up to 20 inches, and wide web presses that are 24 to 54 inches. Superwide presses, 87 inches or wider, are used for preprinting linerboard for corrugated, and some special presses that are even wider are used for printed paper products such as shipping sacks and grocery bags. In Japan and Europe, these wide webs are frequently printed by gravure. In the United States, flexo is commonly used.

Flexo presses for packaging materials seldom exceed speeds of 1500 feet per minute, and at these speeds they are usually limited to one- or two-color line work. Flexo newspaper presses now print 100 lines-per-inch halftones with four-color process work at 2,000 feet per minute, and manufacturers are looking forward to 3,000 feet per minute. There seems to be no reason that packaging presses cannot be made to match this performance, and increasing press speeds have characterized the industry for decades.

Narrow web presses usually have faster changeover. Printing speed is less critical than makeready time for short-run work. For short-run work, the press requires features that will enable the operator to change over quickly. New lightweight plates and plate cylinders enable one operator to change plate cylinders without a hoist.

Wide Web Presses. These presses are used to print flexible packaging, beverage carriers, sanitary food containers, paper sacks and grocery bags which make up the bulk of flexo production. They print four, six , eight or more colors. Wide-web presses show their greatest advantage in high productivity, especially on long runs.

Wide web presses are commonly of the CI or stack configuration (figures 8-7 & 8-8).

Narrow Web Presses There is a major shift toward narrower presses which require lower manning levels, are made ready faster and run faster than the wider presses. Higher speeds are possible because there is less print variation across the web than on wider presses. The print quality obtained on narrow web presses allows flexo to capture some of the web offset package printing market.

The package can often be printed one-up, eliminating the need to make duplicate plates and to separate the diecut blanks at the delivery end. The increased speed and lower crew requirements more than make up for the increased number of impressions required to complete the run. Narrower presses can use in-line rotary diecutting more easily than the wider presses, where diecutting often has to be performed off line to obtain the required quality of cutting. Wide web press manufacturers are, in many instances, developing narrower presses, e.g. 24 to 28 inches wide, as an adjunct to their product lines.

Narrow web press manufacturers have changed their designs to handle folding

Figure 8-7. Wide web central impression (CI) flexo press for printing flexible packaging. (Photo courtesy of Carint S. r. l.)

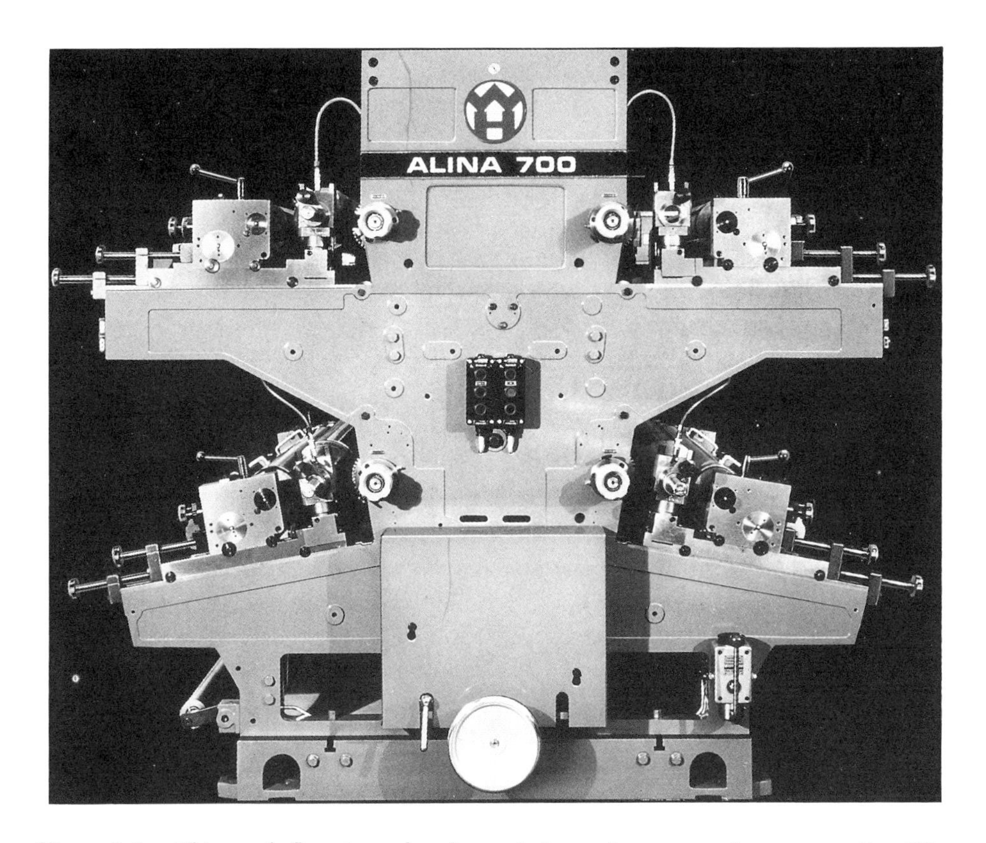

Figure 8-8. This stack flexo press has four printing units, arranged two on a side. (Photo courtesy of Windmoeller & Hoelscher.)

Figure 8-9. Narrow-web in-line flexo press. This machine has eight print stations. (Photo courtesy of Flexo Accessories Co., Inc.).

cartons. Narrow in-line presses, with downstream converting, are taking business from the wider gravure presses that had been sold to the folding carton industry only a few years ago. Narrow web flexo presses are also making inroads into the web offset folding carton market. Most narrow web presses have the in-line configuration (figure 8-9).

Narrow web presses have less ink circulating through the system and waste less on each job and during washup and cleanup.

Labels, notably pressure sensitive labels, are the most common product produced on the narrow web press. Types of flexo presses most commonly used for various products are listed in table 8-I.

In-line Presses. In-line flexo presses (figure 8-10) are designed from modular units that are lined up horizontally in a row. This configuration is the only one usually used for gravure and web offset. The number of colors is rapidly growing. Where one- and two-color presses used to dominate, six, eight and 10 color presses are becoming common. They are used to print multicolor packages and often apply varnish.

Color units on in-line presses, once driven by a common lineshaft, now use electronic line shafts; that is, all stations are individually controlled from a central computer console. In-line presses can be manufactured with whatever number of colors the package printer needs. This type of press can be designed to handle extremely wide web widths since a single frame need not support all colors. Wide-web, in-line flexo presses are commonly found in folding box and multiwall bag plants. Sheetfed corrugated presses have in-line configurations.

Narrow-web in-line flexo presses offer the advantage of quick setup and accessibility for printing pressure sensitive and standard label stocks and for short-run printing of specialized product lines.

The narrow-web press can be equipped to turn the web over and print on both sides, as is sometimes done on labels for glass bottles. It is also used to provide coating or all-over coloring on paper.

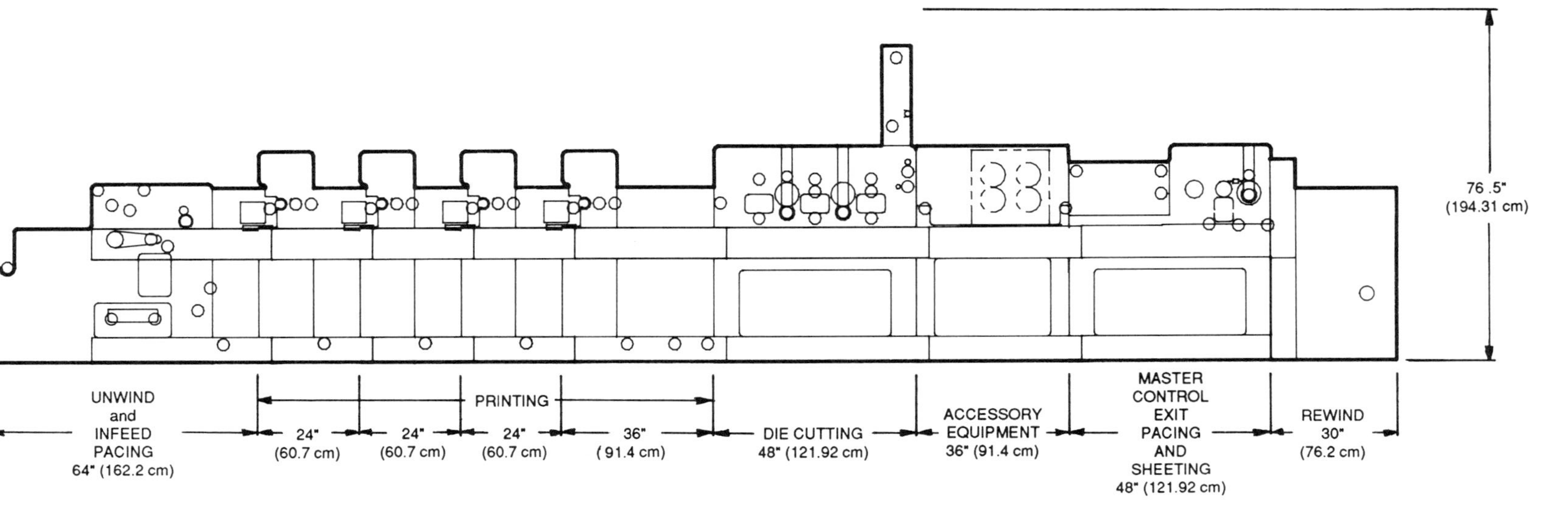

Figure 8-10. Schematic diagram of a narrow web in-line flexo press. Only the basic components are shown. (Illustration courtesy of Mark Andy, Inc.).

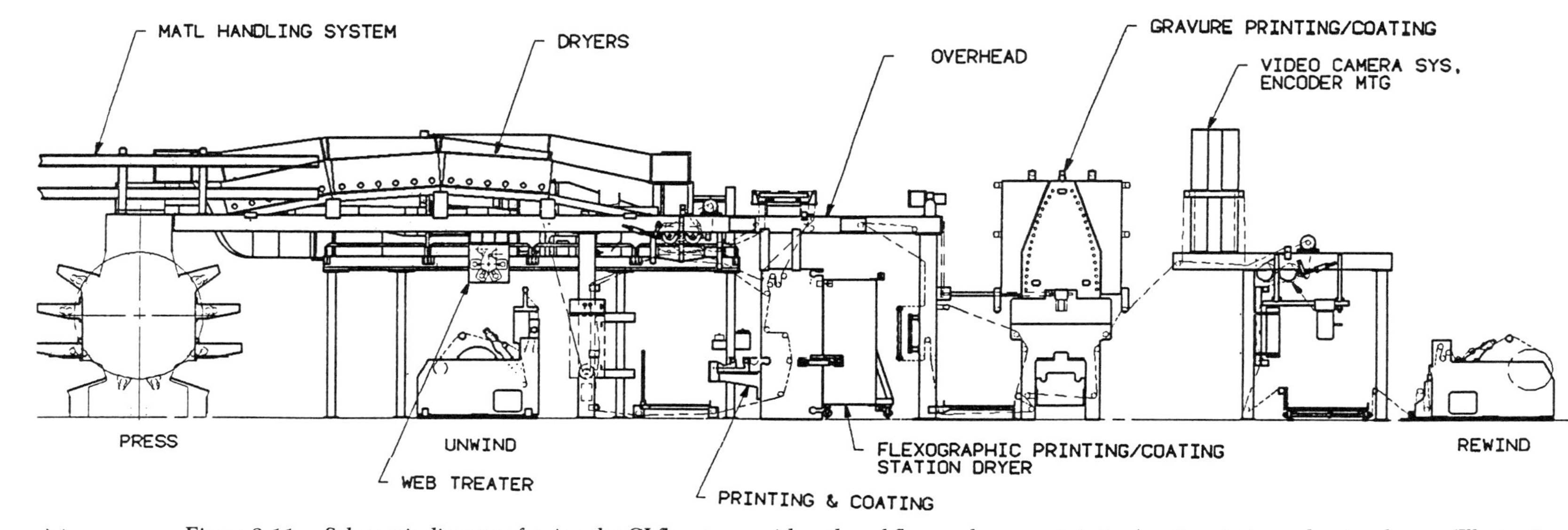

Figure 8-11. Schematic diagram of a six-color CI flexo press with outboard flexo and gravure printing/coating stations, showing dryers. (Illustration courtesy of Paper Converting Machine Co.).

Common Impression Presses. The CI press, sometimes called a drum or central impression press, supports the color stations around a single steel impression cylinder mounted on the main press frame. The substrate is supported by the cylinder, and color register is greatly improved, especially with materials that stretch easily, as with thin plastic films. With in-line presses and stack presses, substrates tend to stretch as they go from one printing unit to the next.

The CI press has two to eight print stations or decks. The most common CI press in use today has six colors (figure 8-11). The substrate is held in place against the impression cylinder, and the color is applied from printing stations, or "decks," around the cylinder.

This ability to hold excellent register has made the CI press the mainstay of the flexible packaging industry. Use of the CI press is promoted by the steady growth in demand for complicated graphic designs and for process printing.

The diameter of the impression cylinder varies with the number of colors, the speed and press engineering. Four-color presses generally used to be 30 or 36 inches in diameter, but impression cylinders up to 60 inches can be run at higher speeds. Six-color CI presses used are found with 83-inch-diameter cylinders and eight-color CI presses with 94-inch-diameter cylinders. However, improved press design has reduced the distance required for drying between colors, so that 60-inch, six-color CI presses are common.

Modern CI features to increase the productive time of the press include: automatic deck positioning systems varying from fully automatic to partially automatic; preregister systems that vary in degree of automation, accuracy and cost; computerized fault-finding systems; robotic or semi-automatic plate cylinder removal systems and robotic reel placement and removal systems.

Stack Presses. Stack presses are used for flexo packaging and sometimes for offset publication printing. The printing units are stacked one above the other (figure 8-12). Each color unit (or "deck" or "section") is driven through gear trains supported by the main press frame. One to four individual print units are mounted on each side of a vertical frame, producing two to eight colors in a single pass. The six-color press is most common.

The structure of the stack press gives accessibility to the color units, facilitating changeover and washup. The structure also accommodates large plate cylinders that make it possible to print large repeats.

It is relatively simple to print two sides. Various web threading arrangements of the stack press allow complete ink drying before turning the web to print the reverse side.

Although the stack press has been used to print every type of substrate, it is not well suited for very thin gauge materials or ones that stretch easily. The stack press is generally restricted to color registrations that require accuracy no closer than ± 1/32 inch (one mm.)

Combination Presses. Combination presses embody CI units and various stacks both upstream and downstream from the CI unit, notably with preprinted linerboard presses, but also with flexible packaging presses as well. Such presses (figure 8-13) often have separate stacks for coating, whether prior to printing or afterwards. The upstream stack sometimes provides a coating or coloring while

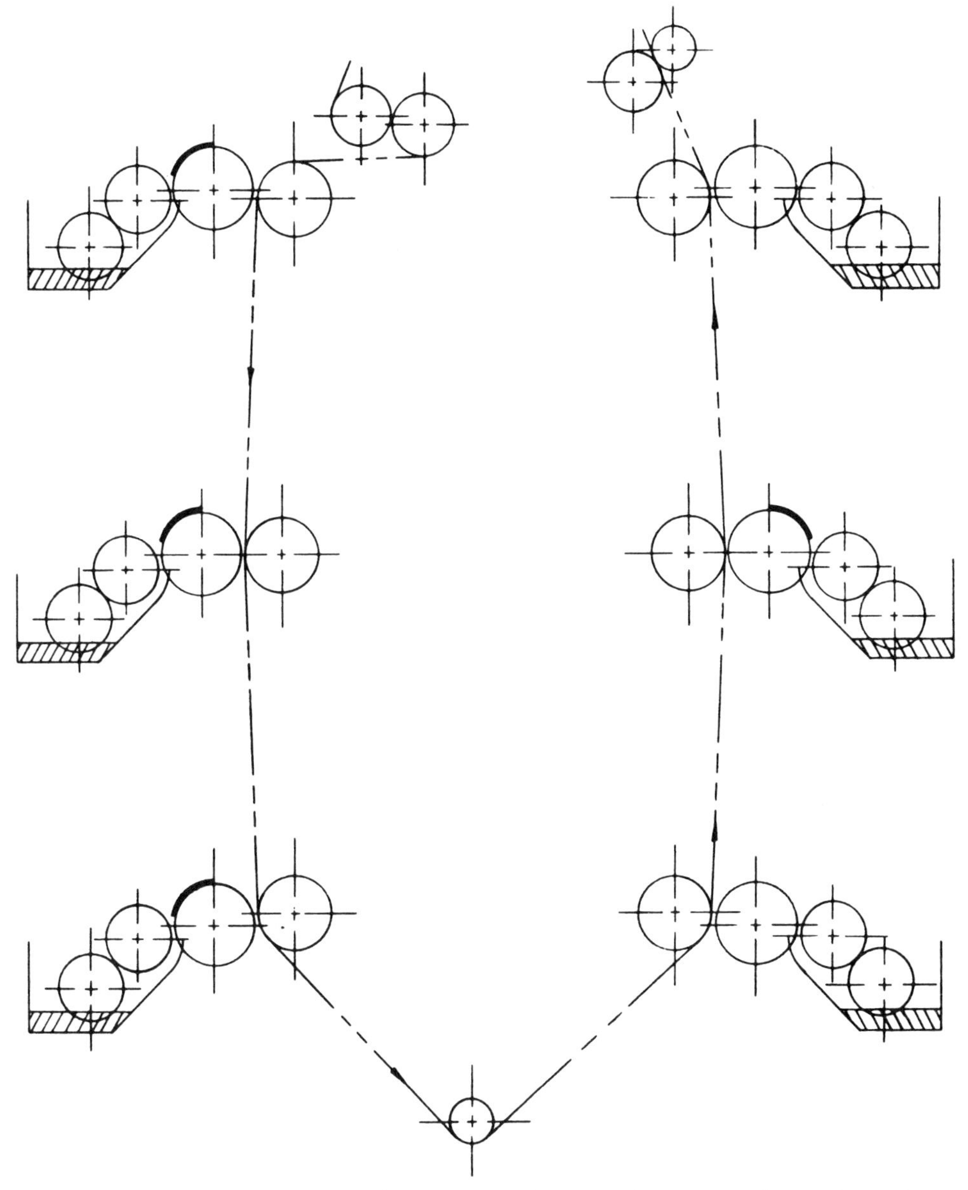

Figure 8-12. Schematic diagram of a typical stack press. Two stations do not show plates, indicating that they are not printing. Dashed line shows substrate travel. (Source: Flexography: Principles and Practices, Fourth Edition.)

the downstream stacks may be used for line printing and coating for gloss, slip and other end-use qualities to complement the process work on the CI drum.

Furthermore, combination presses are found with flexo units on gravure and offset presses, often applying information such as prices or addresses that may be changed during a gravure run or applying coating to a web offset printing operation. Flexo coaters are commonly used on sheetfed offset presses. Flexo plates can be used in place of offset blankets for pattern coating of cartons. Flexo press units are often included to provide solid bars for electronic symbol reading.

Figure 8-13. Combination press incorporating an eight-color CI unit plus two outboard stacks. This unit is used for preprinting linerboard. (Photo courtesy of Fischer & Krecke, Inc.)

Flexo/offset/letterpress/rotary screen machines produce some labels and folding cartons.

Sheetfed Presses. Sheetfed flexographic presses are used to print corrugated board. The stiff board is pushed or pulled into the printing station at speeds as high as 400 "kicks" per minute while maintaining good register. The machines can be adjusted to run many different sheet sizes. Use of sheet cleaners (pneumatic, electrostatic and mechanical) to remove dust from heavy board is very helpful.

Sheetfed corrugated presses can be fed either from the top or bottom of the pile, depending on the feeding devices employed. If fed from the bottom, the weight of the pile must be limited so that it does not create more friction between the bottom blank and the blank above it than the feeder can overcome easily.

Corrugated presses are either top printers that print on top of the sheet or bottom printers that print on the bottom. The sheets may be carried on a belt for top printers; bottom printers generally rely on the plates and pull collars to carry the blank through the press. When the press has a device for pushing the blank through, it is called "timed." When it relies on the plates and pull collars to transport the blank, it is called "untimed."

GRAVURE PRESSES

A gravure press consists of an ink fountain, a cylinder that carries the image, a doctor blade and an impression cylinder (see Chapter 3). The gravure cylinder is a hard, smooth cylinder with the image engraved in it. It turns in an ink bath, and the surface of the cylinder is wiped clean with a doctor blade. This simplicity is one of the great advantages of gravure printing. One result is that gravure cylinders can easily be changed to produce variable cutoff or repeat lengths. Technically the web process is called rotogravure, but there are a few sheetfed gravure presses in the United States, Europe and Japan.

Characteristics of the Gravure Press

The simplicity of the process contributes to its ability to print at high speeds without losing control. Rotogravure presses are fastest printing presses built (figure 8-14).

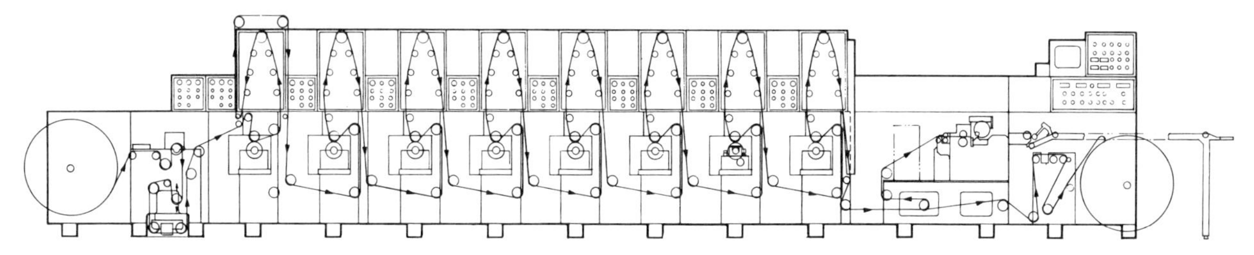

***Figure 8-14.** Schematic diagram of a gravure press, showing the threading pattern of the substrate. (Illustration courtesy of W. R. Chesnut Engineering Inc.)*

High speed and excellent color reproduction make gravure the process of choice for very long run package printing, especially on plastic or coated paper that have the smoothness to accept ink from the gravure cylinder. Flexible packaging benefits from the brilliance achievable with gravure inks, and skin and hair tones on cosmetic cartons benefit from the uniformity of color and structure of the halftone dots.

The gravure inking system requires a fluid ink (one of low viscosity), but the ability of the cylinder to apply a thick film of ink contributes to the brilliance and gloss that are typical of gravure.

The gravure cylinder, made of copper-coated steel, is very heavy, and multicolor gravure presses are usually in-line presses. In-line presses, as noted above, have problems with registration, problems that have been largely overcome with modern electronics.

Wide-web presses (figure 8-15) are popular, but there is a trend to narrower, faster presses in gravure as in other printing methods where they are better suited to short runs. On modern systems, the front end is completely computerized, and the cylinders are being prepared without use of any film. The cylinders for the next run are being engraved while the first job is being run. At the end of

Figure 8-15. Wide web gravure press. (Photo courtesy of North American Cerutti Corp.)

the run, the engraved cylinder is moved into the machine almost as fast as a flexo plate can be changed. The gravure print station rolls in and out of the press. The next job is readied and moved onto the press when the last job is complete.

Most new gravure packaging presses are equipped with trolleys or carts that carry a complete ink fountain, doctor blade and cylinder to allow off-press makeready, reducing downtime between jobs.

Gravure presses are divided into two groups: lightweight presses for flexible packaging, gift wraps, paper and foil labels and decorative films, and heavyweight presses for folding cartons and vinyl sheeting. Some overlap occurs in the midrange so that different materials can be printed on the same press.

There is conceptually no difference between a light press and a heavy press. Gravure board presses are bigger and heavier than presses for flexible packaging because they require bigger, heavier side frames, press roller construction and rewinding for the heavier substrate. Gravure presses are not necessarily heavier than flexo presses. CI gravure presses have been manufactured but are not common.

Web transport through the gravure press relies primarily on web tension control.

Standard gravure units are used to apply primers, and overall coatings, varnishes and adhesives. The unit is heated to apply hot melt or wax coatings.

Although many types of packages are printed by gravure, there are only a few major products. These are listed in table 8-II.

The Printing System

Printing Cylinder. Made of copper-plated steel, the hard, smooth, incompressible surface of the gravure cylinder gives excellent print uniformity, especially advantageous on long runs, but it demands a smooth substrate. Gravure does not print well on rough substrates, such as uncalendered paper, that do not

TABLE 8-II. TYPICAL MARKETS FOR GRAVURE PRESSES AND TYPES OF PRESSES COMMONLY USED

Labels: Narrow web in-line
Folding Cartons: Wide and narrow web in-line
Flexible Packaging: Wide and narrow web in-line

make intimate contact with the cylinder. Even use of electrostatic assist is insufficient to overcome speckle or missing dots caused by a rough sheet.

Inking System. The inking system can be no more than an ink trough with circulating ink in which the cylinder turns, but as with modern flexo presses, modern gravure presses use an enclosed inking system that avoids evaporation of ink from the ink trough or fountain.

Little can be done to the inking system to adjust the color of the printed image. This characteristic makes gravure color highly reproducible, but the cylinder must be properly engraved, and the ink color must be properly controlled.

Stainless steel fountains are required when waterbase inks are used.

Doctor Blade. Gravure uses a trailing blade to wipe excess ink from the gravure cylinder whereas flexo uses a reverse-angle blade that, in effect, shaves the ink from the surface of the anilox roll.

The doctor blade must clean all excess ink from the cylinder roll with a minimum of wear to itself and to the roll. This requires that the blade be rigidly mounted at a controlled angle, normally between 45 and 70°. The blade must have a carefully prepared edge so that it can seat itself easily and wipe cleanly. The blade is usually made from cold, rolled steel but specially coated steel and plastic blades—Teflon or nylon—in thicknesses up to 0.060 inch are becoming popular.

Printers changing from solvent to waterbase inks find that extensive changes are required not only in the doctor blade and doctor blade system but in other components that contact the ink. Conventional doctor blades will not handle the high pigment loading in waterbase inks and the flow characteristics that differ from those in solventbase inks.

Impression Roll. The impression or backup cylinder on the gravure press is a rubber-covered steel cylinder that turns freely with the web. The impression roll squeezes the substrate against the gravure cylinder, making it contact the ink. The impression roll must be set to obtain proper ink transfer, to control the web tension and to move the web through the press.

The impression roll is driven by friction, not by gears, so that the ratio of its diameter to the diameter of the print cylinder is not critical. The diameter of the impression roll is usually less than that of the cylinder.

Electrostatic Assist (ESA). By generating an electrical charge differential between the gravure cylinder and the impression roll, ESA draws ink from the cells, and the problem of speckle or missing halftone dots is greatly reduced, and the quality of the gravure print is significantly improved.

Types of Gravure Presses and Their Markets

Wide Web Presses. Gravure cylinders for folding cartons typically run from 36 to 55 inches wide (figure 8-15). Speeds of folding carton presses typically run up to 700 feet per minute when the web is fed into a reciprocating die cutter. Roll-to-roll they may run at 1200–1300 feet per minute, but these are becoming unpopular because web monitors sometimes fail to pick up critical defects during printing. The trend is to in-line sheeting that delivers a finished product that can be examined immediately. The long-term trend is to rotary cutters that can deliver a trimmed package at speeds in excess of 1000 feet per minute.

Folding carton presses typically have from six to eight printing units and are supplied in either narrow (under 36 inches) or wide web (up to 55 inches) with cylinders of variable cutoff or repeat.

Flexible packaging presses typically have eight printing units, but have been built with as many as 11 units. Web widths range from under 12 inches to 63 inches.

Narrow Web Presses. As in flexo, narrower presses lead to lower cylinder costs and easier handling, and they improve the economics of gravure on shorter runs (figure 8-16). Most flexible packaging presses are designed for roll-to-roll operation.

Cutoff lengths vary from as low as 10 inches to a maximum of 36 inches. Web widths are typically around 36 inches. Printing of folding cartons or labels typically requires seven or eight printing stations. Presses are manufactured that permit the interchange of flexo and gravure units (figure 8-17).

Sheetfed Presses. Sheetfed gravure presses are used for proofing. In the United States, some are used for high-quality cosmetics cartons and for overall coating of sheetfed offset products. They are more commonplace in Europe and Japan.

Figure 8-16. Eight color in-line narrow web rotogravure press. (Photo courtesy of W. R. Chesnut Engineering Inc.)

Figure 8-17. Interchangeable unit for flexo/gravure press. The unit is a flexo/gravure print station. (Photo courtesy of W. R. Chesnut Engineering Inc.)

Offset Gravure. Offset or indirect gravure, in which the ink is transferred from the cylinder to a rubber-covered transfer roll or blanket is used sometimes to print decorated metal surfaces. It is also used when the substrate is not smooth or has an uneven surface.

LITHOGRAPHIC PRESSES

The lithographic press comprises a plate cylinder that holds the plate, a blanket cylinder that carries the offset blanket, an impression cylinder, an inking system and a dampening system. The dampening system is unique to lithography. The inking system resembles the inking system of a letterpress machine and consists of many rolls to work and distribute an oilbase paste ink (one of high viscosity). This complex inking system contrasts sharply with the simple inking systems of flexo or gravure.

The infeed unit to a web offset press is much like that of any web press, but the infeed unit to a sheetfed press differs greatly. Sheetfed presses feed the substrate intermittently. Offset presses are the most popular sheetfed presses for labels and folding cartons.

There are several types of lithographic presses. Because of their popularity in publication printing, a great deal of research and engineering has gone into their development.

Characteristics of Lithographic Presses

Offset lithography is characterized by printing from a smooth or flat plate that separates image from non-image areas by their receptivity to water or ink.

The complicated inking system, the offset blanket, and addition of a dampening system make this the most complicated printing system in use. These difficulties are countered by inexpensive plates that are easily and quickly prepared and mounted, making sheetfed lithography ideal for short-run work. The offset blanket permits the press to print on a wide range of paper and board substrates, and also on sheet metal for three-piece cans.

Web offset presses run at speeds surpassed only by gravure presses. Another advantage of lithography is the stiff or paste ink that can be transferred from plate to an offset blanket to substrate without excessive dot gain. Offset and letterpress inks are much more viscous than flexo or gravure inks. Web offset prints a heavy ink film that produces a glossy print.

Paper transport through the web press relies on the web offset blanket and proper control of web tension. Sheetfed presses rely on grippers located in the impression cylinder.

High waste has been characteristic of web offset printing. Waste results from the need to make ready with the press running on good stock, from the need to trim material unprinted by the gap in the blanket cylinder, and from careless operation typical of so much web offset printing. All of these sources have been attacked, and web offset has become economically useful for printing folding cartons.

Applications of offset, sheetfed and web, for package printing are listed in table 8-III. The ability of lithography to print sharp halftone dots and vignettes has enabled it to retain its markets in printing of top quality labels, folding cartons and litho labels for corrugated.

TABLE 8-III. TYPICAL MARKETS FOR LITHOGRAPHIC PRESSES AND TYPES OF PRESSES COMMONLY USED

Labels: Sheetfed
Litho Labels for Corrugated: Sheetfed
Folding Cartons: Sheetfed or web
Metal Decorating: Special sheetfed press
Carded Packaging or "Bubble Packs": Sheetfed

The Printing System

Plate. Introduction of water permits use of inexpensive offset plates, but it also creates the potential for color variability. Offset lithography has greater color variability than any other printing process. While this may be an advantage in correcting poor prepress work, in the new world of quality management, color control is an added challenge to the printer, and recent press developments help to reduce variability.

Blanket. Offset blankets are textile webs that have a rubber or other polymeric layer laminated on the surface. Manufacture of these products is a highly sophisticated process, and improvements in blanket manufacture have continually improved their performance. Not only is the uniformity of thickness of high importance, but the texture of the rubber surface affects both the uniformity of the print and the manner in which it releases paper or other substrate.

The offset blanket conforms to rough surfaces such as uncoated kraft, and the blanket is even applied to processes other than lithography.

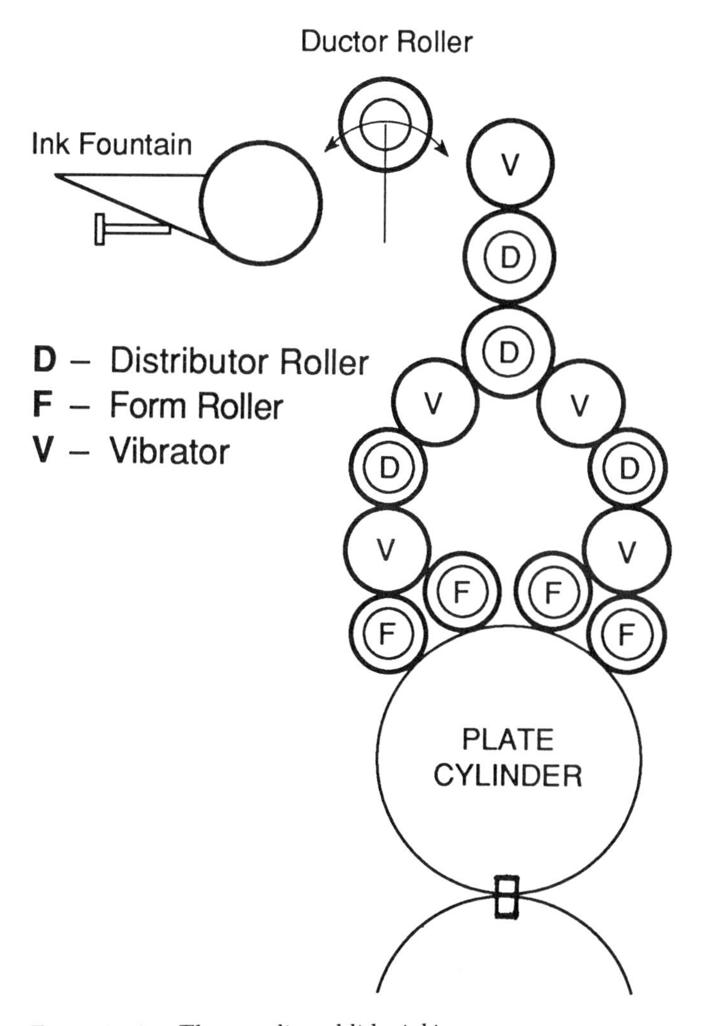

Figure 8-18. The complicated litho inking system.

It takes considerable force to remove a sheet or web from an inked blanket, and this is an important consideration in printing foil or thin plastic.

Plate and Blanket Cylinders. In offset lithography, these cylinders have a gap cut out of them into which the plate or blanket is fastened. This gap does not print, and in web offset printing it creates wasted paper or board. (In sheetfed printing, the sheets are cut no longer than the length of the blanket.) The gap has been made very small on modern presses.

Inking System. Lithography, and letterpress, plates are inked from a roll that carries a thin film of ink, and a heavy plate image can deplete the film on the ink roll. The roll does not transfer enough ink on the next revolution and causes a ghost image in the printing.

The inking system consists of a ductor roll that contacts the ink fountain roller. The ductor carries ink to the first roll in a series of rolls that distribute the ink evenly over the plate. The ink keys and the complicated system of rolls in the inking system (figure 8-18) allow the crew more freedom to adjust the inking than is available in flexo or gravure, but more commonly it causes variability in color, a major problem in package printing. The uniformity of ink feed in flexo and gravure is a major advantage for these processes.

Keyless Offset. This term refers to inking of an offset press in which ink is metered with an anilox roll. In 1992, Europe and Japan had 20 sites operating keyless offset, but only one was successfully operating commercially in North America. One-color printing is successful, but process color is troublesome owing to excessive emulsification of water and ink in a short ink train. Use of keyless offset is largely limited to printing newsprint.

Dampening System. Many types of dampeners have been developed for offset presses. These are commonly divided into intermittent, or ductor feed, and continuous feed types. Intermittent feed units have a ductor roll that picks up moisture from the fountain roll and carries it to the plate dampening rolls. Moisture feed can be controlled by the timing of the ductor. Much interesting engineering has been involved in the development of continuous-feed dampening systems, and these are now used on most new litho presses (figure 8-19).

The presence of water in the ink and on the non-image areas demands attention when offset is used to print nonabsorbent materials. Roll coverings must be easily wet with water but not by ink. Special cloth or paper covers are frequently used over rubber rolls.

Waterless Offset. Waterless offset is accomplished by using a plate whose non-image area is not receptive to ink. It avoids the necessity of controlling ink-water balance and allows printing of very fine halftone screens. Prints can have a high color density and sharpness. Waterless offset requires very close control of temperature on the press and requires special inks and cooling rolls.

Types of Lithographic Presses and Their Markets

Most web offset presses for packaging are 40 to 50 inches (1000 to 1250 mm) in width. Some narrow web label presses are also available.

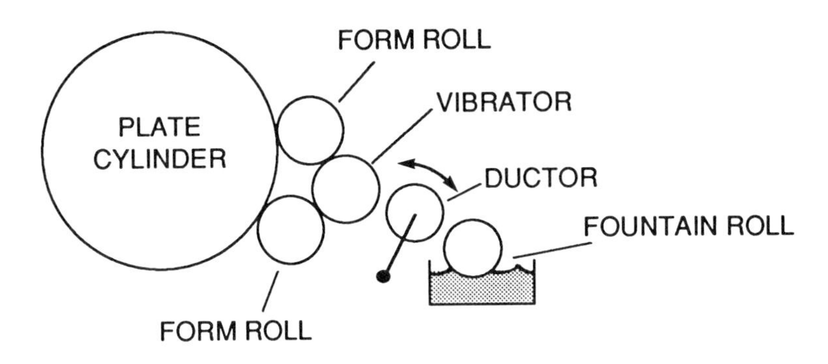

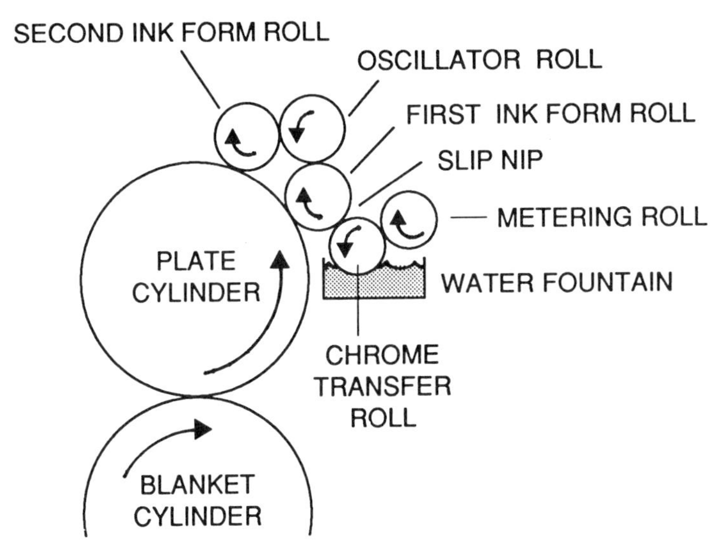

Figure 8-19. Lithographic dampening systems. (Top) Intermittent-feed dampening system. (Bottom) Continuous-feed dampening system.

In publication printing it is important to print both sides of the web, and because the offset blanket is smooth, it is possible for each of two blankets to serve as both image carrier and impression cylinder. For package printing, it is usually necessary to print only one side of the web, and in-line presses are well suited for folding cartons. These presses use an impression cylinder instead of another blanket to back up the web while it is being printed (figure 8-20).

In-line Presses. In-line presses are used when only one side of the substrate is to be printed. The major feature of an in-line press is that each printing unit consists of only one blanket and impression cylinder. Packaging materials are usually printed on one side only although it is possible to print both sides by turning the web between units. Of the packaging materials printed by offset lithography, only folding cartons and labels are commonly printed by web offset.

Variable Cut-off Web Presses. Web offset presses usually have a fixed repeat length or cut-off, which means that all products must be the same size. If various sizes are to be produced on a single press, the trim waste is prohibitive. This works well for standard products such as single-service cereal boxes, but it has limited web offset applications in package printing.

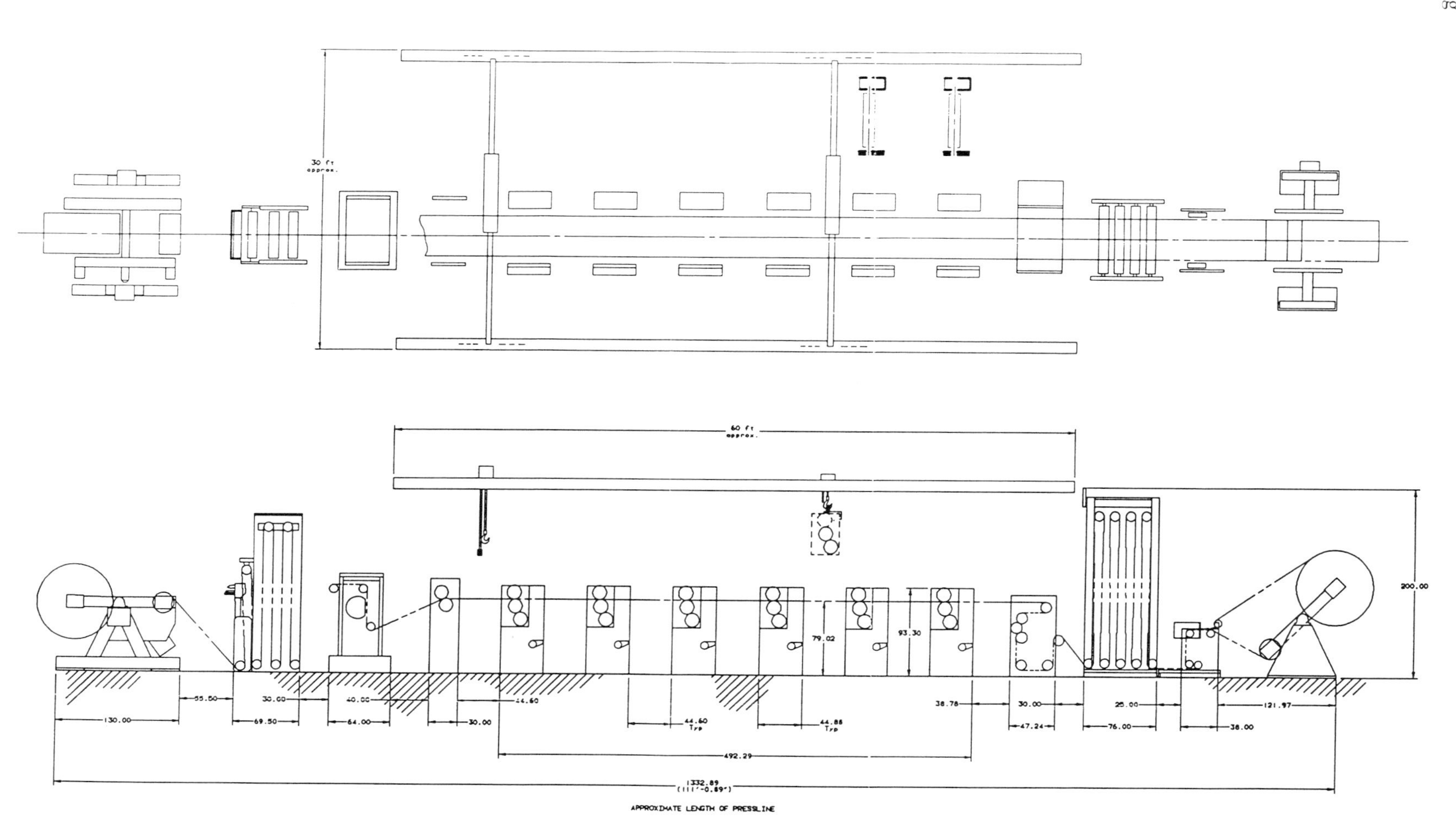

Figure 8-20. Schematic diagram of variable cutoff web offset press showing blanket-to-impression cylinder configuration. (Illustration courtesy of MAN Roland, Inc.).

Figure 8-21. Variable cutoff web offset press. (Photo courtesy of MAN Roland Inc.)

Variable cut-off web presses have been available for years, but changing the entire printing unit to change repeat length proved to be a nuisance. Advances in press design are addressing this problem (figure 8-21).

Metal Decorating Presses. Metal sheets, such as those used for three-piece cans, are printed on sheetfed presses specially designed for metal decorating. These presses are offset presses, usually offset litho but there are also some offset letterpress machines, usually referred to as "dry offset." They may print from plastic or metal litho plates.

Sheetfed Presses. The six-color press is standard for printing offset labels and folding cartons—four units for process color, one for spot color and one for varnish. Seven and eight-color presses offer opportunities for two spot colors and varnish (figures 8-22 and 8-23). Sheetfed litho presses as wide as 77 inches (1956 mm) are in use, although most presses now manufactured are 54 or 60 inches wide.

The principal difference between sheetfed and web offset presses is in the manner of feeding and delivery. In the web press, the paper is moved through the press under tension that is controlled by the web of paper, and it is delivered into a sheeter, folder or rewind. The sheetfed press pulls paper through the press by use of grippers that are placed in the impression cylinder. After the last printing cylinder, the sheet is picked up with grippers attached to a chain, and the delivery system picks up the printed sheets and drops them onto a pile at the end of the press. The only converting or finishing operations normally performed on sheetfed presses is coating. Diecutting, trimming and other operations that may be carried out inline on a web press are always done off-line on a sheetfed press.

The printing pressure is determined by the amount of packing under the plate

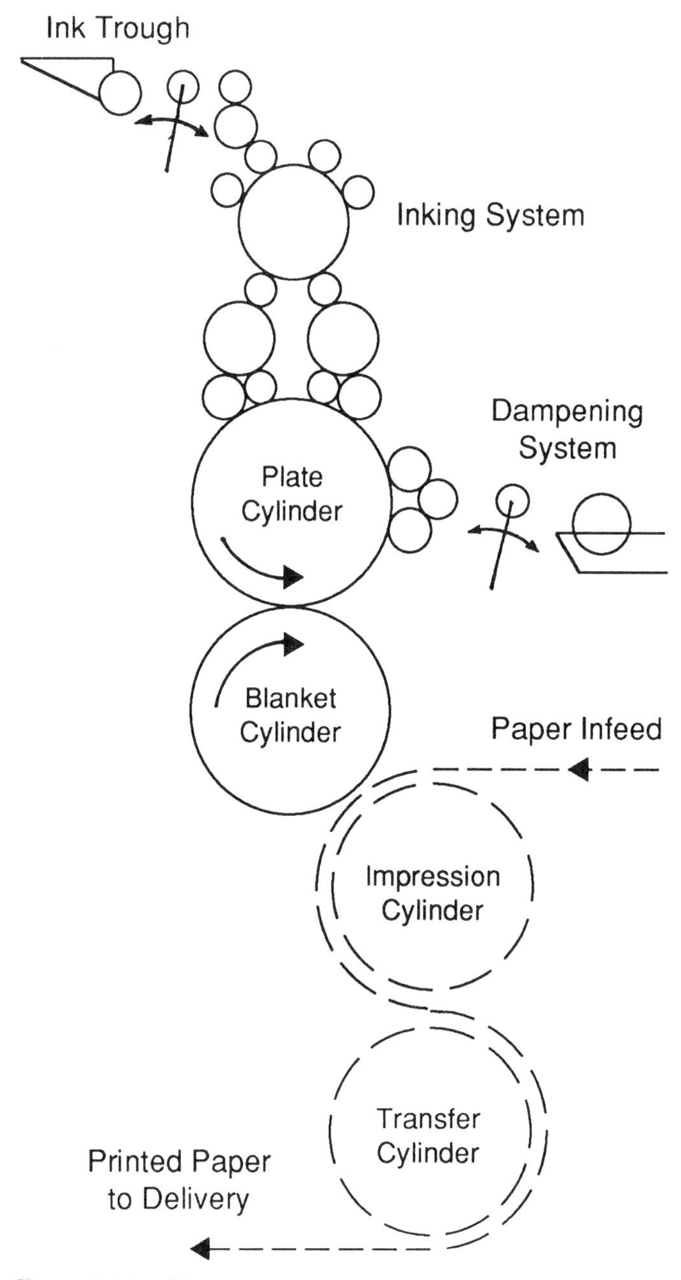

Figure 8-22. Schematic diagram of sheetfed offset press.

and blanket. All cylinders are driven, and it is not possible to change the size of one cylinder.

From the standpoint of capital costs, sheetfed offset presses are about as expensive as web presses, but they are not well suited to in-line operations. Low waste and speed of makeready favor sheetfed presses for short runs. Off-line operations require additional personnel and extra handling that increase costs and delay processing.

Sheetfed offset presses are widely used for printing labels (including litho

Figure 8-23. Sheetfed offset press. (Photo courtesy of KBA-Planeta North America, Inc.)

labels for corrugated) and for folding cartons. They are occasionally used for printing E-flute corrugated.

LETTERPRESS MACHINES

Once the principal method of printing, letterpress has fallen out of favor in both publication and package printing because of the long makeready time required to pack the plate so that it will print with an even impression over the entire sheet, because of the expense of the plates and because the hard plate requires a completely smooth substrate. Coated papers and boards are preferred for letterpress printing.

Offset letterpress, in which the image is transferred from the plate to a blanket and then to the substrate, overcomes the problem of a rough substrate but complicates the press and the process. As with lithography, the complicated process does not bar it from being a commercial success.

Characteristics of the Letterpress Machine

The hard metal or plastic plates characteristic of letterpress produce sharply formed letters. Fine serifs reproduce well, and letterpress is in favor for small pharmaceutical boxes and labels that must carry fine print. Letterpress contributes to good quality color by producing rosettes which, when viewed under a magnifying glass, appear sharp and crisp.

Unlike lithography, letterpress requires no water on the plate so that ink emulsification is not a problem, and printing on nonabsorbent substrates such as plastic offers no special problem. Letterpress has less color variation that lithography.

The principal markets for letterpress printing are now found in packaging. They are listed in table 8-IV.

TABLE 8-IV. TYPICAL MARKETS FOR LETTERPRESS AND TYPES OF PRESSES COMMONLY USED

Market: Press type
Labels: Narrow web
Folding Cartons: Narrow web
Two-piece Cans: Offset letterpress
Foil Stamping, Hot Stamping and Embossing: Platen press

The Printing System

The letterpress machine consists of a plate cylinder and an impression roll, which is a steel roll covered with a hard rubber, much like a gravure impression cylinder.

The inking system consists of many rolls, closely resembling that of an offset press (figure 8-18). A ductor roll carries ink from the fountain pan onto distributor rolls and a form roll that inks the form or image. The system is suited to the stiff paste inks used in letterpress.

Types of Letterpress Presses and Their Markets

Three types of presses have been in common use: rotary presses, in which cylindrical plates print against the substrate which is supported by the impression cylinder; flatbed presses, which are usually sheetfed, roll the paper against a flat plate that carries the image; and platen presses which are "clam-shell" devices in which one plate carries the substrate and the other carries the image (figures 8-24 and 8-25). They are used for embossing and hot-foil stamping. Flatbed cylinder presses are intermittent or stop-action presses with in-line flat die converting. They are used in Japan and Europe for label printing.

Rotary Letterpress. These are web presses with a cylindrical plate that presses against a cylindrical impression cylinder. They are typically built in an in-line configuration, and they are used to print labels and small folding cartons.

Figure 8-24. Flatbed rotary letterpress machine. (Photo courtesy of Gallus, Inc.)

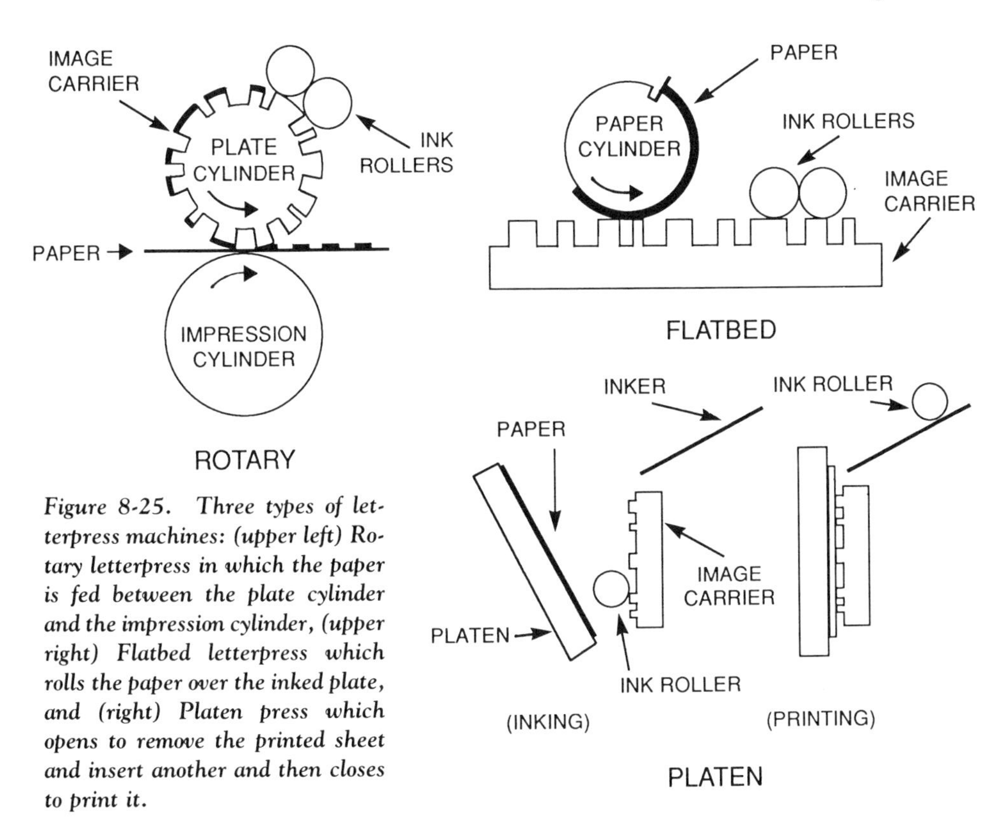

Figure 8-25. Three types of letterpress machines: (upper left) Rotary letterpress in which the paper is fed between the plate cylinder and the impression cylinder, (upper right) Flatbed letterpress which rolls the paper over the inked plate, and (right) Platen press which opens to remove the printed sheet and insert another and then closes to print it.

Some tube stock is printed with letterpress using a "semi-rigid" plastic plate. The process makes it practical to print a paste ink on the tube stock and dry it by UV.

Offset Letterpress. While rigid letterpress plates are suitable for printing on paper or board, they are not suitable for printing on metal. Offset letterpress is used for printing on cans. This process is commonly used to print two-piece cans (figure 8-26).

Platen Presses. Foil stamping, embossing and die cutting are frequently done from platen presses (figure 8-27). They are discussed briefly, later in this chapter. The processes are used for highly decorated boxes, such as perfume and cosmetic boxes.

CI Presses. CI letterpress equipment is used extensively for label printing (figure 8-28).

SCREEN PRESSES

Screen printing is used in a broad variety of packaging and decorating applications. Screen printing can put down a heavier film of ink than any other process. The process yields films of high opacity and brilliance. With fluorescent colors, the heavy film contributes to the permanence of the color. These properties are often desired in labels, cartons and other packages.

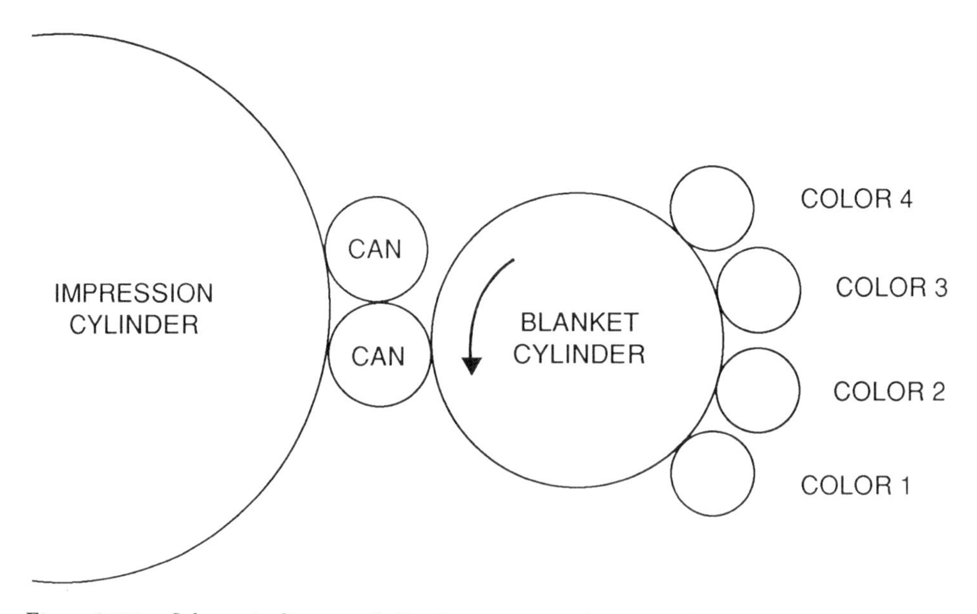

Figure 8-26. Schematic diagram of offset letterpress configuration for printing two-piece cans.

Figure 8-27. Platen letterpress printing press, also used for foil stamping, embossing and die-cutting. (Photo courtesy of Brandtjen & Kluge).

Figure 8-28. CI letterpress machine. (Photo courtesy of KoPack, Inc.)

All screen presses include a frame, to which the mesh carrying the image is attached, and a squeegee to force the ink through the mesh. Preparation of the screen and makeready are relatively inexpensive, and the process is favored for very short runs.

Inking is controlled by the mesh of the fibers used to make the screen and by the shape and operation of the squeegee. Squeegee hardness, speed and drag also determine color consistency. Dot gain is directly related to drag on the squeegee.

Monofilament polyester that is used to make the mesh comes in many diameters including "S" (thin), "T," and "HD" (heavy duty). Low-elongation meshes allow higher tensions to be employed.

Retensionable frames are best for controlling tension (figure 8-29). They allow the mesh to be work hardened to produce best results.

Figure 8-29. Newman stretch frame used to produce uniform tension of the screen. (Photo courtesy of Stretch Devices, Inc.)

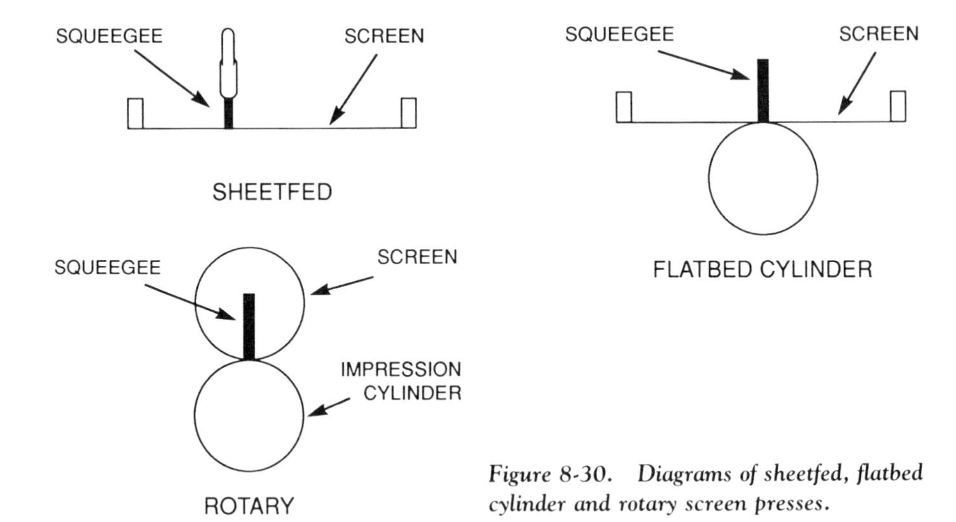

Figure 8-30. Diagrams of sheetfed, flatbed cylinder and rotary screen presses.

Types of Screen Presses and Their Markets

There are three basic designs of screen presses: sheetfed, flatbed cylinder and rotary presses (figure 8-30).

Sheetfed. Round or oval bottles and cylindrical packages are frequently screen printed. The familiar flat screen is used as the stencil carrier. The cylindrical object is positioned beneath the screen, and the screen cylinder is lowered into contact with the container (figure 8-31). To transfer the image, the screen frame is moved under the fixed or stationary squeegee. The squeegee presses the screen against the cylindrical object which rotates as it is printed.

Flatbed Cylinder Press. This press resembles the flatbed cylinder letterpress. The substrate is held by grippers and carried past the printing cylinder on a flat bed. The ink is applied from a cylinder that is rolled against it. The image is carried in a screen mesh attached to a cylinder, and a squeegee on the inside of the cylinder presses the ink through the mesh onto the stock. These roll-to-roll presses have the highest productivity of any screen press, with top speeds of 250 to 300 feet per minute.

Rotary Press. Rotary presses are used to print webs. The screen is held in a cylinder with the squeegee inside. As the web travels under the cylinder, the stationary squeegee extrudes ink through the screen. Rotary machines have top speeds of 130–230 feet per minute (40–70 meters per minute). Most of those used to print packaging materials are used for labels figure 8-32).

NON-IMPACT PRINTERS

These printers are not usually called presses because there is little or no pressure used to generate the image. Ink jet printing operates without any contact

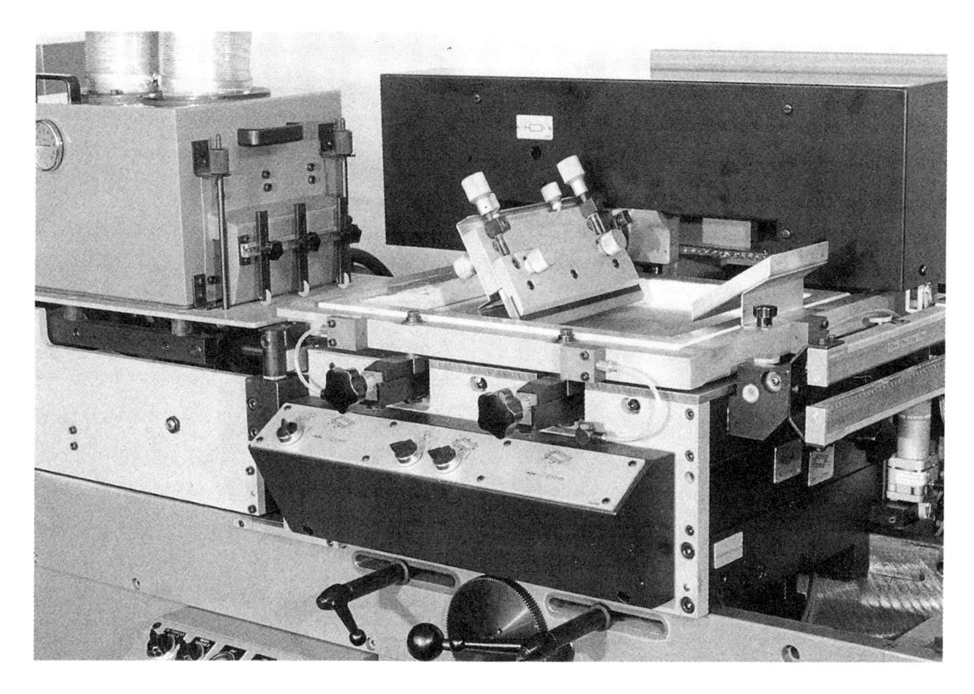

Figure 8-31. Flatbed screen press. (Photo courtesy of Gallus, Inc.)

Figure 8-32. Rotary screen press. (Photo courtesy of Gallus, Inc.)

between the inking system and the substrate. (Pictures of ink jet printers are presented in Chapter 3.) Other processes include electrophotography or electrostatic imaging, which includes laser printing.

Non-impact printers are used to generate proofs of art, design and color separations, but they are also used for in-store labeling and for imprinting batch numbers and codes on cans and bottles.

The most applicable characteristic of non-impact printers is their ability to print a different image every time, even at high speeds. Ink jet printers are used to print batch and control codes while electrostatic and thermal transfer printers are used to imprint in-store and primary labels.

ACCESSORIES AND AUXILIARY EQUIPMENT

Productivity of any press depends greatly on the auxiliary equipment. Some accessories, such as infeed devices, are an essential part of the press. Others are highly beneficial under some situations. Infeed and rewind devices are much the same for every web printing process. Electronic viscosity control is used only on flexo and gravure presses, while ink film control is used only on letterpress and litho presses.

Increasing use of electronic press controls is driven by the demand for greater quality coupled with greater productivity and lower costs. High-speed web printing is impossible without web tension controls; computer controlled consoles reduce crew requirements. Automatic register systems, web viewing systems, ink development and environmental controls are other important trends.

Automatic register, automatic impression and electronic viscosity control greatly improve running consistency. Computer control consoles both control and record many press variables (figure 8-33).

Press auxiliaries are often similar for presses that print by different processes. Sheetfed feeders are an exception. Offset presses use one type, but sheetfed direct-printing corrugated flexo folder-gluers often have a prefeeder at the feed end, and these presses are unique (figure 8-34).

Automatic lifting devices and plate cylinder replacement systems will enable a single operator to replace and reposition the plate cylinder very quickly. A few of the many auxiliaries and accessories are listed in table 8-V.

TABLE 8-V. COMMON PRESS AUXILIARIES AND ACCESSORIES

Control Console	Roll Lifter
Unwind Stand	Diecutter, Rotary or Platen
Flying Paster or Splicer	Dryer
Zero-speed Splicer	Chill Roll
Rewind Stand	Sheet Cleaner
Web Tension Control	Spray Powder Applicator
Demetallizer	Static Eliminator
Ink Feed Control	Web Scanner
Automatic Viscosity Control	
Registration Control including Web Guide	
Plate Cylinder Replacement Systems	

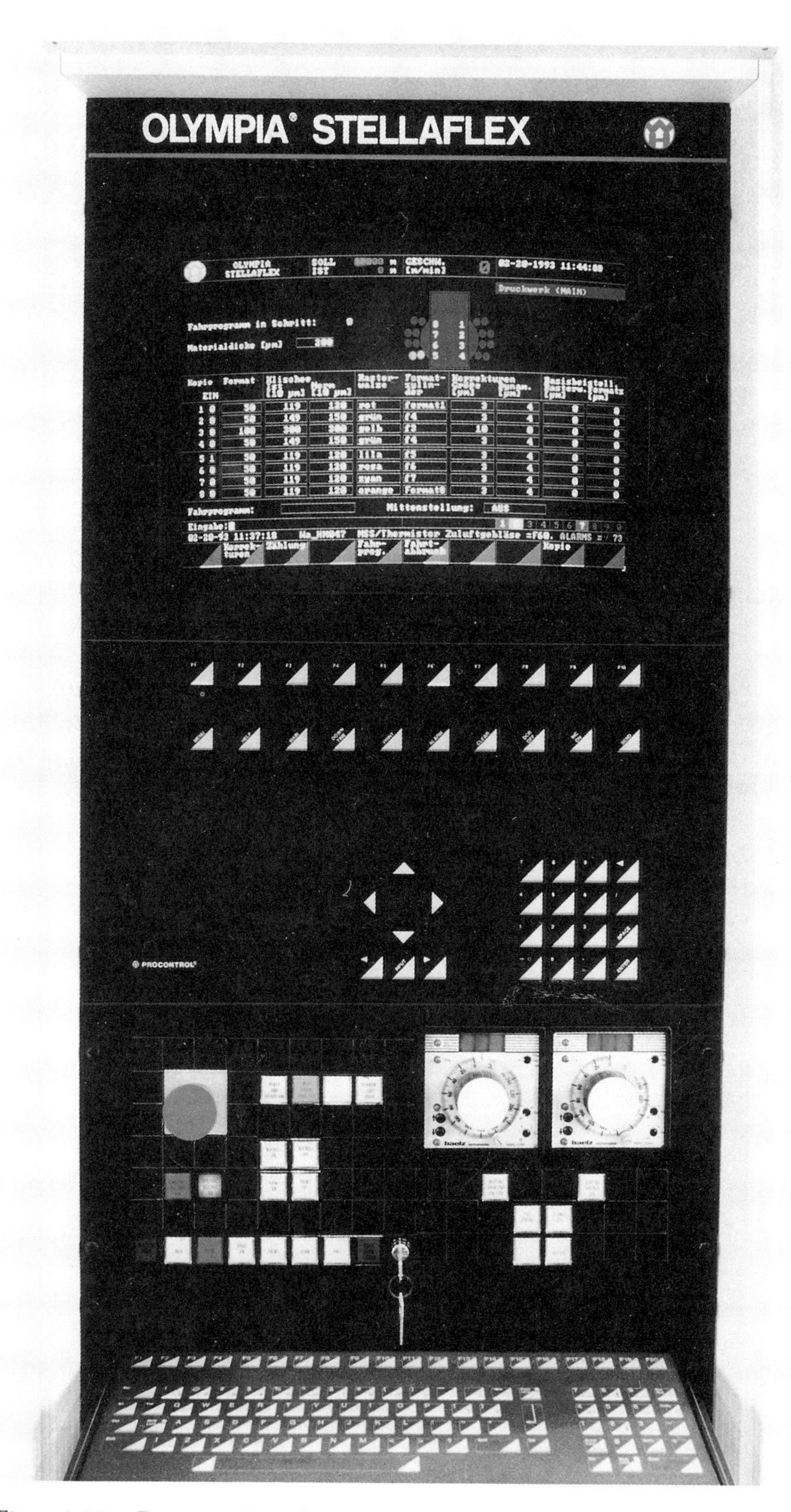

Figure 8-33. Press console on flexo wide-web press. All new machines use dedicated computers and fully electronic systems. (Photo courtesy of Windmoeller & Hoelscher Corp.)

Figure 8-34. Flexo folder-gluer with prefeeder, for printing directly onto combined corrugated board. (Photo courtesy of The Ward Machinery Co.)

Computer Control Consoles

By centralizing press control, it is possible for a single operator to control much of the printing operation, and crew requirements are reduced. Computer-control consoles both control and record many press variables including press settings and conditions, print count, running speed, color and register, ink viscosity, ink key settings, dryer temperatures and a host of other parameters. Offset presses control color by densitometers that can be electronically linked to the inking system, opening or closing ink keys. Flexo and gravure presses control color by controlling the ink viscosity, and there are instruments that read color density and automatically adjust the pigment content of the ink while maintaining its viscosity.

Computers for flexo folder-gluers on corrugated presses also record the thickness of the substrate and the impression of the plate. The settings are put into memory and used for repeat runs, requiring only minor adjustments. New technology is introduced at each major printing show.

Infeed and Rewind Equipment

While the infeed equipment of different kinds of web presses is more or less similar, infeed equipment of sheetfed presses differs greatly from web infeed devices.

Sheetfed Infeed. Corrugated boxes are fed to the flexo press from either the top or the bottom of the pile. Corrugated feeders move the blank through an opening or slit by vacuum or with a high-friction belt or a "kicker" (pusher). Top feeders may use a vacuum as does an offset press, or they may be fed manually. Vacuum transport systems hold the board in place for printing from the bottom (see figure 8-35). Automatic feeding is often stream-fed, but without overlap of blanks.

Two systems, untimed feed and timed feed, are used. Untimed feed relies on the mechanical coordination of feeding and printing so that the board does not stop as it moves through the press. Timed feed brings the board into position where it stops and is then printed. Untimed feed is the faster system.

The most common type of infeed equipment for paper on sheetfed presses

Figure 8-35. Feeding corrugated blanks into a flexo folder-gluer (Photo courtesy of The Ward Machinery Co.)

involves suckers, which are rubber pads that are vacuumized to lift the top sheet from the pile and move it forward and then drop the sheet when the vacuum is released. There are two different ways of feeding paper sheets to the offset press: single-sheet feeding or stream feeding. For faster presses, stream feeding in which the sheets overlap each other, is preferred.

In stream feeding, each sheet is brought to the feed table of the press before the next sheet is removed from the pile. In overlap feeding, movement of the second sheet is begun before the first sheet is free of the pile, with the result that the sheets overlap on the feeding table. Stream feeding requires that sheets be accelerated at a higher rate than does overlap feeding, and the high acceleration sometimes causes sheets to buckle or wave.

On either type of feeder, sheets are advanced to head stops which position the sheet in register. Grippers, located in the impression cylinder, pick up the sheet and draw it through the printing unit.

To feed sheets through a press at speeds in excess of 10,000 sheets per hour, in register and without wrinkling, requires an immense feat of engineering. Corrugated machines generally are designed for "cruising" speeds of up to 15,000 or 18,000 sheets per hour, with some machines capable of 30,000 per hour.

Unwind Stands. The reel stand must be large enough to handle the largest roll of paper or substrate that will be fed to the press (figure 8-36). For paper, this is usually 50 inches although paperboard can reach 84 inches.

Stands for the flexible packaging industry usually are capable of handling rolls up to 24 or 32 inches in diameter. Stands for handling heavier laminates and papers accept rolls with diameters up to 72 inches. Flying splicers for heavy board stock must be capable of handling 72-inch rolls.

Large rolls are very heavy, and roll handling equipment is a necessary part of the press. Some newer equipment is roboticized to bring the roll to the press and mount it.

Splicing of the expiring roll to the new roll is done either by zero speed or flying splicers (sometimes called "pasters"). To make a zero-speed splice requires that both rolls be stopped, the splice made, and the new roll started up. This requires shutting down the press and starting it up again unless there is a festoon

Figure 8-36. Unwind stand at feed end of a narrow-web flexo press. (Photo courtesy of Mark Andy, Inc.)

device that can store enough paper to keep the press running while the rolls are stopped, spliced, and brought back to speed. On a flying splicer, the new roll is brought up to the speed of the expiring web, and the adhesive or tape on the new roll is pressed against the old web. The device automatically cuts off the tail on the old roll.

Rewind Equipment. Rewind equipment is much like unwind equipment except that the reel must be driven (unwinders are usually driven). Both center and surface driven rewinders are used. The center winder drives the shaft, and it must increase in speed as the roll grows. The surface driven rewinder drives the outside of the roll whose speed remains constant. Some single roll rewinds use a combination of center and surface drives.

Web Guides. Web guides keep the web of substrate running in the proper position at both the unwind and rewind positions. They are controlled by an electronic sensor at the edge, and they move the web by placing a small force against its edge. An edge guide maintains the position of one edge, a center guide reads both edges and keeps the center of the web in position.

Sheeters. The web can be delivered in sheets by cutting it into appropriate carton lengths and feeding it to a controlled gathering system. Sheeters are cylinders with knives that cut the printed web after it has been dried.

Tension Control of Webs

It is impossible to achieve good quality and good productivity on a web press that has inadequate tension control.

There are many different types of web-tension-control systems, ranging from manual to highly sophisticated electronic systems. Unwind stands for flexible packaging presses require very sophisticated tension control devices to stabilize lightweight films as they enter the press.

Web-tension control is critical for any web-process machine because it is largely responsible for the quality of the product and the efficiency and productivity of the press. Adding a good tension control system to an older machine in good condition can greatly improve its performance.

Web tension must be controlled at the unwinding station, through the press and at the rewinder (figure 8-37). If the paper or other substrate is not properly wound, control on the press is hampered. Even the best equipped web press will do better with a well-wound roll than with a poorly wound roll.

Inadequate tension control causes many problems. Misregister, web breaks and wrap-ups around driven rolls are only a few. Others include stretching or wrinkling of the web, variation of coating thickness, unwind or rewind core crushing, and waste of time and material. Unfortunately, these problems are often accepted as normal press problems. Anyone familiar with web operation should recognize the relationship between adequate tension control and smooth, efficient operations.

Register Control

On any press, register must be maintained in two directions: across the cylinder or sidelay register, and register around the cylinder or cutoff register. Figure 8-38 shows a typical register mark as printed in and out of register. Figure 8-

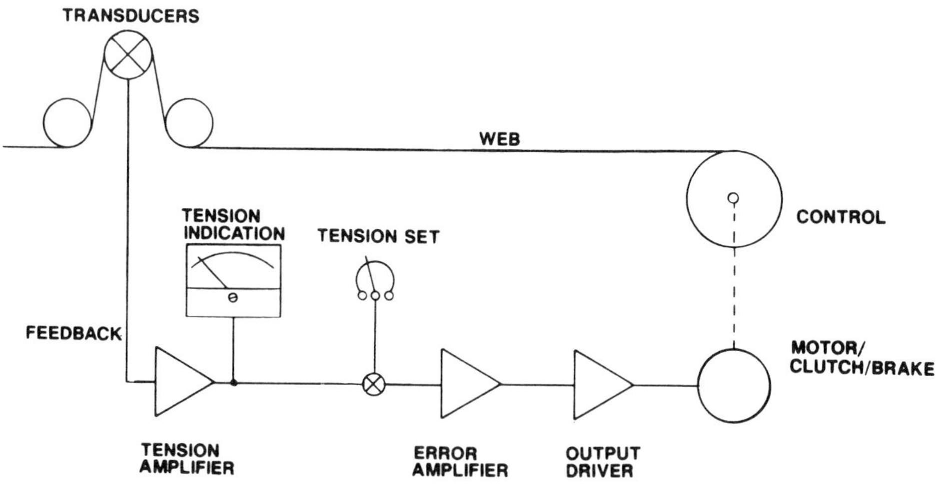

Figure 8-37. Schematic diagram of an infeed tension control on a wide-web flexo press. This closed-loop system provides proportional control to reflect the amount of correction required. (Source: Flexography: Principles and Practices, Fourth Edition.)

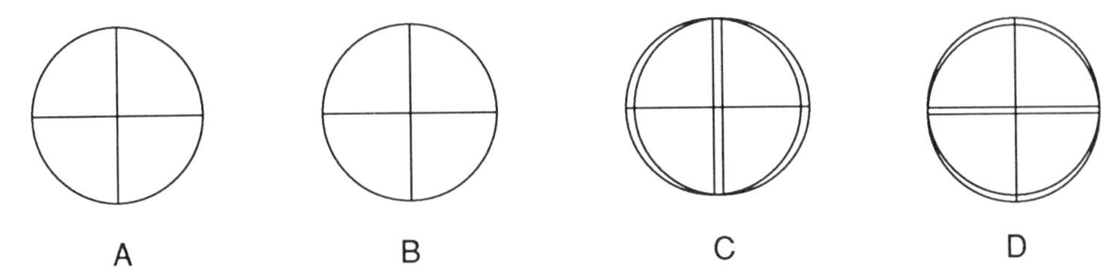

Figure 8-38. Typical register marks showing both in-register and out-of-register printing. (A) is basic register mark. (B) shows in-register printing. (C) and (D) show out-of-register conditions with (C) showing misregister left to right ("sidelay misregister") and (D) showing misregister in press direction ("cut-off misregister"). Other marks are also used for registration.

Figure 8-39. Press-mounted register control device. (Photo courtesy of Quad/Tech.)

38(A) shows the single mark; figure 8-38(B) shows the print of two marks printed in register. Figure 8-38(C) shows side-lay or sideways misregister, and figure 8-38(D) shows misregister in press direction.

Modern presses use electronic equipment to control register (figure 8-39). An instrument detects a mark printed on the image as it passes each print station. If successive marks are out of register in press direction, the electronic system advances or retards the printing plate cylinder to bring the colors back into register. Electronic systems control side-lay misregister by moving the plate cylinder laterally.

Many different types of systems are available for controlling color-to-color register. Register control systems can work from register marks or, by a process of pattern recognition, from a selected part of the image. Some control the position of halftone dots printed on each other, others control the position of the image with regard to the edge or to the cutoff. There are significant differences between control systems for flexo, offset and gravure.

Gravure and flexo need register marks for each unit. In a gravure press there are 30 to 40 feet of web travel between units to allow room for the dryer. Gravure requires control register after (or sometimes on old presses, before) each unit.

Cutoff misregister (misregister in the press direction) is controlled by moving the plate or cylinder in relation to the other colors. Sidelay misregister is often corrected by moving the web or the plate cylinder to one side or the other. The unwind stand can be automatically guided to provide lateral web position control at the press infeed.

Diecutting can also be kept in registration with a print-to-die register system using a reference mark on the print.

A preregistering device permits the press to preregister the cylinders. Another electronic registration system permits a web, printed on a five-color press, to be rerun to produce 10 colors. The system produces a readout of how tightly the registration is being held and gives the operator a real-time measure of deviation from registration.

All systems rely on human judgment. The pressman gets everything into register, then sets the unit to maintain the register.

Polyethylene is very difficult to control, especially when printing thin films. The press must have good infeed tension control, the film must be manufactured consistently and, most important, the roll must be wound consistently. Rolls with baggy edges and rolls with tight and loose spots cannot produce good, economical printing. Large rolls are particularly troublesome. If the dryer temperature goes too high, the film softens. This limits process color work on PE.

Good register requires intelligent artwork. Bad artwork causes all sorts of troubles. Overlaps and skinnies (images on different colors that overlap or fail to butt) cause real trouble.

Register is greatly improved with use of electronically controlled line shafts that are replacing the older mechanical line shafts. Electronic line shafts are computer controlled, with zeroing on each rotation of the cylinders. Some machines read a printed mark on each repeat in a web and adjust the print-to-print accuracy to within 0.0001 inch of the mark. Most manufacturers of modern presses guarantee accuracy to within the size of a halftone dot—0.003 inch.

Dryers

It is possible to print even clay-coated corrugated without a dryer, but the slow drying speed would make the operation unprofitable and prohibit heavy coverage or overprint varnish. Flexo and gravure presses often use interstation dryers (dryers after each printing unit). Because web offset and letterpress inks dry at much higher temperatures, a single dryer is used at the end of the press. (See figure 8-40.)

On sheetfed presses, infrared heaters warm the prints and speed the drying of the ink by accelerating the chemical reaction by which sheetfed inks cure or dry.

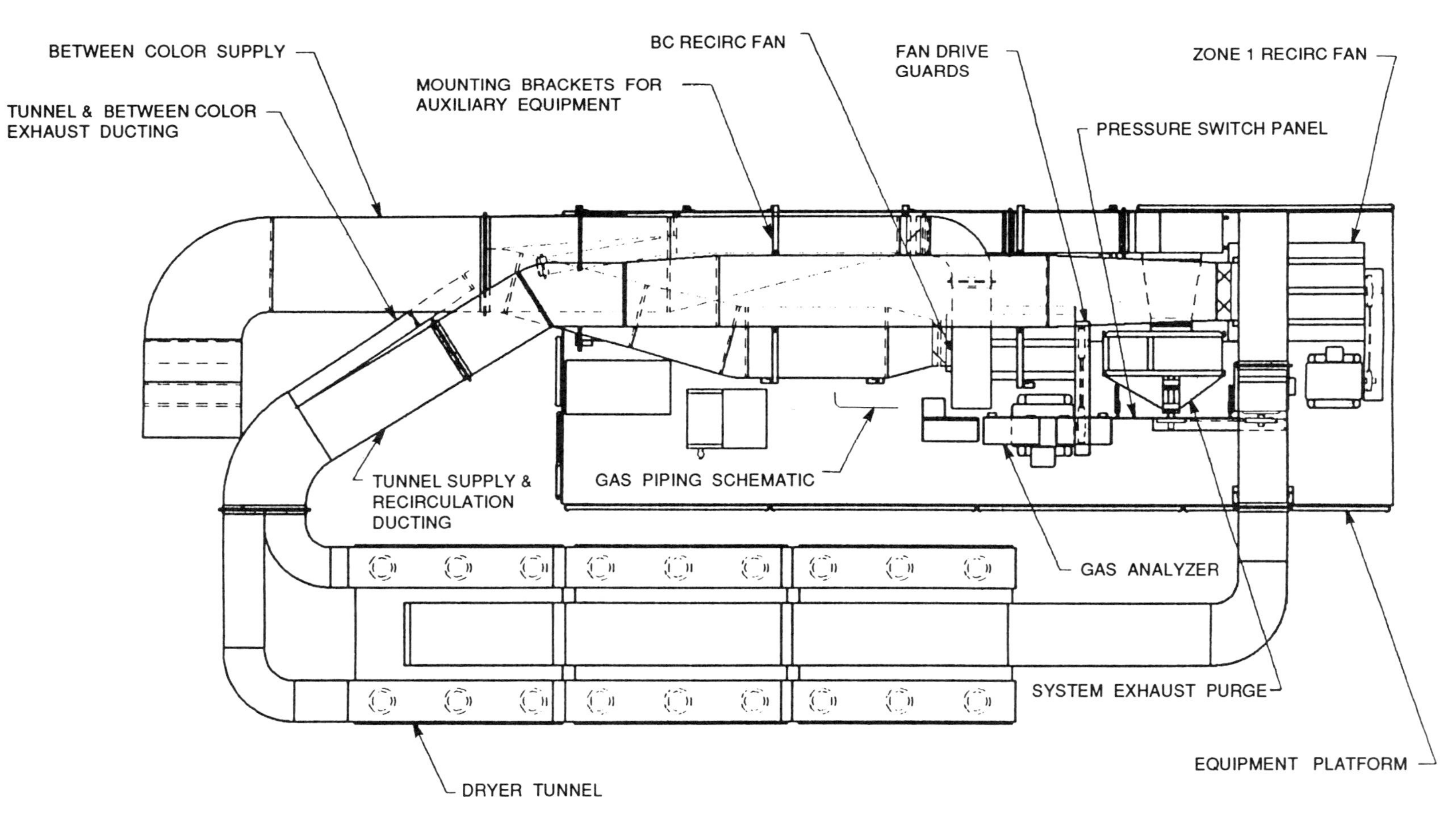

Figure 8-40. Hot air dryer configuration for a flexo press. (Illustration courtesy of Paper Converting Machine Co.)

Hot Air and IR Dryers. Hot air is generated by burning natural gas or even by burning waste solvents recovered from the drying ink. High-velocity hot air (HVHA) is blown against the surface of the web. This removes the ink vapors much more efficiently than does an open flame or other static heater. Infrared heaters are sometimes used to heat the printed web, followed by a jet of warm air.

Dryers are specified for maximum air flow, temperature of drying air, and the web length between printing nips. Sometimes nothing more complicated than an electric fan is used to remove solvent from flexo or gravure inks, and other simple devices are often used. Hot air blowers do not supply a uniform film of air and may provide uneven drying. If solvent recovery is necessary, a more sophisticated dryer is required.

Dryers for waterbase inks must supply more energy to the web than those for solventbase inks because water requires more heat to evaporate. This is partly compensated for by the higher pigment content of waterbase inks, which can therefore be printed with a thinner film.

Radiation Curing. Ultraviolet (UV) and electron beam (EB) radiation supply enough energy to carry out polymerization reactions that convert liquid ink vehicles to solids as the web goes through the press. Although the process is in many ways more expensive than drying of solvent inks, UV and EB inks produce very little volatile organic compounds (VOC), reducing the cost of waste solvent recovery and disposal. The excellent gloss and toughness of the films have made UV and EB inks and varnishes thoroughly established in package printing and decoration.

Steam Dryers. Many gravure solventbase inks contain low boiling solvents that can be evaporated by running the printed web over a steam-heated roll.

Solvent Recovery and Disposal. Air pollution control requirements make it necessary to recover and dispose of solvents from drying inks, which is a costly operation. Many techniques are used for solvent recovery including condensation by chilling and adsorption on carbon or other substrates. The recovered solvent can be used to let down new ink or it can be burned, but sometimes it can be returned to the ink company. The evaporated solvent is frequently burned without recovery, and the heat of combustion helps fuel the dryer (see Chapter 12).

Since waterbase inks usually contain some VOC, they often must be treated like solventbase inks.

Chill Rolls

Dryers used to evaporate water or ink solvent may overheat the substrate, and chill rolls are used to cool the paper and restore it to its original size. Webs coming from an offset press or letterpress dryer are too hot to rewind or fold. The ink is still semifluid, and it must be chilled in order to set it.

Chill rolls are hollow, steel cylinders through which cold water circulates. The number of rolls depends on the type of press on which they are used. The chill roll system at the end of a web offset press usually has three or four chill rolls in it. The water must circulate rapidly to keep both sides of the web at the same temperature.

Chill rolls are sometimes equipped with a remoisturizing spray to replace some of the moisture lost during the drying. The web temperature should be around 90° F (32° C) as it exits the chill roll section. Chill rolls are often required for flexo or gravure printing of packaging materials.

Web Scanning

If the product is sheeted as it comes off the press, individual samples can be examined to be sure that everything is printing properly, but if the product is rewound, problems will be discovered after the press run is complete, too late to do anything about them.

An automatic register system monitors only a small part of the printed image. The entire image still needs to be observed at full production speed, and the wider the web, the greater the need for such inspection, especially on films and materials that stretch easily.

The simplest system is a stroboscope that is timed with the web so that the web is effectively "stopped" by the flashing light. This inexpensive viewing device has severe limitations. At speeds above 25 cycles per second, the stationary effect of the strobe can be lost, and it begins to provide constant illumination. If the speed is too low, the operator observes a brief instant flash that lacks the necessary dwell time to retain the image between flashes. It also rapidly causes operator fatigue.

Other web viewers are based on principles of rotating mirrors. The rotating mirror on a multisided drum follows the web. It is used under ordinary light and eliminates fatigue. The system is suited to web speeds of 100 to 300 feet per minute, but the image is somewhat unstable.

Video technology is replacing these two methods. A video camera with a zoom lens combined with a stroboscopic light source provides high quality images (figure 8-41). The camera can be moved to zero in on any part of the image. The images can be viewed sequentially and compared with a master image to allow close comparison. The system can be computerized to give the operator either a visual or audible warning when register or color go out of specs. The system in the diagram includes an automatic roll flagger and a printer for quality reports.

Ink Control

Viscosity Control. Flexo and gravure presses control color principally by controlling viscosity. When the viscosity of the ink increases, solvent is automatically added to keep the viscosity (and the color strength of the ink) constant.

As solvents evaporate from the flexo or gravure inking systems, the color strength and viscosity of ink change, affecting the color of the print. By monitoring the viscosity of the ink, the printer can add more solvent when necessary, and ink feed is kept under control. This is done manually on many presses, but installation of an automatic viscosity control system that adds solvent as called

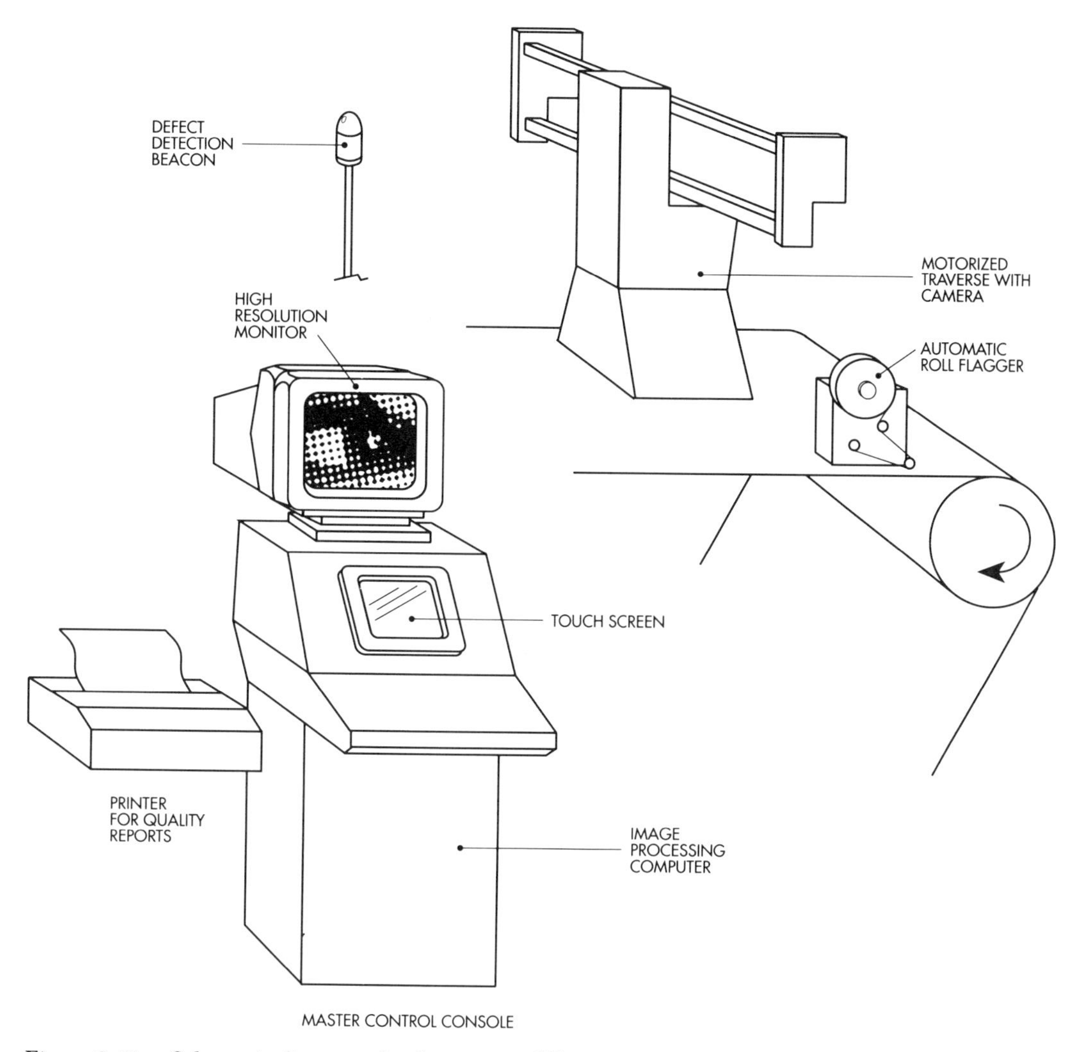

Figure 8-41. Schematic diagram of web scanner. (Illustration courtesy of CC1, Inc.)

for controls the amount of color applied and reduces crew requirements and color variability (figure 8-42).

Ink Feed Control. Letterpress and web offset inks contain high boiling solvents that do not evaporate on the press. However, ink feed must be controlled both before and during the run. The control used to be done manually by adjusting the ink keys or the fountain blade on the ink fountain, but modern presses are controlled at the press console by electrical servomotors.

Adjustment of feed before printing is done with a plate scanner that reads the plate and calculates the amount of ink required to print the image. The ink keys on the press are set automatically but final adjustment is made by operators at the console. A scanning densitometer records the color of the printed image, and adjustments in the ink keys are made from the press console.

Flexo and gravure presses control ink level in the fountain with cutoff switches that maintain a suitable ink level and prevent overflowing or running

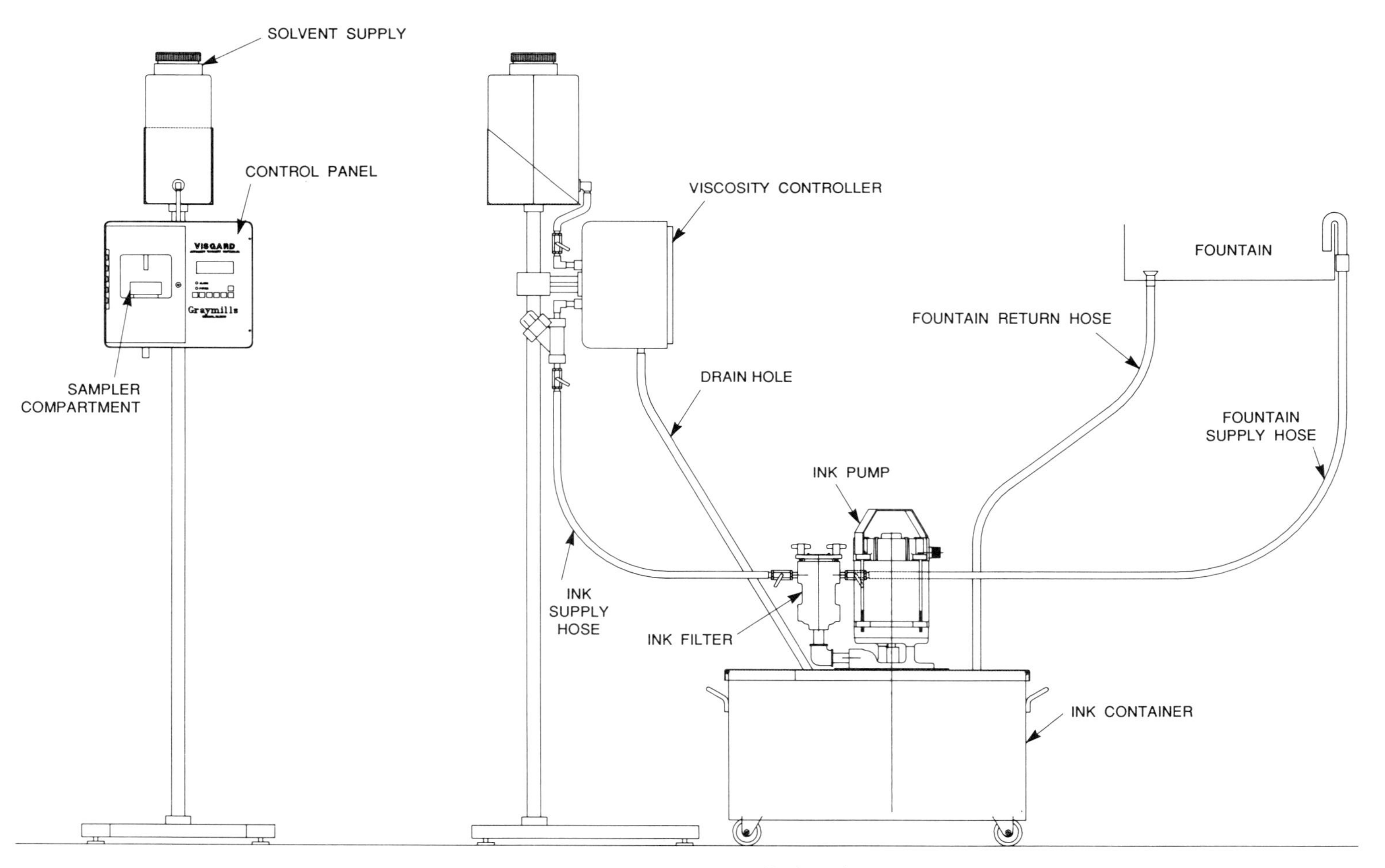

Figure 8-42. ***Automatic viscosity control system (Illustration courtesy of Graymills Corp.)***

dry. Sufficient fluid must be maintained to lubricate the system. Ink is a better lubricant than water. Should a flexo system run dry, the blades scratch and the anilox roll is damaged.

ROBOTS

The robot is making its way into the pressroom where it can change paper rolls or printing cylinders and plates. The robot can pick up gravure cylinders directly from a cylinder storage rack, set them near the printing section, and move them onto the machine, removing and restoring the old cylinders. Once placed onto its station, the cylinder is rotated into register with the other cylinders.

Robots change plates and anilox rolls on a wide-web CI press in less than a minute per plate. They quickly remove and replace gravure cylinders. While expensive, robots pay off in reduced makeready time and labor.

Other Auxiliary Equipment

Sheet Cleaner. A sheet cleaner removes loose dust from board or even from paper and greatly reduces the frequency that the press must be stopped to clean plates. Sheet cleaners include mechanical sweepers, vacuum cleaners, power blowers and combinations. Cleaners are more effective with an anti-static bar to reduce the tendency of the dust to adhere to the board (figure 8-43).

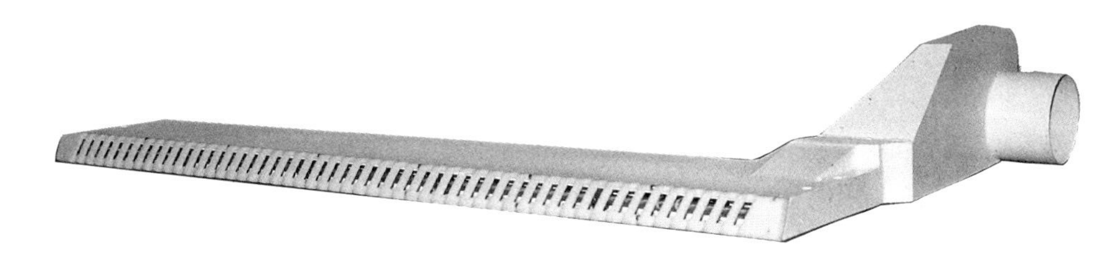

Figure 8-43. This sheet cleaner is attached to a sheetfed press. A web press has such a unit on each side of the web. (Photo courtesy of Baldwin Graphic Systems.)

Spray Powder Applicator. If wet prints or sheets are stacked, the ink on the face will transfer (set off) onto the back of the next sheet. Spraying starch onto the surface of the print puts enough space between the wet prints to prevent setoff. The spray powder applicator places an almost invisible layer of fine starch particles onto the surface of the freshly printed sheet. Spray powder applicators are standard equipment with large sheetfed offset presses (figure 8-44). They are not required when UV or EB inks or coatings are applied.

The powder makes the surface rough, and this is often objectionable for some types of packages such as cosmetics packages.

Static Eliminator. There are several methods of eliminating static electricity. Keeping the relative humidity above 50 percent can be helpful for paper,

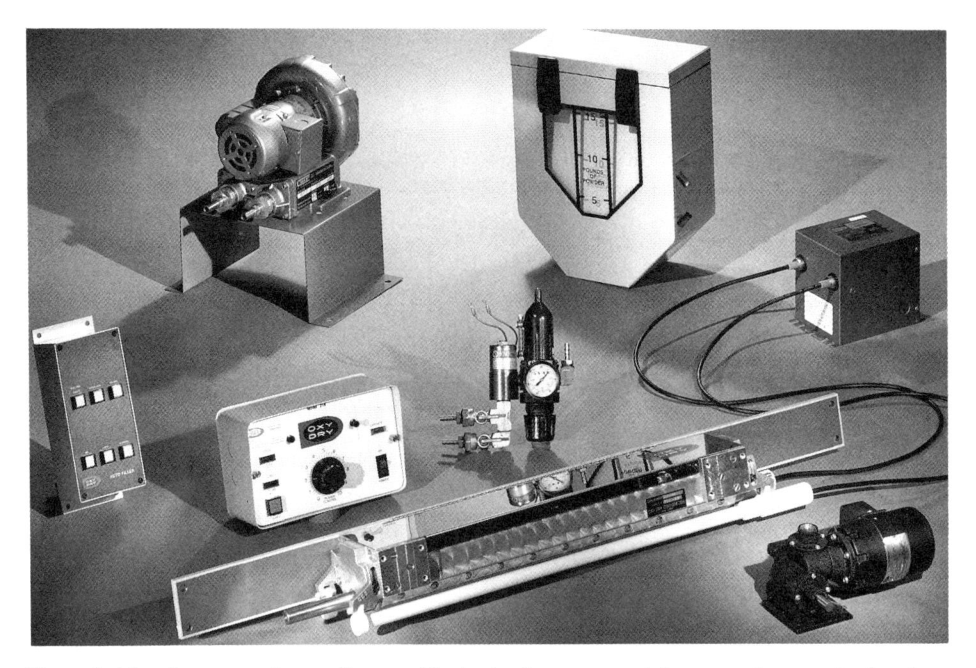

Figure 8-44. Spray powder applicator. Clockwise from upper right: spray hopper, high-voltage transformer, drive motor, sprayer box, controller and control panel, and pump. The regulator/filter and valve assembly are in the center. (Photo courtesy of Oxy-Dry Corp.)

but plastic films present special problems. Many devices are available. A familiar type of static eliminator produces ions to neutralize the ions on the substrate. Most electronic static eliminators place a high voltage on a sharp point close to a grounded shield or casing. This generates both positive and negative ions, and the film absorbs those that are opposite to its charge.

Electronic Impression Control. Electronic impression control keeps the plate cylinder in the proper position for printing. This equipment is particularly useful on flexo presses where control of impression is especially important.

FINISHING EQUIPMENT

A wide variety of finishing equipment is available to convert printed sheets or rolls into functional packages. This section presents several of the finishing operations that can be performed in line with the press and some that are performed on press or off line.

Finishing equipment commonly found on folding carton presses includes cutter/creaser, coater, sheeter, rewinder, rotary die cutter, rotary embosser.

Flexible packaging presses frequently offer reverse coating, laminating, punching/perforating, trimming and slitting. Extruders or laminators are the only major auxiliaries that are designed to precede the printing operation although they usually follow the printing.

Label presses can be equipped with in-line sheeter/jogger/stacker, rewinder, embosser, slitter, perforator, trimmer, coating unit and many other accessories some of which are listed in table 8-VI.

TABLE 8-VI. FINISHING PROCESSES

Coating and Varnishing	Embossing
Extrusion and Lamination	Trimming
Hot Foil Stamping	Numbering
Delivery systems	Windowing
Sheeting	Slitting
Rewinding	Counting
Folder/slotter/gluer	Bundling
Rotary Die Cutting	
Flatbed Die Cutting	
Punching and Perforating	
Scoring and Slitting	

Characteristics of Finishing Equipment

In-line vs Off-line Operations. In designing a process or system, the engineer must often decide whether to carry out a process in line with another process, or whether to break the process into independent operations. There are advantages either way, and one often finds successful companies doing essentially the same job in different ways.

The advantages of in-line operations are reduced handling and operating time. If the printer can put raw paper and ink into one end of an in-line operation and come out with a finished package, time and handling of materials are reduced to a minimum. On the other hand, the entire operation can operate no faster than the slowest component, and if any part of the operation fails or breaks down, the entire operation must come to a halt. The printer/converter should consider it vital to review new equipment periodically as improved technology becomes available.

Modularity. Modular construction means that an in-line operation can be assembled from pieces built to fit together. A common example is an eight-color press built from eight units, each capable of printing one color. Presses can be ordered with any number of units. They becoming larger and longer as better designs are developed.

There is a trend, perhaps led by the computer industry, to make parts in a way that permits the assembly of a press with a wide variety of accessories—coaters, laminators, slitters, trimmers—that are built to fit any press in the manufacturer's line.

Coaters

Coating is application of primers, overprint varnishes and adhesives. Coating procedures are commonly carried out on the press, where they are usually referred to as in-line or finishing operations. The subject of coating is discussed at length in Chapter 10.

There are many types of coaters. A light coating can be applied by any printing process in the same way that ink is applied. Flexo plates, gravure cylinders and offset blankets are well suited to application of either overall coating or spot coatings or varnishes.

Overall coating can be applied with a simple smoothing bar or rod, after the point of application, that lightly wipes the surface of the wet coating. It is an effective way to apply continuous, overall coatings of lacquer. It eliminates streaking or irregular striations known as "crow's feet."

Mixed processes are often used for coating. Offset presses may use an anilox roll to apply a heavier coating than could be applied with the usual offset inking system. In offset-gravure coating, the etched gravure coating cylinder is substituted for the anilox roll.

Folding cartons are frequently given spot coatings, primarily UV or acrylic, that offer eye-catching appeal. Originally applied from an offset press with a blanket cut to fit the design, spot coatings are now being applied from a photopolymer flexo plate that replaces the blanket on a tower coater or add-on coater.

Tower coaters are built into an offset press and are dedicated to the coating process. They often use an anilox roll metering system, but they may also use a two- or three-roll metering system. They are offered by all press manufacturers.

Add-on or retrofit blanket coaters are attached to the last print station engaging the plate or blanket cylinder. These units can be moved up and away from the print station so that the station can be used to print ink or apply a spot varnish. One unit is built so that the last unit can do both.

Finally, spot coatings can be screen printed. The process is slow and expensive, but superior gloss levels can be attained.

Sheeters and Slitters

The printed web must be converted to single units at some point. A sheeter converts it to sheets at the end of the press. Sheeters are usually built with a knife embedded in a cylinder that cuts the web into sheets with each revolution. Slitters cut the web longitudinally. Proper design and maintenance of these operations are critical to satisfactory operation.

Diecutters

There are two types of diecutters, rotary and flatbed. In either case, the cutting must be done in register with the image on the substrate. Diecutting comprises two steps: 1) the actual cutting of the pattern in the material and 2) the removal of the waste material. Also, diecutting may be part way through the laminate to cut a label but not the liner or all the way through the substrate.

Rotary diecutters have the knives placed in a cylinder; flatbed diecutters have them placed in a platen (figure 8-45). Corrugated, narrow web and wide web use different types of cutters. In a typical pressure sensitive label application, the diecutter cuts the labels without cutting the liner or support so that the trimmed part of the label (the matrix) can be removed and discarded, while the printed labels remain fixed to the liner (figure 8-46).

Flatbed diecutters (figure 8-47) are commonly used for folding cartons but less frequently on labels. They can be used in line with the press, but this requires

Figure 8-45. Die mounted in a flatbed platen-type diecutter. (Photo courtesy of Bobst Group, Inc.)

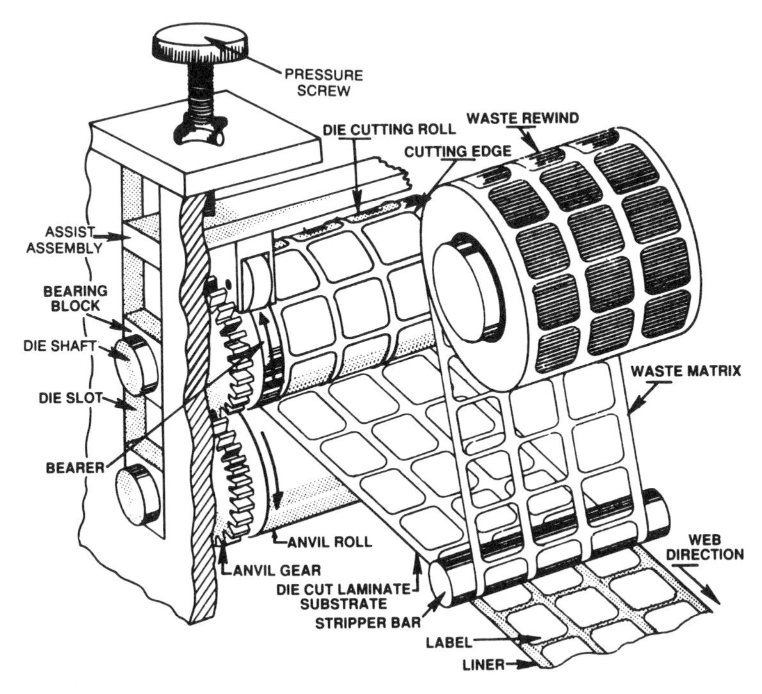

Figure 8-46. Stripping label waste in a label printing operation. (Source: Flexography: Principles and Practices, Fourth Edition.)

Figure 8-47. Sheetfed platen-type diecutter. (Photo courtesy of Bobst Group, Inc.)

that some mechanism be inserted in line to allow the web to flow continuously while the cartons are cut intermittently. The result is a slow process. Production is usually speeded if the cartons can be sheeted and diecut subsequently.

In-line diecutting done on a platen press limits any web press to a maximum speed, currently 500 to 600 feet per minute. It also requires four to six hours makeready. Sheeting can run up to 1500 feet per minute. Rewinding is no problem, so that the product from the web press is usually dried, rewound and fed to the die cutter. It is also practical to sheet in line and feed the sheets to a sheetfed die cutter. Press speed should be up to 900 to 1000 feet per minute from this operation.

An attractive alternative is rotary die cutting, in line with the web press. There are two tooling categories for the rotary cutter: a solid die with the die mounted to the cylinder, and the tongue and groove cutter.

Laminating

Laminating of two plastic layers or a plastic layer to paper or board on a flexo press is shown in figure 8-48. Laminating is often used to seal the printing into a sandwich. Laminating adds important properties to any single substrate, and its use has grown steadily over the years. It is possible to add a laminator to existing equipment to perform in-line laminating, but new presses are often equipped to perform this function.

Some materials, such as polyethylene, require only heat and pressure for adhesion to the substrate. Others require adhesives that must be kept transparent. The laminating equipment must be able to cope with the packaging requirements.

Foil Stamping and Embossing

Hot stamping or foil lamination of labels or cartons is carried out on a heated platen press (see figure 8-27). Webs can be fed through the press on an intermittent basis, but the product is frequently sheeted and run sheetwise.

For hot stamping, the type form of the platen press (see figure 8-25) is heated and a ribbon of adhesive foil is fed over the type form. Where the type presses it against the label or carton, it adheres to produce a gold or silver image. Dyed ribbons can provide other colors, too.

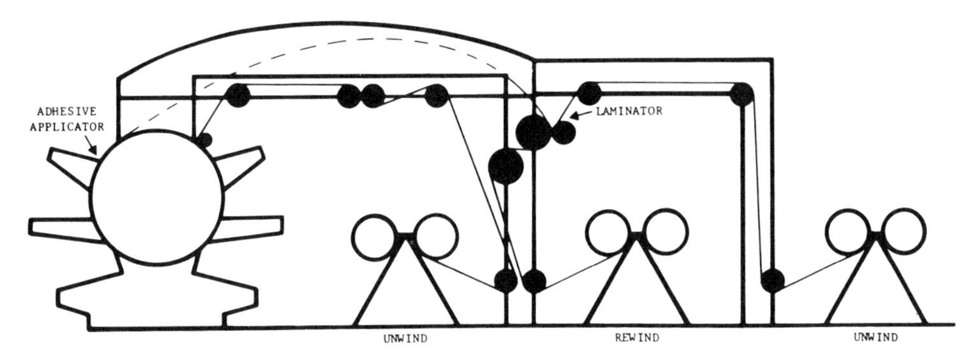

Figure 8-48. In-line laminating on flexo press. (Source: Flexography: Principles and Practices, Fourth Edition.)

Embossing is carried out on a platen press in which an image carried on the form or plate side is pressed into a matching image on the platen.

Demetallizing

Demetallizers are used to remove a metal from film to allow the product to be seen through the film. They are often used with a laminated sandwich, and the window becomes a part of the package design.

PRESS MAKEREADY AND FINISHING EQUIPMENT

The most important factor in controlling makeready time in any printing process is the organization of the process and the training of the crew. This aspect of productivity has been much neglected in the past, but competitive pressures and new business practices such as JIT require that management pay close attention to faster makeready.

Close behind organization and training is proper equipment. Improved equipment and supplies have helped reduce makeready time. Many small things improve the speed of makeready. Improved flexo plates, offset blankets and better backings tolerate greater variability in substrate, but papermakers and film manufacturers are making more uniform paper, board and other substrates. The combination means faster makeready, greater printing speed and better quality.

New press technology is changing makeready. Fast doctor-blade changes on flexo and gravure stations result from snap-in mounting. Sleeves for flexo and roto allow the plate to be fully mounted before makeready (figure 8-49), and robots change plate cylinders and anilox rolls on a wide web CI flexo press in a matter of seconds per plate. A computerized system for control of blade angle on the gravure cylinder speeds gravure makeready. Modern flexo and gravure presses do not have ink pans; ink is pumped into an enclosed system.

Computer-controlled consoles help the crew bring the press to chosen press settings quickly.

In wide-web flexo, the trend is to one-up plates made from single films from the step-and-repeat machine. It would take hours for a proofer-mounter to mount

Figure 8-49. Sleeve for flexo plates. (Photo courtesy of Strachan Henshaw Machinery Inc.)

four colors six up. One boxboard printer said: "We can quickly replace a defective plate. We take the cylinder out, strip the plate, replace the stickyback, and with the help of a pin register system, have the press running again in 18 minutes. If we'd used individual mounts in the old system, it would have been two hours."

Makeready includes getting the plates mounted and the ink and paper onto the press, but it also includes getting the finishing equipment ready for the job. Makeready of diecutters, slitters and trimmers often takes longer than makeready of the printing units.

Roll-to-roll printing requires the least time to adjust delivery. Sheeting requires more time, and in-line die cutting and scoring the most time. Cutting and scoring is often done off line by feeding the printed roll into the cutter. This speeds press makeready but still requires makeready of the cutter/scorer.

Generalizations about makeready time for various printing processes are difficult. For one thing, there is no generally accepted definition of which steps are included in makeready, and the number of steps varies from job to job. The lack of standards is partly responsible for the fact that makeready usually takes more time than necessary. There are great variations in the amount of makeready time for any particular process: the time varies according to the difficulty of the job and how much the new setup differs from the previous job.

Most of all, makeready time depends on the care with which prepress operations are controlled. If everything were planned in advance and kept under control, so that the plates and inks printed a package that matched the proof, makeready would be much faster than it usually is, and printing profits would be greatly increased. Makeready time is a key test of management's ability to organize and control the operation.

On short runs, makeready time may be longer than the press run. This illustrates why makeready procedures are critical. For products like beer cartons or cigarette packages that are put on the press and run for days, makeready time has much less impact on total costs.

Because of the time it takes to make ready on press, many package printers make ready off line on equipment that is an exact match for the press printing cylinder. This is especially helpful if pin registration is used. Optical/mechanical lineup equipment assures precise registration once the plates are mounted on press. The plates are registered on carrier sheets (blankets) that are registered to the press itself.

PRESS MAINTENANCE, CLEANLINESS AND PRODUCT PROTECTION

It has been observed that you can tell what condition the presses will be in when you walk in the front door of the plant. The attitude of management is reflected in everything from the landscaping of the plant to the condition of the presses and the care in protecting shipped products.

Presses that are dirty and in poor mechanical condition do not produce top quality printing. Part of the quality management process is to determine a predictive maintenance program that keeps the presses in top condition. When presses fail to operate properly or break down on the job, printer and customer both lose. In the world of JIT, the customer cannot tolerate late delivery, and a careful study of press maintenance is a major consideration in awarding contracts for package printing.

FURTHER READING

Flexography: Principles and Practices, Fourth Edition. Frank N. Siconolfi, Chairman. Flexographic Technical Association, Ronkonkoma, NY. (1991).

Gravure: Process and Technology. E. Kendall Gillett, III, Chairman. Gravure Education Foundation and Gravure Association of America, Rochester, NY. (1991).

The Lithographers Manual, Eighth Edition. Ray Blair and Thomas M. Destree, Editors. Graphic Arts Technical Foundation, Pittsburgh, PA. (1988).

Printing Technology, Third Edition. J. Michael Adams, David D. Faux and Lloyd J. Rieber. Delmar Publishers Inc., Albany, NY. (1988).

Introduction to Flexo Folder-Gluers. Joel J. Shulman. Jelmar Publishing Co., Inc., Plainview, NY. (1986).

Chapter

9

Printing Inks

CONTENTS

LIST OF FIGURES

LIST OF TABLES

Chapter 9

Printing Inks

INTRODUCTION

The packager needs to know enough about printing inks to ask intelligent questions and to judge the value of the answers. Ink technology involves advanced organic chemistry, physics and a great deal of practical experience. Packagers will rarely have or need this background. The packager needs a technical advisor; usually someone at the ink company can supply the needed help.

The inkmaker is usually selected by the converter, but the packager needs to have confidence in the technical competence and the integrity and reliability of the inkmaker. The packager must be willing to share a great deal of information with the inkmaker to get inks that will provide attractive decoration and meet the many technical requirements. Inks are priced competitively, and the technical support offered by the inkmaker is a large part of the value of the ink.

One of the most important current trends is the trend to waterbase inks for flexo and gravure and the move away from solventbase materials. Another important trend is the rapid acceptance of UV inks for most types of package printing except corrugated. Radiation-curing inks (UV and EB) virtually eliminate the use of solvents and reduce environmental concerns.

Most printing inks are dispersions of insoluble pigments in a liquid that can easily be formed into a thin film and rapidly converted to an adherent solid. The liquid is often called a "varnish." The most common ways of converting the wet ink or the varnish to a dry ink are by evaporation of the solvent from the varnish and by a chemical reaction (polymerization) that ties the varnish molecules together.

It requires a great deal of science and experience to make a printing ink that will successfully decorate a package and keep it attractive during handling.

The packager needs to know about the pigments used to make the inks. Many pigments contain heavy metals. They make excellent printing inks, but they are toxic if they get into food, and they will contaminate landfill when they are discarded. Most of these pigments have been replaced. The package manufacturer is legally responsible for any packaging materials that contaminate ground water, and he must be sure that printing inks used to print packages contain only pigments that conform to environmental regulations.

The packager needs to know about solvents used to make the inks because they may pollute the atmosphere or create a health or safety hazard in the printing plant. They are usually of little concern once the package has been printed.

The following discussion is intended to give the packager a basis for understanding the problems that the inkmaker should be able to solve.

Growth of Printing Processes and Printing Inks

The growth in consumption of printing inks provides a measure of the growth of the various printing processes as shown in table 9-I. Packaging inks comprise 35 percent by value and 45 percent by volume of all printing inks, reflecting the lower cost of packaging inks versus inks used for other purposes.

Table 9-II shows the preponderance of flexography and gravure in the printing of packages, and together with table 9-I, it reflects the dominance of offset lithography in publication printing. Table 9-III shows how package printing dominates flexo and represents a significant portion of the gravure ink market. Offset lithography is well suited for paper and board, but flexo and gravure are usually better suited to the plastic and metal used in packaging. Most offset inks used in packaging are used on folding cartons and labels, and virtually none are used for flexible packaging, as shown in table 9-IV.

Selecting the Best Ink

The printing ink is a key part of any printing process. Each printing process requires its own different type of ink. Printing inks are divided into two groups:

TABLE 9-I. GROWTH IN CONSUMPTION OF PRINTING INKS

	% Annual Growth	
	By Dollar	By Weight
Gravure	4.5	3.0
Litho (Sheet & Web)	10.0	8.5
Flexo	7.0	7.0
Letterpress	−2.5	−7.5
Other*	8.5	2.0
Total	7.1	5.3

* Includes screen inks, but not those used for textiles
Source: Rauch & Associates (1988).

TABLE 9-II. SHIPMENTS OF PRINTING INKS FOR PACKAGING

	$ Millions	Pounds, Millions
Flexo: Solvent	189.3	117.1
Waterbase	172.4	125.0
Gravure	153.6	111.3
Litho	77.7	18.2
Letterpress	27.4	12.1
Screen	59.6	*

Source: NAPIM and U.S. Census of Manufactures 1987

Note: It is probable that the figures for litho are low since commercial printers are heavily involv in the printing of labels and carded packaging.

TABLE 9-III. PERCENT OF PRINTING INKS USED IN PACKAGING

Based on wieght: 25–26% of all inks

Based on value: 31–35% of all inks

	Weight	Value
Flexo inks	95%	95%
Gravure inks	27	37
Offset inks	5	10
Letterpress	1	2
Screen inks	20–30	20–30

Source: Industry estimates.

fluid or low viscosity inks used for flexo and gravure, and paste or high viscosity inks used for offset, letterpress and screen.

In flexo and gravure, inks must have relatively low viscosities to flow quickly in and out of the cells of the anilox roll or the gravure cylinder. Because the printing unit is simple in design, low-boiling solvents can be used. These are easily evaporated, making the ink easy to dry. Flexo inks cannot use some of the stronger solvents that are useful in gravure printing because the plastic or rubber plates and plastic rolls may be attacked or swollen by the solvent.

Offset and letterpress inks must be worked on the press to reduce their viscosity so that they can be printed without tearing the paper. If their viscosities were as low as the inks used in flexo and gravure, they would run or drip off the press. Sheetfed inks dry by a slow, complicated chemical process; web inks dry by evaporation of a high-boiling solvent. If flexo or gravure inks were applied on an offset press, the inks would evaporate and dry before they reached the printing plate. A web offset ink cannot be dried on a sheetfed offset press, nor a sheetfed ink on a web press.

Next, the ink must be selected to suit the substrate. Inks for flexible packaging are different than inks for folding cartons, film or foil. Inks that work well on cellophane are not best for polyolefins.

Printing inks are applied to packages to provide color, so it is obvious that

TABLE 9-IV. DISTRIBUTION OF INK CONSUMPTION FOR FLEXIBLE PACKAGING
(Percent by Value of Shipments)

	1985	1988	Growth Rate
Flexography	56.4%	57.6%	4.5%
Gravure	43.6	42.4	−2.8
Waterbase	16.1	31.1	29.3
Solventbase	83.9	68.9	−2.9
Film	34.2	35.1	4.7
Paper	36.1	33.4	1.1
Laminate	29.7	31.4	5.7
All inks			3.8

Source: GATF, TechnoEconomic Market Profile No. 13: Flexible Package Printing and Converting (1988).

color is a key to printing ink value. The pigment is the colorant used in most inks. Pigments are insoluble and dyes are soluble colorants. Most dyes can be made insoluble by chemical reactions. For most printing applications, dyes are undesirable because of their tendency to "bleed" or to dissolve in water, oil or other liquids that come in contact with the package. Dyes often fade rapidly, making them unsuited to most packaging applications.

Pigment is usually the most expensive component in a printing ink, and the price of the ink can be reduced by reducing the amount of pigment in the ink. This also reduces the color strength of the ink. Reducing the color strength increases the amount of ink that the printer must run, so the lower price may be no bargain. Measurement of color strength is discussed later in this chapter.

Many other properties must be considered in formulating an ink that will perform well on press and on the package after it is printed. The color of the ink is determined by the pigment. It must have the right hue, but many other properties are also required, such as resistance to light and resistance to bleeding or to the product that is to be packaged. Environmental and safety factors must also be taken into account.

From a technical viewpoint, it is optimal to have every ink formulated specifically for each job. This would increase costs and complicate inventory and delivery problems, so an intelligent compromise must be worked out.

In practice, selecting the best ink is a complicated problem. Supplier and customer must work together to set up specifications that will assure that the printing ink provided is the best possible choice. Price is always a consideration.

The design sometimes calls for the same ink to print both a heavy solid and also a light halftone. Line colors tend to be denser and more opaque than process colors; the printer must make a compromise between these conflicting requirements. The ideal solution is to put the two patterns down from two press units, but extra press units are not always available.

It is important to consult with the inkmaker before ordering an ink. Sometimes small changes between one printing job and the next require important changes in the ink formulation. Stock or "off-the-shelf" inks are satisfactory for many packages, but the inkmaker knows the formulation and can make the best judgment.

MANUFACTURE OF PRINTING INKS

The inkmaker buys pigments from one manufacturer and the varnish from another and mixes them with driers, waxes and other additives. High-energy mixers work the inks to be sure that the pigments are thoroughly dispersed. The pigments may be "dry colors" (dried pigments) or "flushes," highly concentrated pigments in an oil base. There is a wide variety of varnishes bearing names such as rosin-ester, cellulosic, polyamide, linseed alkyd and acrylic. Some of these are good adhesives that are dissolved in a solvent, others polymerize to become solid after the ink has been printed. Each has properties suitable for some type of packaging ink. The printer is usually less interested in the formulation than in the performance characteristics.

Although the inkmaker is the expert concerning properties of pigments, vehicles, solvents, driers and additives, he cannot bear full responsibility for formulating the correct ink. Both the converter and the packager must be sure that the inkmaker knows the final requirements of the ink. Converter and packager must give the inkmaker all the job specifications: color, printing process, press, substrate to be used and the product to be packaged. If a varnish or overprint coating will be applied, the ink must be compatible with it. There are special requirements for printing of packaging when the ink or coating may come into

Figure 9-1. Continuous ink mill. (Courtesy of Premier Mill Corp.)

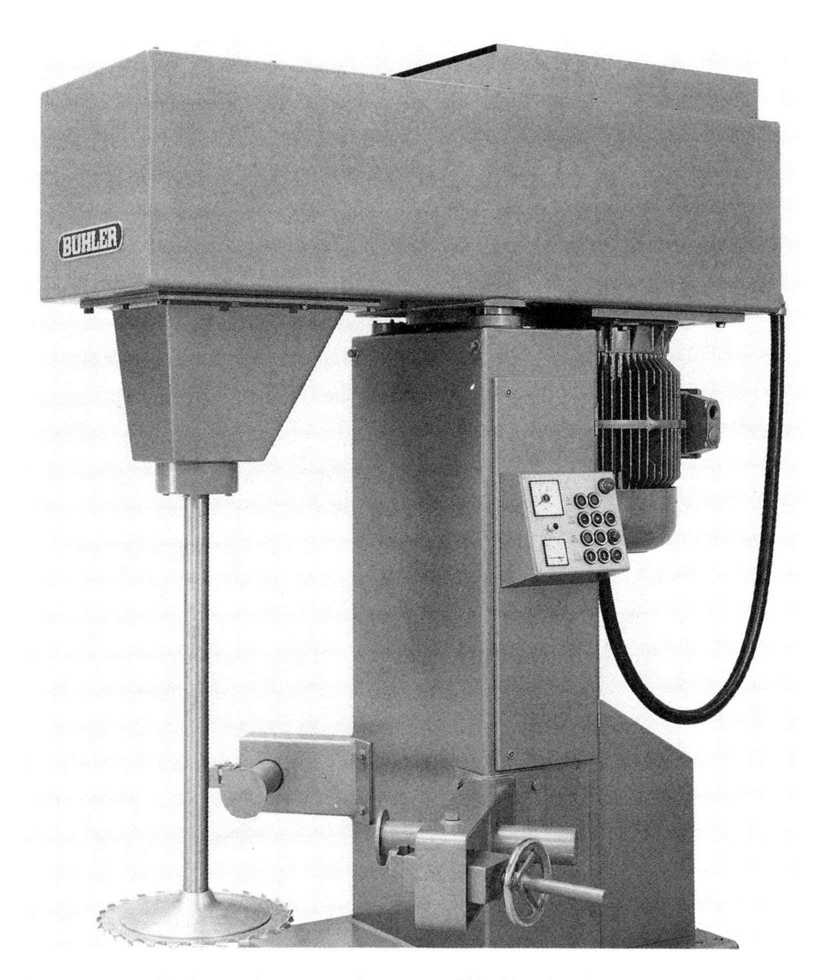

Figure 9-2. High speed mixer. (Courtesy of Buhler Inc.)

direct contact with food products. Oily products may require a different formulation than dry products. Oils, soaps, solvents and other packaged products require special consideration.

Fluid inks for flexo and gravure are easily processed because of their low viscosity. They should be made in closed containers to keep the solvents from evaporating. The inks are now made in continuous mills (figure 9-1) which have beads, steel or ceramic balls, or sometimes sand as the abrasive material that "grinds" the pigment. To keep the pigments from settling out, the inks are sometimes shipped at a high concentration and diluted to the proper viscosity at the press.

Fluid inks are sometimes made from pigment chips. Most of the work required to disperse the pigment is done in mixing the pigment with the resin that forms the chips. No more than a high-speed mixer is then required to make the ink.

Processing of paste inks requires more energy than of fluid inks. If a dry pigment is used, it must be dispersed in the varnish with a high-speed mixer (figure 9-2). Other ingredients are added, and the ink is ground in a three-roll

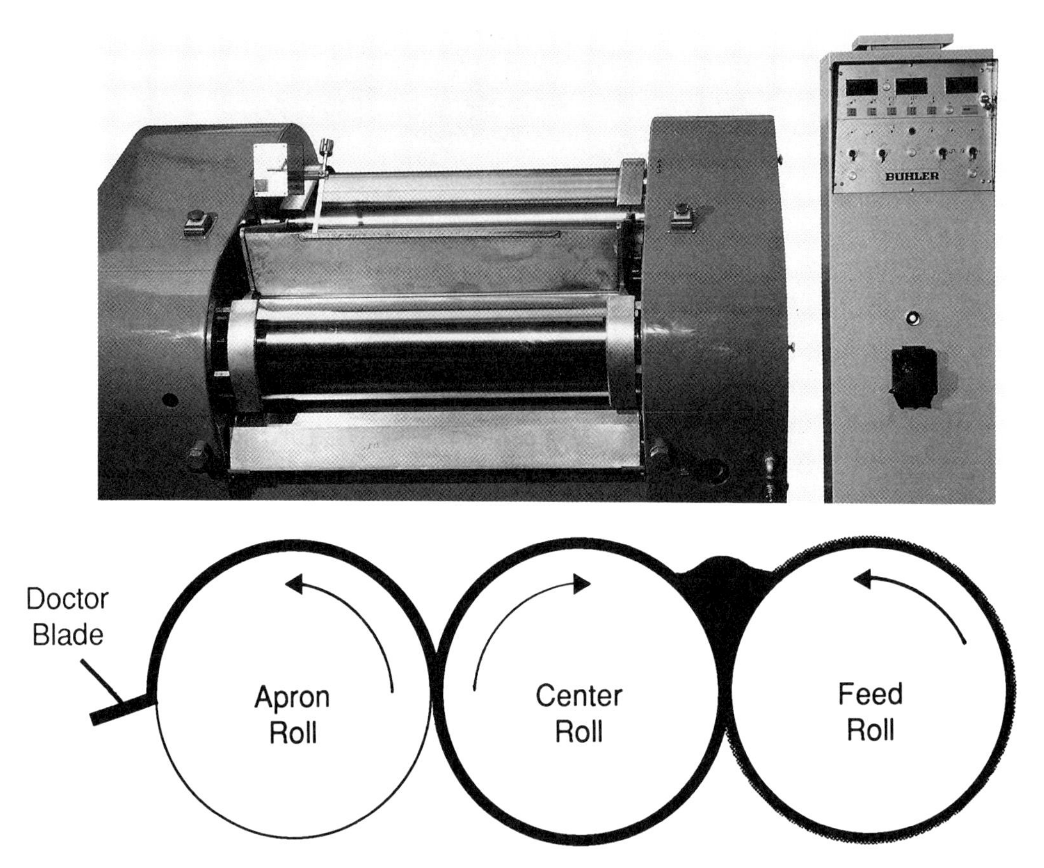

Figure 9-3. Photo and diagram of three-roll mill. The apron roll turns faster than the center roll which turns faster than the feed roll. The speed differential creates a shearing or grinding action. (Courtesy of Buhler Inc.)

mill (figure 9-3) which grinds or removes any grit or undispersed pigment and air bubbles. If a flush is used, the ink is still mixed with a high-speed mixer and ground on a three-roll mill, but the process is faster, requiring less power. Ink made from flushed pigments is less likely to have hard aggregates of pigments in it.

PIGMENTS

Pigments are included in the ink formulation to give ink the desired color. However, pigments and ink are not the only factors determining the printed color. The substrate is also important. An ink printed on art board will print a different color on a plastic sheet and still another color on a metal can. The difference may be slight or great, but it is unpredictable.

Under proper controls, a printing press can match the original color closely, but variability on the press often causes an unacceptable variability in the resulting color.

Pigment Properties

Pigments impart color to inks. The pigment also affects the working properties of the ink and the properties of the printed package. Pigments can be divided into two groups: organic pigments and inorganic pigments. Organic pigments are based on petroleum products; inorganic pigments are minerals that are mined from the earth, or they are made from such minerals.

Many properties are required for a pigment to be useful. Not only is the color (or "hue") important, but the color must resist fading and bleeding in solvent or in packaged products, it must have good transparency (or sometimes good opacity), good color strength, and brightness or purity. Other properties required to make a good pigment include particle size, gloss, durability, refractive index, wettability and dispersibility.

The packager needs to know about resistance to light, heat and chemicals. The packager must be concerned not only with fading, but with darkening or other color changes. Lightfastness depends on the chemical nature of the pigment, but it also depends on the thickness of the printed film, the protective properties of the ink vehicle, the time of exposure and the intensity of the light. Screen printing and gravure, that usually put down the thickest ink films, have an advantage here; lithography, that puts down the thinnest film, has a disadvantage.

Color strength of the pigment, discussed in Chapter 5, determines how much pigment is needed to give the desired color strength to the ink. If a pigment has a weak, washed-out color, a great deal of it is needed to make an ink with sufficient color strength, an undesirable property in pigments.

The ability of a pigment to yield a glossy ink film is important in packaging inks.

Opacity may or may not be desired. If a transparent pigment is desired, as with process color inks, then very low opacity is desired. If the ink is to cover a transparent film, such as white inks, then high opacity is an advantage. Spot color inks often have high opacity so that the colors are unaffected by the background color. Opacity is related to a property called "refractive index" or the power to bend light. Pigments with high refractive index are more opaque than those with low refractive index.

The shade of pigments sometimes varies slightly from batch to batch, and inkmakers usually blend several batches of pigments to make sure that different batches of the same ink all have the same shade. Sometimes, closely related pigments differ significantly in hue or shade, such as red shade or green shade phthalocyanine blue pigments.

Small pigment particle size is preferred. Carbon blacks have the smallest size of any ink pigment, and they show good flow properties. Coarse pigments, like titanium dioxide, are apt to settle out. They may also cause flow problems.

Wettability and dispersibility of pigments refer to the ability of pigments to be wet out and dispersed in the printing ink vehicle. These properties may differ greatly between organic solventbase varnishes and waterbase varnishes.

Manufacture of Pigments

There are thousands of different pigments, but most printing ink is made from one or more of a dozen or so important pigments. Black inks are made from carbon black, mostly in a form called furnace black. Magentas are made from lithol rubine or rhodamine. Rhodamine has a better shade, but rubine is less expensive. Cyans are made from a pigment called phthalocyanine, from which the color is named. Sometimes a reflex or alkali blue is used for blue inks. Yellow inks are most commonly made from diarylide yellow pigments. All of the above pigments are organic pigments, made from petroleum. White inks are made from titanium dioxide, the only inorganic pigment mentioned in this paragraph.

Fluorescent pigments are made by dispersing a fluorescent dye in an inert, insoluble resin and grinding the product to a small size. These pigments work best in thick printing ink films, and they produce the best results when printed by gravure or by screen. Satisfactory results are often obtained by flexo or litho.

Gold inks are made from bronze (an alloy of copper and tin) and silver inks are made from aluminum powders. Like the fluorescent pigments, they are more successfully applied in thick films. Lithography has the most trouble printing gold and silver inks.

Pigments are manufactured by either of two processes. Flushed pigments are now more popular than dry colors, at least in North America. Whether flushed or dry color, pigments are often made from dyes that are manufactured by mixing aqueous solutions of two or more chemicals.

When a pigment is flushed, the filter cake is mixed in a high-energy mixer with the chosen printing ink vehicle. The pigment is wet by the vehicle, and it enters the varnish, leaving the water. The water is poured off, and the flushed pigment is sold to the inkmaker. Flushed pigments, suspensions of pigment in varnish, are free from grit, and they are easily dispersed in the printing ink.

To make a dry color, the filter cake is dried in an oven. The inkmaker can put the dried pigment into any kind of printing ink vehicle he chooses, but the dried pigment is hard to disperse, and it usually contains some grit, requiring that the inkmaker grind the ink carefully. This takes time and electric power. On the other hand, dry colors often yield inks with higher color strength than flushes.

Lithographic, screen and letterpress inks are sold at a strength ready for the press. Paste inks are not readily mixed in the pressroom. Flexo and gravure inks are often sold in a concentrated form and diluted or "let down" to the correct viscosity and color strength at the press. "Proprietary colors," those used for Coca-Cola red, IBM blue, or Kodak yellow, are always sold at the strength to be used on the press.

Pigments for flexo, gravure or screen inks can be milled into nitrocellulose, acrylic or polyamide resins by the pigment manufacturer and sold in the form of pigment chips.

Spot Colors

Spot colors, such as those used for producing a company logo, are usually produced from inks that are mixed, from eight basic colors, to closely specified tolerances. (In addition to the eight basic colors, other specialty colors are available.) They are popularly referred to as "PMS colors" or "Pantone colors" from the name of a leading supplier of standards for mixing pigments to match a wide variety of colors. The colors can be purchased ready mixed, from the ink supplier, but many printers mix them in their own printing plants.

In matching the design color, it is important to use a numerical specification, and the PMS colors are very useful for this purpose. Color descriptions such as "apricot yellow" or "mist blue" are worse than useless—they are misleading.

Heavy Metal Content

Packagers need to know about the presence of heavy metals in the printing inks they use. Inkmakers have been successful in replacing most of the heavy metals formerly used. Lead chromate has been replaced with organic yellows, and when opacity is required, a transparent pigment can be mixed with titanium dioxide. The packager must be concerned not only with the toxicity of the pigment on the package but with potential contamination of the environment when the package is buried or burned.

Pigments based on heavy metals are no longer used in waterbase inks, and they have been largely eliminated in solventbase inks. Nevertheless, some heavy metals do occur as trace contaminants in other pigments. For example, lead shows up as a trace contaminant in titanium dioxide, and it is the packager's responsibility to be sure the ink conforms to governmental requirements. Inkmakers (and all other manufacturers of industrial chemicals) are required by law to supply Material Safety Data Sheets (MSDS) that describe proper handling and any special hazards of each material that they sell.

Dyes

In addition to the pigments used for most printing inks, dyes are used for ink jet printing and sometimes for flexo. Dyes often have chemistry similar to that of pigments except that they are in a soluble form. Sometimes they are significantly different.

Dyes offer strong, brilliant colors, but they are soluble in many fluids and tend to bleed or run. They cannot be used on most packaging materials. They can be used on paper, if there is no chance for water or solvent to cause them to run and bleed. In addition to poor resistance to water, most dyes have poor lightfastness and are useful only on disposable products with a short life, such as paper bags.

Lightfast, spirit-soluble dyes are commonly used for highly transparent, colored foil coatings. The printer and packager must carefully check the bleed resistance of these inks. They may bleed into or through a base coat or topcoat used on the package.

VEHICLES AND VARNISHES

Resins, Oils and Polymers

The terms "vehicles" and "varnishes" tend to be used interchangeably, although in a strict sense varnish means the ink binder (dissolved in a solvent) and vehicle refers to the varnish plus additives such as wax or driers.

Table 9-V lists the resins most commonly used in printing inks and the inks in which they are usually used.

Rosin is a natural resin, obtained from pine trees and from paper pulping. For use in inks, it is chemically modified to improve its printing properties, such as solubility and melting point.

Hydrocarbon resins are manufactured by polymerizing chemicals obtained during the distillation or cracking of petroleum. These low-cost synthetic resins provide good printing properties for offset, gravure, letterpress and screen inks.

Nitrocellulose was one of the first plastics, made by reacting cotton with nitric acid. Nitrocellulose gives tough, brilliant films, and it improves wetting of ink pigments by the ink. It is still used in some flexo and gravure printing inks. Other cellulose esters are used in gravure and screen inks. Methyl cellulose is used as a thickener in waterbase inks.

Shellac is a natural resin that was one of the first binders used for flexo inks, and it still is used for both solventbase and waterbase flexo inks. Shellac is in short supply, and in waterbase inks, it has largely been replaced with acrylic resins. Polyamides used in flexo and gravure inks are produced from natural oils, such as soybean oils. They provide excellent adhesion to polyethylene and other packaging films.

Acrylics are produced from acrylic or methacrylic acid and other monomers. Depending on their chemistry, they can be used in either solventbase or waterbase inks. They provide good flexibility, pigment wetting and gloss.

Long-oil alkyds are made from linseed oil, soya oil, tall oil (from paper pulping), dehydrated castor oil and other vegetable oils. The oil, the alkyd and the reaction conditions all affect the properties of the inks. Long-oil alkyd varnishes

TABLE 9-V. USES OF PRINTING INK RESINS

	Flexo Solv	Flexo WB	Gravure	Sheetfed Offset	Web Offset	Screen	Sheetfed Letterpress
Resin							
Rosin ester	x	x		x	x	x	x
Hydrocarbon				x	x	x	x
Nitrocellulose	x		x				
Shellac	x	x	x				
Polyamides	x		x				
Acrylics	x	x	x				
Long-oil alkyd				x	x	x	x
Ketone resins	x						
Vinyl polymers			x				
Chlorinated rubber			x				

dry by a chemical reaction with air and a drier. They are used extensively in sheetfed offset inks.

Ketone resins are used as modifiers for cellulosic, acrylic and polyamide resins to improve gloss and adhesion in flexo inks.

Vinyl polymers, such as poly(vinyl butyral) yield inks with excellent adhesion to glass, metal and plastics. They are most commonly used in gravure inks.

Vehicles used for ultraviolet (UV) and electron beam (EB) inks are polyesters or low-molecular-weight acrylic materials that polymerize very rapidly under irradiation of UV and EB. Formulations for the two types of inks differ somewhat. Although the resins are relatively expensive, their end-use properties justify their use for printing of some folding cartons, plastics, foils and metals.

Waterbase flexo and gravure inks can be made from emulsified polymers or alkali-soluble polymers that form a film on drying. Most acrylics are not soluble in pure water. They are dissolved with ammonia or an organic base, or they require an organic solvent in order to form a tough film.

UV and EB inks have always tended to take up more water than inks made from conventional vehicles, and chemists have taken advantage of these properties to make water-dilutable resins and to replace expensive, toxic diluents. While they are not useful in offset, the new inks are useful in flexo, gravure, screen, and notably ink-jet printing where they produce a solvent-resistant ink film, something not previously available from ink jet. In screen printing, waterbase UV permits use of a finer screen. The ink film can be as thin as eight to 10 microns. The cured ink films are resistant to temperatures as high as 330-360° F (170–180° C), temperatures much higher than those from most conventional ink resins.

Vehicles for metal decorating may be prepared from long-oil alkyds modified with synthetic resins, but polyester vehicles give excellent adhesion, gloss and abrasion resistance. Polyesters can cure in a minute or less in an oven.

Solvents

Solvents are part of the vehicle. They dissolve the oils, resins and additives to produce varnishes that carry the pigment in the ink and adhere the dried ink to the package. There are several groups of volatile organic compounds (VOCs) used as solvents: aromatic and aliphatic hydrocarbons, alcohols, esters, ethers and ketones. The solvent power of each group is somewhat different, and each chemical within the group has its own boiling point. To make an ink "faster," a low boiling solvent (one that evaporates rapidly) is added. To make an ink "slower," a high boiling solvent is added. "Faster" and "slower" refer to the speed of drying.

Flexo and gravure inks often contain a mixture of solvents. Because one solvent is always more volatile than another, the composition of the solvent changes as the ink evaporates.

A few of the most commonly used solvents are listed in table 9-VI. Isopropyl alcohol is the alcohol most commonly used, but ethanol or propanol are sometimes used. Petroleum distillates used are highly purified aliphatic or alicyclic hydrocarbons. The most common glycol ether is the monoethyl ether of ethylene

TABLE 9-VI. SOME SOLVENTS COMMONLY USED IN PRINTING INKS

Solvent	Gravure	Flexo	Offset	LP	Screen
Water	x	x			x
MEK	x				x
Toluene	x				x
Alcohols	x	x			x
Glycol ethers	x	x			x
Petroleum dist.	x	x	x	x	x

glycol, usually referred to as "Cellosolve," a common trade name, but the methyl ether of propylene glycol is also frequently used. Methyl ethyl ketone is usually referred to as MEK.

Except for some chlorinated materials, organic solvents are flammable, and, under certain conditions, explosive. Different solvents have different degrees of toxicity and volatility, but all have some degree of toxicity and most present air pollution problems. In general, the lower the boiling point, the greater the likelihood that the solvent will reach explosive concentrations in the air. The low concentrations of alcohol in flexo inks helps reduce the hazard.* The toxicity and flammability of organic solvents are driving the development of waterbase and UV inks.

Waterbase inks overcome many of the environmental problems caused by solvents in solventbase flexo and gravure inks. They behave very differently on the gravure press than on the flexo press because of the effect of viscosity and surface tension, on wetting of the gravure cylinder and flexo plate and release of the ink from them. Waterbase inks tend to wet the paper so that it swells or stretches. Waterbase inks also have problems wetting plastic substrates. For obvious reasons, a great deal of research and engineering work is going into the development of waterbase inks and wettable plastics.

Large amounts of waterbase inks are used in gravure packaging, but almost none is used in publication gravure.

Solvents may also be used as diluents or thinners. A thinner is a solvent for the vehicle, and it usually reduces the viscosity. A diluent is a nonsolvent or a poor solvent. Diluents are often added to reduce the rate of evaporation or to reduce the cost of the ink.

Additives

Ink additives—plasticizers, wetting agents, antisetoff compounds, waxes, shortening compounds, reducers, stiffening agents, antiskinning agents, antipinhole compounds and driers—are required to turn the suspension of pigment in a vehicle and solvent into a printing ink with the properties required for good printing and good performance on the package.

* An excellent discussion of the specific properties and hazards of each solvent used in fluid inks is presented in the *Printing Ink Manual*, Fourth Edition. R. H. Leach, Editor. Van Nostrand Reinhold (International). London, (1988). pp 391–392.

Plasticizers make the dried ink softer and more flexible, improving its adhesion to films and other substrates. They often improve abrasion resistance because a brittle film may abrade more easily than a flexible one. Waxes, such as petroleum wax, enhance the slip of the printing ink, improving rub resistance and reducing abrasion and scuffing.

Lithographic, letterpress and screen inks sometimes set off onto the next sheet when freshly printed sheets are stacked in a pile. Antisetoff (anti-offset) compounds such as wax and grease may be added by the inkmaker to reduce this tendency to set off. The sheetfed printer using any of these processes adds a spray powder that helps to keep the paper on the back of the sheet separated from the ink film on the next sheet. If heavily applied, the antisetoff powder makes the printed surface feel and act like sandpaper. It can also improve the skid resistance of the printed package.

Inks that dry by air oxidation tend to skin over in the can or on the press. An antioxidant, that prevents the reaction, will prevent these inks from skinning over in the can when they are applied to the surface of the ink. If they are sufficiently volatile, they will not greatly retard the drying of the rest of the ink.

Aqueous (waterbase) inks are usually formulated with defoamers that may themselves be complex mixtures. By reducing the tendency to foam, they reduce the formation of pinholes in the ink or coating on the printed substrate.

Driers are organic compounds of cobalt or manganese (or sometimes other metals) that promote the oxidation-polymerization reaction by which sheetfed offset, letterpress and some screen inks dry. They are sometimes incorporated into heatset web offset inks to provide a hard film after the prints have been delivered from the press. Sometimes the compounds are suspended in a liquid such as a petroleum solvent. These are called liquid driers. The metals may also be cooked with resins to yield a high viscosity paste that can be added to a printing ink without reducing its body.

FLOW OF PRINTING INKS

Viscosity and Tack

Two terms the packager should be familiar with are "viscosity" and "tack." Viscosity is the resistance to flow; tack refers to the force required to split a thin film. It is rather like "stickiness."

The term "thick" is sometimes used to refer to high viscosity materials like honey or sheetfed offset inks and "thin" to refer to low viscosity materials like water or flexo and gravure inks. However, these words become confused with thick ink films and thin films, and the terms "high viscosity" and "low viscosity" are used to avoid confusion.

Although fluid inks are used for both flexo and gravure, there are enough differences that an ink for one process will seldom work for the other. Gravure inks often contain strong solvents that soften flexible packaging, improving the adhesion or giving the ink a "bite." Such inks will soften flexo plates and cannot be used. Flexo inks must flow into and out of the tiny cells in the anilox roll. Gravure inks must easily flow into and out of the tiny cells engraved on the surface of the cylinder, but the wetting and flow characteristics of inks that work well for flexo are seldom suitable for gravure printing.

Most important, the ink film printed by gravure is usually thicker than that of flexo, and the concentration of pigments in each ink must be controlled to give the proper color density.

The solvent in the ink always has a lower viscosity than the varnish. As the solvent evaporates, the viscosity increases and the pigment becomes more concentrated, changing the color. Monitoring the viscosity of the ink is a convenient way to keep the color strength of the ink constant on the press and on the package being printed.

Viscosity is measured to check whether the correct amount of solvent has been added to a flexo or gravure ink. The viscosity is also measured during the run to check on the evaporation of solvent from the ink on the press. As the viscosity increases, the amount of colorant delivered by the press increases. Since the color strength of the ink also increases when solvent evaporates, the two factors compound each other.

With flexo inks particularly, a slight increase in viscosity will not cause noticeable color change on the printed substrate. Unfortunately, the cost of the ink used can be doubled before a noticeable change shows up in the printing. Control of viscosity of flexo inks thus becomes more than a matter of appearance of the finished package.

Modern flexo and gravure presses are now built with closed ink fountains or chambers so that evaporation is largely eliminated. They have electronic instruments that continuously measure and record the viscosity, and they are often programmed to adjust solvent to maintain constant viscosity. Economy and uniformity of color throughout the run are greatly improved by these two steps.

Monitoring the ink viscosity in the pressroom gives the printer valuable control of the process. There are several ways of measuring viscosity. Two of the most common are the viscometer and the Zahn or Shell cup. Table 9-VII compares readings obtained by these different methods for various efflux cups.

The viscosity of waterbase inks depends on the pH of the ink and the viscosity changes if the pH of the ink changes. It is vital to monitor both the viscosity and pH of these inks.

The viscosity of ink decreases rapidly as it becomes warm. If the temperature of the sample to be tested is not kept constant, the viscosity measurement becomes of little value.

All printing inks exhibit a property called "thixotropy." The viscosity decreases as the inks are worked on the press, and it recovers when the inks are no longer worked. Thixotropy is especially important with paste inks. It is the purpose of the complicated inking system on litho and letterpresses to reduce the viscosity of the printing ink at the moment it is printed on the substrate. When the mixing or working is stopped, the ink again increases in viscosity. This is one of the ways in which printing inks "set."

Trap

"Tack" is the stickiness of a wet ink film, and "trap" is its transfer. Control of tack is important to assure the proper transfer of ink and to maintain good trap and good color uniformity. The ink with higher tack pulls the ink with the lower tack from the plate or substrate. If ink on the plate or blanket pulls ink from the substrate, the behavior is referred to as "back trapping."

As the printing ink dries, the force required to split the ink film increases, and a dry printed ink film traps the ink from the next printing plate better than a wet film. This gives an advantage to printing processes in which the ink is dried after each unit. Where wet-on-wet trapping is used, ink must be formulated to transfer properly.

In flexo and gravure inks, the effect of tack on trapping is not clear. Because these inks often dry between print stations, either with dryers or by absorption into the paper, the question of tack of the liquid ink is not a major problem. Nevertheless, proper formulation of flexo and gravure inks is just as important as that of any other ink.

With the paste inks used in litho and letterpress, tack is related to the viscosity. The body of the ink must be properly adjusted if the ink on the plate or blanket is to transfer to the wet ink on the print. Try to visualize a peanut butter and jelly sandwich. The jelly will spread on the peanut butter, but if there is a thick layer of jelly on the bread, the peanut butter will not "trap" over the jelly.

In gravure and flexo printing, printing ink films are frequently dried between stations (called "dry trapping"), which usually gives better trapping and better color control than wet trapping. The ability to maintain constant color is one mark of a printer who has the process under control.

HOW PRINTING INKS SET AND DRY

One key to successful inks and coatings is that they must flow readily until they are on the package, and then they must stop flowing—they must dry. An ink's drying is as important as its color. If the ink fails to dry, it is useless.

An ink is dry when its viscosity has increased enough to allow the print to go on to the next operation—such as cutting, folding or rewinding. There are many other definitions of drying, but this functional definition is useful. The term "setting" refers mostly to offset, letterpress and screen inks which may increase significantly or substantially in viscosity—resistance to flow—without becoming dry. When the ink has set, the printed board or labels can be stacked or jogged, but not trimmed.

There are several different mechanisms (modes) by which printing inks set and dry. We'll look at various packaging inks, and discuss the mechanism by which each dries.

Setting

The absorption of low viscosity vehicle or solvent raises the viscosity of the ink remaining on the surface, and in the case of quickset lithographic inks it substantially increases the rate at which the inks set and dry.

The thixotropic nature of paste inks (litho, letterpress and screen) substantially increases the viscosity and setting of these inks once they have been printed.

Evaporation

The most common drying method in package printing is evaporation of the solvent. Drying of fluid inks is usually a simple matter of evaporating the solvent from the varnish and leaving the dried resin to bind the pigment to the surface of the package. If solventbase inks are used, the printer must capture the organic solvent and dispose of it.

Most packaging inks dry very easily. Flexo and gravure inks dry when the solvent is evaporated. Problems with flexo and gravure inks arise when the varnish has "poor solvent release." Some resins, like nitrocellulose, dry more easily than others, such as some of the hydrocarbon resins. Resins with poor solvent release tend to remain soft or sticky after the print has passed through the dryer.

The solvents in flexo and gravure inks evaporate at relatively low temperatures, much lower than the temperatures required to dry offset or letterpress inks. Because the solvents evaporate or boil at such a low temperature, it is relatively difficult to condense them for recovery, and it is more economical to replace them with safe products. The most popular solvent is water—inexpensive and generally safe to use. Water is not as powerful a solvent for the binders used in printing inks, and there has been a great deal of work in recent years to develop resins that work well with water.

One widely used way to solve the problem of poor solvency of water is to use some organic solvent like alcohol in the waterbase ink. This not only does a better job of dissolving the resin in the ink, but it helps to wet packaging materials such as polyolefins. Chemists have recently overcome most of these problems so that solvent-free waterbase inks are now available that will print on any substrate that has been properly treated.

Water has a higher heat of vaporization than organic solvents, and more heat is required to dry waterbase inks than solventbase inks. Although aqueous inks on paper usually dry by absorption, a dryer (or oven or heater) is used to dry preprinted corrugated liner, plastic film and most printing on paper. Newer sheetfed corrugated presses may have interstation dryers, but presses without ovens have to be run very slowly.

Drying of waterbase inks can be speeded by addition of small amounts of alcohol.

Absorption

Solvent is absorbed from inks that are printed on paper or board. This is often enough to dry the ink. Simple line work using waterbase flexo inks printed on paper usually requires no further drying. Even waterbase coatings on corrugated or paperboard often dry upon standing for a few minutes. Some flexo presses are now equipped with small blowers to drive off the moisture.

All inks, including flexo and gravure inks, are absorbed to some extent when they are printed on paper. For some products, such as paper bags and corrugated boxes, the absorption of the ink into the substrate is all the drying that is required.

Oxidative Polymerization

Drying of oilbase sheetfed inks involves a complicated chemical reaction initiated by air and a catalyst or promoter that is called a "drier." Called "oxidative polymerization" (or "oxidation-polymerization"), the reaction occurs with drying oils such as linseed-alkyds or soya-urethane varnishes. The process ties the molecules of varnish together so that they cannot flow—they are solid. A chemical reaction in which many small molecules react to make one large molecule is called "polymerization."

Radiation Polymerization

The packager also needs to know that ultraviolet (UV) or electron beam (EB) inks cure by radiation.* Cured UV or EB inks have much better resistance than other inks to abrasion and to packaged products such as oils, fats and solvents. They can be formulated either as fluid inks for flexo and gravure or as paste inks for offset, letterpress or screen. These inks also cure by polymerization that is initiated not by air and a drier but by UV or EB radiation.

Curing by UV or EB radiation is significantly different from oxidative polymerization; the driving force for the polymerization comes from radiation. Unlike the oxidative polymerization, radiation-initiated polymerization is very fast, and is complete in a fraction of a second while the print is on the press, not in hours as with sheetfed litho inks. The press itself must be shielded from radiation so that the ink does not cure on plates or blankets. UV damages the eyes, and workers must be protected. Electron beams are even more hazardous and require extensive shielding.

UV inks have better chemical and weather resistance than conventional inks. The inks have a shelf life of 12 months if they are stored at 77°F (25°C) (lower temperatures give longer life). Presses require appropriate rubber rolls, anilox rolls and doctor blades. They should be equipped with rolls and blankets of EPDM rubber. UV inks and varnishes are used for a variety of packaging applications and are being rapidly introduced into flexography. Modern formulations cause less skin irritation than the early ones.

UV flexo inks under development offer an alternative to solvent- and waterbase inks, providing enhanced print quality, lower costs and compliance with environmental regulations. UV inks and coatings are prepolymers and monomers that harden to form an abrasion-resistant, glossy finish at line speeds up to 450 feet/minute (140 m/min). They perform well on a variety of substrates, notably on porous and unevenly coated materials. They are suited to process-color printing. Because they do not dry on the press, fewer wash-ups are required. Conversion to UV printing is relatively simple on in-line presses, but it requires more than simply hanging a UV lamp on the press.

---

* Although the terms "drying" and "curing" are sometimes used interchangeably, the word "drying" is usually used for evaporation of waterbase and solventbase inks and for oxidation-polymerization drying of sheetfed ink. Polymerization by high heat or radiation is usually called "curing."

Catalytic Curing

Catalytic polymerization is used on metal cans, and sometimes on glass. For example, melamine formaldehyde polymers can be baked to give a hard, resistant film of ink or varnish on beverage cans and bottles. The process usually takes several minutes, and UV and EB radiation are becoming the preferred method of creating hard, scratch resistant coatings on a hard surface.

INK TESTING

The packager requires much more complete testing of the printed ink than does a publication printer because of the wide variety of final uses to which the printed package may be subjected.

If the packager needs details of testing procedures, they are available from the GATF text *What the Printer Should Know about Ink*, and in greater detail from ASTM.* Standard tests are published by ASTM in their *Annual Book of Standards* and their *Paint Testing Manual*. This section mentions the tests that are useful for packaging inks. Standard tests avoid the errors that often occur with home-made testing procedures.

If the packager does not have a testing laboratory, it is wise to establish close relations with an analytical laboratory and a consultant who can supervise sampling, select suitable tests and carry out those that are too complicated or expensive for the packager to perform in his own plant.

Sampling

One of the most important aspects of any testing program, and one that is overlooked by people untrained in testing, is the sampling program. If the sample does not truly represent the material to be tested, the test results may be meaningless or, worse, misleading. The ink often arrives in several drums or different shipments, and the tests must give results that are representative of the entire run. The packager's product consists of many boxes or cans or labels, and the amount of variability to be permitted is one of the important properties that the packager must determine. The results can be no better than the sampling procedure.

In-Plant Tests

The testing program should be designed to meet the packager's needs. Some tests should be run on every batch of incoming ink; some should be run periodically on selected samples, while others can be run by a service laboratory as needed. For liquid inks, viscosity should be run on every shipment. It is probably wise to check the color of every shipment against a standard using a wire-wound

* Graphic Arts Technical Foundation, 4615 Forbes Ave., Pittsburgh, PA 15213. (412) 621–6941. American Society for Testing and Materials, 1916 Race St., Philadelphia, PA 19103. (215) 299-5400.

rod (Meyer rod) or an anilox hand proofer. Opacity can be judged if the ink is drawn over a black bar printed on the test paper.

For paste inks, a drawdown on a swatch pad gives a great deal of information about color, gloss and ink body. A fineness-of-grind test is a quick, convenient way to be sure that the ink is free from dirt or grit and well ground to disperse ink pigments.

Many folding carton printers use a Sutherland rub tester routinely to evaluate the rub resistance of the printed inks.

Preparing the Print

Ink testing usually involves testing the printed ink on the substrate. The test print can be made on a printing press, but cheaper and faster methods are also useful. The simplest is a hand drawdown with a spatula for paste inks, or a Meyer rod for liquid inks (figure 9-4).

The wire wound rod gives consistent results only in the hands of an expert, and, in fact, it can be manipulated to print a thicker or thinner film as desired. There are mechanical/electrical devices for making consistent drawdowns.

For greater speed, convenience and reproducibility, the anilox hand proofer (figure 9-5) is useful for liquid inks, and the Quickpeek tester (figure 9-6) is useful for paste inks. More sophisticated devices for printing test samples of flexo or gravure inks are the Genik web proof tester and for offset and letterpress inks the Prüfbau printability tester or the IGT printability tester. The biggest laboratories use small production presses to generate test prints.

Optical Tests

The above tests provide information about the color, gloss and drying of the ink. The tests also generate samples for testing the properties of the dried ink on the substrate.

Color. One of the most important properties of ink, its color, is measured on the print. This is one reason why the test print must be prepared carefully.

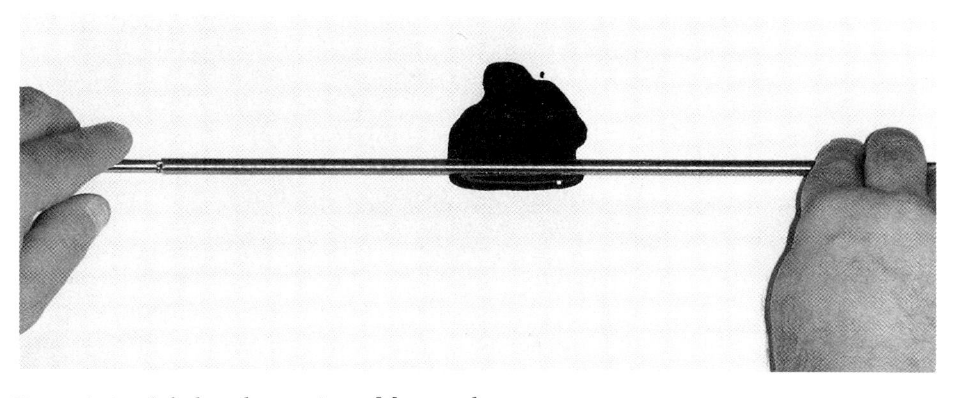

Figure 9-4. Ink drawdown using a Meyer rod.

Figure 9-5. Anilox hand proofer made by Pamarco.

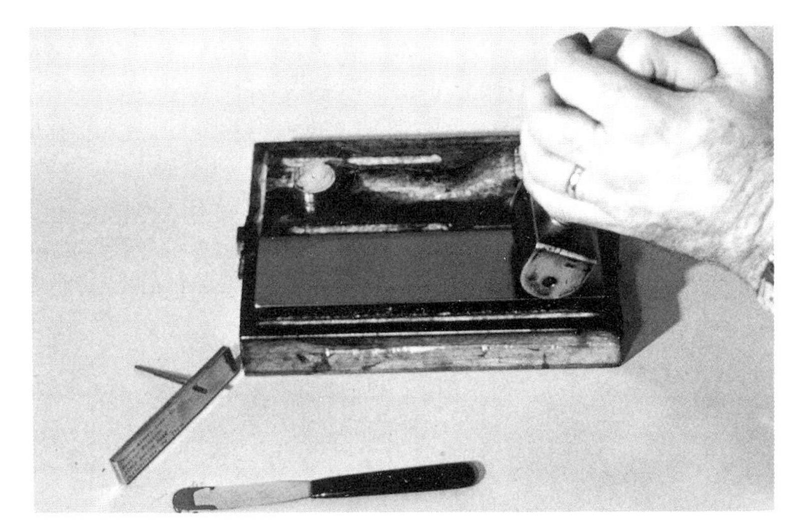

Figure 9-6. Ink rollout using a Quickpeek tester.

For the very simplest tests, two inks can be compared by placing a spot of each ink on a drawdown pad and drawing the two inks across the pad with a wire-wound rod for fluid inks or a spatula for paste inks. Visual comparison is an excellent way to judge the match of two colors. The colors should match both in undertone and masstone—in the thinnest ink films and in the full thickness of a solid.

Actually, fluid inks as well as paste inks can be examined by making a drawdown with a spatula. Prints are made by holding the ink knife or spatula upright and pressing down hard for the first half of the drawdown (which creates the undertone), then bringing the knife down to a 30-degree angle and pressing firmly but not so hard for the second half (which creates a masstone). Gravure inks should also be examined for middletones. Occasionally, an ink will match satisfactorily in the masstone but not in the undertone, or vice versa.

There are several systems for generating a standard set of matching colors. The Pantone Matching System (often referred to as "PMS") produces spot colors for commercial printing from several standard colors, matched to given formulas. Standard ink colors used for corrugated printing have been referred to as "GCMI colors" (Glass Container Manufacturers Institute, now called the Glass Packaging Institute).

Color Strength. The value of a printing ink is directly related to its color strength—the greater the color strength, the less the amount of ink is required to produce the desired color on the substrate, and the better the ink mileage. Accordingly, color strength is an important property to be tested. The pigment is the most expensive component in most inks, and the inkmaker can always reduce the price of the ink by using less pigment. Printers who buy on price alone cannot tell whether they are getting value for their money. Only by testing can the ink buyer judge which of two inks has the better value.

Measuring color strength requires an analytical balance and a special white ink. The bleach test is often performed more conveniently by a testing laboratory than by the printer. Two samples to be compared are weighed on the analytical balance and diluted with 50 parts of white tint base. After careful mixing, the two bleached inks are drawn down side by side on a piece of paper, and the color strengths are compared visually. The test is repeated, adding more bleach to the stronger ink until a match is obtained. This gives the printer a direct numerical comparison of the color values. The monetary value of the two inks is closely related to the color value.

Opacity. Opacity of two inks can be compared by drawing them down on one of the test swatches, used by ink manufacturers, which have a black bar across them. The test ink is placed near the standard ink and a spatula or ink knife is used to make a drawdown on the two inks. The more opaque ink hides the black bar better than the other. The eye does a good job comparing the two inks; if numbers are required, an instrument must be used.

Quantitative measurement of opacity is made with an opacimeter, an instrument that measures reflected light. To measure the opacity of a printed white film or paper, place the sample on a white block and adjust the instrument to read 100. Now place the sample over a special opacity cup lined with black fabric and again read the instrument. The reading is "percent opacity."

Color Density and Spectroscopy. One of the most important properties of the printed package is the print density, the absorption of light by the printed ink film. This is a property of the ink, the printing press, and the substrate itself. It is measured with a reflection densitometer (figure 9-7). Control of optical density is essential to production of packages or labels that will match when they are placed side-by-side on the shelf. It is especially important if two different lots of product may be displayed next to each other. Optical density is discussed in Chapter 5.

Densitometers can be used to measure the amount of light absorbed by a printed ink film. The numbers can be plotted on a GATF Color Triangle or Color Circle to give useful information in judging hue and grayness of an ink. Spectrophotometers and colorimeters measure the color of the ink. They are much more expensive than densitometers, and they are used for a complete analysis of printing ink color rather than in routine laboratory control.

Every pressroom, every packaging print buyer and every art director should have a reflection densitometer. It is necessary for establishing numerical quality standards instead of relying on opinions or guesses or slippery descriptions of color. If quantitative densitometric or spectroscopic measurements of color are to be made, the print must be prepared under carefully controlled conditions.

While the eye does a good job of comparing the color or gloss of two different prints side-by-side, it cannot assign a numerical value to the prints, nor can it remember, even for a few seconds, the exact shade or gloss level.

Gloss. Gloss is measured with a gloss meter, an instrument that printers sometimes have, and one that is available in most ink labs. Once again, however, if two inks have been printed side by side on a substrate, the eye does a good

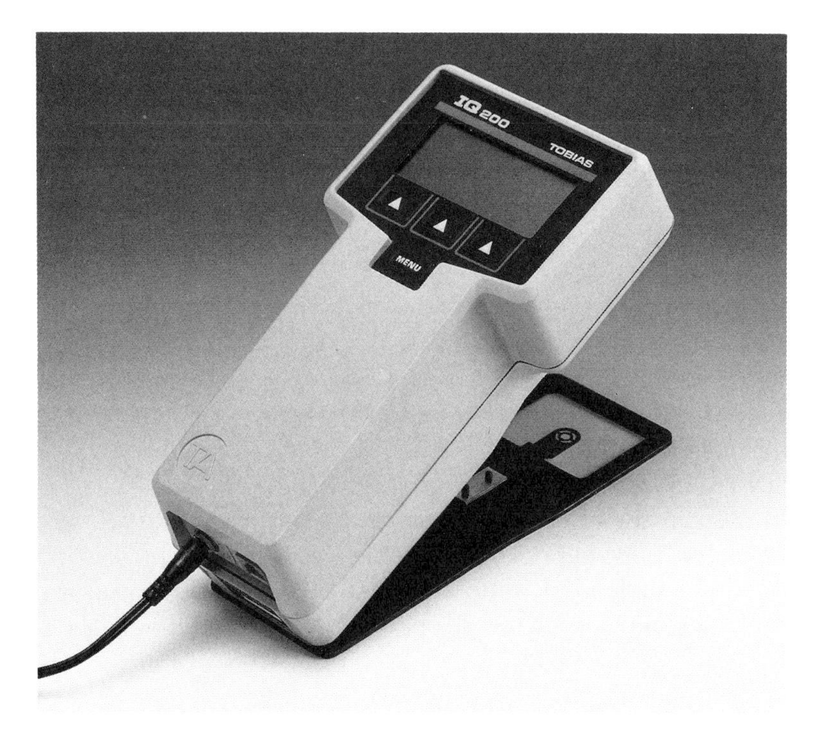

Figure 9-7. Hand-held reflection densitometer. (Courtesy of Tobias Associates, Inc.)

job in judging which ink has the higher gloss. Since the substrate affects the gloss of the print, it is important to test the ink on the substrate to be used.

Working Properties

Drying. Drying is as important a property of an ink as its color. Web inks for flexo, gravure or litho can be tested with an instrument called a "Sinvatrol Tester" that carries a fresh print through a heated oven (figure 9-8). Both speed and temperature are controlled, but the conveyor speed can be varied from 0 to 100 feet per minute (0 to 0.5 meter per second) and the temperature can be varied from 100° F to 600° F (38° C to 315° C).

Drying of sheetfed inks that dry by oxidation-polymerization is usually judged by making a drawdown on the paper to be printed and checking the dryness of the ink by rubbing every half hour with a finger.

Viscosity. Viscosity is an important property of all inks. Viscosity is the measure of resistance to flow. Rate of flow on the press governs printability.

The most popular instruments for measuring the viscosity of flexo and gravure inks are efflux cups, specifically the Zahn Cup, the Shell Cup and the ISO (or DIN) Cup. The Zahn is the oldest instrument and the most popular, but the Shell Cup gives better accuracy and greater precision (more repeatable results). The DIN gives the best of the group. The Shell and Zahn cups are shown in figure 9-9, and the ISO (or DIN) cup is shown in figure 9-10. The cup is dipped

Figure 9-8. Sinvatrol Tester for measuring ink drying time. (Courtesy of Flint Ink Corp.)

below the surface of the ink, filled and lifted from the ink, and the time for the ink to flow from the cup is measured with a stop watch. Table 9-VII presents a viscosity conversion short for various efflux cups.

Measurement of viscosity of paste inks on press is not particularly significant. Owing to their high viscosity, paste inks cannot be measured with efflux cups. Measurement of viscosity of paste inks is largely limited to the research laboratory.

Flash Point. The flash point of an ink is of concern to all printers. A sample of the ink is placed in a special closed cup and slowly heated. The cup is opened at regular intervals, and a small flame is played over the surface of the liquid. When the liquid reaches the flash point, the flame ignites the vapors over the surface of the liquid. This test is seldom performed in the printing plant, but the inkmaker will report the flash points of any ink upon request.

Explosions and fires are a constant threat in plants using solventbase inks. Explosion-proof motors and electric switches are obviously required as are suitable handling and storage of the inks. Safety is discussed further in Chapter 13.

Tack. The tack of paste inks is especially important in the printing of process colors where inks must trap over wet films. The best color uniformity is

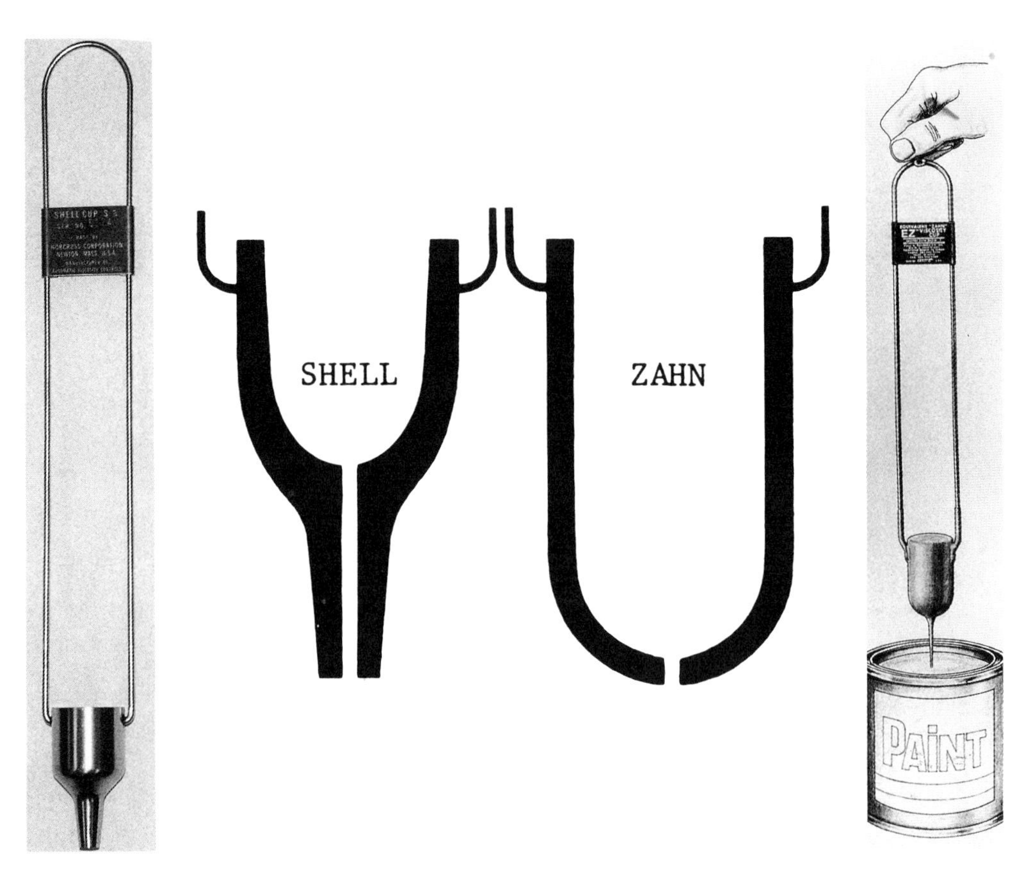

Figure 9-9. Photos and cross-sections of Shell and Zahn cups. (EZ cup picture courtesy of Paul N. Gardner Co., Inc.)

TABLE 9-VII. VISCOSITY CONVERSION CHART FOR VARIOUS EFFLUX CUPS

	Efflux time in Seconds						
Viscosity Centipoise	Shell #2	Shell #2½	Shell #3	Shell #3½	Shell #4	Zahn #1	Zahn #2
7.5	18.3					30.5	
10.0	22					32	
15.0	30.4	19				35	
20.0	39	25				38	18
25.0	47	30	18.6			42	19
30.0	56	35.5	22			45	20
40.0		46.6	28.6	20		52	22.5
50.0		57	35	24.6		60	25
60.0			42	29.2	18	68	28
70.0			48	34	21		30
80.0			55	38.4	24		34
90.0				43	27		38
100.0					30		43
125.0					37		53

Source: Norcross Corp.

Figure 9-10. Photo and sketch of ISO or DIN cup. (Courtesy of Paul N. Gardner Co., Inc.)

obtained when the inks have the proper tack sequence. The higher tack ink should be printed first, and it will pull the next ink from the plate or blanket. Ink tack is measured on an instrument called the "Inkometer." This instrument is found in most printing ink manufacturing plants (figure 9-11).

Fineness of Grind. This property is usually measured with paste inks as a precaution against poorly ground ink or ink that has dirt particles in it. The test uses a fineness-of-grind gauge, a tool-steel block with a highly finished surface (figure 9-12). Two or three parallel channels have been ground into the surface.

Figure 9-11. GATF Inkometer for measuring ink tack. (Manufactured by Thwing-Albert Instrument Co.)

Figure 9-12. Fineness-of-grind test.

They are one mil (0.001 inch or 0.025 mm) deep at one end of the block and taper to zero depth at the other. Beside each channel are marks indicating graduations of 0.1 mil in depth. A doctor blade is supplied with the instrument. When a dab of ink is placed at the deep end and drawn towards the shallow end, scratches appear where the solid particles can no longer flow under the doctor blade. As a rule of thumb, an ink that causes more than two scratches at 0.4 mil or 10 scratches at 0.2 mil should be reground. Poorly ground inks do not behave well on press and do not give the highest gloss. They can contribute to wear of printing plates and gravure cylinders.

Misting. Flying or misting of inks can also be observed with an Inkometer. A piece of white paper is placed at the back of the instrument. It is removed, observed and replaced after each ink is tested. This undesirable characteristic of letterpress and web offset inks has been largely overcome by recent developments in ink formulation.

End-Use Properties

Abrasion Resistance. The resistance of a printed package to abrasion is very important to the packager. The most widely used abrasion test is the Sutherland rub tester (figure 9-13). The printed sample for the test is mounted on the base of the instrument, and the paper or board to be rubbed is mounted on a metal block (there are two blocks: one weighs two pounds, the other four pounds), and rubbed against the clean substrate. The instrument rubs the clean substrate back and forth across the printed sample, counting the number of rubs on a dial. Paper cartons are commonly tested by giving 50 rubs with the two-pound block and 80 rubs with the four-pound block.

A new tester, introduced in the 1980s, is reported to give much better cor-

Figure 9-13. Sutherland rub tester. (Courtesy of James River Corp.)

Figure 9-14. Gavarti Comprehensive Abrasion Tester. (Manufactured by Gavarti Associates, Ltd.)

relation with actual performance in the field, and to require less time to run. It is the Gavarti Comprehensive Abrasion Tester (CAT) (figure 9-14). It is useful for measuring abrasion resistance of folding cartons, corrugated boxes, plastic films or sheets, metal foil, and steel or aluminum sheets. The amount of scuffing or rub off can be measured with a densitometer.

Several other instruments have been used to measure abrasion of printed samples.

Adhesion Tests. Adhesion of ink to the printed sample is especially important on flexible packaging. Adhesion of ink to paper and board is not usually a problem. Several tests are commonly used: wrinkling, fingernail-scratch, "Scotch" tape and, with metals, manual bending. The wrinkle test is readily performed in the office or at pressside. A piece of the printed film is held between the thumb and forefinger of each hand and rubbed back and forth rapidly about 10 times. The printed sample is examined for disruption of the ink.

The "Scotch" tape test is run by firmly pressing a 4-inch piece of cellulose tape onto the printed sample, saving a small edge that is turned under so that it can be grasped. The tape is pulled rapidly away from the print, and the amount of ink remaining on the print is observed. The "Scotch" tape test can also be used on printed glass or metal. Ink formulation affects adhesion, but the surface treatment of the film or foil is more important (see Chapter 10).

Adhesion of ink over bends in printed metal can be judged by manually bending the printed metal. Adhesion is affected by the flexibility of the ink as well as by its adhesive power.

Scratch Resistance. Scratch resistance is observed by placing the printed film on a pad or cushion and quickly drawing the back of the fingernail over the film. The film is inspected for scratch marks.

Caps and crowns are often scratched during handling, and scratch resistance can be tested in laboratory hoppers designed to duplicate production conditions. There are also tests to measure the resistance of printed beverage cans to abrasion by jostling in ice-filled chests.

Blocking. Flexible packages must not stick together when they are stacked. The problem is referred to as "blocking." A test is easily conducted in the laboratory by stacking printed film and placing the stack under a weight in an oven at 120°F (50°C). If many tests are to be performed, it may be wise to purchase a block-point tester.

Heat Sealing. The ability of a package to be heat sealed is of critical importance on the packaging line. Heat-seal ability is measured by sealing a printed sample and observing the package afterwards. This can be done with a flat iron, but a heat-seal tester is more suitable for laboratory use.

Lightfastness. Unfortunately, the cleanest, brightest pigments often have the greatest tendency to fade or discolor under exposure to light. The testing of lightfastness is complicated by the fact that relative humidity and the composition of the ink vehicle affect the results. The simplest test is to place printed samples in a window and observe the results. Prints should be placed about a foot from a south window. For outdoor exposure, samples can be placed on an inclined board facing south. The samples should include unprinted stock so that any effect of sunlight on the substrate can also be observed.

For quantitative and controlled measure of lightfastness, a machine called the Fade-Ometer (figure 9-15) is used. Printed samples in a light cabinet are exposed to a controlled amount of light, and there are no cloudy days to alter the amount of exposure received by different samples.

A similar tester, the Weather-Ometer, is available when exposure to light and moisture are both important.

Other Performance Tests. The ink must affect neither the odor nor taste of packaged foods. Odor and taste are observed by panels of people chosen for the purpose because different individuals do not observe taste and odor in the same way. Organizing and training a taste or odor panel requires the services of an expert if the results are to be significant.

Traces of volatile solvents can be detected and analyzed by use of gas chro-

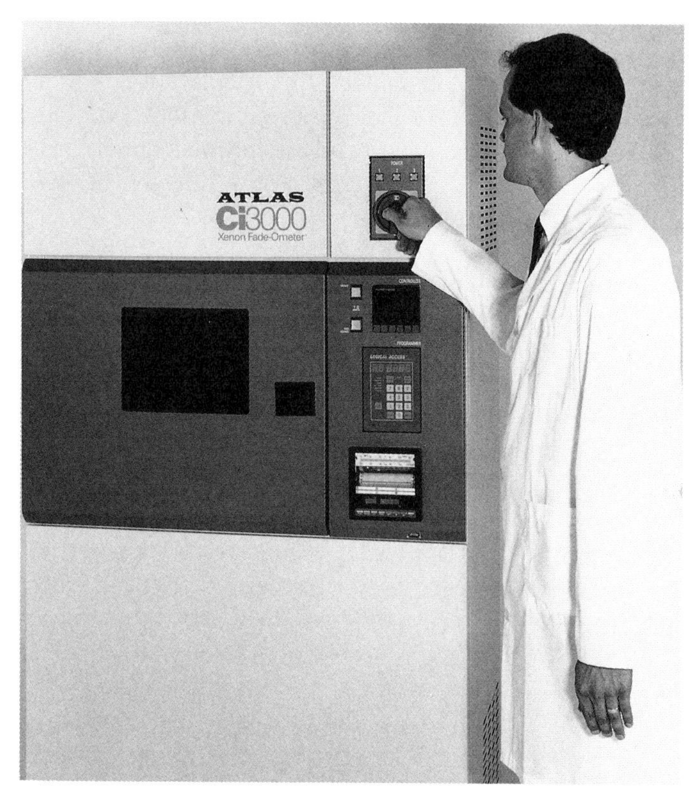

Figure 9-15. Atlas Fade-Ometer. (Courtesy of Atlas Electric Devices Co.)

matography, an analytical technique that is commonly used in the chemistry laboratory.

Three tests of water resistance are commonly performed on printed packages: resistance to bleeding in water, adhesion of the ink when the package is frozen and thawed, and resistance to boiling water. Bleeding is easily observed by placing the sample in contact with a wet white paper towel or tissue paper for several hours. Freeze-thaw resistance is measured by placing the sample in the freezer, then checking ink adhesion after the sample has been removed and thawed. Boiling resistance is measured by putting the package in boiling water, removing and observing the adhesion of the ink.

Water vapor transmission rate is not ordinarily a function of the ink. It is an inherent property of the film, foil or substrate, but it is affected by coating (see Chapter 10.)

Inks used for laminated substrates must adhere to both the base sheet and the laminate, but the printed stock must not block in the roll coming from the press. If the package is laminated in line, blocking is not usually a problem. The bond strength of the laminate may be observed by placing the laminate between two pieces of "Scotch" tape, or quantitatively by using a tensile strength tester.

SPECIFICATIONS FOR PRINTING INKS

Packagers or converters who practice Total Quality Management (TQM) will develop meaningful specifications for the printing inks they purchase. This is not a simple task. Desired properties sometimes cannot be measured accurately, and common tests often give little information about the performance of the ink. There is no point in writing specs that do not assure proper performance.

The cost of the effort in writing useful specs should be returned in increased productivity, reduced waste and packages that help gain greater consumer acceptance.

Specs are written in cooperation between the supplier and the customer in order to measure and control those properties that are important in the decoration, performance and disposal of the package, always at an affordable cost. The inkmaker can usually provide technical expertise, but sometimes it is wise to contact an independent consultant or authority who can help design a testing program that will give good results for the least money.

The above discussion of testing of inks should be a guide for developing meaningful specs for printing inks. Different types of inks and different applications require different specs. Properties that should be considered for specification include: color and color strength, opacity, heavy metal content and appropriate end use properties for all types of inks including product resistance and performance during printing, filling, shipping, storage and handling by the consumer. Specifications for fluid inks include volatility, viscosity and flash point. Specifications for litho, letterpress and screen inks include drying time and open time (resistance to drying on the press), yield value, fineness of grind, shelf life, tack and tack stability.

After specs have been agreed to, the packager must set up some sort of testing program. The testing program should be developed together with the specs. Not all properties must be tested on every sample, but some ink testing is required to be sure that the specs are serving the purpose of improving communications between supplier and customer and of assuring that inks will always meet the requirements of the job.

When problems involving ink are encountered, the inkmaker should be consulted. The price of the ink includes at least some service when problems arise.

INK CONTACT WITH FOODS

Printing inks must not come in direct contact with food.

The U.S. Food and Drug Administration (FDA) has not issued any regulation on the subject of printing inks. Printing inks have not been considered to be food additives if the printing is separated from the food by a packaging material or coating that serves as an effective functional barrier. On the other hand, printing inks that come into contact with foods will probably become food additives, but it is impractical to formulate any printing ink from approved food additives. Not only do most of the components of printing inks not appear on the FDA's list of GRAS (generally recognized as safe) materials, but inks are not made from food grade raw materials nor are they handled as food grade materials.

Packagers can usually overcome this problem by coating the printing ink with a material that is "FDA compliant," that is, published on a list of GRAS or other acceptable materials.

FLEXOGRAPHIC INKS

Flexo inks are fluid inks. They are volatile and easily dried, but they also evaporate in the press. To reduce evaporation from the press, enclosed inking units are being supplied with modern flexo presses.

In general, flexography gives good color consistency throughout the run, but evaporation of the solvent from solventbase inks or changes in the pH of waterbase inks can affect ink feed and print color.

Control of the drying temperature is important with flexo as it is with all web inks. If the temperature is too high, the ink film or even the substrate may soften; if the temperature is not high enough, the ink does not dry and the printed roll may block or stick together. Hot ink films can be cooled on a chill roll, but it is often possible to avoid using a chill roll if inks are properly specified and the dryer temperature is properly controlled. Solventbase inks can be made to dry faster or slower by addition of "fast" (low boiling) or "slow" (high boiling) solvents.

So that the pigments will not settle out rapidly, these inks are often shipped and stored at a higher viscosity than that used on the press. The inks are "let down" or diluted to running viscosity at press side. On modern presses, the running viscosity is automatically monitored to control both the color strength of the ink and the amount of ink fed to the press.

Polyamide inks in alcohol solvent give excellent adhesion, gloss and flexibility on polyolefin films, nitrocellulose and polymer-coated nitrocellulose films and aluminum foil. A polyamide cosolvent ink (alcohol and aliphatic hydrocarbon) gives good resistance to water, acid and alkali. Nitrocellulose inks have excellent solvent release and resistance to grease and heat.

The sparkle of the aluminum surface is retained by using highly transparent inks, made of transparent pigments and vehicles.

Vinyl films often contain plasticizer which can bleed back into the ink, softening it. The package printer should consult the ink manufacturer and the film manufacturer before trying to print on any vinyl film.

Flexographic inks (and gravure inks) can be printed on the underside of a clear film. This gives protection to the ink and retains the clarity and sparkle of the film. Similar results are obtained by laminating a clear film over the print. Inks for laminating must adhere to both layers, and they must have good solvent release so that solvent is not trapped between the layers.

Waterbase inks are not new to flexography, but the first ones provided little gloss and had poor rub resistance. New binders provide excellent properties and have made waterbase inks increasingly popular with flexographers. They reduce safety problems in the pressroom and environmental problems outside the plant.

Many waterbase inks contain small amounts of organic solvents to improve pigment wetting, solubility of the ink binder and the formation of a tough film when the ink dries. Advances in polymer chemistry and ink formulation have made possible flexo inks that are free of organic solvents.

Waterbase inks have penetrated flexo printing more rapidly than gravure printing. They are always used on paper and often on film or laminates. Waterbase inks have many advantages over solventbase inks: they wash up with water, the do not swell most flexo plates, they are virtually odor free, they reduce fire hazards and insurance costs, and they are less affected by fluctuations in the price of petroleum. Under good circumstances, they nearly match solventbase inks in printability, gloss, rub and heat resistance, shelf life and ability to dry at press speeds.

Control of pH is critically important in controlling the viscosity of waterbase inks. Control of viscosity is essential in maintaining ink performance and controlling print color.

Waterbase inks also have some disadvantages: they require more power to dry and they are more likely than solventbase inks to have trouble wetting plastic surfaces. Like solventbase inks, they develop good water resistance when they dry, and once they are dry, they resist washing from the press. Waterbase inks sometimes cause problems on improperly treated polyolefin films and polyolefin coated paper, and in order to solve the problem the package printer must determine whether the film or the ink or both must be corrected.

Waterbase inks for use on packaging films and foils present a challenge to the ink formulator. Increased chemical knowledge and experience in surface science have produced better ink formulations. Dedication and assistance from experts in various fields, from plastic engineers to surface scientists, enable ink companies to provide improved solutions to packaging needs.

One way to improve graphic quality is to use high-solids, high-viscosity printing inks. Reverse-angle doctor blades on the anilox rolls make it possible to use higher viscosity inks than with the older roll-metering system. The higher viscosity provides better holdout on paper and board, and the printed ink film is therefore denser and more opaque. The applied ink film is thinner and easier to dry. Pigment levels are increased so that the thinner ink film has the same optical density as the film from conventional waterbase flexo inks.

First used in Europe in the mid 1970s, these new waterbase flexo inks run around 50 percent solids, twice that of conventional flexo inks. The solids content of black ink runs somewhat lower and white ink somewhat higher than the solids content of colored flexo inks. (The solids content includes the binder and the pigment.) Colored inks contain both colored pigment and extender pigment, formulated to give the correct color strength with the anilox roll to be used. If the anilox volume runs higher than eight billion cubic microns (bcm) per square inch, considerable extender is used; below five bcm, little extender is used. These variables—binder content, colored pigment and extender—can be manipulated to improve performance or to decrease price, and close cooperation between printer and inkmaker is required for optimum results.

The high-viscosity inks avoid striations, a common flexo problem caused by insufficient resin in the ink. The metering system and printing plates must be appropriately selected. The inks have many advantages: higher gloss, better color consistency throughout the run, higher color intensity, cleaner color, better trapping, better coverage and elimination of striations, faster drying and reduced dot gain.

The high-viscosity inks contain more pigment and have a higher price per pound than the older inks, but experience has shown that much less ink is needed to achieve good color densities. Actual ink mileage increases dramatically, and the advantages are attained with net ink costs close to those of "conventional" flexo inks.

UV flexo inks are rapidly becoming popular. They contain no solvent and thereby solve many environmental problems. As noted above under radiation-curing inks, they offer superior resistance to chemicals such as solvents and products and to abrasion and other mechanical damage.

Use of laminating inks is a growth area in flexible packaging. Laminating inks must not only provide appropriate visibility, but they must be compatible with the laminant. They are printed on the underside of a transparent plastic, and they become the filling in the sandwich of that plastic and its partner that is laminated to it. They are being used for plastics laminated to corrugated. Laminating inks enable the packager to avoid direct contact with the product. The lamination protects the ink from solvents and other products or physical damage.

Troubleshooting Flexographic Inks

Every printing process and ink has typical problems. Poorly printed flexo is readily identified by halos and striations. Many other problems are shared with other printing inks.

Control of printing pressure is critical in flexography. Insufficient pressure causes mottle or ragged solids; excess pressure causes halos around the print. Halos are also caused by overfeed of ink. If the plate is not uniform, it is impossible to create even pressure throughout the print. Specs on thickness variation of the plate must be considered.

Plugging of the flexo plate is sometimes related to ink formulation or plate composition, but it is usually caused by dusty or linty paper.

As in every printing process, there are many causes for color variation—variations in the ink, wearing of the anilox roll, and evaporation of the solvent from the press. In printing of process color, failure to control printing pressure causes variable dot gain that results in color variations.

Pinholes on nitrocellulose coated films can be reduced by adding an anti-pinhole agent that reduces the surface tension of the ink, allowing a smooth pinhole-free coating to be obtained. Pinholes from use of waterbase inks result from foam which may be corrected by preventing the ink from falling from an open line, by adding a filter to the ink line, or by adding defoamers.

Inks that dry too fast, sometimes cause picking of paper or board.

Striations are caused by poorly formulated ink and failure to control the press.

Poor adhesion is corrected either by treating the substrate or by reformulating the ink, sometimes both.

Good troubleshooting guides are presented in FTA's *Flexography: Principles and Practices*, Fourth Edition and in the GATF text by Eldred and Scarlett.

GRAVURE INKS

Like flexo, gravure uses fluid inks. Gravure can use stronger solvents than can flexo or offset because the printing unit has no plastic or rubber plates or rolls. Solvents, such as toluene, Cellosolve, and methyl ethyl ketone improve the adhesion of inks to plastics and films. On the other hand, they are more highly toxic than the alcohols often used for flexo inks. When used, special precautions must be taken to prevent people from being exposed to them.

Viscosity is related to solvency. Strong solvents produce varnishes of lower viscosity than do poor solvents.

Gravure cells deliver a relatively thick ink film which tends to give better gloss than is normally obtained by flexo or offset, especially when printed on paper or board. The thick film is a clear advantage in printing gold or silver inks and fluorescent colors. Gravure golds are usually brilliant and nontarnishing.

Like flexo inks, gravure inks printed on the back side of a transparent plastic film produce a bright, sparkling package.

Waterbase gravure inks reduce fire hazards in the pressroom and cause less environmental pollution, but they require special formulation in order to flow into and out of the cells in the gravure cylinder.

Although solventbase inks have a strong hold on the gravure industry, much research and engineering work is being done with waterbase inks. Waterbase inks release differently from gravure cylinders than solventbase inks, and replacing them with waterbase inks requires adjustments in the cylinder manufacture as well as modifications to the inks and plastic film on which gravure may be printed.

In 1990, 65–70 percent of packaging gravure inks were solventbase, the rest waterbase. but waterbase inks are growing in popularity as the technology improves and as environmental agencies insist on reduction or elimination of VOCs in inks.

During printing, the solvent may evaporate from the gravure ink pans or troughs, concentrating the pigment and changing the color, especially with inks based on the more volatile solvents. Evaporation is greatly reduced on modern presses that have enclosed or chambered ink fountains (like modern flexo presses). The problem can be avoided by checking the viscosity of the ink (which changes as solvent evaporates) and replacing solvent as required. It is well controlled with a continuous sensor and automatic viscosity control.

With flexible packaging—including plastic films, glassine and vegetable parchment—flexibility of the ink film is needed; nitrocellulose and polyamide inks do well. Adhesion to aluminum foil requires that the foil be clean and free of grease.

Gravure inks are often classified according to their composition. Class A and Class B inks are used for publication printing. Class C inks are those based on nitrocellulose. Class D are polyamide inks. Class E are alcohol soluble inks, usually containing alcohol-soluble nitrocellulose. Chlorinated rubber inks, soluble in toluene, are Class T, and waterbase inks are Class W.

Troubleshooting Gravure Inks

The most typical problem in rotogravure is skipped dots, often referred to "speckle" or "snowflaking." This is usually caused by an uneven printing surface. The problem is greatly diminished by using electrostatic assist. Occasionally, missing dots are created by ink viscosity that is too high or ink that dries on press. This problem can be corrected by adding solvent to the ink. In waterbase inks, foam can cause bubbles that create missing dots. A defoamer may correct foam problems, but redesign of the pumping system may also help. The problem can also result from broken dots on the engraved cylinder.

Gravure gives the least color variability of any printing process. Most variations in color result from changes in the ink, evaporation of ink solvent from the press, or variations from one batch of ink to another.

Static electricity on the web can cause fuzzy prints. Ink crawl or mottle occurs when the ink does not wet the substrate properly or when the ink viscosity is too low.

Blocking of printed film or paper is caused by insufficient drying of the ink. It may be corrected by increasing the temperature of the dryer or using a faster solvent. Occasionally, plasticizer in a vinyl film may migrate into a dried ink film, softening it.

For food packaging, freedom of odor is critical, and ink solvents must be selected that will leave no trace of odor. The gas chromatograph is the standard instrument for detecting and identifying traces of solvents.

LITHOGRAPHIC INKS

Although both are lithographic inks, formulations of sheetfed and web offset inks are very different because they dry by different mechanisms. Both groups of inks must work with water. It has been said that lithography depends on the fact that oil and water do not mix, but in offset lithography, they must mix to a controlled extent. If the ink is too resistant to water, it will not transfer properly to the plate. If it is not sufficiently resistant to water, it will emulsify water, causing weak, washed-out prints, or specks of ink in the non-image area. A test of the emulsification of an ink by stirring it with water yields useful information about its ability to work with water (figure 9-16.) In the figure, ink A will take on water too readily, inks B and C may be expected to perform well, while ink D will not transfer properly to the plate.

The ink carries tiny droplets of emulsified water to the plate. Ink that accepts water too readily produces weak, washed-out prints, or it emulsifies in the fountain solution and colors the non-image areas. Ink misting on high-speed web offset presses has been reduced by development of improved ink formulations.

Because the presses run so much faster, web offset inks must have lower tack than sheetfed inks. Offset inks are sometimes tacky enough to pick bits of paper from the surface of the paper or board.

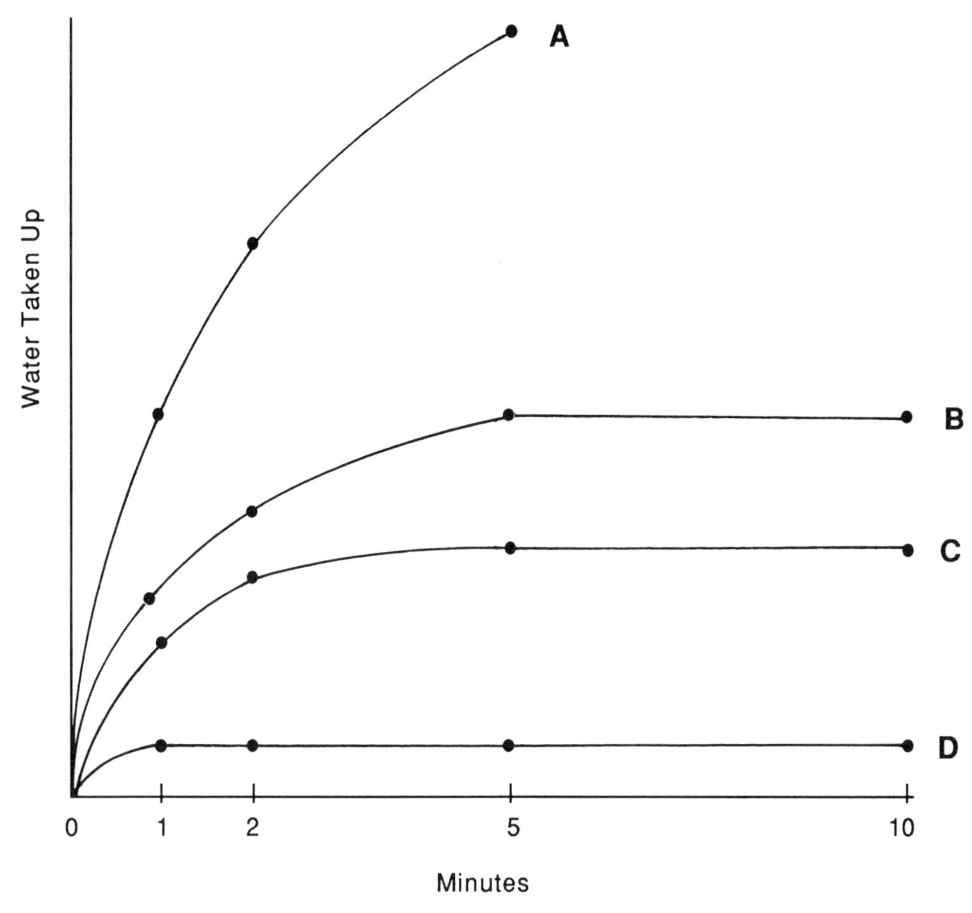

Figure 9-16. Graph of Surland test showing differences in water takeup by different inks.

All offset inks contain some hydrocarbon solvent—sheetfed inks in order to promote quick setting and web inks to dissolve the resin. In sheetfed printing, the solvent is largely absorbed by the paper or board on which it is printed; in web offset, most of the solvent evaporates and must by captured to prevent discharge to the atmosphere.

Offset inks for metal decorating contain specially reactive polymers that cross-link upon baking. UV and EB inks are especially useful with metal decorating because they replace long, slow, drying ovens with fast curing procedures.

Inks for printing on plastic or plastic-coated paper are sometimes called "collodion" inks, in reference to the viscous solution of nitrocellulose used to coat paper (or cuts or scratches on one's skin.) They are formulated with little or no solvent. If the proper ink is used and if dampening is properly controlled, it is possible to print plastic by lithography, although it is not practical to feed thin plastic films through an offset press.

Although waterbase inks are not used in litho, it is possible to use them, either by a process called "reverse lithography" that uses a hydrocarbon as the "dampening" agent or by using a driographic plate.

Troubleshooting Lithographic Inks

Sheetfed offset inks are plagued with drying problems (slow drying, offsetting, blocking, mottle and ghosting) when the printer does not take adequate precautions. There is usually no cure for the problems because they are not observed until long after the job has been printed, but strict attention to ink formulation, proper selection of the ink, control of spray powder and stacking of the printed sheets help prevent the problems.

Color variability is one of the greatest problems of offset lithography. The ability to vary the color is sometimes used to compensate for poor prepress work, but it is far better to do the prepress work properly and to maintain close control of color. Variations in ink feed and water feed are important causes of color variation. As mentioned earlier in this chapter, color sequence and trap sequence should be selected and maintained to provide good color uniformity on offset presses. The inking system sometimes contributes to color variations, especially mechanical ghosting. The thin printed film can lead to mottle, when the ink unevenly penetrates non-uniform board.

Variations in color that match an image on the previous print are called ghosts. They are characteristic of the offset inking system and are aggravated by the thin ink film printed by litho.

Web offset problems are generated by lack of control of the dryer (often referred to as an "oven"). If the dryer is not hot enough, the ink remains soft after going over the chill rolls; if the dryer is too hot, the paper or board is dried out and embrittled.

All offset inks are troubled by contamination with paper—picking, piling, paper dust and debris. The problems are reduced by selection of good paper, but if the ink is too tacky, it can contribute to the problems.

Setoff from the surface of a freshly printed sheet to the back of the next one in the pile is controlled by spray powder and by proper ink formulation. Low color strength requires a thicker ink film that has greater tendency to set off than a thin ink film.

Ink in the non-image area—catch-up, scumming, and toning or emulsification—are unique to offset lithography. "Catch-up" or "dry back" occurs when not enough dampening fluid is fed to the plate. Scumming can arise from many different causes, but the result is that the plate prints ink in an area that is supposed to be clean. Too much water or poor ink can cause ink to be emulsified in the water, where it is carried to the non-image area of the plate.

Dirt in the ink causes problems with any printing process, but because offset ink films are so thin, dirt is more troublesome in this process. Not enough ink is fed through the nip to wash away even the finest bits of dirt. These often print a donut-shaped defect commonly called a "hickey."

LETTERPRESS INKS

Letterpress inks strongly resemble offset inks. Both groups of inks are paste inks, both are thixotropic requiring a press with many ink róllers to work the ink before printing. Sheetfed inks dry by oxidative-polymerization; web inks dry by evaporation. Two differences are significant: letterpress inks do not need to work with water, and since the press prints a thicker ink film than the offset press, the color strength of letterpress inks must be lower. To maintain ink body while reducing color strength, the inkmaker sometimes adds a transparent pigment, such as clay, or a transparent varnish to the ink. These additives should be avoided in offset inks where they interfere with proper ink transfer from plate to blanket.

The letterpress printer can adapt offset inks for letterpress printing by adding varnish or transparent white ink. It is preferable, of course, to get the inkmaker to formulate the ink properly for the press.

As with other ink systems, UV letterpress inks are becoming very popular because they provide instant drying and require little or no solvent. UV inks also provide an especially tough ink film when dried.

Letterpress inks used for packaging are printed mostly on labels. UV formulations overcome the difficulties inherent in packing and shipping pharmaceutical products, where a recall of product may result from scuffed labels. It also has enabled the elimination of partitions for pharmaceuticals packed in plastic containers.

Troubleshooting Letterpress Inks

Letterpress and offset both use paste inks, and they both print from a multiroll ink distribution system. Many of the problems are the same. Mottle results when the ink is not stiff enough to hold out on the surface of a paper that varies in absorbency. Mechanical ghosting is largely a result of the press design, but increasing the ink film thickness or using a more opaque ink can help to cover it up. Scuffing of printed labels on shipped cans or bottles can be reduced by adding a tough wax to the ink. Excess heat in the drier can aggravate scuffing by making the ink film brittle. UV or EB inks have highly scuff-resistant properties.

Ink setoff occurs when the ink is too soft or the paper or paperboard has high ink holdout. Careful ink formulation will reduce both setoff and mottle. Setoff is usually controlled by the use of spray powder, as in offset printing.

SCREEN PRINTING INKS

Screen inks belong to the family of paste inks, along with offset and letterpress inks. They are sometimes referred to as pastes, especially if a clear varnish is being used. Screen is the most versatile of printing processes, and screen inks are the most difficult to describe because of their wide variety. The first screen inks used oxidative-polymerization drying, like sheetfed letterpress or offset inks, but screen printing can be accomplished with many other kinds of inks: evaporative drying inks (like web offset), waterbase inks, UV and EB curing inks, catalytic curing inks for metal decorating that dry in a few seconds at a high temperature, and vitreous enamels that can be fused onto the surface of glass bottles or even onto metal surfaces.

The thick film promotes the brilliance of colors and opacity, distinctive characteristics of screen printing. Screen is the preferred process for printing fluorescent inks and gold and silver inks (bronze and aluminum) because of the thick ink film that can be applied.

Application of a film of ink through a screen makes it possible to print on a great variety of packaging materials: paper and paperboard, plastics of all sorts, metals, glass and ceramics, wood and wood composites, and textiles. Each substrate has somewhat different ink requirements, and screen can apply almost any sort of ink.

Even more than sheetfed offset inks, screen inks are specially formulated for a particular job. Ink mileage varies greatly depending on the mesh of the screen and the surface of the product to be printed. Rough surfaces require a great deal more ink than do smooth surfaces. The press itself must be taken into account; for example, a sheetfed screen press requires a different ink formulation than a rotary screen press.

One characteristic of screen inks that differs significantly from other printing inks is "length." A long ink will produce a long string of liquid when a spatula is dipped into the ink and withdrawn. Most inks must be "long" in order to transfer well from roll to roll. Screen inks must be short and buttery so that long strings do not form when the screen is pulled away from the print, forming fuzzy edges around the printed letters or form. The viscosity of the screen ink must be low enough to allow it to be squeegeed through the mesh but not so low that it will smear or give a fuzzy print.

The screen ink must be compatible with the material used to make the screen. If the material is a plastic monofilament, it must not be dissolved or softened by the ink.

Troubleshooting Screen Printing Inks

Drying problems of screen inks are treated like drying problems of other processes. If an oxidative drying ink is used, it must be formulated with the proper amount of drier so that it will dry in a reasonable time without drying in the screen.

Poor adhesion may result from improper ink formulation or improper treatment of the surface of the substrate. Polyolefin films should be flame treated or corona treated; aluminum foils must be clean and free from oil. If the ink solvent is too strong, it may attack the printed film causing it to pucker. Screen ink films are sometimes so thick that they can form ridges or waves if the velocity of an air dryer is too great. The problem can be overcome most easily by reducing the air velocity or by printing a thinner ink film.

SPECIALTY INKS AND TONERS

Many inks are formulated for specialty application. In package printing, ink jet inks are used to print bar codes and batch numbers. A major requirement is that the inks must not clog the tiny nozzles in the ink jet printer. Use of dyes instead of pigments reduces clogging. In addition, the ink must not dry in the nozzle. Ink jet inks resemble writing inks more than they do printing inks. A typical ink jet ink contains a solution of a dye in water or glycol. Hot-melt inks can be applied with special drop-on-demand printers.

Hot-melt inks are a small but important specialty. They may be applied by impact or non-impact devices (e.g., ink-jet). Hot-stamp applications include plastic and metal decorating, fabric label printing, pressure-sensitive label printing and graphic arts. The hot-melt ink must be formulated to operate in the printing process for which it is intended and the image must meet all performance requirements. Accomplishing both of these goals is a real challenge in designing hot-melt inks.

For medical supplies that must be sterilized by high pressure steam in the autoclave, it is useful to apply an ink that changes color at high temperatures. A glance at the package quickly shows whether or not the contents have been sterilized. These inks can be printed by any printing process.

The word "toner" is applied to two different materials for printing. In printing inks, a toner is a dye or pigment used to change the hue of an ink, notably a dark blue dye that changes black inks from a brown shade to a blue ("blacker") shade. The word toner is also applied to the black powder used to generate the image in electrostatic duplicators and laser printers.

Toners for electrostatic duplicators are complex materials that have controlled electrical properties and soften after they have been printed. These materials may be used in copying or printing devices that use light to generate a bar code or other symbol on any package.

FURTHER READING

Flexography: Principles and Practices, Fourth Edition. Frank N. Siconolfi, Chairman. Flexographic Technical Association, Ronkonkoma, NY. (1991).

The Lithographer's Manual, Eighth Edition. Ray Blair and Thomas M. Destree, Editors. Graphic Arts Technical Foundation, Pittsburgh, PA. (1988).

The Printing Ink Handbook, Fifth Edition. National Association of Printing Ink Manufacturers, Harrison, NY. (1988).

What the Printer Should Know About Ink, Second Edition. Nelson R. Eldred and Terry Scarlett. Graphic Arts Technical Foundation, Pittsburgh, PA. (1990).

The Printing Ink Manual, Fourth Edition. R. H. Leach, Editor. Van Nostrand Reinhold (International), London. (1988).

Chapter

Coatings: Decoration and Protection

CONTENTS

LIST OF FIGURES

LIST OF TABLES

Chapter

10

Coatings: Decoration and Protection

INTRODUCTION

Coatings are both decorative and functional. They serve to enhance the appearance and feel of a package and to protect its contents. Coatings are used not only to promote sales but also to protect the package and the printing on the package. They may even help the package meet legal requirements by protecting required printing from scuffing and partial obliteration.

Coatings must perform quantifiable and defined functions. They must be formulated to meet specific end-use requirements. Clarity, absence of color, good adhesion, product resistance, and resistance to rub and abrasion are almost always necessary. Specialty coatings offer many other properties. Colored coatings add color to the list of properties that can be achieved. Although they are not usually considered coatings, foil stamping, holograms, and other specialties are mentioned in this chapter.

Market research studies show that the visual impacts of labels and packages greatly influence the decision to buy. The package is a "silent salesperson." Ink and coatings have replaced sales clerks in promoting products sold in supermarkets and discount stores. It is widely believed that "gloss" sells, and increased glossiness is a prime reason for coating, although protection of printed data is significant for pharmaceutical labeling.

Coatings are classified by their function or by the type of material used or by the method of applying them. We may speak of primers, topcoats and barrier coats, or of varnishes, lacquers and aqueous coatings, or of gravure coatings and laminates.

In many ways coatings are like printing inks without pigments, and much of the technology in Chapter 9 applies to coatings. They can be applied by the same methods used for inks: from a printing plate or an engraved cylinder, from an offset blanket or through a screen. In addition, they can be applied by methods that are not used for applying a pattern: by extrusion and by lamination.

Coatings are applied both in line with the printing press and by off-press treatment of the previously printed sheet or roll.

Printing presses are frequently equipped with one or more stations for applying coatings in line with the printing. These coating stations may or may not use the same printing method that the press uses. Gravure presses may use flexo stations to apply a topcoat; offset presses may use anilox rolls to apply aqueous coatings at the end of the press.

Gravure and flexo presses have been running solventbase coatings for years. Aqueous flexo coatings are widely used on preprinted linerboard. Dryers on the last station of gravure and flexo presses are built to dry solventbase coatings and inks from the preceding stations. A few dryers have been built to cure solventbase or waterbase catalyzed (or "catalytic") coatings at temperatures of 200–275°F (95–135°C). These are also capable of drying waterbase coatings at near the design speed of the press.

Coating technology is evolving rapidly, led by better control of coating thickness, more versatility in coater capabilities and growth of aqueous or waterbase coating technology. Methods of coating are shown in figure 10-1.

Markets for Coating Materials

Sales of coating materials (by weight) exceed the volume of printing inks. Since most publication printing is uncoated, this suggests that more coating than printing ink is applied to packaging, an observation that confirms what you see on the grocer's shelf. The large volume of coatings is required because ink covers only part of the surface of the printed form while coatings usually cover the entire surface, and coatings are often applied at two to three times the thickness of ink.

Coatings are used for purposes other than enhancing the appearance or protection of printing. For instance, coatings on corrugated boxes are used to prevent stacked boxes from slipping; multiwall bags are coated with non-slip coatings to enable them to be handled on large pallets without strapping; frozen food cartons are coated both inside and out to provide a moisture barrier. Frozen food cartons are not usually printed, but are overwrapped with a printed and coated paper. Milk and juice cartons are coated on both sides but are printed on only one side.

Definitions

Nomenclature used for coatings is often not standard, a common problem in the graphic arts. Coatings include primers, barrier coatings, varnishes, aqueous

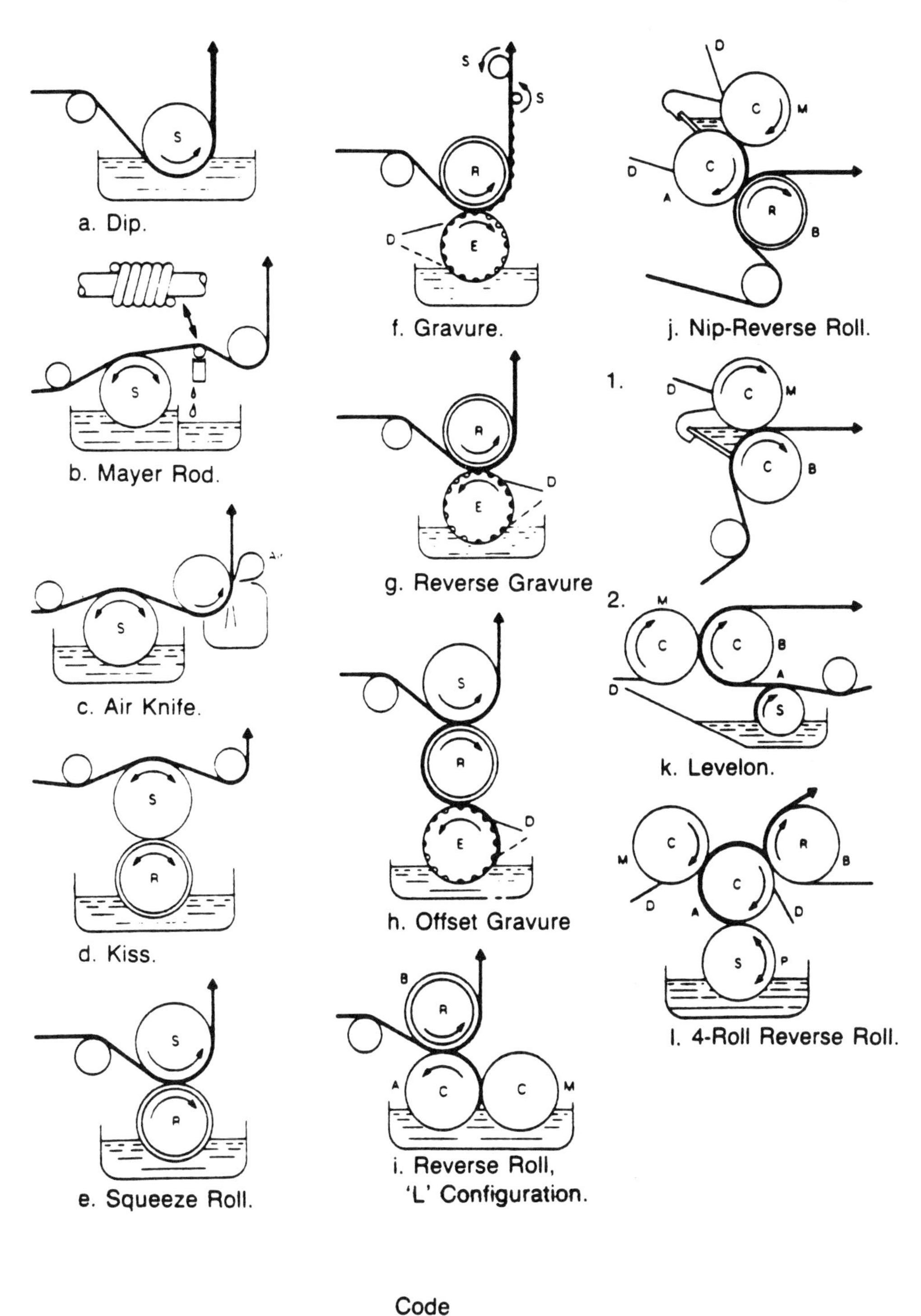

Figure 10-1. Methods of coating. (Source: Modern Plastics Encyclopedia, 1988.)

coatings, ultraviolet (UV) and electron beam (EB) coatings, lacquers, catalytic coatings and hot melts. This chapter also includes laminates as coatings. Varnishes include full-coverage varnishes, spot varnishes, and dull and gloss varnishes. Varnishes are sometimes distinguished from coatings, but in this chapter, varnishes are considered to be some of the materials used for coating.

Primer coats, that are often applied to improve adhesion, are also called undercoats. Actually, primers and undercoats are not the same. It is possible to apply a primer coat over a print, as on a flexible packaging film that has been reverse printed. The primer over the print improves adhesion to another film. In this case, the primer coat is also an overprint coating.

Overprint coatings are usually called "varnishes" by offset printers, even if the material is not an oil-base varnish. Metal decorators call them "top coatings." Flexographers call them "lacquers."

The word "varnish" is probably used for overprint coatings, because the first overprint coats were oil-based varnishes applied from a letterpress plate. Lacquers are solventbase solutions of any film-forming material applied as an overprint coating. The term "lacquer" is taken from the paint industry which used solutions such as shellac in alcohol to coat unfinished surfaces.

Coatings are usually clear, but the term "coating" is sometimes used for a pigmented coating applied to provide a tint or metallic gloss to the background of the package.

CLASSIFICATION OF COATINGS

Coatings may be classified according to their **type**: varnish, lacquer, aqueous coating, ultraviolet or electron beam. They may be classified according the **method of application**: flexo or gravure coatings, extrusion coatings or laminates. They may be classified according to **where they are applied**: primers or underprint coatings, topcoats or overprint coatings. Coatings are also classified according to their **function**: barrier coatings, non-slip coatings, etc.

Primers are applied chiefly to improve adhesion of inks or other coatings. Overprint coatings improve gloss and protect the printed surface from scuffing or scratching and to maintain its good appearance. Barrier or functional coatings protect the contents of the package from air or from loss of flavor or moisture.

Coatings often serve more than one function. Primers are applied mostly to improve the adhesion, gloss and holdout of inks or a subsequent coating, but they may also confer some resistance to the transmission of oxygen, moisture vapor or oil.

If an aqueous coating is applied in two coats, the first serves largely as a primer, the second mostly as an overprint coating.

Primers (Undercoats)

Primers improve the graphic effects or the adhesion of inks and overprint coatings to flexible packages and metal cans. For example, a primer or undercoat

TABLE 10-I. PERFORMANCE OF PRIMERS

Type of Primer	Adhesion to			Resistance to		
	Paper	Metal	Plastic	Heat	Moisture	Chemicals
Shellac	Poor	Exc	Poor	Poor	Poor	Poor
Organic Titanate	Good	Good	Good	Fair	Fair	Fair
Polyurethane	V. Good	Exc	Exc	Exc	Exc	Exc
Polyethylene imine	V. Good	Good	Exc	Exc	Fair	Poor
Ethylene acrylic acid	Exc	Exc	Fair	Fair	Exc	Good
Polyvinylidene chloride	Exc	Fair	Exc	Good	V. Good	Fair

Source: Paper, Film & Foil Converter, November 1986

on flexible packaging, foil or tinplate improves the opacity of a white coating, overcoming a gray tint. An acrylic or nitrocellulose may be applied to aluminum foil, a thermoset to steel cans, or a flexible primer to film to improve adhesion of inks.

A tough, brilliant wrapper/label for batteries is reverse printed on plastic then primed (here, the primer coat is an overprint coat). The primer promotes adhesion to the aluminum foil and improves appearance. Table 10-I summarizes some primers used on packages.

Overprint Coatings and Varnishes

Overprint coatings protect the printed package and improve visual impact and sales of the product. The oldest applications of overprint coatings (or top-coats) were varnishing and lacquering of the printed package to provide gloss. This, like modern coatings could be done either on line or off line. Coatings are now used for many functional purposes: to improve resistance to scuffing, rubbing and handling of the package. They protect the products in the package by supplying barriers to moisture, oils and fats, oxygen or sometimes light. They function as adhesives, and they also serve as primers for subsequent coatings.

Oil-based varnishes are applied as a liquid, usually from a printing press. They dry by a chemical reaction just as oil-base printing inks do. In offset lithography, oil-based varnishes have long served to improve gloss and protect the print. When a higher level of gloss was required, the printed sheets were sent to a finishing plant for application of a solventbase coating or lacquer. Around 1980, UV coatings began to replace the solventbase coatings. They provide higher gloss. UV coatings are applied at 100% solids. They contain no volatile organic solvents to pollute the atmosphere and therefore require no investment in solvent recovery or afterburner systems. Since these UV applications are trapped over dry litho inks, there is no loss of gloss as happens when a UV coating is wet trapped over litho inks.

Gravure cylinders and anilox rolls are commonly used to apply coatings to web packaging products: labels, folding cartons and flexible packaging.

Web litho coaters came on the scene in the mid '80s. Originally these were dry trap coaters, installed after the press oven and chill rolls. They made it possible for the web litho plant to apply a coating on line over dry ink.

Functional Coatings

Functional coatings are used for such purposes as the reduction of the loss of moisture and flavor of materials in the package and to prevent damage to the contents by substances in the environment, mostly water and air. Functional or barrier coatings protect the packaged products by providing a barrier to oxygen, moisture, oils, soaps, fumes and solvents.

Oxygen barriers, such as polyvinylidene chloride (PVDC), (often referred to by the common trade name "Saran" of Dow Chemical Co.) can be applied as an aqueous emulsion by using a "gravure coater," which is a name given to an anilox roll applying the coating to an offset blanket. The PVDC is typically applied in line with the printing, overprinted, then laminated. Laminating may also be done in line, but because of its low speed, it is usually done off line.

Some coatings are applied for specialized barrier properties. A coating can be used to block oxygen yet permit nitrogen to pass through, making is possible to use nitrogen sterilization of a finished product package. Other coatings can selectively block other gases while allowing the desired gases to pass through the packaging material.

A hot polyethylene coating can be applied in line after the dryer. The web is then put over the chill roll and rewound. For food items such as food trays or plates, the sequence may be: print, extrude, form.

Coatings may even be used on the inside of packages. Moisture-vapor barriers are applied to the inside of boxes of frozen foods to prevent freezer burn that results when the product is dehydrated by intense cold.

REASONS FOR COATING

Protection of Print

Overprint coatings protect the print on the package against scuffing or defacing. In the case of pharmaceutical labels and cartons, they may help the packager meet legal requirements concerning labeling and information about the contents of the package.

Gloss

Gloss is always an important property of topcoats and varnishes. Gloss plays a major role in impulse buying, and the coating is of critical importance in determining gloss.

The packager usually wants to achieve high gloss, but low gloss or a matte finish is sometimes desired for special visual effects. Dull varnishes, produced by adding a filler such as silica to a conventional oil varnish, are often applied in a pattern by the printing press.

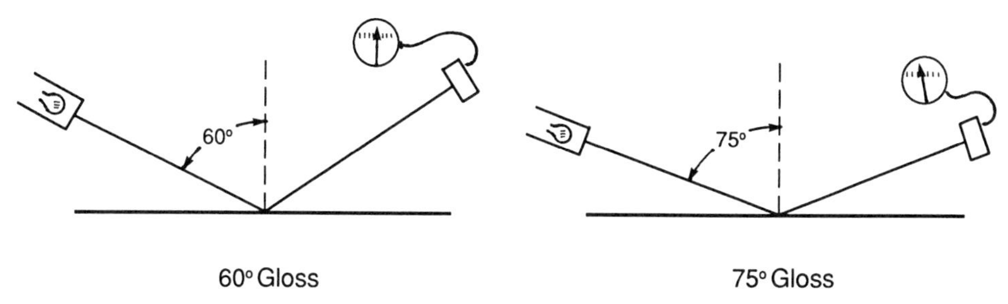

Figure 10-2. Gloss angle.

Gloss is measured with a gloss meter and is reported as "% gloss," that is, the percentage of the incident light that is reflected at the angle at which the determination is made (figure 10-2). Gloss varies with the angle of illumination (the higher the angle, the higher the gloss), and the angle must be reported. Gloss is most commonly recorded at 60°; gloss at 45° will be lower and at 75° higher than at 60°. Gloss in excess of 80 percent (at 60°) is usually considered good, and modern coatings can often achieve gloss values of 90 to 95 percent.

Many factors affect gloss: the coating, the smoothness of the substrate, coating, holdout of the stock, and the color of the print beneath the coating. To determine the gloss level of a particular paper or board and varnish combination, gloss readings are usually made on coated unprinted paper or board. The best gloss results from careful selection of both stock and varnish.

Decoration: Spot Varnishes

Coatings and varnishes applied in a pattern are referred to as "spot varnishes." They are mostly decorative, but they may also protect important parts of the design. Spot varnishes can be applied with a litho plate, with a blanket cut to fit the pattern or by adding a flexo unit or other printing unit to the press. Flexo plates have the advantage of being much cheaper than offset blankets.

Adhesion or Release

Adhesives are important in forming packages and can be applied by a press. Additional adhesive is applied to the setup box or bag on the filling line. The design of printing and coating must take into account the gluing requirements. It is often difficult to glue coated surfaces together.

Sometimes, the package requires a release coating. Pressure sensitive labels must be placed on a sheet with good release properties. Varnishes adhere well to ink but poorly to wax, silicone or Teflon. Actually, adhesion depends on the chemical nature of both materials that come into contact with each other. If the materials are very different chemically, adhesion is usually poor. Adhesion or release depends on proper formulation and the nature of the surfaces to be adhered.

Oil and Moisture Resistance

Overprint coatings also provide resistance to handling with greasy fingers. Oil resistance can be numerically specified. The term "squalene resistance" is used to express the resistance to a test oil called squalene that simulates the oil found in fats and human skin.

Paper and paperboard are especially permeable to water vapor. The resistance to water vapor can be significantly increased by coating with a polyvinylidene chloride latex or by laminating with a PVDC film. This retards frozen foods from drying out and moisture-sensitive packaged goods from becoming damp.

Product Resistance

Overprint coatings provide resistance to handling, but they must also provide protection against the product in the package. Labels for cooking oils require oil resistance; soap wrappers should be resistant to soap; and packages for fatty foods should provide barriers to fat. Most foods contains fats or oils.

A common application of barrier coatings is to milk and orange juice cartons where the coating keeps the product from leaking through the carton. The effectiveness of the coating as a barrier depends on the structure of the carton as well as on the coating itself.

Barrier Properties

A major purpose for applying functional coatings is to provide a barrier for moisture, air, tastes and odors and other materials that may enter or leave the packaged product. Modern barrier materials are sometimes designed to meet very specific needs such as a block to nitrogen gas but not to oxygen.

Slip and Nonskid Coatings

Coatings can have a low coefficient of friction, which is associated with high slip and a pleasant, smooth feel or "hand." However, coatings with a high coefficient of friction are sometimes applied to packages to improve skid resistance so that bags, for instance, can be stacked high without slipping off the pile. Coefficient of friction is adjusted by the coatings manufacturer who formulates the coating to produce the desired amount of slip.

Nonskid coatings are normally applied to multiwall bags at the last station of the printing press. For corrugated, however, the coatings are applied in the converting operation and are not strictly part of the printing process.

Nonskid coatings are also useful in preventing plastic bags from slipping past each other when they are stacked. The skid resistance of printed plastic bags is measured by placing one on top of the other on a board, and determining the angle to which the board can be raised before the top bag begins to slip. The test requires great care if the results are to be significant.

Faster Processing of Packaging Materials

Aqueous (waterbase) acrylics or UV coatings or both are often applied over wet litho ink (wet trapping) to folding cartons on the litho press to speed the manufacturing process.

Aqueous coating of wet offset inks eliminates setoff and blocking and reduces the need for spray powders. This enables the sheetfed printer to process folding cartons or corrugated at the end of the run without waiting for the ink to become fully dry. Coatings may actually slow the final drying of the ink, but the ink dries eventually underneath the coating. In addition, of course, the coating increases gloss and protects the printing.

Coating of wet litho inks is usually successful with aqueous coatings, but UV coatings sometimes adhere poorly to wet ink and give variable gloss, or reduced gloss after coating, especially over heavy films of ink.

There is a significant loss of gloss (5–15 points) when UV coatings are wet-trapped over litho inks. In the worst case, the UV coating does not spread on the wet ink, and it fails to produce a film. When a wet film breaks and forms beads on the surface, it is said to "crawl." If a film crawls, a dry film does not form and the coating provides neither gloss nor protection.

If wet trapping of an overprint coating is to be successful, the inks must be properly formulated, free of wax, silicone, polytetrafluoroethylene (commonly referred to by DuPont's trade mark "Teflon") or other slip agents. The coating must also be formulated to make it wet trap. However, wet trapping usually diminishes the gloss. UV coatings give higher gloss if applied over dry ink.

UV coatings over conventional inks have rapidly replaced UV inks. Now, presses with 9 or 10 stations can give a double bump* of aqueous coating over the wet ink, achieving both high gloss and quick delivery of the pile. For the premium buyer, printers developed the practice of wet trapping a waterbase coating over the litho ink, then taking the loads thus printed and coated to another press for UV coating.

Blister Packs

The paperboards used for blister packs (or carded packages) are produced with aqueous acrylic or modified acrylic coatings. They can also be produced with UV adhesives. They require a heavy coating, and this requires a good dryer. Blister packs can be made by wet trapping the coating over the wet ink if the line has sufficient drying capacity at the end. The product is inserted under the "blister" on a special finishing machine.

APPLICATION METHODS

Coatings are applied by a wide variety of techniques. They are applied by flexo, gravure, offset and rotary screen presses; by laminating machines and by

* double bump = two impressions at two stations of the printing press.

curtain coaters. Combinations of these methods are common. Coatings are applied in line with the press and also off line. Because of the thickness or weight of the film that is applied, special drying equipment is often required.

At one time, varnishes were applied on a "varnishing machine," but conventional varnishes are now mostly applied in line on a sheetfed or web offset press. On short runs, cylinder screen presses offer an effective and efficient means of applying varnishes.

A catalog lists more than 30 suppliers of coating equipment, coaters and dryers. Waterbase coatings are supplied by at least 17 companies. Coating has become so important, not only in packaging but also in publication and commercial printing, that printing presses have been largely redesigned to apply coatings as well as ink.

Flexo Coatings

Properly adapted, the flexographic printing press is capable of carrying out several types of coating applications. Figure 10-3 illustrates the threading arrangement for different coating operations.

If UV coatings are to be run on a flexo press, it is necessary to replace any rubber or plastic rolls that are not resistant to UV coatings.

Flexo presses require low viscosity coatings. Strong solvents such as toluene

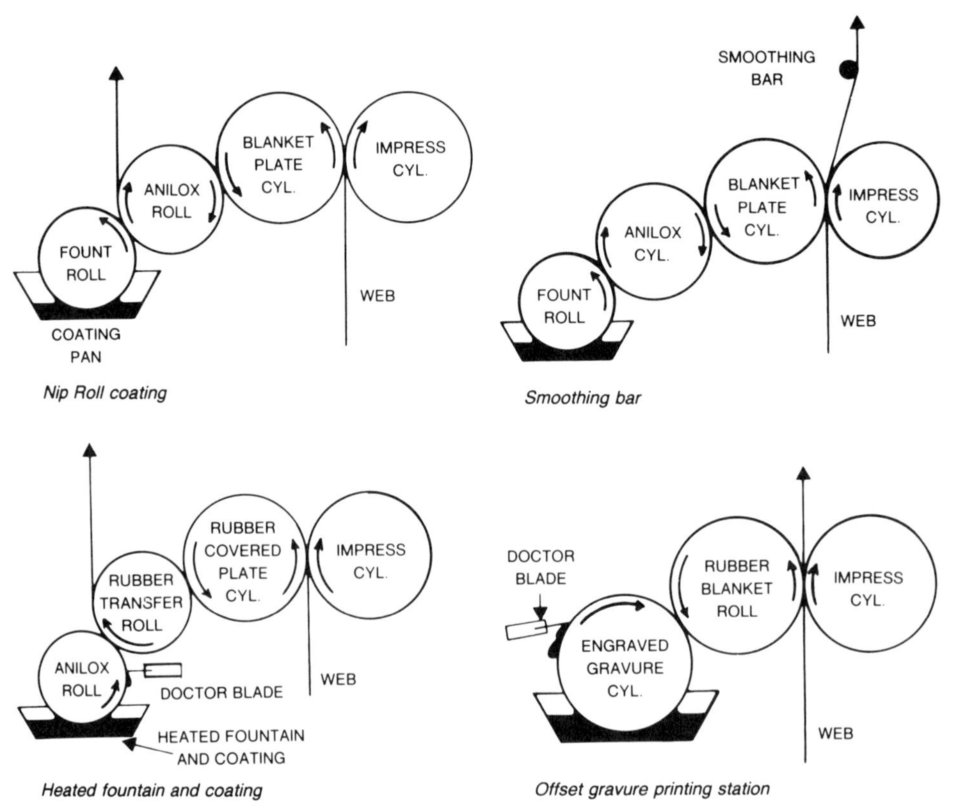

Figure 10-3. Flexo press coating arrangements. (Source: Flexography: Principles and Practices. Fourth Edition.)

and methyl ethyl ketone attack flexo plates and rolls. Aqueous coatings are greatly preferred because most flexo presses are not equipped with solvent recovery systems.

Gravure Coatings

No special design changes are required to apply coatings from a gravure press. Gravure presses handle low viscosity coatings, and they are useful for lacquers or solventbase coatings containing strong solvents since they usually have solvent recovery systems, and since (unlike flexo and offset presses) all parts of the press are resistant to strong solvents. Waterbase coatings are commonly applied to packaging materials on gravure presses.

The term "gravure coating" is sometimes applied to coating from the blanket of an offset press which is equipped with an anilox roll to meter and deliver the coating.

Offset Press Coatings

Varnishes have a higher body or viscosity than most other liquid coatings and are applied in line, through the ink train of the offset press. Aqueous coatings have lower viscosity, and they are applied to the blanket through the dampening system of the press or off line with a roller coater, gravure coater or by other means.

A coater/dampener or a blanket coater can be retrofitted on most presses. The coating is applied to the printing plate, which in turn applies the coating to the blanket, and thus to the sheet (figure 10-4). There are two film splits in the coating from the applicator to the sheet, which limits the thickness of the coating that can be applied to the sheet. Where the replacement of spray powder with moderate gloss or moderate scuff resistance is the reason for coating, this thin film is no drawback.

A blanket coater, as the name indicates, applies the coating directly to the blanket. Since there is no film splitting from plate to blanket, one less coating film split is involved, and it is possible to put more coating on the sheet and to get higher gloss and scuff resistance.

Web Offset Coatings

Excellent results are obtained by combining wet trapping with a dry trap coater added after the chill rolls. A wet-trap waterbase primer is completely dried at full press speed. The web exits the press oven with better surface holdout than when it entered. A high-gloss waterbase topcoat is now applied and dried at normal press speed to yield a high-gloss, scuff resistant coating.

Two press configurations are possible: 1) with no capital expense, the web can be rethreaded back through the oven or go through the second oven if there is one; 2) with a capital outlay, a vertically mounted shortwave infrared (SWIR) drying system can be installed before the dry-trap coater.

Presses are being designed to provide space after the plate coater on the last

Figure 10-4. Epic Delta coater/dampener. (Courtesy of Epic Products International Corp.)

print station to mount a SWIR dryer to dry the waterbase primer applied there before the sheet enters the tower coater.

Screen Coatings

Coating with a flat-bed or rotary screen press offers no special problems. Unless the press is equipped with solvent disposal equipment, it cannot handle lacquers or solventbase varnishes. Screens require high viscosity coatings, usually varnishes or ultraviolet coatings.

Smoothing Rod

Coating with a "smoothing rod" is another effective way to apply continuous, overall coatings of lacquer or low viscosity aqueous coatings. The rod can be attached to any flexo or gravure press. It is especially effective in eliminating streaking or irregular striations known as "crow's feet." This adaptation is commonly used to apply adhesives and consists of a simple smoothing bar or rod, after the point of application, to lightly wipe the surface of the wet coating. This wiping action levels the surface of the lacquer in direct proportion to the angle of the web over the bar and the closeness of the bar to the point of application.

Mixed Coating Processes

Presses designed to print packages often use two or more printing methods. A gravure printing press may have a flexo unit to apply coating at the end of the press. More and more offset presses are purchased with special coating units in line.

In the offset gravure coating method, the etched gravure cylinder transfers the coating to a rubber-covered transfer or offset roll. The transfer roll applies the coating to the web. To apply a patterned varnish or coating, the patterned gravure cylinder must be of the same diameter as the transfer cylinder. For an all-over coating, the offset rubber blanket roll can be any diameter.

Offset gravure is used for varnishes with viscosity of 200-800 centipoise (cp). Below 200 cp, we would use direct gravure or flexo. Flexo can go up to 400 cp. Roll coaters are used for higher viscosities. Offset gravure coatings are heavy viscosity, high solids such as UV coatings, and a very thin coating may be applied.

Special Equipment

The drying equipment is of special importance for coating because coatings are usually applied at higher film thicknesses than inks. Dryers that satisfactorily dry inks may be inadequate for coatings, requiring the press or coater to operate at an uneconomically slow speed. Additional drying capacity can be economically added with infrared dryers or blowers.

Press attachments, such as unwind stands (used for, among other things, laminating), UV and EB curing systems, stackers and conveyors are used to build a press for package coating. Instruments such as viscometers are also required.

Off-line coatings are applied with equipment dedicated to coating, such as specially converted offset presses, cylinder screen presses and laminators.

Printing presses are increasingly being purchased with units especially built for coating, and most press manufacturers now sell coaters as separate units for inline operations. Coaters can be ordered together with the new press or as retrofit units.

Extrusion Coatings

Extrusion coatings are often applied to carton board at the board mill, to provide special properties required for packaging soap, milk, cosmetics or frozen foods. These are really a type of laminating, in which an extruded film (often polyethylene) is laminated to the board. If the polyethylene is properly treated, it will adhere to the board without use of an additional adhesive.

Extrusion coatings offer several advantages for folding cartons. They can be glued and printed, they can be given a variety of finishes (gloss, satin or matte) and they can be colored to enhance the appearance of the carton or to hide product residues.

Curtain Coaters

Curtain coating is a type of extrusion coating where the coating is extruded perpendicularly to the horizontal web. Curtain coaters are special converting equipment in which a curtain of the coating material flows by gravity or under pressure through a slot and is laid on a substrate that passes underneath. Curtain coaters may be able to apply many coats at one pass. They are used for making corrugated boxes almost impervious to moisture (e.g., containers for lettuce and other produce) and are occasionally used in coating other packaging materials. Speeds are slow.

Curtain coaters are used to apply multiple coats, such as polyvinylidene chloride (for oxygen barrier) between two layers of polyethylene (figure 10-5) or another functional coating can be added, plus another layer of polyethylene. The thickness is controlled very precisely at low speed. It requires good control of viscosity and temperature.

Other Coating Methods

Corrugated is also waxed by dipping it into a tank of melted wax. The wax not only coats the surface, but some gets into the fluting to give good moisture resistance to the structure. This gives a box that is useful for packaging wet lettuce or other produce. The board must be printed with wax-resistant inks before it is waxed. Corrugated can also be waxed in a tunnel where hot wax is blown on it.

LAMINATING

Laminating is a special method of coating. Laminating is used because no single material provides the optimal balance of properties for every application: e.g., barrier properties, heat sealability and durability, needed for the many end-use applications in today's packaging market. Laminating creates a beautiful surface on a package and supplies a barrier to moisture, oxygen, light, odor and flavor for specific packaging applications, especially in food packaging.

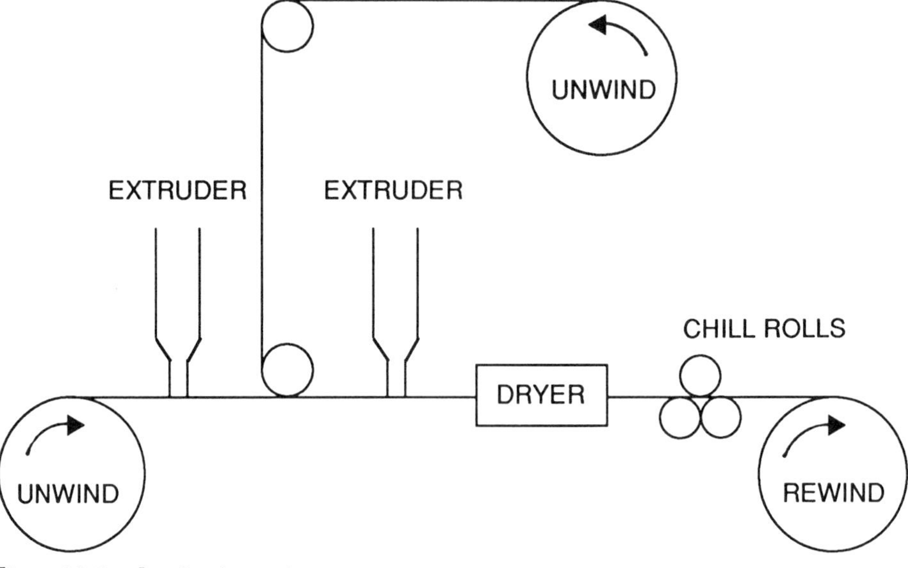

Figure 10-5. Laminating and coating on a curtain coater.

A host of products: spices, fats, oils and acids, attack or penetrate many packaging materials. Dry food seasonings such as chili powder tend to penetrate and delaminate the plies. The package must resist deterioration by the packaged product, as well as light, moisture and oxygen during storage and exposure to high and low temperatures during shipping and handling.

Proper pretreatment, primer coat, drying time and surface wettability must be considered to ensure successful laminating of films.

Flexible packages often comprise multiple plies of paper, film or foil laminated together with an adhesive system. A survey by *Paper Film & Foil Converter* magazine (Nov. 1986) showed that 86 percent of flexible packaging firms engaged in coating or laminating had laminating equipment, 85 percent used a gravure applicator, and most converters used both solvent and waterbase adhesives. Only three percent of the respondents used EB curing, while 15 percent used UV curing. For the future, 70 percent of the respondents were planning to use waterbase adhesives.

Lamination applies a preformed film to the substrate. Extrusion lamination involves the extrusion of a melted polymer directly onto the substrate. Lamination adds tensile and tear strength to the functional properties obtained by combining coating and substrate (paper, board or film). Coatings do not usually add strength.

Laminating adds a web to the original web. An adhesive is applied to one of the webs, and a nip-roll (ironing roll) is used to seal the two materials together.

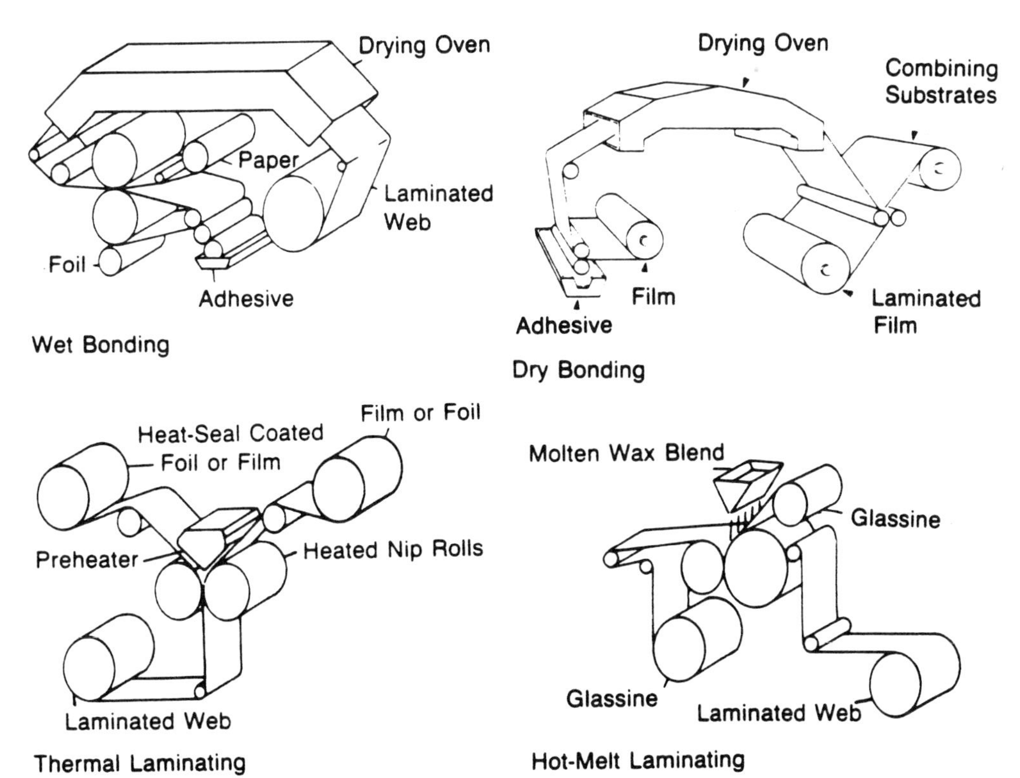

Figure 10-6. Methods of laminating. (Source: Modern Packaging Encyclopedia, 1988.)

Off-line laminating is performed on special machines that apply films to paperboard and other packaging materials.

There are four basic methods of laminating: wet bonding, dry bonding, thermal laminating and hot-melt laminating. These are illustrated in figure 10-6. In wet bonding, substrates are combined prior to going through the drying oven. In dry bonding, the combining of substrates occurs after the drying process. This affects the choice of adhesive.

In thermal laminating, heat is the laminating agent, being applied just before, and also as the substrates pass through, the nip rolls. For hot-melt laminating, where heat is also a laminating agent, a molten wax blend is applied where the substrates are combined.

Film laminating is often performed on special laminating equipment (see Chapter 8), but it can also be done on a thermal combiner (heat and pressure) added to the exit end of a flexo press. Adhesives are applied at the last printing station of the press, dried and combined with the second web. A rubber-covered nip roll supplies laminating pressure to a heated roll at 175° F (80° C). If the adhesive is to be applied by flexo, the rubber transfer roll (plate cylinder position) must be solvent-resistant and have good release properties.

A self-contained laminating machine can be installed in line with the flexo press. It is independent of the press except that it and the press use the same rewind, and the electric drives of the two machines are connected and employ necessary web tension controls.

Materials commonly laminated to packaging materials are high density polyethylene (HDPE), low density polyethylene (LDPE), polypropylene (PP) and polyethylene terephthalate (PET), which is often referred to as Mylar, the familiar trade name of DuPont.

The same economic forces that are pushing the expansion of laminating in-line with flexo printing are also leading to the development of various in-line coating processes such as:

- overall heat seal coating
- thermal stripe heat seal coating
- coadhesive latexes, both overall and registered
- overprint lacquers that may or may not be heat sealable.

COATING MATERIALS

Varnishes are usually oil-based materials used for coating or to make printing ink. The term is sometimes used for other overcoat materials. Aqueous coatings are either solubilized acrylics or thermosetting (or catalytic) coatings supplied in aqueous solution. Ultraviolet and electron beam coatings contain no solvent; they are applied as liquids, then cured with suitable radiation. Lacquers are solutions of resins in organic solvents; they are applied as liquids but become dry when the solvent evaporates. Hot melts are melted polymers that solidify when they cool. They require special machinery for application.

Coating materials used in packaging are summarized in table 10-II.

TABLE 10-II. COATING MATERIALS USED IN PACKAGING

Purpose	Material	Purpose	Material
Gloss	Litho varnish Aqueous coating UV Coating EB Coating Polyurethane Wax	**Water Resistance**	Litho varnish UV coating EB coating Polyethylene Polypropylene PVDC Waxes
Protect Print	Litho varnish Aqueous coating UV coating EB coating Polyurethane	**Oil, Fat, Grease Resistance**	Aqueous coating UV coating EB coating PVDC Nitrocellulose Shellac
Barrier to Oxygen	PVOH PVDC PET		

(Abbreviations: UV = ultraviolet, EB = electron beam, PVDC = polyvinylidene chloride, PVOH = polyvinyl alcohol, PET = polyethylene terephthalate.)

Varnishes

The advantages of varnishes are that they are processed like ink, produce patterns easily and give good gloss and good moisture protection. Their disadvantages are that because they dry slowly, they cannot be applied in a thick coating. They require spray powder, and it may be necessary to rack the delivery. Varnishes tend to yellow with age, particularly when they are exposed to fluorescent light.

Conventional oil-base varnishes are formulated to control gloss, adhesion, slip and other properties. Applied by sheetfed offset, they generate all the problems of stacking sheets coated with a heavy film of ink. Drying takes several hours, but it should be complete in eight to 10 hours or less if the varnish is properly formulated and applied. The gloss of a well-varnished folding carton may be as high as 85 percent.

Carton printers often run four-color process work with varnish on two- and three-color presses by running the cartons through the offset press two or three times. By the time the pile begins to fill, the weight of the load forces the bottom sheets together and may cause picking, setoff and blocking. Preventing these problems requires the judicious use of spray powder which sometimes reduces gloss and visual effectiveness. Spray powder gives the sheet a rough, abrasive, sometimes sandpapery feel and reduces scuff resistance. Carton buyers do not like the rough feel that embedded spray powder gives their cartons. Spray powder tends to collect on blankets and plates, interfering with image transfer. It also fouls the air, causes expensive press wear, maintenance and downtime, and it is expensive. If there is no sheet cleaner on the press, the pressman must stop every few hundred sheets to wash blankets. He throws away 20 sheets at each startup because of color variation.

On critical jobs, the pressman may rack the loads every eight or 10 or 12 inches to prevent offsetting. Varnishes are being replaced with aqueous coatings and UV coatings.

Waterbase acrylic coatings, formulated to wet-trap over litho inks, can reduce or entirely eliminate the amount of spray powder needed, particularly if they are dried before the sheets land in the pile.

A great deal of conventional varnish is still used, but the aqueous and UV coatings show less tendency to yellow and are superior in gloss, clarity and resistance to rub, abrasion, grease and oils. They require no spray powder. Yellowing that does occur can be counteracted with optical brighteners.

Aqueous Coatings

The growth of aqueous or waterbase coatings has paralleled or even exceeded the growth of waterbase inks. Aqueous coatings are readily applied by any printing process, including offset. In 1975, aqueous coatings had less than two percent of the market, but by 1990, that share approached 40 percent. They will undoubtedly dominate the market in years to come.

Aqueous coatings are often applied as the sole coating material to folding cartons, and sometimes to corrugated boxes, or they may be used as primers to improve the adhesion of UV coatings to lithographic inks on cartons and labels.

Modern aqueous coatings are usually based on acrylic resins. Developed as floor coatings, the acrylics provide packages with a coating that has the gloss and toughness of floor coatings. They can be applied to packaging materials by flexo, gravure or offset.

Aqueous emulsions are used to provide barrier coatings. PVDC provides an excellent barrier to grease and moisture. Waxes and polyethylene applied as emulsions are used as water barriers. Polyvinyl alcohol can be applied from aqueous solution. The following discussion applies mostly to the aqueous topcoats that are classified as thermal drying or thermoset (also called catalytic or "cat") coatings.

Thermal drying coatings dry by evaporation of water. No minimum surface temperature is required. If the moisture is not evaporated it will soak into the board. Thermoset coatings must be heated enough to cause a chemical reaction to occur, and they can be dried or cured at or near full press speeds. The coated paper or paperboard is generally repulpable. Although the gloss is excellent, the best UV coating will still provide a higher gloss than the best aqueous coating.

The printer will achieve best gloss if the waterbase acrylic coating is applied in two coats (double bump), drying the first coat before applying the second. The first coat serves as a primer. This technique is often used on folding cartons.

Waterbase acrylics give 60–70 percent gloss on one pass when applied in line, 80 percent on two passes. The maximum gloss reaches 85 percent or even 90 percent when two coats are applied by conversion (that is, off line). They come into the carton plant at 50–55 percent solids (45–50 percent water) and are applied as is or diluted to suitable viscosity for the particular method of application. Almost without exception, thermal drying waterbase coatings employ acrylic or modified acrylic film formers. Typically, they are one-component systems that dry when water evaporates from the film or is absorbed by the substrate.

The acrylic resins used in aqueous coatings are brought into solution with ammonia or an organic base. When the water evaporates, the base evaporates or is neutralized by the substrate, leaving a water-resistant coating.

Thermosetting coatings are shipped at 80–85 percent solids and applied from a solution containing about 70–75 percent solids. The gloss usually runs around 70–75 percent, but it can reach 85 percent, or, on a double-bump coating, as high as 90 percent.

A good thermosetting coating will have better scuff resistance than a typical UV coating. It is waterproof, and resistant to alcohol, detergent and grease. In layman's language, what we have here is Formica with enough flexibility to be scored and folded without cracking.

Thermoset aqueous coatings are two-component systems. Somebody in the press area has to stir a catalyst into the coating. This requires attention to detail. Thermoset coatings have low pH in contrast to the alkaline aqueous acrylics. They cure by a heat-induced chemical reaction that requires a temperature of 200–275° F (95–135° C), depending on the coating.

Thermosets arrive in the plant with 75–85 percent solids. The solution is usually let down to 65–70 percent solids with water to overcome its high viscosity. The formulation should have a pot life of at least 24 hours at room temperature after the catalyst has been added.

Surface temperature (not oven temperature, which may be 100-200° F higher) must be 200–275° F (95–135° C). About 200–250° F is as high as most presses are equipped to go.

Packages with acrylic coatings are easier to repulp than are UV/EB coated board. Acrylic coatings are generally considered to be biodegradable.

The chemistry of the waterbase acrylic and "cat" coatings is well developed, and the drying technologies are well known. Waterbase acrylics satisfy most ecological requirements although some contain 5–10 percent isopropanol to speed the drying. Waterbase thermoset coatings satisfy some ecological requirements.

Waterbase acrylic coatings are less costly than UV coatings yet they provide good gloss and good abrasion resistance. They require no special curing equipment, and they contain no toxic photoinitiators. Board coated with aqueous coatings is repulpable; board coated with UV or EB varnishes requires deinking equipment that is generally unavailable in North America.

With a wet trap waterbase primer applied on press as an undercoat, feel, gloss and depth perception are comparable to UV coating at lower cost. The dried coating can be hot foil stamped or imprinted, it does not pick up fingerprints, and it has write-and-wipe ability. However, to achieve these properties, it is important to choose a good stock.

Comparison to Lacquers. Waterbase coatings are slower to dry than solventbase lacquers and require more energy, but waterbase coatings are replacing lacquers because they give good gloss, scuff, water resistance, slip and slide angle, while eliminating volatile organic compounds (VOCs).

Comparison to Varnishes. Unlike the press varnishes they replace, waterbase acrylics don't yellow with age, so they hold their appearance better on the supermarket shelf and they can be formulated to give better rub resistance to withstand the scuffing rigors of the transport and distribution cycle. Board freshly coated with aqueous coatings can be piled high in the delivery. Such board has little or no odor.

If an aqueous coating is to be applied by offset, the press must be equipped with a coater/dampener, blanket coater or inline free-standing tower coater.

It may be worth considering replacing a UV coating with an aqueous coating if UV scrap is not repulpable using current technology. Many buyers state that whatever they buy must be biodegradable and repulpable, and waterbase coatings generally fit that description. They can be made FDA compliant for direct food contact.

Cost. Because of variations in the solids content of various coating materials, the price per pound should be used only in calculating the cost on an area basis: per thousand square feet or per thousand cartons. Any salesman can lower the price of the raw material by adding more water. Aqueous coatings are less costly than UV coatings on the basis of the area coated.

Manufacturers are trying to learn how to ship just the solids portion of the formulation. This will eliminate shipment costs for the water in the coating formulation, as well as concern over shipping at or below freezing temperatures.

Gloss. Gloss depends to a great extent on the nature of the substrate. Film and foil usually give a very high gloss when coated. On paperboard, gloss achieved by acrylic coatings is determined in part by holdout of the board surface, by film thickness, and by the color and character of the ink beneath the coating. The higher the solids content of the solution at the time of coating, the higher the gloss of the dried film.

However, given the same board surface and with all other factors equal, gloss depends on the skill of the coating formulator. The best acrylic will have lower gloss than the best UV, but higher than the best oil-base varnish. Gloss of the thermal drying coating is about the same as that of the thermoset coating.

Gloss is always increased by applying the coating in two layers, provided the first coat is dried before applying the second coat. It is also increased by applying the coating at the highest possible solids content.

Table 10-III gives the level of gloss that can be expected if a coated paperboard is coated with a sheetfed offset press. The numbers can be varied by applying a heavier coat—as with a gravure press—and by drying between coatings, which is impractical on an offset press. (A few sheetfed presses do have UV curing lamps between units.) Dry trapping gives higher gloss than wet trapping. However, any wax or silicone in the first coat makes it impossible to apply an additional coat.

Problems of Aqueous Coatings. Drying problems are the most frequent complaints about waterbase coatings. Assuming correct coating formulation, drying may be affected by, among others, humidity, stock and ink. The most common difficulty, however, may be insufficient drying capacity. It is surprising how often this problem exists.

TABLE 10-III. TYPICAL GLOSS LEVELS ON PAPERBOARD WITH A SHEET-FED OFFSET PRESS

	In Line		Off Line	
	Single Bump	Double* Bump	Single Bump	Double* Bump
Varnish	65%	**	85%	**
Thermal Aqueous	70	**	75	90
Catalyzed Aqueous	70	**	85	90
Catalyzed Solvent	75	**	85	92
UV/EB	90	92	93	95

* Assumes drying before application of second coat.
** Cannot be dried on press.
(Various sources)

UV and EB Coatings

The first waterbase coatings didn't come close to the gloss of UV coatings, and where ultra-high gloss is required, UV early established itself as the preferred coating. Now, under the best conditions the gloss of the best waterbase coatings is approaching that of UV coatings.

UV/EB coatings enhance and protect the printed product. UV light or EB radiation is used to solidify the liquid applied to the substrate. A solventbase lacquer provides excellent gloss, but UV coatings avoid the use of solvent. (Recovery or afterburner systems cost a quarter of a million to one million dollars.) There is no loss in gloss when UV is applied over dry ink.

The coatings are applied on the press in the same way as ink, or they can be applied offline with a coater. They cure in a matter of seconds.

Viscosities of UV/EB liquids range from 50 cp to 1800 cp. The lower viscosity liquids are suitable for flexo and gravure, the higher ones for web or sheetfed offset. Electron beam radiation is more energetic than UV light and cures faster. UV coatings (like UV inks) require the addition of a photoinitiator to achieve satisfactory polymerization rates during production.

EB can achieve 99 percent conversion at 1500 fpm (7.5 m/sec). UV attains 90 percent conversion in line, 95 percent when applied and cured off line. The unconverted material plasticizes the film or soaks into the sheet.

EB coating of flexible packaging is limited mostly to producing aseptic packages or to highly pigmented films. Radiation penetrates opaque inks as thick as 20 mils. EB coatings show no postcuring, and the actual coating cost is five to 10 percent less than UV after paying for the initial equipment investment.

UV/EB coating can be run in line, at press speed, producing gloss of 87–92 percent (60°) when the equipment permits dry trapping. When cured, the coatings have good scuff resistance. Off-line applications may yield gloss of 95 percent or even higher.

UV coating systems have several advantages over waterbase coatings. With greater solids content, the coating has better film properties, e.g., faster cure, fewer wet or tack problems. UV systems are nonflammable. They require no special storage to prevent freezing. Emission of solvents on curing is essentially zero. UV coatings generally have somewhat better gloss than waterbase coatings; EB can achieve even higher gloss.

UV and EB coatings are based on low molecular weight acrylic polymers, often called oligomers, that are further polymerized by exposure to UV or EB radiation after coating on the substrate. UV or EB coatings contain the same acrylic resins that are used in UV or EB inks.

Advantages of UV/EB Coatings. UV/EB coatings dry by radiation. They are essentially 100 percent solids—all the coating becomes a solid film, and no solvents have to be recovered and disposed of. They have the highest gloss and are very hard and resistant to abrasion and solvents. Resistance to alkali makes UV/EB coatings great for soap and detergent boxes. They dry instantly and require no spray powder. They show no tendency to yellow, and they have excellent resistance to blocking. When applied over dried inks, they have excellent adhesion.

Disadvantages of UV/EB Coatings. UV/EB coatings require expensive application equipment. The coatings are expensive, and curing requires a lot of electrical energy. The uncured coatings are not completely free of odor. UV and EB radiation are hazardous and require proper precautions and shielding.

It is often difficult to determine optimum cure. Because UV coatings contain photoinitiators, they are not compliant with FDA requirements for direct contact with foods. One of the most common problems with UV systems is curl caused by shrinkage of the curing film.

UV and EB coatings are difficult to glue, and if they are used, glue lines should be left uncoated.

Early UV coatings frequently caused allergies or skin irritation, but in recent years manufacturers have succeeded in greatly reducing the amount of irritating materials in the UV varnish. Nevertheless, some workers may still find them irritating.

Recycling poses special problems. While it is possible to repulp stock coated with UV or EB varnish, it requires a flotation process that is not widely used in North American repulping mills. UV/EB coated stock can be used in low-grade recycled materials or blended off in higher grade recycled board, but much of it gets exported to countries with suitable recycling equipment.

Comparison of EB with UV Coatings. EB has some advantages over UV. EB coatings are less costly than UV coatings (which require an expensive photoinitiator), and the high energy output of the EB generator permits higher press speeds than UV. The beams can penetrate pigments, permitting cure of coatings under ink or film.

EB coatings show the best gloss obtainable in a liquid application system. However, when gloss levels get above 90 percent, the package becomes susceptible to unsightly finger-printing. EB coatings have very high rub or abrasion resistance, and can be given a durable matte finish. They exhibit superior curing on foil and plastics.

Users report that EB has the following advantages: quality-consistent product, high processing speed, cost effectiveness, low energy consumption, no solvent emission, lack of heat generation.

Disadvantages for end users include the high initial capital equipment and a higher amount of downtime and maintenance inherent in the system. EB equipment is much more costly, partly because EB radiation generates X-rays, and it is more difficult to shield workers from EB than from UV.

Lacquers

Thermoset Solventbase Coatings. Analogous to the thermoset aqueous coatings are thermoset solventbase coatings. These are dissolved in toluene or methyl ethyl ketone and applied by gravure at 20–30 percent solids. They are cured on press. They provide excellent gloss and protection of the surface from rub and from oil and moisture. Gravure presses usually have the required solvent recovery systems, and they contain no rubber or plastic rolls that swell with the strong solvents used.

Nitrocellulose and Shellac. The earliest lacquers were made of nitrocellulose or nitrocellulose dissolved in alcohol or hydrocarbon solvents. They still find some use although aqueous and UV coatings now provide better adhesion, excellent gloss and freedom from volatile organics (VOCs). When used as primers or barrier coatings, nitrocellulose lacquers provide resistance to oils and animal fats, including handling with greasy fingers.

Waxes and Hot Melt Coatings

Folding cartons and corrugated boxes have been waxed with petroleum wax for decades. Wax can be applied in emulsions or from hot melts. In recent years, petroleum waxes have been supplemented or replaced by low molecular weight polyethylene waxes or by ethylene copolymers.

Hot melt wax coatings impart gloss and moisture resistance. They are roll-coated onto paper to make waxed paper that replaces the polystyrene clam shell that was used previously to serve hamburger sandwiches.

Wax emulsions—petroleum wax—and polyethylene emulsions provide gloss and moisture resistance to cartons for frozen foods and milk. Heavy films are applied. They are used on a great number of packages so that the sales volume of these wax emulsion coating materials rivals the volume of all decorative coatings.

Other Coating Materials

Materials such as polyurethanes and polyethyleneimine (PEI) are used as primers on metal and polyester to promote an excellent bond of coatings to foils, cans, paper and paperboard, film or composites. Polyethyleneimine provides superior adhesion between ink and substrate. Lamination bonds up to 1000 grams per inch can be obtained for such applications as retort pouches and boil-in bags. Building the pouch may include printing over a primer (PEI), and applying an extrusion laminate like polyethylene with a polyvinyl alcohol (PVOH) oxygen barrier and a polyvinylidene chloride (PVDC) topcoat.

Polyvinylidene chloride finds extensive use in providing grease, oil and oxygen barriers.

Colored Coatings

Coatings are usually thought of as clear and colorless, but sometimes the coating is colored to provide a tint to the entire surface of the package. In this application, a coating is much like an ink except that the color strength is usually very low and the coating covers the entire surface, forming no pattern.

Corrugated is sometimes coated to give bright colors. It can then be printed or labeled.

Adhesives

Packaging is one of the largest consumer of adhesives. According to Skeist, almost 30 types of adhesives are employed in 35 different applications. Adhesives are a type of coating, but they have little relationship to graphic arts and decoration and are mostly outside the scope of a book on printing.

Adhesives are grouped by end use: corrugated manufacture, folding carton assembly, film lamination and label adhesives; or according to type: hot-melt, pressure sensitive, solvent base, aqueous solutions and dispersions.

A Frost & Sullivan report (1987) predicted U. S. sales of $3.0 billion (2.6 billion pounds) of specialty adhesives in 1991 of which packaging was predicted to consume about 900 million pounds.

Solventbase adhesives were first used for film laminations, both inline and offline. Some markets and substrates still require solventbase formulations that cannot be matched with waterbase adhesives. However, market share for waterbase adhesives in flexible packaging is growing.

Adhesives in common use currently include polyvinyl acetate, epoxies, acrylics, cyanoacrylates, urethanes and solvent cements, in addition to starches, dextrins and glues.

DRYING AND CURING

There are many different ways of drying and curing coatings. Solventbase lacquers and aqueous coatings are dried in line with high velocity hot air, or sometimes with infrared (IR) radiation. Aqueous coatings sometimes dry adequately merely by absorption into the board.

Solvents must be captured and disposed of, but water can be discharged to the atmosphere, provided it is not contaminated with other gases.

Natural varnishes dry slowly by air oxidation in the presence of a catalyst or accelerator called a "drier." UV and EB varnishes are cured by radiation generated by special UV lamps or EB generators. Hot melts are set by chilling on a chill roll. Catalytic coatings are cured by heating the coated product.

A combination of high velocity hot air and an IR dryer is often the most efficient way to dry a waterbase coating. An IR dryer helps dry the ink beneath the coating, preventing potential problems caused by wet ink. Infrared dryers are classed as short wave (SWIR), medium wave (MWIR) or long wave, depending on the wavelength of the radiation. Short wave is the most energetic and SWIR dryers are usually favored for in-line drying.

SWIR dryers have a half second on/off cycle compared to 30-45 seconds for MWIR dryers to cool and reheat. This delay in cooling may cause paper to catch fire on the press. The MWIR dryer is less costly to operate than the SWIR dryer. An air sweep is usually necessary to dry the coating at full speed on the press. If there is no air sweep, the evaporated moisture follows the sheet, making it impossible to get the last of the moisture out of the coating. A hood is desirable to evacuate the moisture vapor and any odor coming off the drying coating.

It is important to distinguish air temperature from web temperature. In one instance, a flexo coater was unsuccessful in curing a new ultra-high gloss and scuff-resistant catalyzed waterbase coating with a 14-foot air oven on the press. Despite an oven temperature of 450° F (235° C), the surface temperature of the web reached only 168° F (75° C) at 125 fpm; and the lowest temperature required for curing coatings of this type is 200–225° F (95–105 ° C). Four banks of powerful SWIR dryers behind the oven cured them at 600 fpm, and the dryer was taken off the press.

TREATMENT OF FILMS: WETTABILITY AND ADHESION

When a liquid is applied to a surface that has low surface free energy, it will not wet the surface. It crawls into rivulets or droplets. In order to apply a coating, especially an aqueous coating, the substrate must be treated to increase its surface free energy. This is done with a flame treatment, a corona discharge or a chemical treatment.

Many packaging films require corona treatment (electrical discharges) to promote adhesion with waterbase adhesives. Problems of initial adhesion and cracking occur if they are not properly treated. Polyolefin films are particularly difficult to glue with waterbase adhesives. Sometimes adhesion occurs initially, but the adhesive causes the substrate to degrade and become brittle.

Wettability of a low energy material such as polyethylene is improved by a corona treatment or a chemical treatment that increases its surface free energy. Initial corona treatment may degrade upon storage for several weeks. Chemical treatment is permanent.

Wettability or "dyne level" of the film is easily determined by applying drops of "dyne solutions" to the film. (Surface tension is measured in terms of dynes of force, hence the name.) These are solutions of varying surface tension. When the surface tension of the film is equal to or greater than that of the liquid, the liquid will wet or spread on the film. If the film has low surface tension, aqueous adhesives or coatings will not wet or adhere to the film (figure 10-7).

COATING SPECIALTIES

Metallized Papers and Films

Metallized paper is widely used for labels, and metallized films are used for labels, decals and flexible packaging. They have eye-catching appeal. Labels are used for liquor, wine and beer bottles, paint cans, soap wrappers and personal

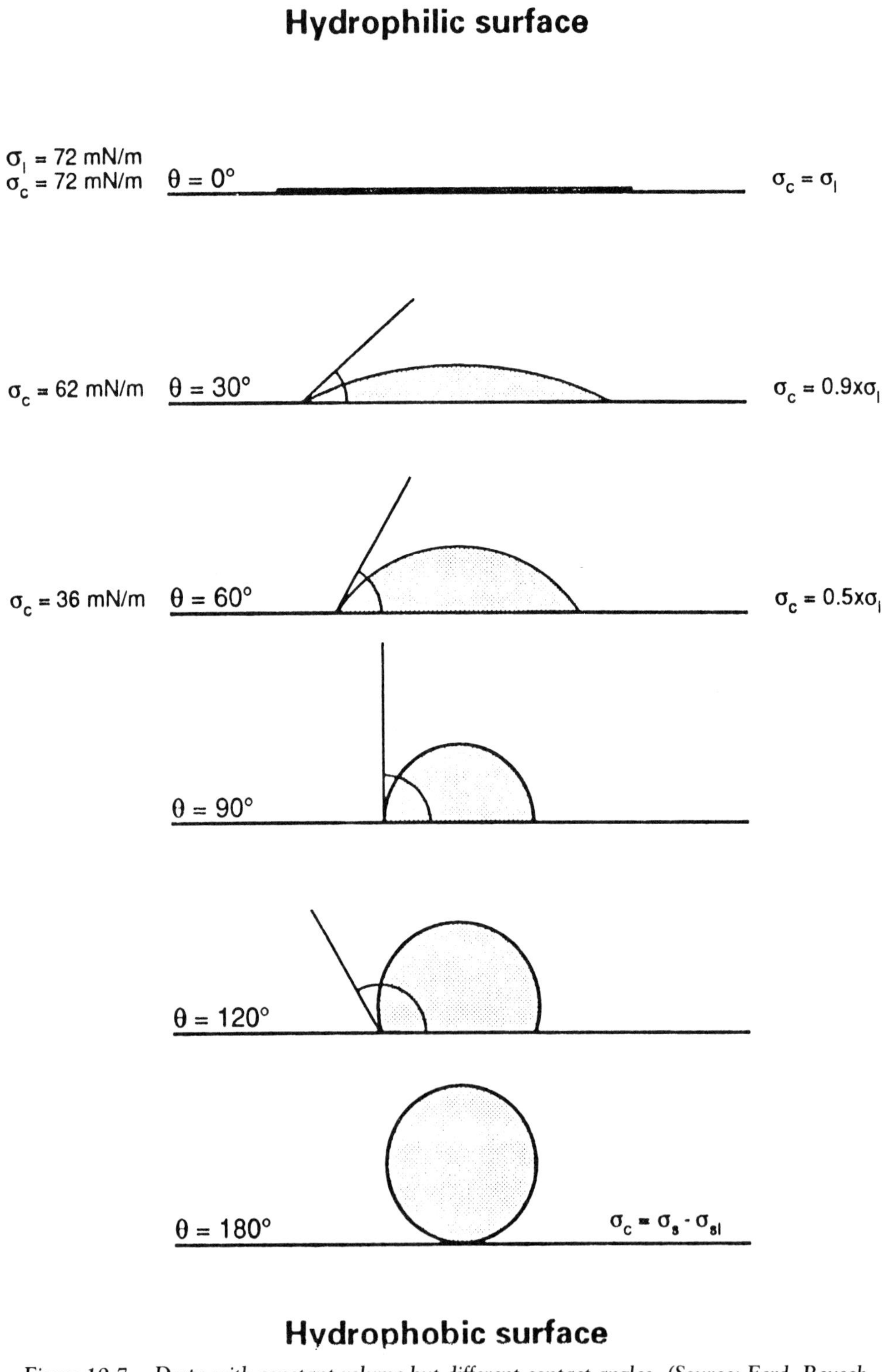

Hydrophilic surface

σ_l = 72 mN/m
σ_c = 72 mN/m θ = 0° $\sigma_c = \sigma_l$

σ_c = 62 mN/m θ = 30° $\sigma_c = 0.9x\sigma_l$

σ_c = 36 mN/m θ = 60° $\sigma_c = 0.5x\sigma_l$

θ = 90°

θ = 120°

θ = 180° $\sigma_c = \sigma_s - \sigma_{sl}$

Hydrophobic surface

Figure 10-7. Drops with constant volume but different contact angles. (Source: Ferd. Reusch AG.)

care products. Other uses for metallized paper include gift wraps, pressure-sensitive laminates, bags and even battery parts.

Unlike foil laminates, metallized papers lie flat during printing and resist moisture-induced curl. The flatness increases production speed by eliminating press jam-ups, misfeeds and wrinkles.

The coatings have a surface that is wetted by both solventbase and waterbase inks, but the printer should consult with the inkmaker to be sure the formulation is suitable for metallized paper.

Films are metallized either in a vacuum metallizer that deposits a thin layer of metal, usually aluminum, on the film or paper or by a liquid that is radiation cured. Metallized films, often polyester or oriented polypropylene, are not only visually attractive but they improve the insulation of the package by reflecting infrared radiation and heat. The metallized film is so thin that it has little effect on flexibility and other mechanical properties.

When metallized films are laminated with the metallic film inside, printing characteristics of the metallized film are irrelevant. Metallized films sometimes undergo a change within a few days so that inks will not adhere. They should be printed as soon as possible after coating.

Holograms

Strictly speaking, a hologram is a truly three-dimensional image, generated with a laser beam and a diffraction pattern that can be captured in photographic film. However, it is possible to produce a flat image with the appearance of three dimensions in aluminum or plastic film. These images are usually referred to as holograms. Pieces of aluminum or plastic with three-dimensional images impressed on them can be adhered to cans, bottles or other packages to produce startling effects.

Bronzing, Foil Stamping and Embossing

Bronzing is carried out by printing a pattern on the label or package using varnish, then dusting it with a silver (aluminum) or gold (bronze) bronzing powder, drying, and cleaning the bronze from the unvarnished areas. The metallized area does not have high gloss. This process is often used with liquor labels.

Prestige labels carry crisp, highly glossy, visually striking metallic images. Gold or silver adhesive sheets are placed on the labels by hot stamping on a platen press. If the press is equipped with the proper die, the label can be embossed at the time that the hot stamp is attached. A roll of material carrying the gold or silver finish on a backing that carries a heat-sensitive adhesive is fed to the press, together with the label, and the hot platen is pressed against the label.

FURTHER READING

Flexography: Principles and Practices, Fourth Edition. Frank N. Siconolfi, Chairman. Flexographic Technical Association, Ronkonkoma, NY. (1991).

Flexible Package Printing and Converting. Techno-Economic Market Profile Number 13. Nelson R. Eldred, Editor. Graphic Arts Technical Foundation, Pittsburgh, PA. (1988).
Folding Carton Printing and Converting. Techno-Economic Market Profile Number 12. Nelson R. Eldred, Editor. Graphic Arts Technical Foundation, Pittsburgh, PA. (1987).
Label Printing. Techno-Economic Market Profile Number 7. Nelson R. Eldred, Editor. Graphic Arts Technical Foundation, Pittsburgh, PA. (1986).
Handbook of Adhesives, Third Edition. Irving Skeist. Van Nostrand Reinhold, NY. (1990).

Chapter

11

Bar Codes

CONTENTS

LIST OF FIGURES

Chapter

Bar Codes

BAR CODES AND THE PACKAGE PRINTER

Bar codes are one method of inputting information to the computer and thus play an important role in electronic data interchange (EDI). Bar codes make it possible for computers to route and sort packages and other transport items. Bar codes are used in retail stores, warehouses, in invoicing, inventory and shipping and in many more applications. Symbols on shipping containers provide information on product routing and the source of the product, making it possible to recall a bad product.

Package printers are involved with bar codes in two ways: bar code symbols must be printed precisely and accurately on the packages produced for customers, and they expedite management and control of materials in the package printing plant. Raw materials coming into the printing plant carry bar code symbols that are important in handling rolls of plastic, rolls or skids of paper, plates and other supplies. A bar code is required if the printer is to handle the paper efficiently and keep inventory costs to a minimum. In North America, paper carries a bar code called a TAPPI roll identifier or a sheetfed paper identifier, which is different than the Universal Product Code (U.P.C. or UPC)* used in the supermarket.

Bar codes must be printed within the specified limits or the entire communication system collapses—the packager's system and the retailer's system. Accordingly, a misprinted bar code can be very costly. Printers are often held

* Properly abbreviated U. P. C., the more compact UPC is usually used.

responsible for the accuracy of the original bar code symbols, or film masters, as well as for their reproduction, and customers sometimes fine printers for printing bar codes out of specs. A second or third offense can disqualify the printer from further consideration as a supplier. Errors and their sources are discussed below.

Codes and Symbols

A bar code is a system of information embodied in computer software that enables suitable equipment to generate or read a symbol, the actual printed image.

Each bar code system uses a different "language" or symbology. While the software for bar code readers can generally be adapted to read several different symbologies, the nature and structure of the information is not always obvious. The UPC and the EAN (European Article Number) have a different number of digits in the symbol, and the EAN symbols are generally not acceptable in the United States.

Function of Bar Codes

Bar codes are part of a management information system that is becoming part of manufacturing resource planning, a necessary part of JIT and of Total Quality Management (TQM), whether by the package printer, his customer or his customer's customer, the retailer.

Management information systems (MIS) compile, analyze and distribute information necessary for managing an organization's activities. Any manufacturer engaged in JIT and TQM needs the types of information that are produced by use of bar codes and computers: placing orders, receiving and storing, inventory control, tracking work-in-progress, invoicing and shipping materials, components and products. Thousands of different transactions must be recorded and analyzed to develop a comprehensive data base.

One function of bar codes is nicely pictured by Roger Palmer in *The Bar Code Book*; "The receiving dock at most manufacturing plants can be a hectic place. Material often arrives in spurts, and there is no uniform package size or labeling format. Expediters are rummaging through the area, disrupting the receiving staff as they search for the 'hot' items that are needed to pay off their shortage list. Each packaging slip needs to be reconciled with the outstanding purchase order field, either with manual paperwork or via a computer terminal. During peak periods, significant backlog can develop in the receiving area.

"If packages that are received have been premarked with a bar code by the vendor, transaction time can be significantly reduced. If the vendor is fully integrated into a mutually agreeable electronic data interchange (EDI) system, your computer already knows what packages to expect at the receiving dock. If your vendor marks the packages with the unique bar coded serial numbers, a simple scanning operation at the receiving area can update the computer as product is received."

Even if EDI is not being used, packages that are received without bar codes can be labeled with bar code symbols generated in the system as the clerk looks up the purchase order number from the packing list and enters it into the computer. The computer generates a bar code that guides the package onto the proper cart or trolley and gets it delivered to the proper manufacturing location or into the storeroom.

In the shipping department, exactly the reverse occurs. The bar code is generated together with the shipping label, and if the customer has agreed to an EDI system, the shipping information is entered into his computer as the product is shipped.

EDI includes the communication of specified data between the computers of two companies or institutions involved in a business transaction. It can, in fact, be automated to the point that new orders are placed automatically whenever inventory reaches a predetermined level. EDI typically gives the purchase order number and delivery instructions, and the supplier generates the serial numbers to be encoded into the bar code and placed on each master carton or pallet to be shipped. Because of its speed and accuracy, the bar code has become an essential part of this communication system.

Bar Codes and the Computer

Of the many methods of data entry to the computer (magnetics, machine vision, magnetic ink character recognition, etc.) bar codes are preferred for packaging and shipping, inventory and similar operations.

The bar code reader converts the symbol into bit streams of ones and zeros—the fundamental logic of digital computers. Three of the most important methods for getting data into a computer are manual key entry, optical character recognition and bar codes. Bar coding is much faster and more accurate than manual key entry or optical character recognition. The Graphic Communications Association reports that manual key entry will result in about 10,000 errors per three million characters, optical character reading about 300 errors, and bar coding only one. The actual error rate may vary considerably, but the numbers give a feel for the order of accuracy achieved by the three methods of data entry. Accuracy of data input is crucial because if the data has errors in it, management must work with flawed data.

Bar codes make it possible for the computer to deliver real-time information to management—information that is compiled, analyzed and transferred to the manager's computer with electronic speed. Bar codes promote real-time inventory and make it possible to place orders automatically whenever inventory reaches some predetermined level. In fact, without that kind of information, JIT operations become ineffective. Real-time management is possible only when managers have the benefit of the speed, accuracy and reliability of a computerized management information system.

The first effect of a computerized management information system fed with computer-readable information is that the quality of the information is greatly improved as errors are reduced. The second effect is that all data are current because the computer is updated whenever anything in the system changes.

Bar codes have been in use since the early 1970s, and they are still evolving and being developed. The best known bar code is the Universal Product Code since it appears on most of the packages of food and other household items that we buy.

History of Bar Codes

One of the earliest and best-known bar codes is the Universal Product Code (UPC), commonly used in handling groceries and other retail items. By 1955 the grocery industry realized that automated checkout* was possible, and about 1970 a committee was organized representing various sponsoring major retailers and manufacturers. They commissioned McKinsey & Co. to work on a universal product code. One of McKinsey's actions was to retain the Graphic Arts Technical Foundation to determine the precision of reproduction of original artwork compared to the finished package found in the grocery store. In 1973, the U. P. C. and Uniform Product Code Council were announced. Within a few years, the UPC symbol began appearing on printed packages, and bar code scanners appeared at the checkout counter in grocery stores.

Bar codes are now used wherever computers are used to record or control the movement of parcels or packages: in retail and department stores, in the post office, at the airport, in the warehouse and in shipping and receiving departments.

STRUCTURE OF BAR CODES AND THEIR SYMBOLS

Symbology is the language; it is the set of rules or specifications for the bar and space widths and their arrangement. The "symbol" is the arrangement of the bars and spaces that encode the data. The "code" is the actual data contained. It can be a serial number; it can be a number representing a particular packaged product; it can contain numerical or alphabetic characters or both. The UPC/EAN and Interleaved 2-of-5 codes are all numeric. Code 39 is an alphanumeric code.

A symbology standard specifies the bar code structure. It describes how information is encoded into the particular arrangement of bars and spaces of varying widths that make up a bar code symbol. The symbology also specifies printing tolerances, optical properties, spots, voids and edge roughness. The width and sharpness of fine bars and the accuracy of spacing are critical.

SPECIFICATIONS FOR PRINTING BAR CODES

Every symbology has an associated set of specifications that must be adhered to during the printing process. Failure to comply with the published specifications will result in the creation of symbols that give unacceptable low read rates (sometimes called the "first read rate") and high substitution error rates if they can be read at all. Printing specifications include limits on dimensional tolerances, spots,

* The first suggestion for bar codes was developed in the early 1930s. It followed shortly after the first supermarket was opened, but it was originally suggested for retail checkout control and inventory. Michael J. Cullen started the first supermarket in Brooklyn, N. Y. His firm, King Kullen Grocery Co., Inc., furnished the artwork used on the cover of this book.

voids, edge roughness, reflectivity and contrast. The bar widths and space widths, as well as the overall width of the symbol, must be within specified tolerances.

As the printed symbols become smaller, the specifications become increasingly more stringent, and package printers generally try to avoid printing reduced-size symbols. Various techniques are available.

Printing tolerances vary according to the width of a bar code symbol's narrowest elements—both bars and spaces. Larger symbols with wide bars and spaces tolerate more deviation, but they take up more space on the package.

Specifications of Bar Width and Uniformity

Specifications differ for each code. UPC has different specs than Code 39, etc. These are established by the organization that publishes the code or application standard, or by ANSI.

A symbol whose average dimensions are within specification can still prove to be unreadable if there is excessive edge roughness, or spots or voids. A spot may be read as a bar, while a void may be read as a space. Defects are covered by American National Standards Institute* (ANSI) MH10.8M-1993 and by ANSI X3.182 "Bar Code Print Quality—Guidelines."

Current specifications of individual symbologies such as Code 39, Interleaved 2-of-5 and Codabar can be obtained from AIM USA,** the leading manufacturer's trade association for vendors and suppliers of automatic identification products, services and supplies.

If a symbol is printed with dimensions that are at the extremes of the published printing tolerances, there will still be adequate margin for the reader to decode the symbol accurately. If symbols are beyond published tolerances, it might still be possible to read the symbol, but the read rate and substitution error rate will suffer. The fact that a symbol is readable does not necessarily mean it is within specs.

The direction of printing the symbol is important. In flexo, when the symbol is mounted with the bars perpendicular to the axis of the cylinder, some distortion occurs, and this must be accounted for in preparing the film master. When the bars are parallel to the axis of the cylinder, much less distortion occurs.

Symbols printed by offset or letterpress usually require no special precautions, although each process has problems that can cause the bar code to print out of specifications.

When possible, press direction in the paper or board should run parallel to the direction of the bars because tests have shown that the percentage spread or gain of the bars is less when the symbol is run parallel to the press direction than across the press. However, when good press control is exercised, direction has been shown to have little effect on the results.

The original specifications for the UPC code were drawn without apparent regard for the specific characteristics of gravure, and this caused severe problems

---

* American National Standards Institute, 11 West 42 Street, New York, NY 10036. Phone (212) 642-4900.

** AIM USA, 634 Alpha Drive, Pittsburgh, PA 15238. Phone (412) 963-8588.

in the gravure printing community. The bar code readers then in use had trouble reading the scalloped bars printed by gravure, and many gravure printed packages were not acceptable. The situation has been remedied, partly by increased knowledge on the part of gravure printers and partly by better scanning equipment. Sometimes, it has been necessary to add flexo or letterpress equipment to print the symbols on packages printed by gravure.

Reflectance and Print Contrast

Several optical properties are usually specified. Since the reader detects this difference in reflectance between the printed bar and the unprinted space, best results are obtained when the contrast is great. The standard, specified in ANSI MH10.8M-1993, depends on the application, and the grade for labels is different from that for natural kraft. It is likely that UCC/EAN will adopt a similar standard. The reflectance of the ink must be low. This means that a black ink is usually preferred. The hue of colored inks is important. Since most bar code readers use a red light, the symbol must not be printed with red ink because it fails to absorb red light, and unless the ink has other colors in it in addition to red, the symbol will not read. Serious problems have occurred when this simple concept was ignored.

Blue, green and black are the best colors, but a determination of the reflectance of the ink with a densitometer is a worthwhile precaution.

The contrast, that is, the difference in reflectivity of the bars and spaces under the illumination of the bar code reader, is one of the most important optical properties. A dark background or an ink that does not absorb sufficient light creates problems. Dark spots in the spaces, or voids and roughness in the bars may be misinterpreted by the reader, and they must be kept within tolerances (figure 11-1) to avoid erroneous readings.

The wavelength of light is also specified. Inks and substrates have different reflectivities at different wavelengths. It is possible to print labels that are in-

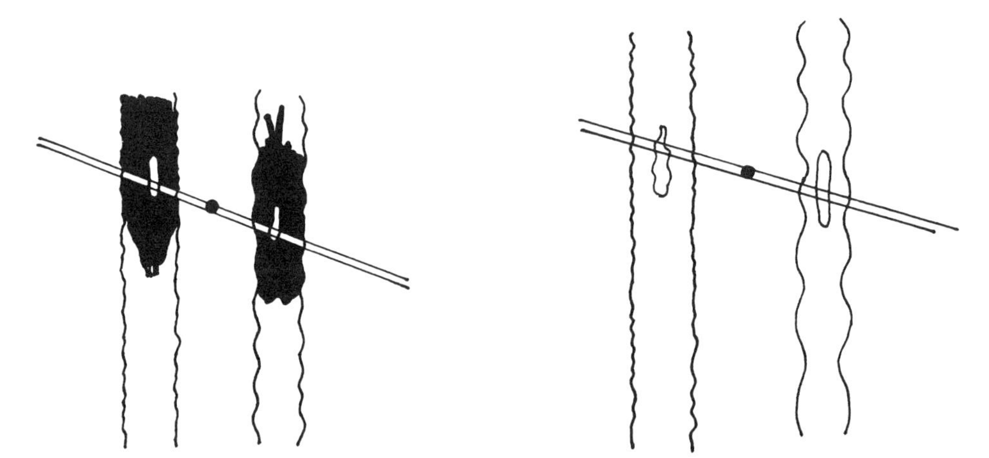

Figure 11-1. Effect of specks, voids and edge roughness on readability of bars and spaces. Any of these may cause a misread on the reader.

visible to the human eye if they have differing infrared reflectivities and are read with an infrared reader, such as security access cards.

Highly glossy inks may also create reading problems, as do some overwraps and laminates. Matte finishes are easier to control. Printing a clear varnish over the symbol may protect the label, only to cause a misread when the reader detects light reflected by a glossy coating. A transparent wrapper over the printed symbol tends to reduce contrast by increasing reflection.

Size

Variability in the width of the bars and spaces must be controlled so that the bar code reader can recognize every bar or space that has the width of "X" plus or minus the specified variability as 1X. The bar code reader must also be able to recognize bars and spaces with widths that are multiples of "X" (figure 11-2). When "X" has a high value, poor control of width has less effect on the readability, but it takes more space to print the bar.

From the point of view of the package design, it is desirable to keep bar codes as small as possible. However, the smaller the symbol, the more problems there are in reproducing it accurately. Each printing process has some variability, and the thinnest bar and space must be wide enough to read properly when the printing process has the maximum variability within tolerance. This tolerance is established for each bar code size, based on the reading equipment and printing process. It can also be established for each press, ink, substrate combination and reported to the manufacturer of the film masters, who modifies the film master with a Bar Width Reduction or Increase (BWR/BWI) to accommodate the growth or loss of a particular printing press and process.

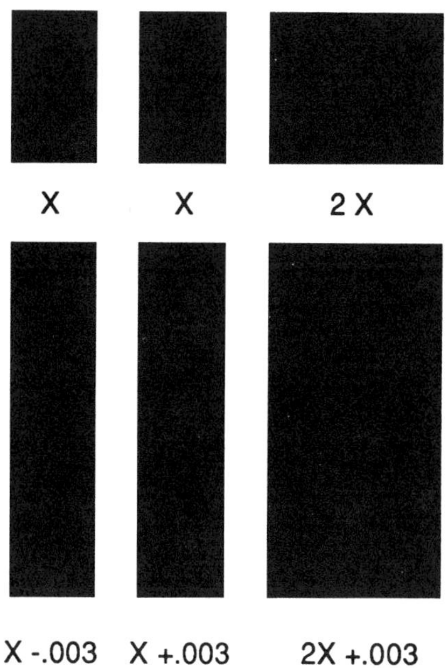

Figure 11-2. Effect of variability of bar width on bars printed within specs.

Character density is also related to the "X" dimension. More information can be crowded into a given space with a narrower "X" dimension. Since narrow bars are most affected by small variations in width, the increased character density requires careful control of printing.

Of the conventional package printing processes, flexo is thought to produce the greatest variability in bar width. The high dot gain of flexo probably contributes to this variability.

Just over one inch (2.5 cm), including required white space on the left and right of the symbol, is required for the smallest UPC-A product code symbol.

There are several means of truncating or reducing the size of a bar code symbol. However, this can significantly reduce read rates, particularly at the supermarket counter, but also with hand-held scanners, bringing fines or returns from the retailer. Truncated symbols should be avoided wherever possible. The UPC has a zero suppression Version E that is about half as wide as the regular version.

Position on the Package

Placing the UPC on a retail package is more than a matter of design and good taste. The symbol doesn't add to the artistic effect in any case. The old specs called for the symbol to be placed on the natural bottom of all products, that is, the side of the package that the shopper tends to put on the checkout counter. This proved to be the side of the package that gets dirtiest. In any case, the printer or converter has little say in the matter of design.

Since the symbol can be read in any direction, its position does not affect the ability of the bar code reader to read it.

Check Digits or Modulo Check Characters

The substitution error rate (SER) is controlled by the use of check numbers or digits. A check digit (or modulo check character) is an additional character that is based on a mathematical formula using the other characters in the code. (In the UPC symbol, it is the digit at the far right.) The bar code scanner reads the bar code, performs the calculation and compares the scanned check character with what is calculated. If there is a match, the results are sent to the computer; if no match, the scanner will not enter the bar code. Code 39 has many checking features built into the code and its optional check character is rarely used, except in health care.

Other Specifications

Although the package printer is indirectly involved, he must be aware that the specifications of the symbol also specify certain characteristics of the bar code reader. Different symbologies require readers with different apertures. A scanner with a large aperture may not be able to read symbols with a small X dimension, while a reader with a small aperture cannot tolerate significant spots and voids.

The bar height must be specified together with the amount of free space at each side of the printed symbol. The smaller the height, the more difficult it is to get a good read.

ERRORS: READ RATE, SUBSTITUTION ERROR RATE

A successful bar code system allows data to be collected rapidly, economically and accurately. The bar code error rate is very small if a secure symbology is used, if the print quality is high, and if the reading equipment is properly designed and in good working condition. Statistically, the best symbols and systems will occasionally make a mistake; error rates may be one in million or even one in 100 million if everything is within specs.

All symbologies are subject to the introduction of errors. Two characteristics are most important: the substitution error rate (SER), often referred to as the error rate, and the read rate (RR), sometimes called the "first read rate." The SER refers to the number of symbols that are read incorrectly. RR, expressed as a percentage, refers to the percent of successful reads per read attempt. For a given technology, there is usually a strong correlation between RR and SER although exceptions can occur. Well-printed symbols have a high RR and a low SER. Bar codes that are well printed and read with good equipment have a read rate in the order of 90 percent or better and a very low substitution error rate. The read rate must be 90 percent if the reader is to achieve a 99 percent read rate by the second try.

Quality is not obtained through inspection; it must be built into the system. The poorer the printing job, the greater the number of substitution errors and the lower the read rate. Employees, customers and others begin to lose confidence if the read rate falls much below 90 percent. A higher rate is much to be desired.

PRINTING THE BAR CODE SYMBOL

Securing the Film Master

The bar code symbol is generated off-site or on-site. It is critically important that it be properly specified. Because of the cost and complexity of the equipment used to generate film masters, printers usually purchase the masters from a trade house or from an organization specializing in this technology.

The film masters usually arrive at the platemaker's camera department in the form of photographic negatives or positives. Most bar code symbols are produced by computer generation. Large quantities of identical or sequential numbers can be produced on site with the use of special on-demand printers.

Specifications vary according to the type of symbology, size of the symbol being used, the plate, paper, ink and press, and other items listed in the section on specifications. They should be clearly indicated on the order. Ideally, the amount of adjustment in the master symbol should be based on the fingerprint (a printed bar code or control bar) of the press on which the symbol is to be printed. Lacking a fingerprint, the trade house will apply what they consider to be suitable adjustments in the master. Some trade houses also supply flexo mats and printing plates already imaged with the bar code symbol.

Sources of Error

Printing of good quality bar code symbols requires close attention to many factors. First of all, the package printer must obtain a copy of the appropriate specifications to understand the requirements and be certain that the printing is within specifications.

The negative film master must not be used as a master negative. It must be stripped into the film. Copying the film master onto another film changes the dimensions of the bars and creates printing errors. Film master negatives can be provided emulsion up or down, and the correct one must be ordered.

To remain within specs, the printer must pay attention to the printing plate, press operation, substrate quality, ink quality, print quality and protection of the print. The printed symbol should be free from dirt or grease which can cause reading errors.

Paper or board causes the fewest problems when it has a matte surface and high reflectance. Unbleached kraft, commonly used on corrugated, has a rough surface and low reflectance and causes problems that require special attention. Printing of large symbols helps as does a smooth white coating on the surface.

In flexo printing of corrugated, stripping in small spots of rubber into a larger plate is customary for size changes, product number changes and price changes. Attempts to strip a new bar code symbol into an old plate usually fail because the bar code does not print with the same pressure and the symbol is squeezed far out of specs. A bearer bar should be placed around the bar code as this "box" helps in evening out the pressure throughout the printed symbol.

In gravure, a flexo station is frequently used to print the bar code. The flexo station, in fact, is useful for inserting all sorts of variable information into the gravure image.

If the bar code is printed on clear plastic that is to contain a dark product or an uneven or mottled background, the printer should coat the plastic with an opaque white ink to provide proper reflectivity. Ink should not smudge or spread. It should give good contrast with the substrate. If the printed symbol is apt to be marred during shipping or handling, it may be covered with a coating or a laminate. However, the coating should not have a high gloss that can interfere with the reading. One advantage of UV ink is that it can be printed with a matte finish, avoiding the reflections that can compromise a symbol.

The above are typical problems of quality control. Like all quality maintenance procedures, printing of symbols within specs requires constant education of personnel, both operators and staff. These problems will be largely eliminated by fingerprinting the press, a subject discussed in Chapter 6.

To function properly, all equipment must be maintained in good condition. All crews must be trained and retrained, and the training must be kept up to date.

One text relates the story of a successful bar code printing program that began to have far too many errors. A study showed that when the project leader moved on, he was replaced by a new operations manager. The new manager found a cheaper source of paper, and the change in substrate resulted in symbols that were no longer printed within specifications.

VERIFICATION

Bar Code Verifiers

The fact that a bar code scanner can read a symbol does not necessarily mean that it is printed within specs or that other bar code scanners will be able to read it. Out-of-spec symbols can sometimes be read. To be sure that the bar is printed within specifications requires the use of a device called a "verifier" (figure 11-3). Several commercially available instruments can be used to automatically determine if a symbol's dimensions are in compliance with published printing specifications. They scan the printed symbol, quickly measuring the contrast, reflectance, width of each bar and space, and supply a printout of the results (figure 11-4). Verifiers typically have an overall precision of better than one mil

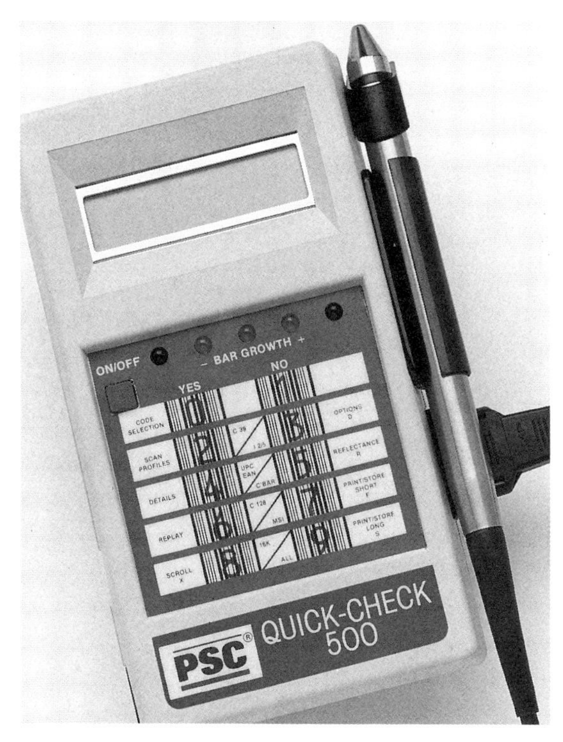

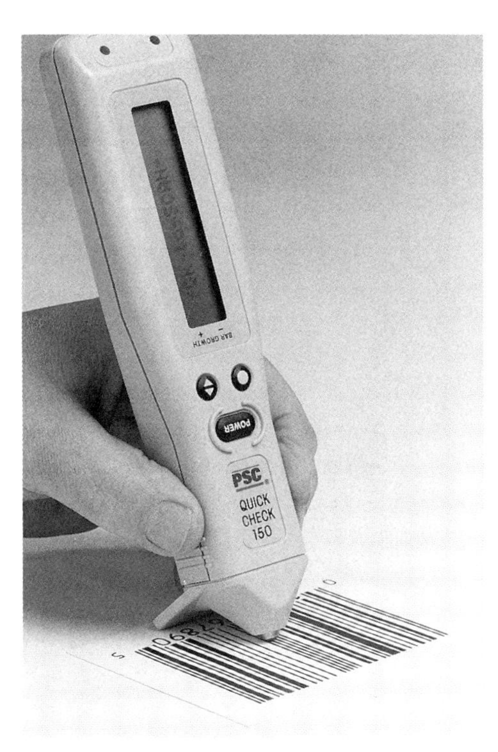

Figure 11-3. Bar code verifiers. (Courtesy of PSC Inc.).

(0.001 inch or 0.025 mm). Commercially available instruments range from simple units with handheld scanners to totally automatic machines that evaluate scan patterns and provide a printout of all data.

An optical comparator with 50X magnification in the hands of a skilled operator will determine the bar width and uniformity, but optical comparators and densitometers do not measure all of the parameters required by ANSI X3.182. It is the exact location of the edge that causes the most problems for symbols with relatively small "X" dimensions.

More complete results are obtained with a verifier. Verifiers measure bar width variability, edge roughness, print contrast ratio and other key properties. Measurements must be made at the specified wavelength. Verifiers are relatively inexpensive ($900 to $15,000). Printing a faulty symbol can result in loss of a customer or a fine that is far more expensive.

Many web presses are equipped with automatic verifying equipment that reads the symbol on the fly. This is backed up in the lab after the ink has dried. The on-press reading can be adjusted to anticipate the gain that will occur when the ink is dry.

The bar code verifier should check such items as:

- Widths of bars and spaces
- Print contrast ratio
- Freedom of the spaces from ink and specks
- Freedom of the printed bars from voids and white specks
- Bar edge roughness (If the edge is rough, it may cause a misread.)

\\\ Quick Check 500B \\\

660 nm, 06 mil Scanner

UPC-A: 100% Mag. Factor

0-1234509876-5

Avg Bar Err = .09X OK!
Refer. Decode Passes <A>
Decode Margin = 88% <A>
Symbol Total = 671 "X"

Prnt Contr Sig = 98% OK!
Reflect (Light) = 95% OK!
Reflect (Dark) = 01% OK!
Symbol Contrst = 94% <A>
Global Thresh Passes <A>
R (min) / R (max) = 01% <A>
Modulation = 73% <A>
EdgeCont (min) = 69% <A>
"Defects" = 02% <A>

Message Length = 11 OK!
Check Charctr Passes OK!

Traditional Tests <PASS>
Bar Growth : IN TOL
"P C S" : OK
Print Quality Grade: <A>
Formatting Checks <PASS>

LASERCHEK REPORT

NAME:

DATE:

UPC#: 0 12345 67890 5

% DECODE: 100%

UNITS: .001 INCH

GUARD BARS
MIN MAX
09 17

LEFT	CENTER	RIGHT
13	13	14

DYNAMIC REFLECTANCE
BACKGROUND: 76%
BARS: 07%
PCS MIN: .68
PCS ACTUAL: .90
CHARACTER BAR-WIDTH DEVIATIONS

TOLERANCE: +/- 4.0
RED: TOLERANCE WARNING

CHAR	BAR1	BAR2
0:	0	+ 1
1:	0	+ 1
2:	+ 1	− 1
3:	− 1	0
4:	+ 1	0
5:	+ 1	+ 1
6:	0	+ 1
7:	+ 1	+ 1
8:	+ 1	+ 1
9:	0	+ 1
0:	0	+ 1
5:	+ 1	0

Figure 11-4. Printouts from bar code verifiers: PSC 500B on left and SCAN Laserchek 2811 on right. These printouts show the conformity or lack of conformity of the bar code to the specs. (Courtesy of PSC Inc. and SCAN Newsletter, Ltd.).

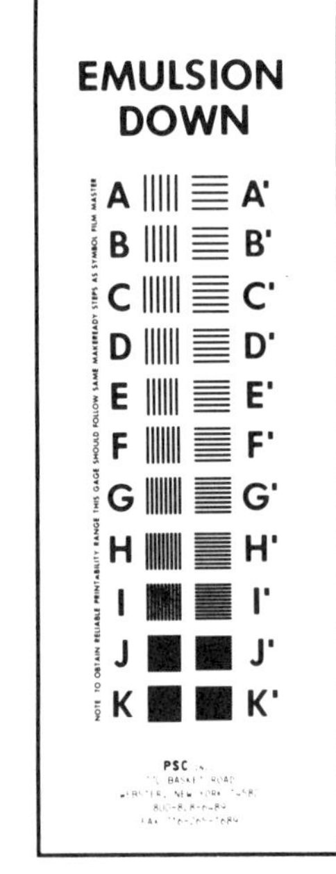

Figure 11-5. Printability gauge. The gauge gives visual indication of the resolution of the press system. (Courtesy of PSC Inc.).

- Adequate quiet zones (unprinted space before and after the
- symbol.)

Pass/fail depends on many things. The fact that a symbol can be read by a scanner does not guarantee it is within specs. Curvature causes many problems. An infrared reader is not apt to be affected by dirt and grease on the surface that often affect readers that operate with visible wavelengths.

Printability Gauge

A printability gauge (figure 11-5) can be printed together with the symbol to show the printer/converter how precisely the press can reproduce fine details and how much gain and slur occurs. The results can be used in ordering film masters that will produce a symbol printed within specifications.

BAR CODES

The Morse code used for telegraphy is familiar to many people. It makes use of only two bits of information: a dot and a dash. It makes no use of the space between the dots and dashes.

Most bar codes make use of the spaces between the bars as well as the bars themselves. In most symbologies, the widths of the bars and the spaces are both variable. These are read by the bar code reader and transformed into computer bits, the 0's and 1's of computer language, as shown in figures 11-6 and 11-7.

The width of the narrowest bar used in composing a bar code is commonly called "X." Every bar and space must be a multiple of "X." In Code 39 this multiple is sometimes called "N." In the UPC, bars and spaces have widths of 1, 2, 3 and 4, and the symbol has bars and spaces of four different widths.

Retail applications drove the early technological development of bar coding. The bar codes most commonly used in the supermarket are the Universal Product Code (UPC), both version A and version E (figure 11-8), or the European Article

SYMBOL

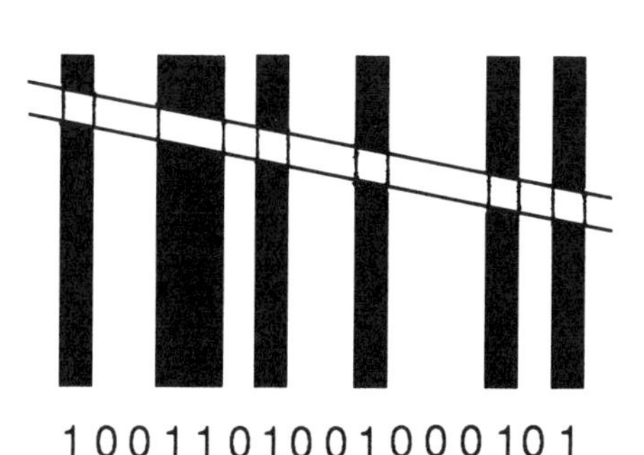

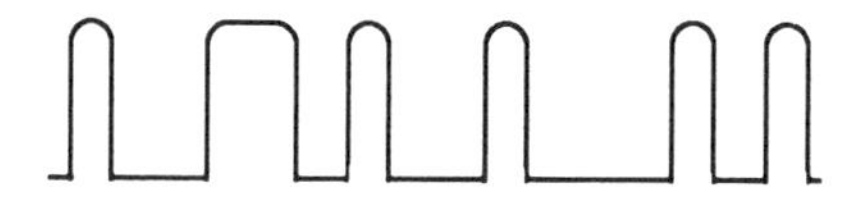

Figure 11-6. How the bar code symbol produces electrical signals that generate a computer response.

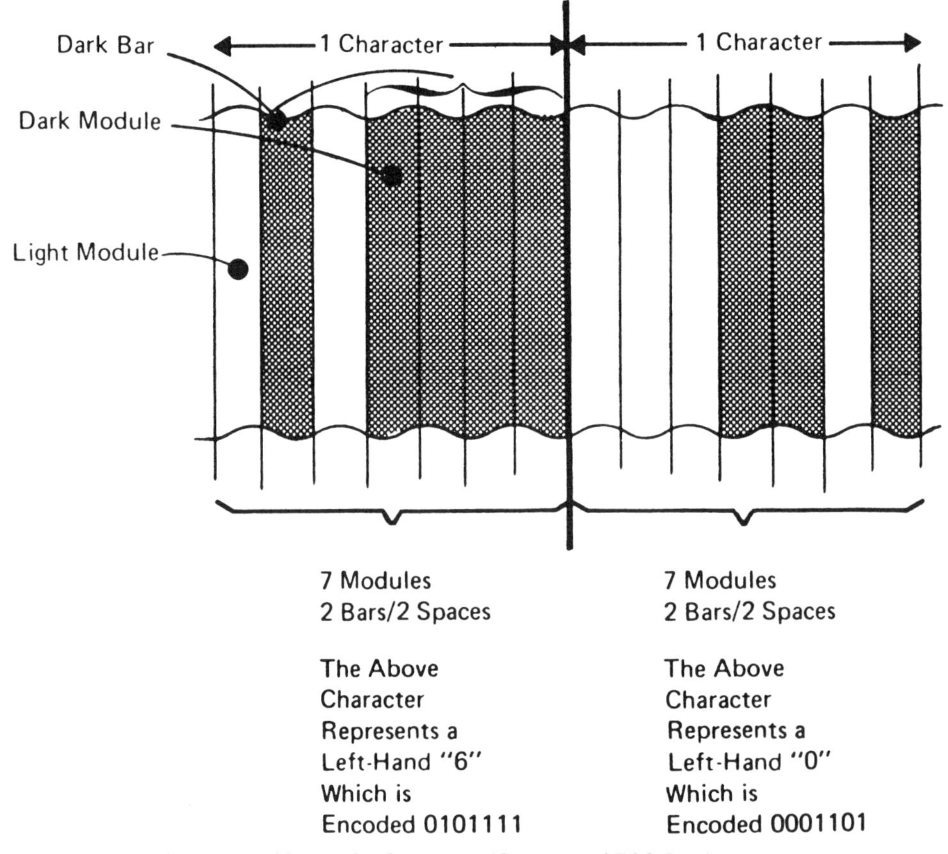

Figure 11-7. Structure of bar code character. (Courtesy of PSC Inc.).

Figure 11-8. UPC symbols. On the left is Version A, on the right is Version E.

Numbering system (EAN). In retail applications, the UPC/EAN symbol is pervasive.

Other familiar bar codes are Code 39, sometimes referred to as 3 of 9 code, Interleaved 2-of-5, Codabar and Code 128. In industrial applications, Code 39 is dominant, but Interleaved 2-of-5 and Codabar are also used. Postnet is often seen on mail, and the International Air Transportation Association Code (figure 11-9) is used to confirm that every bag carried on an airplane has an associated passenger. Military Standard 1189 (Code 39) was adopted in January 1982.

I 2/5 AIRLINE

Figure 11-9. IATA baggage tag symbol. (Courtesy of Matthews International).

UPC and EAN Codes

UPC version A has four "fields." At the extreme left is the "number system character." The number identifies the category into which the product fits (groceries, National Drug Code, random weight item, etc.). That number and the left field of five numbers comprise the six digits that identify the vendor. The right field of five numbers identifies the specific product (figure 11-10). The fourth field, a number at the right, is a number that checks or verifies that the numbers in the code have been properly encoded, printed and read.

The UPC also is printed in Version E, a condensed bar code. Version A has 12 digits. Version E has six numbers which can be expanded to 12 following certain rules; it is often used on small-label items such as chewing gum.

Figure 11-10. Structure of the UPC symbol. (Courtesy of PSC Inc.).

The UPC is administered by the Uniform Code Council (UCC).* The UCC publishes the specifications for printing the UPC and issues numbers to each company that wishes to sell coded products. Other bar codes have similar organizations that monitor them.

EAN is a variation of the original U.S.-based UPC adapted for the international marketplace. It too occurs in two versions: the EAN-8 which carries eight digits, and the EAN-13 carrying 13 digits. EAN is monitored and issued by ICODIF** (formerly the European Article Numbering Association) in Brussels. It is much like the UPC symbol. The first field (that contains a single number in the UPC symbol) of the EAN contains two numbers—the country number. The number of digits assigned to the vendor varies between countries. Because the symbol carries 13 digits, UPC scanners normally cannot handle the EAN symbol, but adding a leading "0" to the UPC code makes it readable by an EAN scanner.

UPC has been accepted nationwide in the U.S. and Canada, and the EAN internationally. It is a successful international computer input language. For all its success, the UPC/EAN code has some limitations. It can encode only numerals, no letters, and it can be misread somewhat more frequently than Code 39.

UPC-SCS Code

The UPC/EAN code gives very poor readability when it is printed on the rough, absorbent surfaces of corrugated boxes, and the Fibre Box Association (FBA), together with the UCC, developed the UPC/SCS (Shipping Container Symbol or sometimes standard case symbol or standard pack) for use on corrugated boxes. The UPC-SCS system is based on the Interleaved 2-of-5 Code which is entirely different from the UPC code.

The UPC/SCS symbol is shown in figure 11-11. The black border, called a

Figure 11-11. UPC/SCS shipping container symbol showing bearer bars. (Courtesy of Matthews International).

---

* Uniform Code Council, 8163 Old Yankee Road, Dayton, OH 45458. Phone (513) 435-3870.

** ICODIF, Rue Royale 29, B-1000 Bruxelles, Belgium. Phone 32-2-218-7674.

"bearer bar," helps to keep the plate on the printer-slotter from punching into the corrugated and broadening the bars in the symbol. It also helps to level the plate. If bearer bars (or a bearer box) are not used, outer bars of the symbol will serve as bearers and distort more than the inner bars. Other symbologies are used on corrugated containers such as ITF-14 and SSC-128.

A UPC symbol relates more closely to the contents that will be sold at retail. Printing the UPC code on the corrugated container makes it just as necessary that the packager use the correct shipping container as it is to put the proper label on the can, bottle or box.

EMBARC: Tappi Roll Identifier and Sheetfed Paper Identifier

Several associations—Technical Association of the Pulp and Paper Industry (Tappi), American Newspaper Publishers Association (now the American Newspaper Association), American Forest and Paper Association, Canadian Pulp and Paper Association, National Paper Trade Association and the Magazine Publishers Association—have cooperated in developing a bar code that all could accept. The result was called EMBARC (Electronic Manifest BAR Code) or the Tappi Roll Identifier. By 1990, most paper mills and most large printers were using the system. The success of the Tappi Identifier has led manufacturers and dealers of other graphic arts materials such as film, plates and ink, to devise another bar code, GIBC (Graphics Industry Bar Code.) This helps printers computerize job, inventory and production control systems.

The Graphic Communications Association* is the registrar for the Tappi codes and GIBC, which is based on Code 39, to control paper and board shipments and handling. Figure 11-12 shows a Tappi Roll Identifier Symbol.

Since the early 1980s, bar codes have been used to control and keep track of rolls, skids and cartons of paper. The label on the package contains information about the date of manufacture, the plant and paper machine, and the position on the main roll from which the roll or skid was cut. Through EDI the information makes it possible for the customer to keep track of the scheduled shipment date and the shipping route. When the paper arrives, the bar code helps keep track of inventory and guides the roll or skid to the correct printing press or to storage. The bar code reduces handling and is helpful in identifying sources of printing problems.

Code 39

Code 39, sometimes referred to the "3 of 9 Code," has bars and spaces of two different widths (figure 11-13). The code can be very long, but it can encode a great deal of information. Code 39 has the advantage that it is an alphanumeric code, whereas most bar codes can encode only numerical information.

The U. S. Department of Defense and other government agencies use a form of Code 39 called Logmars (Logistics Marking and Reading of Symbols). It is

---

* Graphic Communications Association, an affiliate of Printing Industries of America, 100 Daingerfield Road, Alexandria, VA 22314-2888. Phone (703) 519-8160, FAX (703) 548-2867.

Figure 11-12. Tappi Roll Identifier. (Courtesy of Graphic Communications Association).

required of suppliers to the DOD. The Health Industry Business Communications Council (HIBCC) also uses Code 39 (along with UPC and I-2/5). Code 39 is the standard for most industries that do not use UPC coding.

Interleaved 2-of-5

Another common bar code is the Interleaved 2-of-5, sometimes called I-2/5 or ITF . Bars and spaces are arranged so that the bars represent one numeral and the spaces another. Each group of bars and spaces thus yields two numbers.

Because of its configuration, the I-2/5 Code requires that an even number of characters be encoded. If an uneven number of digits must be encoded, a leading zero is used to produce the final number of even digits.

Codabar

Codabar is the language used in libraries, blood banks and air parcel express applications. It can encode numbers 0 through 9 and the characters $: / . , + , – . Codabar has two bar/space widths. Blood banks use ABC Codabar which is a specific application of Traditional (fixed-font) Codabar.

Postnet

Postnet (figure 11-14) is the familiar code used by the U.S. Postal Service.

Figure 11-13. *Code 39 symbol. (Courtesy of Matthews International).*

Figure 11-14. *Postnet symbol.*

The code has very low information density—a long code carries relatively little information.

GENERATING BAR CODE SYMBOLS

Printers usually purchase the bar code film masters from a trade house, an organization specializing in this technology. On-site printers can be used to produce large quantities of identical symbols or sequential numbers. They are usually used as on-demand printers.

Sometimes bar code symbols must be generated that contain variable information, as for example if the symbol carries a purchase order number, variable weight, serial number or other information unique to that package. For such bar code symbols, the conventional printing processes are impractical. Conventional printing processes generally do not respond rapidly to the computer. Computer-drive numbering machines are available for this "crash printing."

When bar code symbols must be generated at the place of use, an on-site printing capability is required. Symbols can be generated by thermal, thermal transfer, laser, ion deposition, ink jet, dot matrix, impact or drum printers, often called "formed font impact printers," designed explicitly for printing bar code symbols on tags and labels as in the produce or meat departments of the supermarket (figure 11-15). These printers can produce images in direct response to the magnetic information on the computer disc.

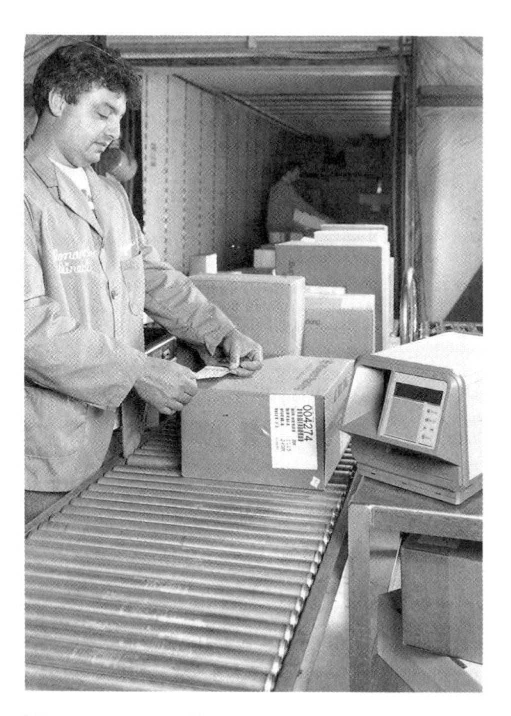

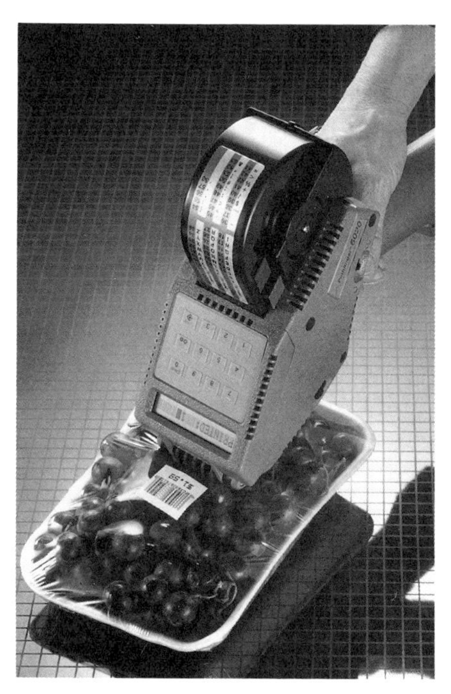

Figure 11-15. Bar code printers for secondary labels: (left) shipping carton labeling system; (right) portable bar code printer for grocery labeling. (Courtesy of Monarch Marking Systems, A Pitney Bowes Co.)

For other applications laser etching, screen printing, stencil spray painting and hot stamping are also used to print bar code symbols.

Xerographic printers generate the image by charging a semiconductor drum in the dark, discharging the non-image areas with light, then toning the charged, image areas, transferring the toner to paper and fusing the toner particles.

Ion deposition is a fast method for adding high-quality variable information to tags and labels. It can be added to a standard narrow web press to print bar code symbols or human-readable characters onto paper or plastic. The machine deposits ions on an image drum, and the image is developed with toner and transferred to the substrate in a manner resembling xerography. The method is fast, simple, economical and offers the printer a substantial mark-up in prices.

Thermal printing has been used for many years to generate hard copy for calculators and office machines. It has now been improved to the point that the process can generate good quality bar code symbols. All thermal printers use the same basic principle. A highly reflective or light colored substrate (typically paper) is impregnated with a clear coating that changes to a dark color (usually black, sometimes blue) upon exposure to heat. The image is not formed by burning but by a chemical reaction in the coating. The image forms rapidly at high temperatures. If the temperature is below 140°F (60°C), the coating takes more than five years to darken. At 400°F (200°C), the chemical reaction will occur within a few milliseconds. A small electrical print head contacting the substrate forms the image. Print heads have a limited life and are considered as consumable items.

Thermal transfer printers have many similarities to direct thermal printers. Like a dot matrix printer, the thermal print head presses a ribbon against the substrate. When heated, the special ribbon releases coloring materials and forms the image.

Electrostatic printers generate the image on a semiconductor drum with an electrostatic generator. They use a toner to develop the image, much like xerographic printers.

Magnetographic printers generate the image similarly to an ion deposition printer, except that they use magnetic images and toners instead of electrostatic images and electrostatic toners.

FURTHER READING:

The Bar Code Book, Second Edition. Roger Palmer, Helmers Publishing, Inc., Peterborough, NH. (1991).

Automating Management Information Systems. Vol. 1. Principles of Barcode Applications. Vol. 2. Barcode Engineering and Implementation. Harry E. Burke. Van Nostrand Reinhold, New York, NY. (1990).

Behind Bars: Bar Coding Principles and Applications. Peter L. Grieco, Jr., Michael W. Gozzo, and C. J. (Chip) Long. PT Publications, Inc., Palm Beach Gardens, FL. (1989).

Chapter

12

Technical Challenges to Management: Quality, Health and the Environment

CONTENTS

LIST OF FIGURES

LIST OF TABLES

Chapter

12

Technical Challenges to Management: Quality, Health and the Environment

In this chapter we do not attempt to give a complete statement of the package printer's responsibilities for quality, health, safety and the environment. Rather, we shall attempt to outline quality requirements and opportunities and to give an understanding of the reasoning behind the laws and regulations and some sources of help. These laws are often complex, and the manager of any manufacturing plant is responsible for knowing the law and following it.

The chapter discusses health and safety in the plant and the effects of environmental considerations upon the choices involved in printing a package.

NEW MANAGEMENT TECHNIQUES

Two new management techniques, Total Quality Management (TQM) and just-in-time manufacture and delivery (JIT), affect every aspect of packaging. In fact, they affect all manufacturing processes. They bring technology, leadership and organization together with training and improving crew skills to alter the relationship of quality, service and cost, making it possible to increase profits in a competitive economy.

TQM is not a project that management can assign to a group or a specialist, it is a way of doing business. TQM establishes a policy of continual improvement in all operations. It employs the tools and procedures of Statistical Process Control (SPC) and Statistical Quality Control (SQC). Because management and the public have misunderstood what is involved, these names mean more or less the same thing. A successful TQM process involves all of these concepts. TQM requires involvement at all levels of the company and in every department. Without leadership and commitment of top management, the quality process will not succeed.

TQM is the control of operations to assure that all products and services are within specifications ("zero defect manufacturing"). It is closely associated with JIT, which is impossible without TQM. No longer can manufacturers afford the luxury of remaking or working over faulty products or of producing overruns to assure production of sufficient salable material. This requires that suppliers of raw materials also have certified quality control programs, and the effect grows to include all manufacturers. Package manufacturers and package printers are finding that their customers or their customers' customers require that they operate a certified quality control program.

JIT is forcing shorter runs, and this requires faster makeready. We cannot afford a one-hour makeready on a press run of one to two hours. New technology will be developed; old technology requiring long makereadies will be discarded. Better procedures will be developed, and crews must be trained to minimize the time between printing the last item on one job and the first on the next. JIT favors printing processes like flexography and lithography that can be put on press and printed in the least time; it challenges rotogravure that has longer prepress times. JIT will stimulate new printing technology that can produce quality printing in a minimum time.

TOTAL QUALITY MANAGEMENT AND PACKAGE PRINTING

Quality control at one time meant putting an inspector at the end of the line to find and discard or repair unsatisfactory product. This way of doing business is unreliable and prohibitively expensive. Before and during World War II, methods for improving the quality of manufactured products were developed in the United States, but these techniques were most successfully applied in Japan. It is ironic that in order to remain competitive, Americans have had to learn from the Japanese how to apply American techniques of quality management.

International Standards

Total Quality Management is becoming a universal concern. There are international standards for establishing a quality control operation, such as ISO

9000 (International Standards Organization procedure 9000). These standards comprise five documents (9000 to 9004) that provide definitions, concepts and procedures and summarize how to select and use standards. They embody one rationale of the many different national approaches in this sphere.

Companies that adopt these standards buy supplies only from companies that operate under a certified quality program. This is increasingly common in Japan and Europe, but many North American companies (big, medium and small) are now beginning to insist that suppliers provide proof of their conformance to international standards of quality management. Packagers may find that without a certified TQM program, they will be barred from bidding on work from many large customers. Packagers and printers must work with their suppliers to meet established quality standards.

National Standards

These standards have been adopted by the national standards organizations in most countries. In the United States, they are identified as ASQC QC90 to QC94. The British have incorporated them into British Standard 5750. The Canadian designation is CAN-CSA-ISO-9000–9004. Although other countries also have national standards, they usually conform to ISO-9000–9004. National standards often have added concepts that are of special relevance to that nation.

Forces Supporting TQM

Many package printers have certified TQM programs, perhaps because certification is being pushed by the world's leading manufacturers. These industrial leaders are putting pressure on their package suppliers because of the importance of packaging on their sales.

Ultimately, the question is: "What is the actual cost of the job?" not "What is the price of the raw materials?" or "What is our wage scale?" A package that adds one or two cents in printing costs but results in increased sales is a bargain. Printing ink or paperboard that is priced a few dollars below a product that meets specs is more costly than the higher priced product if it causes the press to be shut down for a few minutes or if the printed package loses sales for the product. Lack of makeready procedures and waste control procedures have far greater impact on cost than do wage scales. The real problem is that the price is on the invoice, and wage rates are a matter of record. Quality and its effect on sales are more difficult to measure and are accordingly ignored.

Resistance to TQM

One major restraint on the acceptance of TQM is the conflict with most current business practices. TQM requires that the purchaser develop close relations with the supplier and must not purchase on price alone. Says one American manufacturer of folding cartons: "We must break old traditional ways of doing business. We visit our customers' operations, and they visit ours. We are working on a five-year basis. We've become partners with our customers. The old practice was to put everything out for bids every two years."

The organization with all the power concentrated at the top is very effective in a crisis, but it deals poorly with day-to-day problems such as determining the causes and limits of variability in a process. It deals poorly with correcting problems of manufacture where they actually occur—on the manufacturing floor. In modern business, mid-level officers interfere with the flow of communications between the shop floor, where problems occur, and the corporate leadership, which often winds up with poor understanding of the causes of problems and appropriate management action.

The Tools of TQM

TQM is a process of placing responsibility for quality at the level that can take the most effective action for quality—the manufacturing floor. This means that operators must be given tools (statistical skills and computers), the training to use them, and the authority to make effective decisions immediately. Says one manufacturer of folding cartons; "The industry is changing fast. We have a plant-wide Total Quality Involvement program. If you're not ready to involve people, the program will not work. The third shift has full authority to shut down a scheduled job if it is off specs and they cannot reach qualified help." (The third shift also must have numerical specs, the required instruments and training to use them.)

The tools require use of statistics. It has been found that workers can learn the required statistics quickly when they are shown how statistics and a calculator help them produce better product, faster and more easily. Standard deviations, bar graphs and histograms are not difficult to learn and understand, once workers see how they help them in their work.

Workers do take pride in their work. When management makes it clear that doing the job right cannot be sacrificed in order to meet a production deadline, everyone soon learns to meet both objectives.

The Impact of TQM

The first results of a TQM process are happier employees and lower costs, coupled with higher profits, because it actually costs more to do things wrong than to do them right the first time. TQM leads to better understanding of the processes in the plant, lowered barriers between departments, and it makes everyone's job easier—the manager's as well as staff and line workers. The manufacturer works with the suppliers to make sure that all materials received at the plant meet quality requirements.

TQM affects not only manufacturing, it affects the entire organization. One manufacturer of corrugated containers found that its TQM program not only increased product quality while decreasing waste, but it reached the front office, reducing from eight percent to two percent the number of invoices that contained errors.

The entire organization must be directed towards conformance to standards and specifications. These must be set reasonably to meet both the customers' needs and manufacturing capabilities of the packager and printer. Specifications that are set tighter than the system can meet actually increase variability. If specs must be tightened, it requires a study of the system and correcting the equipment and procedures that cause variability.

Setting meaningful specifications on manufactured products and purchased materials is a difficult and never-ending activity, one that requires close cooperation between customer and supplier. It cannot occur without committed support from top management.

TQM can be applied to any manufacturing or service process; it has much the same effect on printing as it has on the manufacture of corrugated boxes or flexible packaging. Any package manufacturer or printer who does not understand and apply modern principles of TQM is going to find the competition growing more and more formidable and customers increasingly demanding.

Package buyers must also understand and practice TQM. They must work with the package converter to develop specs that assure that purchased packages perform properly on the filling line, during shipment and on the store shelf.

Getting Help

Where can a printer or package manufacturer get help? The first place to turn is to your suppliers. Several of the world's largest companies that supply the printing and packaging industries have leading edge TQM programs and are willing to share them with enthusiastic customers. Your supplier is the first place to seek help. You set up partnerships with your suppliers and your customers.

In addition to a supplier who will help you set up your TQM program, many qualified and certified consultants are available. There are several organizations that have programs guided specifically to the printing industries:

Rochester Institute of Technology, T & E Center, P. O. Box 9887 Rochester NY 14623–0887. Telephone (716) 475-5000.

Graphic Communications Association, Division of Printing Industries of America, 100 Daingerfield Road, Alexandria, VA 22314–2804. Telephone (703) 519-8100.

JUST-IN-TIME

"The core philosophy of JIT is the elimination of waste: idle time, waste movement, waste of overproduction, waste of readjustments, defects, and returned goods, set-up times...JIT provides for the cost-effective production and delivery of only the necessary quality parts, in the right quantity, at the right time and place, while using a minimum of facilities, equipment, materials and human resources."*

JIT strengthens the competitive advantage. It increases cash flow while reducing inventory and working capital. It facilitates design changes and product modification so that the company responds more easily to changing market con-

* *Just-In-Time.* Chris Voss and David Clutterbuck. Springer-Verlag, New York, NY. (1989).

ditions. The product must be produced at highest speed, for least cost, while producing the quality that the customer expects, including prompt delivery.

JIT eliminates waste—anything that does not contribute directly to the value of the product: wasted time, wasted raw material, and off-spec or poor quality product. Waste includes excess or buffer inventories to cover for defective material in the production lines. Waste includes labor hours spent producing unnecessary or defective parts, reworking poor quality products, and time invested in setting up the machine before it starts operating. JIT forces the printer to reduce makeready times.

JIT and TQM

JIT is either the next step after TQM or it can be developed in parallel with TQM. JIT cannot exist without a successful TQM program, because JIT forces the manufacturer to prevent the production of off-spec products. Most manufacturers, printers, converters and packagers are overloaded with large inventories, massive amounts of paperwork and poor ways of solving problems.

Overall production time, from receipt of the copy to delivery of the finished product is a prime consideration. This includes prepress processes, handling of raw materials and finished goods, not to mention makeready of the press. Overall production time includes time involved in producing waste and off-specification product.

Like TQM, JIT involves the entire organization, and it reflects top management's attitudes and leadership. It involves training of the entire staff and the establishment of good attitudes and communications. JIT works well in industries of all sizes: small companies can move faster, large ones have more clout with suppliers. Like TQM, JIT depends on the delivery of material that meets specifications, and purchasers must insist that suppliers have certified TQM programs.

JIT appears to favor in-plant preparatory processes as opposed to obtaining color separations and plates from trade houses. If color separations and other prepress services are to be obtained from a trade shop, a great deal of cooperation and planning is required—typical of the type of cooperation and planning required if JIT is to function between departments within the shop.

Reducing Inventories

Buffer inventories are required if delivery to the assembly line or printing press is not controlled and organized. The customary buffer inventory, which increases working capital, is an expensive way of avoiding the solution of problems of quality, organization and delivery. The most expensive inventory is the outdated inventory that must be discarded.

Not only must inventory be reduced, but the manufacturing process must be simplified in order to reduce setup and makeready times and to quickly detect problems and force immediate solutions.

One visitor to Japan observed that one of Tokyo's biggest printers had only a 24-hour supply of paper on hand. American publication printers are comfortable with an inventory 60 to 90 times that large. Manufacturers will have to make long-term commitments to suppliers and must depend on them in order to meet production goals. This sort of relationship is unfamiliar to many American businessmen and will take a long time to establish.

Resistance to JIT

Where management fails to gain commitment of everyone, doubters remain who will resist actively or passively. As in the U.S., experiences in the United Kingdom and Europe vary from successful to failed. If you are competing against a Japanese company, you are competing against JIT. While there have undoubtedly been failures in Japan, JIT has made its greatest progress there.

As with TQM, JIT is unfamiliar, and most people, including managers, resist the unfamiliar and are reluctant to change familiar practices. Getting rid of large inventories and eliminating the buffer of a few extra packages or reels of paper is unfamiliar.

Effect of JIT on Package Printing

The long makeready of the letterpress machine, the costly and lengthy production of gravure cylinders, the high waste of the web offset press do not by themselves determine what process will be used, but they are important considerations. The process of bringing the offset press "up to color" (achieving ink-water balance) requires time and wastes paper. Plate scanners and densitometers that control the inking system are designed to speed the process, but flexo and gravure presses have an inherent advantage here. The lengthy process of proofing and gaining customer approval of the proof is a great place to cut unnecessary time and costs. All of these sources of waste can be reduced by application of the principles of TQM. Dealing with a JIT customer requires that the problems be addressed.

Cooperation between customer and supplier is essential to produce the most satisfactory product. Application of TQM and JIT often produces a low-cost product, but ultimately it is the satisfactory fulfillment of requirements that determines profitability and market share.

JIT means shorter lead times, smaller order sizes and stricter quality requirements. It requires application of all known techniques to the quality-cost-service triangle: better technology, better management techniques, and better training of personnel.

Like all other manufacturers, printers, converters and packagers are rethinking their operations. They must not buy the cheapest ink or board or plates, but should purchase those products that will allow them to meet the customer's specifications at the lowest cost. The manufacturer's goal is not the cheapest raw material or the lowest wage scale but the greatest profits.

LAWS AND REGULATIONS CONCERNING HEALTH AND THE ENVIRONMENT

The great interest in health and the environment of the last two or three decades is related to crowding resulting from a rapidly expanding number of people, to growing industrialization, and to the rapid change of technology. We are learning a great deal about ourselves and our environment, and this knowledge has contributed to our increasing life expectancy.

A century ago, new materials were introduced slowly to industry, and although there are many examples of death and poor health resulting from unsafe practices and misuse of industrial chemicals, workers usually were familiar with the hazards of their craft.

Now, customers, workers and the public at large are concerned about safe handling of new and unfamiliar materials that are coming into use with increasing frequency. People are often more concerned about unknown hazards than they are about well-known ones, for which safeguards often already exist. In addition, groups concerned about the environment are attracting public attention. Finally, the amount of space available is strictly limited in which to bury the burgeoning waste heaps that we generate, and there is concern about the long-term effects of buried waste.

Problems of health, air pollution, toxic wastes and recycling may be created by printing or decorating the package and by disposing of unused stock. Every packager, like every other manufacturer, is required by law to take certain actions to protect workers, customers and the environment. Many of these laws are based on common sense. Laws are often enacted because special knowledge of hazards is required if new products are to be handled safely. No one would think of eating or drinking printing ink, but other safety practices must often be taught.

Because these laws and regulations come from governmental bodies and agencies, they are governed by politics as well as by science. State and federal regulations are sometimes in conflict, and they often emphasize different aspects of safety and environmental control.

There are five U.S. federal government agencies responsible for establishing environmental and safety regulations that affect the packaging industry. They are the Environmental Protection Agency (EPA), which is concerned with the environmental effect of releases of chemicals to the air, water, and land and with hazardous waste management; the Food and Drug Administrations (FDA), concerned with the regulation of food additives, product tampering and labeling; the U.S. Department of Agriculture (USDA), concerned with the packaging of meat and poultry products; the Consumer Products Safety Commission (CPSC), concerned with the safety of products other than those under FDA or USDA jurisdiction; and the Occupational Safety and Health Administration (OSHA), concerned with the use of chemical substances and employee protection in the workplace. Some Federal laws that deal with health, safety and the environment are mentioned briefly in table 12-I.

In addition, many state, county and local agencies are concerned with environmental problems. When there are different standards between local, state or federal regulations, the most stringent limitation generally applies.

TABLE 12-I. MAJOR HEALTH, SAFETY AND ENVIRONMENTAL ACTS

CAA. The first Federal Clean Air Act was enacted in 1970. It has since been amended so that it now covers emissions from automobiles, emissions from stationary sources like printing plants, and control of acid rain.

CERCLA. Comprehensive Environmental Responsibility, Compensation, and Liability Act, enacted in 1986, often referred to as the "Superfund," provides for the clean-up of abandoned hazardous waste sites.

CWA. The Clean Water Act regulates the discharge of pollutants into surface waters.

EPCRA. The Emergency Planning and Community Right-to-Know Act often referred to as "Right-to-know," is Title III of SARA. It requires that manufacturers make public disclosure of chemical information concerning their products, and it requires development of plans to handle emergencies.

EPA. There are many Environmental Protection Acts—regional, state and federal. Manufacturers are required to conform to regulations from the federal agency, agencies in their own states and often from regional authorities.

HSWA. The Hazardous and Solid Waste Amendments to RCRA set up a timetable for banning landfills, and they set requirements for underground storage tanks.

OSHA. One of the best known laws, the Occupational Safety and Health Act set up the Occupational Safety and Health Agency that regulates health and safety practices in the workplace. Employers are required to follow all the rules and regulations set up by OSHA.

RCRA. The Resource Conservation and Recovery Act requires special handling in the disposal of hazardous wastes that are generated, treated, stored or distributed by manufacturers, including printers.

SARA. The Superfund Amendments and Reauthorization Act of 1986 expanded the coverage provided by CERCLA.

TOSCA (also referred to as TSCA). The Toxic Substances Control Act provides that the EPA gather information on chemical risks in the environment.

Manufacturers of toys and packages for toys are running into a "toy test" that involves extracting the paint or ink from the painted toy or printed package with hydrochloric acid in a test for heavy metals. This is a more rigorous test than is required for testing the safety of most packaging materials.

Outside the U.S., most national governments have agencies with responsibilities similar to those in the U.S.

THE SAFETY PROGRAM

OSHA requires that all printers and other manufacturers have a written safety program for plant personnel that covers all activities in the plant. This includes shipping and office work in addition to manufacturing operations. Keystones of the safety program are a written safety plan, safety training, and the assignment of responsibility for safety to one manager or department. Management should be thoroughly familiar with the requirements of the Occupational Safety and Health Administration (OSHA).* Management needs to be aware of state

* The printer should contact the Association for Suppliers of Printing and Publishing Technology, Inc. (NPES), 1899 Preston White Drive, Reston, VA 22091-4326; Phone (703) 264-7200; for copies of the American National Standards for Press Control B65.2

OSHA regulations in addition to Federal regulations. Whichever regulations are more restrictive or severe apply in the plant.

A 1988 survey by the National Safety Council showed that printing industries are about average in the number of industrial accidents. Hazards in the printing plant are little different from those of any other manufacturing operations: electrical hazards, mechanical hazards and falls produce injuries, while back and muscle strains occur when proper equipment is not available or people fail to use proper procedures. Chemical hazards include corrosive, toxic, flammable and carcinogenic substances, and wastes and emissions that cause environmental problems. Injuries in traffic accidents, whether on the job or while traveling to or from work or elsewhere, are common to all kinds of work and should receive attention in the safety program.

Safety equipment such as the nip guard, shown in figure 12-1, reduces the hazards of the work place. Such equipment is often required by OSHA.

Even when reasonable efforts are made to keep hazardous materials out of the eyes and off the skin, accidents do happen, and an eye wash fountain (figure 12-2) and a safety shower (figure 12-3) must be readily available in most printing plants.

Noise is hazardous to the human ear. The simple precaution of requiring ear protection doesn't always work. Ear plugs and ear muffs are something of a nuisance, and workers often fail to wear them. The workroom can be designed to reduce noise, and machinery can be designed to generate less noise. OSHA regulations specify the volume of noise to which workers may be exposed, and hearing tests are required to monitor changes in the worker's hearing.

Although none of these hazards are unique to printing, hazards of inks,

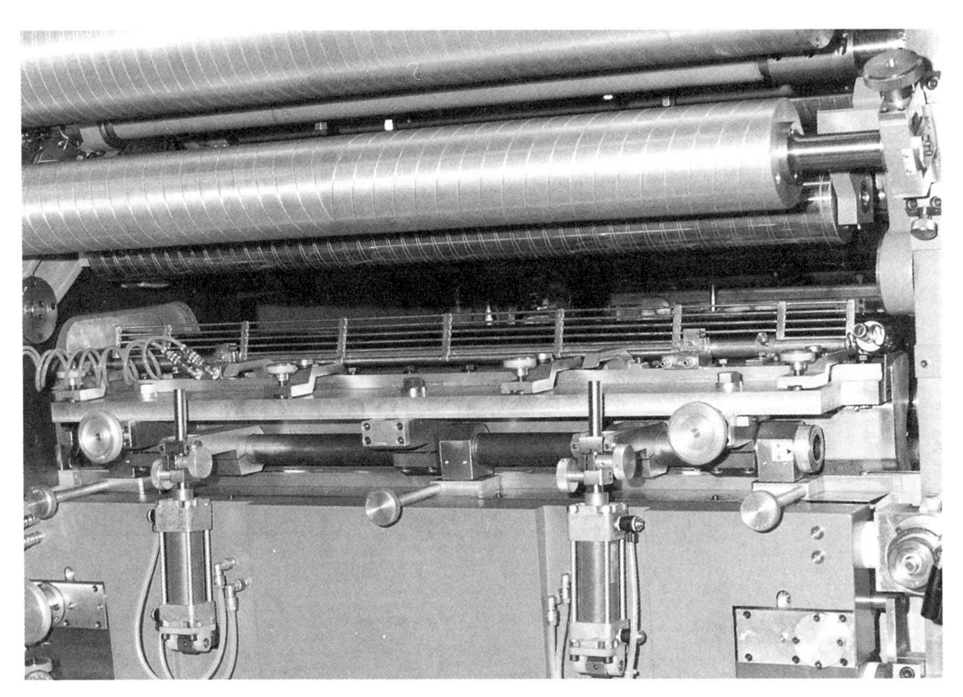

Figure 12-1. Safety guard on gravure press. (Courtesy of North American Cerutti Corp.)

Figure 12-2. Eye wash fountain.

Figure 12-3. Safety shower.

solvents, coatings and lithographic fountain solution call for some special discussion.

There are probably more chemical hazards than mechanical hazards in the preparatory section of the printing plant where photographic developers, plate washes and gravure cylinder etches present significant chemical hazards. In the pressroom, mechanical hazards with moving equipment and machinery may be more serious, but inks, coatings, press washes and ink solvents often present chemical and fire hazards.

CHEMICAL HAZARDS

Printing inks and packaging materials are chemical substances. In fact, all materials and substances are chemicals. Air, water and food are chemicals. When misused, all can be dangerous. For example, water causes drowning and electrical fires; air at high pressure causes the bends (caisson disease) in deep sea divers, and most fuels will explode when mixed with air. Excessive eating causes both acute and cumulative health problems.

Safe Handling and Packaging

There is, therefore, no such thing as a safe substance or chemical, but there are safe ways to handle any material. Materials as corrosive as sulfuric acid and caustic soda are handled commercially in tank car quantities every day. Drain opener, laundry bleach and household ammonia, although familiar substances, can be deadly, and many home accidents occur every year from their misuse. Security packages are designed to reduce the hazards of household chemicals, materials that provide both challenges and opportunities for the packager. The development of tamper-evident packaging is another example of use of packaging design to promote safety.

Hazardous Substances

A hazardous substance, then, is a material so designated by a governmental agency. It is listed as hazardous because special precautions are required when handling it. Hazardous chemicals include carcinogens, flammable or explosive materials, irritants such as solvents that remove fat from the skin and leave the hands dry and itchy, corrosive materials that burn or eat away the eyes or skin, and materials that are toxic if inhaled or swallowed. The skin is a good barrier to most toxic materials.

Eyes are always more sensitive than the skin; goggles and gloves are recommended when handling corrosive materials or chemicals that may be irritating. Sometimes, one person is very sensitive to an irritant that does not affect most people. This sensitivity is called an allergy.

Chemicals Used in Printing

Isopropyl alcohol has been widely used as a solvent for flexo inks and in the fountain solution of the offset press, where it improves the operation of the press. However, since it is toxic (it is much more toxic than grain alcohol) and since it reacts in the air with sunlight to cause irritating products, its use is being curtailed.

The chemicals used in the etching of gravure cylinders are highly toxic and must be handled with respect. Most of the chemicals used in photography and the processing of photographic images are also highly toxic. These include such chemicals as silver nitrate, hydroquinone, formaldehyde and sodium thiosulfate ("hypo"). Printing inks, photographic developers, and press and plate washes and other materials used in printing are mixtures of chemical compounds. They are hazardous if handled improperly.

TLV and PEL

Because everything is toxic if enough is taken, we are concerned with numbers such as the "Threshold Limit Values" (TLVs) and the "Permissible Exposure Levels" (PELs) that indicate safe levels of exposure to hazardous materials. The TLV is a term established by the American Conference of Governmental Industrial Hygienists (ACGIH) to indicate the toxicity of a wide variety of chemicals. The numbers are changed occasionally as we learn more about specific materials. The TLV is the concentration (in parts per million) in the air to which most people can safely be exposed for eight hours a day, year after year, without adverse effects; the lower the number, the more toxic the chemical. PELs are set by OSHA and therefore have the force of law.

Industrial workers and managers usually cannot tell from a list of contents on the package what precautions are necessary. These materials bear names that are often unfamiliar even to chemists. How is the printer to know when gloves or goggles are required? Therefore, the directions for proper handling are more important than the list of contents.

Material Safety Data Sheets

To assure that substances are handled safely, government agencies require manufacturers to supply Material Safety Data Sheets (MSDS) whenever a product requires special or unusual handling. The employer must make these MSDSs available to employees who handle materials that may be toxic, explosive or corrosive. Use of gloves or goggles is mandatory when specified by the MSDS.

Flash Point

Solvents in most inks and press washes are flammable. They can be explosive under certain conditions. Solvents with high boiling points (or low volatility) are less likely to reach explosive concentrations in the air in the plant than are low boiling, volatile solvents. The flash point is a number that indicates how hot a fluid must be heated in order to start an explosion. If the flash point is lower than 100°F (38°C), a solvent is usually considered hazardous.

Carcinogens

There are a great many materials that will cause tumors. They are called "carcinogens." Some materials are very active: a few exposures to a small amount of material will cause a tumor to grow. Others cause tumors only after large doses are received repeatedly over many years. Few of the materials used in manufacturing packages or in printing inks are of concern as carcinogens.

One chemical that is of great interest is dioxin—actually a mixture of closely related chlorinated dioxins. It occurs with "Agent Orange" and in smoke from incinerators. Dioxin is generated in incredibly small amounts during the bleaching of pulp. Dioxin may appear in barely detectable amounts in bleached pulp and in papermill waste, and it can sometimes be detected in sanitary food board.

Confusion about dioxin arises from several sources: the material is 5000 times more toxic for guinea pigs (the usual test animal) than for the closely related hamsters, and toxicity in humans has not been quantified. This earned dioxin the reputation of being one of the most toxic materials known, but people have lived through industrial accidents that have released large amounts of dioxin into the atmosphere. The only apparent effect is a skin rash called "chloracne." Other effects on human health remain a subject of controversy.

We don't know exactly how toxic dioxin is to humans, and we have not yet learned what other effects incredibly small amounts may have on the environment, including fish and wildlife.

As our analytical techniques improve, we are able to detect increasingly smaller quantities of these materials—parts per trillion (one ounce in 25 million tons). One chemist has observed: "As our instruments and techniques continue to improve, we are finding that everything is everywhere."

Unfortunately, in a great many cases our knowledge of toxicity or long term effects of exposure to chemicals, even some familiar ones, is as sketchy as our knowledge of dioxin.

Handling Hazardous Chemicals

Chemists and other workers who routinely handle corrosive, toxic and irritating chemicals are familiar with the use of face masks, fume hoods and other means of protection. Use of gloves and goggles, is, unfortunately, not very common practice with printers. It is difficult to change habits and practices even if it is legally required and makes good sense. (Workers in printing plants, as well as other plants, in Japan, wear gloves—white gloves.)

The packager needs to know that some inks may contain pigments with heavy metals or other materials that may contaminate ground waters around landfills. It is the responsibility of the inkmaker to avoid using such pigments in the inks, but it is the packager's responsibility to avoid using them on the package. These have been largely replaced with less hazardous pigments, but the printer must still ask the inkmaker about the safe disposal of both printed ink and waste ink. The package buyer must also be sure that the package is safe for disposal in a landfill. Suppliers of chemicals (including ink) are required to furnish a Material Safety Data Sheet (MSDS), and the purchaser is required to make that information available to any employee who asks for it.

Electrical Hazards and Fires

Solvents and volatile organic compounds (VOCs) that are found in ink or used to wash the press are sources of fires and explosions, and sparks can set them off. Therefore, electrical equipment should always be grounded, and sparkproof electrical switches should be used. A ground wire should connect the solvent drum to the container when drawing solvent to wash the press (figure 12-

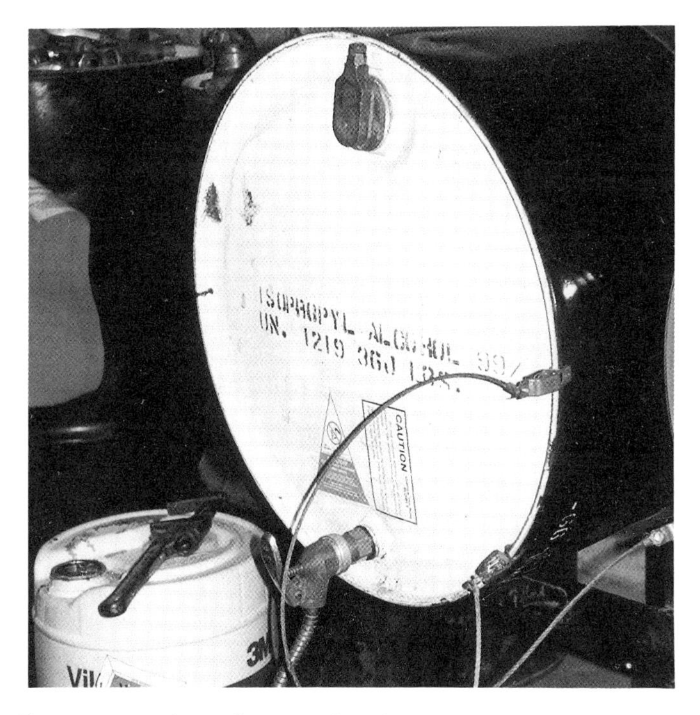

Figure 12-4. Grounding wire for solvent drums and cans.

Figure 12-5. Safety solvent can.

Figure 12-6. Safety can for disposing dirty ink cloths.

Figure 12-7. Fire extinguisher.

4). Special attention should be paid to the volatility of the solvents used to wash the press or to make the ink.

Solvents used in the printing plant should be kept in safe containers (figure 12-5). These containers have a special wire shield inside the pouring spout to reduce danger of sparks from static electricity.

One special hazard of the printing plant is the spontaneous ignition of wiping cloths wet with solvents or inks, especially paste inks—litho, screen and letterpress inks. Wiping cloths that contain ink or solvent must be placed in specially constructed steel cans (figure 12-6) and cleaned at frequent intervals.

An elemental and basic fire precaution in the printing plant, office or home is a fire extinguisher (figure 12-7). The fire extinguisher must be placed in a convenient location and be inspected and refilled on a timely schedule.

Storage of Ink

Because solventbase or petroleum-base inks are flammable, they should be stored away from heat and flames or electrical sparks. The fire department should be consulted concerning the suitability of the area where inks are stored. Insurance companies are also good sources for information regarding safety. Even sheetfed litho, screen and letterpress inks (paste inks) that have relatively little solvent should be stored safely. Waterbase inks are not flammable. The ideal, of course, is to practice just-in-time manufacturing, and to keep the amount of stored ink at a minimum.

In addition to safety considerations, a good storage plan includes convenience and concern for the storage life of the ink. Solventbase inks must not become too hot, and waterbase inks must not be allowed to freeze.

Dust and Fumes

Even inert dust is a form of air pollution, and federal regulations limit the level to which workers may be exposed. A common source of dust in the lithographic pressroom is offset spray powder. Several devices are available to limit the discharge of dust to the pressroom atmosphere. A hood over the press (figure 12-8) is useful for eliminating dust from the press. Such hoods, mounted over a press, can also reduce the amount of vapor in the atmosphere of the pressroom.

INKS AND COATINGS

Three types of pollution must be considered by the packager: air pollution, water pollution and solid waste including litter. Air pollution is caused by solvents from inks, press and plate washes, and, for the lithographer, fountain solution additives. Water pollution can be caused from chemicals in the camera department or pressroom. Many of the problems come from inks and coatings.

Inks for sheetfed printing usually contain small amounts of solvents. These are of low volatility, and they are absorbed into the surface of the package when the ink dries. The characteristic odor of sheetfed inks comes mostly from byproducts of the chemical reaction that occurs when sheetfed inks dry.

Web inks for flexo, gravure and heatset web offset dry mostly by evaporation. Unless it is water, the evaporated solvent cannot be discharged to the atmosphere. Waterbase inks on paper and paperboard dry partly by absorption. This accounts for the great interest in waterbase inks for gravure and flexo. Most, but not all, waterbase inks contain some organic solvent, usually alcohol. Safe disposal of the vapor depends on the solvent content of the ink and the amount of ink used.

When solventbase inks are used, the solvents must be recaptured or burned. Modern technology allows ink solvents to be recaptured and burned to heat the dryer on the press or even to heat the plant in winter (figure 12-9). However, the lower the boiling point (the higher the volatility) of the solvent, the harder it is to recapture or to "condense." Low boiling points make the ink easy to dry, but they also make the ink unstable on the press.

When ink solvents are discharged to the air, they form "smoke" or plumes that disturb the neighbors, but they also react with the air to create irritating chemicals that cause "smog." Discharge of solvents to the air is therefore strictly regulated.

Water pollution was not a major problem with printing until the introduction of waterbase inks. Since most waste-water discharges from the printing plant go to the sewer, the printer must follow the rules of the sewage authority. Solvents and waterbase printing inks are the materials that most obviously require attention. The printer must not wash down the press with large amounts of water, allowing it to flow to the sewer. Like solventbase inks, waste waterbase inks must

Figure 12-8. Hood over the press to reduce fumes or spray powder dust.

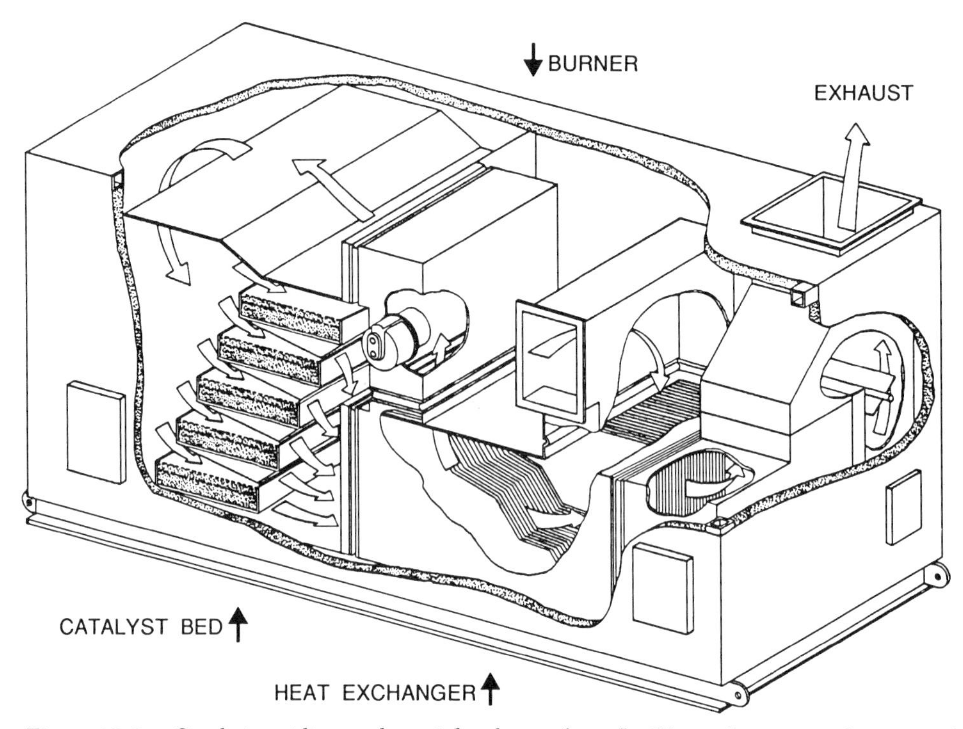

Figure 12-9. Catalytic oxidizer to burn ink solvents from flexible packaging or other printed webs. The gases are oxidized in a catalyst bed, and the heat is recovered to fuel the unit. The burner provides additional heat when needed to maintain the combustion process. Temperature control minimizes discharge of carbon monoxide and other toxic gases. (Courtesy of TEC Systems, Grace Emission Control Products.)

be disposed of according to guidelines. Attempts at excessive dilution are often considered attempts to evade the regulations.

In most areas, small amounts of waterbase inks can safely be dumped into the municipal waste stream, insofar as toxicity is involved. However, color is often objectionable and unless the amount of color is small, the effort to dilute it sufficiently cannot be successful. Regulations concerning discharge to the sewer are usually enforced by the local sewage authority.

Ink Solvents

Ink comprises solvents, pigments and binders. Coatings involve solvents and binders. All have environmental significance and problems.

The most environmentally troublesome components of inks and coatings are the solvents that evaporate when the ink is dried. Organic solvents are flammable, toxic and often irritating to the eyes and lungs. Because waterbase inks (or water-dilutable inks) contain relatively small amounts of VOCs (volatile organic compounds), they are growing in popularity, and some solvent-free waterbase inks are available. But even waste waterbase inks are not generally permitted to be flushed down the sewer.

EB and UV inks usually have only small amounts of solvent in them. It is not necessary to recapture them if good ventilation is provided.

Some highly toxic solvents have, in the past, been used in solventbase flexo inks, but the most common organic solvent now is isopropyl alcohol, the same alcohol used in lithographic fountain solutions (but now being replaced). It is considerably more toxic (intoxicating) than grain alcohol. No one should work with high-speed, precision equipment without full control of his reflexes and judgment.

Alcohols and other volatile organic compounds (VOCs) must be condensed, rather than discharged to the atmosphere. It is possible to purchase or build an exhaust hood for the press, but it is expensive to condense the alcohol by chilling, with activated carbon or by other means. Gravure presses are sometimes totally enclosed. The money spent building and operating such a system will buy a lot of ink.

Ink pans or troughs on modern flexo and gravure presses are enclosed, reducing by as much as 95 percent the evaporation of ink before printing. Sometimes covering the ink pan is all that is needed to bring a printer into compliance by eliminating a major source of unwanted effluents.

Common gravure solvents are much more toxic, and even greater efforts are made to capture them. Toluene, commonly used in gravure, is collected and usually used in the pressroom to dilute the ink or returned to the inkmaker to make more ink.

One of the important features of organic solvents for flexo and gravure inks is that they evaporate easily, allowing high-speed press operation. However, solvents that evaporate easily are difficult to condense.

The solvents used in lithographic inks tend to be less toxic than those commonly used in gravure. In web offset, petroleum solvents are usually burned, but it is possible to recycle the heat for drying the printed web. Furthermore, all organic solvents are flammable, and at the proper concentration in air, they are explosive. There have been several serious fires caused by ink solvents in printing plants.

Because of the ink-water relationships, waterbase lithographic inks are impractical,* but the industry is turning to waterbase flexo and gravure inks to reduce problems of venting VOCs to the outdoors or to the pressroom.

Heavy Metals

Although many compounds containing heavy metals are highly insoluble (and therefore not immediately toxic), there are many reasons to avoid use of heavy metals. Most printing ink pigments containing heavy metals have been replaced with other pigments free of heavy metal.

Resins

Resins, such as alkyds or phenolics that are used to make printing inks, are of relatively low toxicity. (In spite of the fact that phenol, from which phenolics

* Waterbase lithographic inks have been printed using a hydrocarbon solvent to protect the nonprinting areas of the plate, or waterbase lithographic inks could be printed from a dry plate, such as the Toray plate.

are made, is highly toxic, it is not toxic once it has been chemically linked in the resin or plastic.) Nevertheless, it is not recommended that anyone eat or drink printing inks or eat lunch without washing the printing ink from their hands. UV varnishes used in ink and coatings contain irritating chemicals and toxic activators that should be treated with respect. However, once cured, they are benign.

SOLID WASTE

Solid waste has become an item of social concern, and these are reflected in legislation. Unfortunately, there is a great deal of misinformation that is widely held such as: (1) there is more plastic in landfill than paper (2) packaging increases the amount of landfill and (3) plastics in landfill never degrade. These beliefs can be disputed. There is actually about five times as much paper as plastic in landfill. By reducing waste of agricultural and other products, packaging actually reduces the amount of garbage and other wasted solids, and the ability of plastics or paper to degrade depends on the conditions in the solid waste disposal facilities.

Disposing of Waste Ink

Printers used to dispose of ink by sending it to the landfill in cans, or if it was waterbase ink, flushing it down the drain. Printers were once encouraged to send waste ink back to the inkmaker, but inkmakers have found that the wastes are sometimes contaminated with other refuse or waste, and they cannot dispose of them. Accordingly, American inkmakers no longer accept waste ink from the printer. Disposal of the waste now becomes the printer's problem, although inkmakers will assist with technical advice.

Federal and state regulations are increasingly restrictive. Local sewer authorities are cracking down on anything that pollutes or appears to pollute the water stream. Printing ink sent to a landfill can create future problems because the printer who sent it there becomes liable for payment of at least part of any subsequent cleanup expense. The Superfund regulation specifies heavy fines. Alternatives to landfill and sewer include recovering and recycling the ink in the printing plant or paying a specialty waste handler to dispose of it.

Greater care in estimating means less waste ink at the end of the job. The easiest and cheapest way to minimize disposal of used ink is to use a minimum of ink in the fountain. However, there must be enough ink left in the fountain to maintain the required color on the last item printed.

After the press is cleaned, cleaning cloths must be cleaned, and waste solvent must be disposed of.

It is easier to dispose of waterbase inks. If the quantity of ink is small, it may be, in most cases, washed down the drain. The press is washed down with water and cleaning cloths can be washed out in the sink. For larger quantities, the water can be evaporated, and the remaining ink can be made into a cake and disposed of in solid landfill. This requires special equipment.

Newer press designs reduce waste by reducing the amount of ink needed in the fountain or by replacing the fountain with a device that meters ink directly to the ink train to give equal—and in some instances superior—results.

Landfill and Waste Disposal

Studies in the United States show that paper and paperboard make up the largest volume of landfill. Plastic is another major component of solid waste found in landfills.

There are several ways to dispose of solid waste: bury, burn, recycle, dump into the sea or convert it to compost.

Burning has many advantages and some problems, but people do not want an industrial incinerator located near their homes. Burning releases the sun's energy that the living tree captured during growing, and it releases the carbon dioxide to grow more trees. But it creates air pollution problems. Paper can be burned and the heat recovered as steam, so that the wood is effectively used twice. Some plastics also contain high heat potential and could be burned effectively for generating heat, but special conditions must be met.

Recycling gets much public attention, but less than 10 percent of America's post-consumer solid waste is as yet recycled, and only about 10 percent is burned.

Even burying at sea has been found to pollute the environment. The garbage is found washed up on the beaches and in the bodies of dead sea creatures.

Composting uses microorganisms to convert paper fiber into products that may be useful in the farm and garden, but it is still an experimental procedure.

The United States produces almost 200 million tons of solid waste each year, and we are running out of land on which to bury it—and the United States is not alone in this regard. The space that remains grows more expensive year after year. Packaging accounts for 30 percent of this solid waste. Burying the waste tends to preserve it, and 40-year-old newspapers that are still legible have been dug up from landfills.

Recent and projected generation and disposal of solid waste has been reported by the U. S. EPA. Data are summarized in Table 12-II.

Recycling

A great deal of confusion arises regarding the ability of various printed products to be recycled. Part of the problem is that while it is possible to recycle almost any printed product in the laboratory, industrial operations that can handle the process are often lacking. We occasionally read that a scientist has developed a new recycling process to recover additional kinds of printed products, but that is of little use to the consumer. It is not only necessary to know what can be done but what commercial facilities are within practical shipping distance of the source of the waste.

As landfill costs increase, the cost of disposing of waste will exceed the cost of recycling. Currently, newspapers and aluminum cans are the most profitable items to reclaim. Studies have shown that the cost of curbside recycling programs

TABLE 12-II. GENERATION AND DISPOSAL OF MUNICIPAL SOLID WASTE IN THE UNITED STATES

	Millions of Tons							
	1960	1965	1970	1975	1980	1985	1988	1995
Generation	87.8	103.4	121.9	128.1	149.6	161.6	179.6	199.8
Recovery for Recycling	5.9	6.8	8.6	9.9	14.5	16.4	23.1	38.8
Recovery for Composting	0.0	0.0	0.0	0.0	0.0	0.0	0.5	9.5
Materials Recovered	5.9	6.8	8.6	9.9	14.5	16.4	23.5	48.3
Materials Discarded*	81.9	96.6	113.3	118.2	135.1	145.2	156.0	151.5
Combustion								
Energy Recovered	0.0	0.2	0.4	0.7	2.7	7.6	24.5	45.0
Energy Not Recovered	27.0	26.8	24.7	17.8	11.0	4.1	1.0	0.5
Discards to Landfill**	54.9	69.6	88.2	99.7	121.4	133.5	130.5	106.0

* Does not include residues from recycling/composting processes.
** Does not include residues from recycling, composting or combustion processes. (Totals may not add due to rounding.)
Source: U.S. Environmental Protection Agency

for newspapers and aluminum cans is about half the cost of collection and disposal in landfills.

Development of recycling centers and the shipment of reclaimed materials to manufacturers for reprocessing promotes the growth of recycling. Without doubt, increasing amounts of all waste materials will be recycled.

Recycling of Plastics

Packagers are being pressured to reduce the amount of plastics going to landfill. A trend to ban the use of plastics in packaging appears to be developing in Europe. Recycling is an excellent way to respond. The recycling networks for plastic have not developed as well as those for recycling aluminum or newsprint, but they are developing.

Plastics can be recycled if they are not mixed or contaminated. Plastic waste in the printing plant can usually be recycled. Postconsumer waste is usually sent to landfill, but it need not be. If the plastic is thermoplastic, such as polyethylene, polypropylene or nylon, it can easily be reused, provided it has not been printed. If it has been printed, it can be sold to a reprocessor who can upgrade or reclaim it. If the plastic is thermosetting, such as phenol-formaldehyde or urea resins, it can be ground and used as filler in other plastic articles.

Information on recycling can be obtained from the Institute of Scrap Recycling Industries.*

* The Institute of Scrap Recycling Industries, 1627 K St. NW, Suite 700, Washington DC 20006; Phone (202) 466-4050, incorporates the former National Association of Recycling Industries, the Institute of Scrap Iron and Steel, the National Association of Waste Materials Dealers, the National Association of Secondary Material Industries and the Rubber Reclaimers Association.

Recycling of Paper and Board

In the early 1990s, about 30 percent of all paper and board in the United States was recycled, and the percentage has been growing together with a growing per capita consumption of paper. In other words, use of recycled paper is growing faster than use of all paper. One problem in the collection and recycling of paper is the wide fluctuation in the price of waste paper that makes dealing in waste paper a highly speculative business. Another major restriction is lack of mill production capacity. Although the number of mills making recycled paper and board continues to grow, most states have more paper to recycle than recycling mill capacity to handle it.

Most of the recycled fiber comes from newspapers and corrugated board. About 40 percent of used corrugated is recycled in the United States, with the amount reaching 60 percent in some metropolitan areas.

About 70 percent of recycled paper and paperboard winds up in boxboard. Recycled paper is combined with virgin fiber to manufacture corrugating medium and linerboard. Although few folding cartons are recycled, about 50 percent of folding cartons are made from recycled board. There is also a 3.7 million ton export market for secondary fiber. Almost no printed paper packaging is recycled.

The amount of waste paper going into a paper manufacturing process is at least 10 percent greater than the amount produced because some of it is lost in the process. If the paper is to be deinked, losses may range from 15 to 40 percent.

Several processes are used to recycle waste paper, depending on the type of paper. If the waste paper consists of unprinted trim from SBS board, cards or paper plates, it can be converted to a bright sheet with minimum processing and loss of weight. Old newspapers must be treated to remove ink, increasing the processing cost and loss of fiber. If clay coated papers are recycled, losses may amount to 50 percent of the waste paper. The process must handle and dispose of the large amounts of clay.

Printed paper can be reprocessed without deinking, but the product will have lower brightness than paper made with bleached virgin fiber. Even if it is to be coated, the brightness will be lower than if the base sheet were brighter. Products made from un-deinked paper are low-priced products such as egg cartons, flower pots and roofing paper. Un-deinked paper is coated to make shoe boxes and boxes for suits and dresses. In other words, un-deinked paper is used where protection is needed without display quality.

Recycled paper is losing its bad reputation. Waste is classified as mill broke (waste in the mill), preconsumer (waste at the converter or printer), and postconsumer. Preconsumer waste includes butt rolls, rejected unused stock, trimmings, envelope cuttings, bindery waste, printed paper discards and overruns. Postconsumer waste is paper recovered from the ultimate user: old newspapers, corrugated boxes recovered after use, and office waste. When the waste paper is mixed, its recovery is difficult and its value is greatly reduced.

There are many grades of recycled paper. Recycled paper has been improved in appearance, texture and printing performance. Although recycled paper usually has a lower brightness than virgin paper, opacity and flatness are usually better. In 1990, at least 34 mills in the United States produced recycled writing and printing papers.

Deinking of Paper

Deinked papers are widely used for packaging materials, notably folding cartons and corrugated. Mixed waste can be made into egg cartons or chipboard without deinking.

Paper can be deinked by one of three processes: by washing, by flotation and by dispersion, which is a combination of the two.

For washing, the stock is dumped into water, dispersed with a high-speed mixer, screened, and deinking chemicals are added to assist in the physical removal of the printing ink. Centrifugal or cyclone cleaners may be used to remove heavy contaminants such as sand or grit. The stock is beaten or refined depending on the mill equipment and washed or sprayed with a large amount of water, while the contaminants go to the waste pond.

In flotation, the contaminant particles must be larger, so the amount of refining is less. Flotation works only with petroleum-base inks, because the floating particles must be insoluble in water. Foaming agents and "collectors" that render the contaminants hydrophobic are added. Air injected into the stock causes bubbles. The ink residue attaches itself to the bubbles that float to the top of the tank for skimming and removal.

Dispersion is a combination of washing and flotation. The object is to keep the contaminants in large particles and float them away. The stock is concentrated to 30-percent consistency (almost dry) and dispersed in a double-faced disk refiner. This brushes the almost-dry pulp and breaks up any remaining contaminants. By adding detergent and washing, almost all the ink is removed from the waste paper.

Another problem remains. After getting the ink out of the printed waste, it must be disposed of. The best way is to flocculate it and collect it. The relatively small amount of solids that are collected are sent to landfill.

Papers printed by heatset and sheetfed inks can be mixed, and varnished pages can be redispersed and recycled. Waste paper from the office is usually contaminated with plastic, rubber bands, pressure sensitive tape and other undesirables that are difficult to remove; it is a poor source of recycled fiber. Laser image papers cannot be successfully deinked if they constitute more than around five percent of the waste. Flexo waterbase inks also create some special recycling problems that cause a large loss of fiber.

Recycling of UV or EB coated stock is technically feasible, but its commercial practice is strictly limited because the amounts of stock are not large enough to justify building special facilities. Sometimes such stocks can be "bled off," that is, mixed with large quantities of conventional waste stock and used that way. The value of conventional waste paper is greatly reduced by contamination with UV or EB inks or coatings.

Recycling of Metal and Glass

Recycling of aluminum cans now exceeds 50 percent, part of it encouraged by taxes or laws requiring a deposit on each can sold. Recycled aluminum is about 90 percent cheaper to process than new aluminum ore, and recycling networks have been established. These cans are used for beverages. Less than one percent of steel cans, used for canned foods, are recycled in the United States, although this quantity is rising with the advent of municipal waste separation and collection programs.

Of all the materials used in packaging, glass is the most permanent or least degradable. Silicon dioxide is the principal constituent of glass, and in the form of sand, sandstone and quartz, silicon dioxide is one of the most common materials on earth.

A major problem in recycling glass is the high purity required in the raw material. The glass must be free of many contaminants, small quantities of which can ruin a batch of glass. Colored glass, for example, cannot be recycled for the production of clear bottles. Some recycled glass is used to make glass building blocks.

Manufacture of glass requires the addition of cullet (recovered glass) to the melt, but most of the cullet comes from the glass manufacturer as in-house defects and breakage.

Using Recycled Materials: Paper and Board

Recycled paper has many desirable qualities. It is smoother and often more opaque than paper made from virgin pulp. Undeinked recycled board can be coated with white paper or with pigment coatings to produce a fine packaging material that will serve for many applications in place of more costly SBS stock. Recycled corrugated can be coated in different ways to provide excellent printing properties. The board is usually not as strong as virgin board, but the printing properties may be better. Use of recycled paper and board for packaging is discussed in chapter 2.

Recycled paper is also used to make roofing paper, sides of gypsum board, panel board and other products.

Using Recycled Materials: Aluminum, Plastic and Glass

Aluminum cans are recycled to make aluminum foil. Recycled glass is used for structural glass blocks, fiberglass and packaging fill (dunnage). Unprinted or deinked plastic can often be spun or extruded into fibers or films, or, when its quality is unsuitable for this, it can be used as a filler or packing material. Where the purity of the stock to be recycled cannot be guaranteed, the final product must be used in an application that does not call for critical or pressurized use as do beverage cans or food jars. Recycled stock is unsuitable for milk cartons or paper plates where impurities might contact food for human consumption.

To avoid problems with long-lived chemicals that may occur in used glass containers (for example, if the householder uses a soft drink bottle to mix toxic lawn chemicals), all recycled glass is crushed and refired. None is merely washed and returned to the market. High processing temperatures eliminate further potential problems with such chemicals.

Source Reduction

Source reduction is supported by some well-organized environmental groups, but as new laws are enacted, packagers find new ways to meet the consumers' desires for better and more colorful packages that offer convenient use of the product. Packaging manufacturers also have their lobbyists at federal and state legislatures to see that their interests are well represented so that marketing, technology and politics all play a part in the future of packaging.

Packagers must be concerned about reducing litter because much of the litter in our cities and along the roadside comes from packaging. The public reacts by supporting legislation that restricts use of packaging, and this directly affects the packager. One approach is to use packaging materials that are easily degraded. Yet, it is not practical to eliminate the most resistant of all packaging materials: metal (particularly aluminum) and glass. Deposit laws have been enacted in some states to cope with metals, plastics and glass—all of which are recyclable.

Because environmental groups view packaging as a source of solid waste, they lobby for reductions in the amount of packaging used. This political trend seems likely to continue.

Waste minimization, a major aspect of just-in-time manufacturing, can become an important means of reducing the amount of material sent to landfill from the factory, converting plant and printing plant. It is far better to process and sell raw materials than to convert them to waste and landfill.

RAW MATERIAL SUPPLY

Of all the packaging materials, only paper is truly renewable. Most of the pulpwood used for paper is grown on tree farms, owned either by the wood-process companies or by small farmers. Some of it is from byproduct sawdust produced in saw mills. The supply tends to vary, according to demand, but wood has now replaced cotton as the chief agricultural crop in the southeastern United States.

Plastics come from petroleum which is a nonrenewable resource, but supplies are sufficient to last for many decades. It does, however, seem likely that other sources of plastics will be developed by the time petroleum is depleted. Of the other materials, supply of sand for glass is effectively inexhaustible, although not all sand is of a suitable grade for glass. Aluminum ores and steel ores are plentiful, although as the best ores become exhausted, the cost of obtaining pure metal increases.

WHAT THE PACKAGER AND PACKAGE PRINTER SHOULD DO

It takes a large manufacturer to afford a staff who can keep up to date on all of the health, safety and environmental issues. One source of information for small and medium sized packagers and package printers is the Environmental

Conservation Board* of the Graphic Communications Industries (ECB). The ECB supplies information and guidance on the ever-increasing legislation and regulations that the printer must follow. The ECB also answers questions relating to federal, state and local requirements and represents the printing industry at legislative sessions considering new legislation or regulations. For printers of flexible packaging, the Flexible Packaging Association** is also helpful.

Laws and regulations are going to continue to grow more complex owing to a combination of increased industrial activity, increased emissions from highway traffic, and increasing public concern about health and the environment.

As noted earlier, each printer should have a formal, written safety program with responsibility formally assigned to a manager. It must include all activities in the plant. There are OSHA standards for a safety program, and the any printer who does not have one should immediately contact OSHA or the ECB. The elements of the safety program include proper warnings and labeling of cans, containers and equipment, and the training of employees in the safe use of equipment and materials.

The safety program need not be a financial drain. In fact, a good safety program should increase profit, not only by reducing insurance costs, but as a result of employee training—leading to increases in both safety and productivity.

The safety program is also a selling point for any manufacturer. Competent suppliers have a written safety plan and plan for control of hazardous substances. Lack of such a plan is sufficient reason to decertify the company as a supplier. The approved supplier concept goes hand in hand with control of all types—safety, pollution, JIT and ISO 9000.

THE FUTURE

Without question, as our world grows more crowded and as we consume more and more goods, environmental problems are going to require more attention. As we conquer one problem, another appears to come to the surface. Legislatures and regulatory bodies will continue to create more and tighter restrictions that will require additional pollution controls and record keeping. More than one environmental regulator has said that the ultimate goal will be zero emissions, both in the plant and to the atmosphere. How close we can approach this goal remains to be seen.

There is still much to be learned about improving safety in the workplace and protecting and improving the environment. New regulations will be based, partly, on new technology: improvements in the printing processes and in printing equipment.

---

* The Environmental Conservation Board of the Graphic Communictions Industries (ECB), Care of the Association for Suppliers of Printing and Publishing Technologies (NPES), 1899 Preston White Drive, Reston VA 22091-4326; Phone (703) 264-7200. The ECB is supported by Printing Industries of America (PIA), Graphic Arts Technical Foundation (GATF), Association for Suppliers of Printing and Publishing Technologies (NPES), National Association of Printers and Lithographers (NAPL) and several private companies.

** Flexible Packaging Association, 1090 Vermont Avenue N.W., Suite 500, Washington, DC 20005. Phone (202) 842-3880.

To meet the demand, scientists and engineers are finding better ways to monitor and control pollution. These include automation and redesign of the press, so that less ink is allowed to evaporate from the fountains.

Recycling of paper and other packaging materials is a growth industry, but one that has many problems. The shortage of space for landfills is an important incentive to keep recycling programs growing. Recycling programs will become more convenient and profitable.

Waterbase inks and varnishes are replacing solventbase inks in all processes except lithography. New pigments have largely replaced older pigments that contain heavy metals.

The future of safety and the environment is highly dependent on political forces, and political forecasts are hazardous in themselves. The public sees empty packages in the high mountains of waste, and politicians are taking drastic steps to respond to public pressures. In Germany, recycling of packaging is already mandated by stringent laws. Other laws restrict the use of packages and packaging materials. Similar laws are proposed in other nations, and it seems safe to predict that tight regulation of the amount and types of packaging and strict recycling laws will rapidly become adopted throughout the world.

Packagers, designers and package manufacturers must remain alert to these used and recycling laws and regulations as they are to health and environmental regulations.

FURTHER READING

Quality, Productivity, and Competitive Position. W. E. Deming. MIT Center for Advanced Engineering Study, Cambridge, MA. (1982).

Quality Without Tears: The Art of Hassle-Free Management. Phillip B. Crosby. New American Library, New York, NY. (1985).

Quality Control Handbook, Fourth Edition. J. M. Juran, Editor. McGraw-Hill, New York, NY. (1988).

Quality and Productivity in the Graphic Arts. Miles Southworth and Donna Southworth. Graphic Arts Publishing Co., Livonia, NY. (1989).

Questions and Answers on Quality, the ISO 9000 Standard Series, Quality System Registration and Related Issues. U. S. Department of Commerce Publication NISTIR 4721. National Institute of Standards and Technology, Gaithersburg, MD. (1991).

How to Implement Total Quality Management. Miles Southworth and Donna Southworth. Graphic Arts Publishing Co., Livonia, NY. (1992).

Just-In-Time Manufacturing: A Practical Approach. Arnaldo Hernandez. Prentice Hall, Englewood Cliffs, NJ. (1989).

Just-In-Time. Chris Voss and David Clutterbuck. Springer-Verlag, New York, NY. (1989).

Chemistry for the Graphic Arts, Second Edition. Nelson R. Eldred. Graphic Arts Technical Foundation, Pittsburgh, PA. (1992).

Packaging and the Environment. Susan E. M. Selke. Technomic Publishing Co. Inc., Lancaster, PA. (1990).

Packaging in Today's Society, Third Edition. Robert J. Kelsey. Technomic Publishing Co., Inc., Lancaster, PA. (1989).

Chapter

13

Choosing a Printer or Graphic Arts Service

CONTENTS

LIST OF FIGURES

LIST OF TABLES

Chapter

13

Choosing a Printer or Graphic Arts Service

Service, quality and price are still the principal standards in selecting a printer or supplier. Financial strength is also important. Each printer has one or more special niches. These factors are among the most important in selecting a printer or graphic arts service.

TYPES OF PRINTERS AND GRAPHIC ARTS SERVICES

The package printer and the packaging buyer face different problems. The package printer looks for vendors of paper or film, inks, plates and color separations. The packaging buyer is more apt to look for a completed box or bag in addition to print quality and suitability of the process to the product. The package printer is interested in printability of the packaging material while the packaging buyer is more interested in its barrier properties, transparency, strength and color.

Package buyers do not usually buy all forms of packaging. Just as most printers do not print many types of packages, only a few large diversified companies buy a wide variety of packages. A packager who buys flexible packaging may not buy labels, cans or tubes.

A packager may buy gummed or pressure sensitive labels but not both. The packager who buys both types of labels may choose one supplier for pressure-sensitive labels and another for ungummed sheet labels. Printers of labels for canned vegetables or chewing gum often do not produce pressure sensitive labels, nor do printers of pressure sensitive labels usually produce gummed labels. A packager requiring both kinds will probably wind up with two different suppliers.

Printers

Most package printing is done by the printer/converter who may need to purchase color separations, printing plates and typesetting. The larger packagers usually have their own design departments and often do, or contract separately for, all of the prepress work. If not, the printer/converter selects the prepress contractors and develops a working team that includes the customer, artist/designer, separator and platemaker. Even small packaging buyers require a team effort, but they may leave the selection of team members to the printer/converter who will bear ultimate responsibility for the finished package.

In this day of Total Quality Management and Just-In-Time manufacture and delivery, the buyer tries to have as few suppliers as possible. At most, the packager should consider using only a specialist printer for each different type of product; e.g., one for labels, one for flexible packaging, one for corrugated and one for general commercial printing.

The greatest service the print buyer can obtain is often a good technical representative who understands the needs of the job and can suggest better, faster, easier, less costly ways to achieve the packager's goal. The tech rep and the supplier's designer are key members of the team that plans the package. The prospective buyer must carefully evaluate the type of service the company is able to furnish.

Specialty Printers. Most package printers are specialists whose equipment is geared to a specific type of product. For instance, label printing and metal decorating are different industries with different equipment and different customers. The best and latest equipment for printing and converting labels is seldom suitable for manufacturing printed folding cartons. As a rule, printers of offset labels do not print screen or gravure labels. Pressure sensitive labels are printed on presses specially equipped with appropriate waste stripping equipment.

Leading package printers normally maintain an art department to convert supplied artwork into material suitable to be printed. Too many times a package design is executed by a designer familiar only with offset printing. Such a design, as noted in earlier chapters, will not produce a good job when printed by another method. The design must be reworked for the process to be used, at some cost penalty to the package customer.

Staff designers make minor (or major) revisions to artwork sent in from advertising agencies so that the copy can be printed by flexography or by gravure. Design work may also be requested when the basic design of a new product is identical to existing products with only typesetting or layout changes needed.

Buying from a Distant Printer. Often a specialty printer is too far away to visit. For most package printing such as primary labels, flexible packaging and preprinted liner, printers compete in national and even international markets. Bulky items such as corrugated boxes and cans are usually not shipped long distances, and the printer/converter will be nearby, usually closer than 200 miles (300 km).

Buyers of commercial and package printing routinely send specifications, mechanicals and art over long distances, using mail, couriers, fax and telephone. They send proofs and receive shipments over the same distances. They may travel personally to visit the printer only for a press check.

Although nearby printers are convenient, it is more important to find a printer (or a trade house) that has the required equipment and expertise. Commercial printers are located in every city, but specialists in packaging are not. Because of shipping costs, every major city has corrugated converters, but one with the equipment you require may not be nearby. Distant printers are often represented locally by brokers or sales representatives in daily contact with the shop. Dealing with a distant printer limits close supervision, but once you have established good specs and learned how to communicate effectively, close supervision becomes unnecessary. In any case, you will have checked out the printer carefully before beginning to work together.

For large orders, the advantages of securing a supplier who understands your needs and has the equipment and staff to meet them overcomes any disadvantages of distance.

Graphic Arts Services

The day is long past when typesetter, color separator, platemaker, printer and diecutter were all separate companies. It is now common to see many of these operations under one roof. Photocomposition, platemaking and diecutting have become so highly automated that a single company can usually own the necessary equipment and hire skilled people to be sure it is properly operated. With the web equipment used to produce most packaging, these operations—such as diecutting, windowing, slitting and stacking—are done in line with the printing.

The preparation of high quality, electronic color separations is so complex and the best equipment is so expensive that specialty companies frequently perform the service. Only big companies can afford the overhead and capital expense involved in keeping the equipment busy. Nevertheless, more and more package printers, especially label printers, have installed equipment that performs operations of design through platemaking in a single unit on a continuous basis. With such equipment in house it is feasible to make up a family of designs for little more than the cost of a single basic design. Furthermore, scanners and other electronic systems not only assist CAD but generate digital data needed for laser platemaking.

The package printer can buy color separations locally or from an overseas firm. While color separators in Asia can often supply color separations at a lower price than domestic suppliers, it is difficult to communicate regarding adaptability of the color separation to the needs of the job.

Color separations from overseas, and in fact, many produced locally, are often made to the separator's standards instead of the printer's requirements. It takes time and trouble to print test patterns for each press-substrate-ink combination that is required to give optimum information for the color separator. Many printers and package buyers accept less than optimal separations to avoid this cost (see Fingerprinting—Chapter 6). Printers who do standardize the color separations for their operations usually work closely with a single color separator.

The printer still secures most typesetting and design services from the trade house, but computerized design and layout are beginning to bring much of this work into the printer's shop.

LOCATING PROSPECTIVE SUPPLIERS

With all of the sales representatives who call on you, you probably already know most of the printers and trade houses that can supply the printing or printing services you need. If you wish to know of others, your package design artist or one with a commercial design firm can supply names of printers and trade houses to recommend, companies that he/she has worked with over the years. A color separator will know which printers can do your kind of work, and your printer is a source of recommendations for a color separator or typographer.

A few well-placed telephone calls to experienced packaging buyers can help you identify more printers and trade houses suited to your needs. Knowledgeable printing buyers attend local and national conventions and trade shows. You may even find printers and trade houses in the classified section of the telephone directory, sometimes listed under the products in which they specialize and sometimes along with general printers. They also advertise in periodicals and trade magazines oriented to potential buyers of their products. Color separators and other trade houses advertise in the printing trade press.

There is no single source of names of package printers, but part of the need is met by the *Directory of Packaging Sources** that lists many label printers and printers of flexible packaging, plus many non-printers and converters as well. For corrugated boxes, folding cartons, rigid boxes and fiber cans the appropriate source is the *Official Container Directory*** (OCD). No suitable listing yet exists for other types of packaging in the United States. The OCD, which includes Canadian sources, is the only packaging directory for Canada. Such directories are available in some other countries.

WORKING WITH YOUR PRINTER OR SUPPLIER

The manager's responsibility is not to get the lowest price but to secure a package that will ultimately produce the greatest profit. The manufacturer must

* JPC Directories, Inc. P. O. Box 488, Plainview, NY 11803. Phone (516) 822-6861.

** Official Container Directory, Advanstar Communications, One East First St., Duluth, MN 55802. Phone (218) 723-9200.

learn the difference between the lowest cost and the lowest price. A low-priced folding carton that shuts down the filling line is not inexpensive; ink that fails to match the required color becomes very expensive; low-price color separations that require extensive reworking and delay are costly, and a plastic film that splits does nobody any good.

Will your supplier give you new ideas about markets and new technology? Your supplier, as well as your customer, can be an excellent source of ideas. Suppliers can improve your marketing ability if they are good marketers. They should improve your package design and function. The supplier you want may be able to recommend a better way of doing things, a method of packaging that will improve marketing, or suggest operations that will be far less costly.

The question of how often a packager should review relations with a supplier is a major business decision. The modern way is to develop a long-range relationship between customer and supplier through a quality-development contract.

The old-fashioned way of doing things is still the most common: the purchasing agent sends out for bids every year or two or for each major job. This has proven to be more expensive than to develop close relationships with a printer or supplier who can provide a product that meets your quality requirements and who can deliver just-in-time.

To develop JIT capability, we must have suppliers with TQM procedures. You cannot produce on-spec product with off-spec raw materials. If you are a packager, there is no way to repair a poorly printed job. If you are a printer, you cannot expect top performance from ink, substrate or plates that do not meet your requirements. In neither case can you operate your equipment while your supplier is trying to locate something to replace off-spec products you rejected.

The packager must be careful to specify performance requirements, not production methods. Specifying production methods creates higher costs. When results are specified, the supplier has more options.

Years ago, the United States Postal Service requested bids on a paperboard tray to hold and ship mail. Their initial specifications were production specifications, and they agreed to pay for shipping from point of production to various delivery points. The initial bids were unacceptably high. A delegation from the Fibre Box Association met with officials of the Postal Service and recommended that the specifications be changed to performance specs and that the industry practice regarding delivery be accepted. The new bids opened the door to many more suppliers and the final cost to the Postal Service was a quarter of the original bid. The result is the familiar brown, printed corrugated tray seen in every U. S. Post Office.

Certified Suppliers

In the early 1990s, there were very few printers with certified TQM programs. TQM or statistical quality control (SQC) or statistical process control (SPC) is being driven largely by the world's largest manufacturers who insist on dealing with suppliers that have a certified quality program. Most of them require vendor certification programs that conform to ISO-9000 to ISO-9004. These five standards describe different levels of certification. If you have international customers, you will need to be certified under the international program, not a national one such as the American ANSI/ASQC Q-90 or the British BS 5750. If you are selling to a customer with a TQM program, you have an excellent source of help in setting up a certified program. Your customer may go so far as to specify the supplier of the ink or substrate or color separations.

A manufacturer or supplier can be certified by a qualified certifier, but manufacturers who practice TQM and ISO-9000 have their own procedures for certifying their suppliers. Ultimately, the package buyer must certify the supplier's TQM program although some major manufacturers of printing supplies can help their customers get started.

The concepts of TQM demand close relationships between supplier and customer, and buyers of package printing or graphic arts services will want to be especially careful in selecting a supplier to work with. It's like choosing a partner. It is of utmost importance that the supplier be able to meet quality needs and remain a competent supplier for at least the next five years.

If you are a printer/converter selling to one of the world's leading manufacturers, your customer will help you become certified. Modern management practice is to retain as few suppliers as possible. It takes a great deal of time to be sure that suppliers know your requirements and that specs are properly written. As a buyer, you do not want to spend this time developing close relations with many suppliers or printers, and once this relationship is established you will not need many suppliers or printers.

Furthermore, color separations, printing plates, inks and substrates must be designed to work together if they are to deliver the highest quality printing available for the money. The more suppliers there are, the harder it is to be sure that all parts of the system work well together. Whenever the packager changes any part of the system, other parts may require changing in response.

One of the advantages of TQM is that by reducing waste time and materials and spoilage, costs go down as profits go up. JIT further reduces costs by reducing inventories and the need for excessive working capital. These cost reductions should be shared between supplier and customer, and in the biggest operations they are frequently scheduled into the contract. This consideration should not be disregarded even in small or medium-sized operations.

The printer or trade service must not only be able to meet appropriate quality requirements but must be able to meet reasonable JIT requirements.

For most customers, service is the most important consideration. The size of a printing company is a poor guide to the quality of work it can do or the service it can render. Shops with small presses as well as those with large presses may do showcase work. Large and small shops may turn out work that is very disappointing. A company with a certified quality management program will turn out consistently good work.

Sales Reps and Customer Service Reps

After you select your regular printer and trade services, you rely on their familiarity with your needs and the requirements of your product. With well-written specs, you can trust them to take care of your printing needs.

Your contacts with the printer or trade house are your sales representative and a customer service representative. A good sales rep makes sure that jobs are done properly and on time. He or she works to raise quality and cut costs and will pick up and deliver copy and samples. The sales rep answers technical questions and shows how a job might be done better, cheaper or faster.

A good sales rep monitors estimates to keep costs down, and works with production crews to assure that problems are avoided or solved quickly. They keep quality up and check with production supervisors to keep jobs on schedule. Part of their job is to represent the customer's interests at the plant. As Beach, Shepro and Russon summarize their function: "Good representatives make everything go smoothly from planning to delivery. Take your rep to lunch once in a while just to say thanks."

Sales reps are away from the shop most of the time, so larger shops also have in-house customer service reps. Customer service reps support the sales force in representing the customer. You call them with technical questions or to learn about the status of a particular job.

Specifications

When you choose a printer, it is your responsibility to match your expectations to the shop's ability to produce. This is the object of specifications. To verify that the product meets your specs, you measure the specified qualities. Qualities that cannot be measured make poor specs.

Useful specifications are not dreamed up by a purchasing agent or a committee. An experienced purchasing agent may be able to draw up a fair list of specs, but the most useful specs are developed between supplier and customer. Some specifications may unnecessarily increase the price of a job with little or no benefit. Others may be impossible or impractical to fulfill. Discuss your needs with your supplier to be sure that specs properly describe the work you want done.

Price, Cost and Quotations

We repeat a central theme of TQM: "It is not the price of a product that matters, it is its cost." The buyer without useful specs cannot tell whether a product with a lower price represents a better buy or not. For example, it is possible to take 10 pounds of ink, add one pound of solvent and sell the product for five percent less. If you have a color strength specification on the ink you know that the lower-priced product is more costly because you must use 10 percent more of it.

In current practice, final costs are often different from quotations because customers ask for changes as the job progresses. If you are late with mechanicals or change the schedule, your printer may have to pay overtime and will pass that cost on to you.

Getting the Team Together

At the start of the job, before the package is designed and the specs are drawn up, it is important to get your whole team together—your marketers, the design artist, the trade services and the printer—to be sure that the design is going to work. The printer may be able to point out that the job can be controlled on a 30-inch press but not on a 42-inch press, so that it should be designed for a 30-inch press right from the start, or recommend a step-and-repeat format instead of a one-up because he has recently installed electronic prepress or platemaking equipment. If the buyer works closely with suppliers, he/she will know what new equipment is available. With an appropriate interchange and lack of recriminations, someone will come up with ideas that make the job more effective and less costly.

Brokers

For medium or "commercial" quality printing and color separations, a broker may be able to get you a good bid. A good broker knows the industry intimately and puts that knowledge to work for his client. Brokers know the capabilities of a great many printers and color separators and should be able to obtain the printing or service at a good price.

For the highest quality work or if you are doing JIT manufacturing, the interposition of another party between buyer and supplier may interrupt or slow communications.

EVALUATING THE PRINTER'S CAPABILITIES

Dependability and convenience are key factors in selecting a printer, factors that should be checked out before starting negotiations. Suitable equipment is vital for the work at hand and anticipated in the near future.

Equipment Lists

You also should have a list of the printing equipment necessary to print your package. A review of Chapter 2 may suggest some items to include on the list of required equipment. Be sure that the prospective printer's press equipment is suitable for printing your packaging. Most printers have an equipment list that they will be glad to furnish to prospective customers.

Most printers have some prepress equipment: cameras, desktop computers that control color scanners, composition and image assembly, step-and-repeat machines, platemaking equipment. For some flexo work, reproportioning must be built into the design so that after stretching the plate around the cylinder, it will overcome the effects of the stretch. Special equipment easily introduces the required reproportioning. Labels for unusual shaped packages require special reproportioning, too.

Many package printers have design departments. If you do not have your own designer, either on staff or as a subcontractor such as an advertising agency, you may want to use the printer's design department. This can be especially cost-effective since the staff designer knows the capabilities of the equipment in the plant and will design for that equipment. No conversion is needed (at extra cost) to adjust the design so that it will run properly on existing presses.

After you decide how much equipment and service you need, find out whether your printer can furnish it. If not, find out how the printer will get this work done.

Information about maximum and minimum sizes and the number of units is relevant to presses of any brand. Brand names are less important than the condition of the printing equipment, how well it is operated and how much waste is generated.

In addition to printing presses, you need to know what kind of prepress and finishing equipment a printer has. How much is new? New equipment is usually more productive than old equipment, but old equipment that is in good shape and updated with state-of-the-art quality assurance equipment such as register controls, automatic densitometers and viscosity controllers will give better results than new equipment that is poorly managed and operated. A trip through the printing plant will give more information than 100 samples.

The printer's converting equipment is fully as important as the printing equipment. There is an immense array of converting equipment, some of it highly specialized. It is convenient to find that your printer has all of the prepress, press and converting equipment to do the job, but if a special piece of equipment is required to make your package more effective, you may want to work out a long-range agreement with your printer. If you are a large volume packager, you may even have a contract with your printer to buy special equipment for the manufacture of your special packages. Or it may be a good idea to find a printer who has the equipment in the first place. You may even purchase it and place it in the printer's plant. Or your printer may purchase special equipment and install it in your plant to make final printing or converting operations after the package is filled. This is often done for corrugated manufacturers where a secondary printing and joining operation is done after the product is inserted into the package.

References and Other Customers

As a part of your effort to choose a printer or graphic arts service, you will ask for the names of some of their customers. Perhaps you know some already. Since they will probably not select names of people with whom they have had unsatisfactory relations, you need to probe fairly carefully to find out whether the supplier in question will be able to give you the kind of service you are looking for.

Ask whether the quality of work met their expectations. Customers are especially helpful for evaluating how much service a printer offers and whether they can keep on schedule.

When you contact a printer's other customers, check in some detail concerning the printer's performance. Every printer has loyal customers and others who have left after a single unsatisfactory transaction. Try to find out whether the person you are interviewing is a typical customer.

Evaluating the Printer's Products

Ask your printer for samples of his products. It is particularly helpful if you are able to compare the printed package with the proofs. Since the printer will provide you with "customer samples," it may be a good idea to go to the supermarket or the retail warehouse to find examples of printing that were used commercially. Better yet, try to secure samples from two or three different parts of the country and compare quality. How well do they match the samples the printer gave you?

Evaluate the samples for those print qualities that are important to you. The following properties are usually important:

- Uniformity of color from package to package
- Uniformity of color across and around the package
- Color register
- Freedom from spots, broken letters and mottle
- Sharpness of image, absence of moiré
- Ink adhesion (thumbnail or crinkle test)
- Rub resistance

Curiously, printing that passes a few simple tests and observations usually passes more sophisticated testing as well. This is because something seriously wrong with the printing usually shows up in several different ways. Nevertheless, before signing a contract, you will list properties important for your packaging that are to be specified and controlled. But be careful; these tests can be expensive. It is a common error in product testing to test for properties that need not be specified.

One simple visual test for color variability is to take a package and cut a round hole in it about one-half inch (one cm) in diameter. Place your template sample directly over other packages so that the hole exactly fits over the same part of the image underneath (figure 13–1). Under good illumination, this will

Figure 13-1. Template for detecting small color differences between samples.

quickly show even small variations in color. Variations in flesh tones are especially objectionable, but also hard to control. One useful spec for color uniformity is to prepare two packages showing the maximum amount of color variability that can be tolerated.

Quality Control

If you have a TQM program, you will look for a supplier with a certified TQM program. In the early 1990s, there were few printers with TQM processes. You may have to help your printer set the process in motion.

A tour of a prospective printer's plant can be very instructive. You will quickly see how clean and well organized the plant is. You can easily determine how much attention the printer pays to "operating by the numbers" by noting the presence of control instruments and how they are used.

If you visit a prospective printer, keep your eye out for auxiliary control equipment and instruments such as densitometers and register control systems on the press that tell about the printer's concern for keeping variability under control. The printer who insists his crew can do excellent work with little or no instrumentation will not do as good work as a printer whose crews who are well trained and well equipped. Modern densitometers show up throughout the printing plant—camera department, platemaking and pressroom. Most of them are computerized and help control all parts of the process through feedback loops.

Are the presses equipped with viscosity control systems for the ink? If not, how is viscosity controlled?

Does the printer you are investigating use print quality control targets on his products? These targets help control ink density, gray balance, slur, dot gain and many other print variables. They have been standard in high quality printing for years, and they are being used with increasing frequency on packages (figure 13–2). They can be trimmed from sheets of labels or folding cartons and hidden on the flaps of boxes, but for some types of packages, they cannot be hidden. New quality control equipment can be set to read areas in the printed image, getting around the problem of finding appropriate areas on the package to place a quality control target.

You may also be interested in your supplier's testing program. Some common pieces of testing equipment for packages are listed in table 13-I.

Suppliers' Training Programs

What sort of training programs does your prospective supplier have? Do all employees participate? How often do employees get to attend training sessions? Explore this question carefully, because a well-developed training program is an essential part of TQM. Suppliers sometimes offer training programs for their customers' personnel.

Financial Strength

Check the financial strength of your supplier just as carefully as you check the financial strength of your customers. It takes time, money and effort to build a close business relationship, and you don't want to invest that effort in a company that won't be around in another year or two.

Vision Statement, Mission Statement

Ask to see the printer's or supplier's Vision Statement and Mission Statement. Are their goals and principles compatible with your needs?

Personnel

Before you make any commitments, get acquainted with key personnel. You will, of course, know the sales rep and very likely the president or owner. You

Figure 13-2. GATF Star Targets on a package printed by offset lithography. The arrows point to the various targets. Tears near the targets were created when the package was opened. Targets are placed in a location which will not be seen by the consumer.

TABLE 13-I. TESTING EQUIPMENT FOR PACKAGE PRINTING PLANTS

Densitometers, computerized and manual for checking film and print densities	Bar code reader and verifier
Color viewing cabinet or room	Glue trim detector
Micrometer for gauging paperboard	Rub tester
Moisture meters	Tear tester
Basis weight scales	Burst tester
Stiffness tester	Gloss meter

should also get acquainted with the customer rep. It is very helpful if you can get acquainted with some of the crew to learn the attitudes of the people who will be doing your work. It seems that companies with happy customers also have happy employees.

TRADE CUSTOMS

Most industries have practices concerning issues such as credit, delivery and insurance. Trade customs are legally binding only if the case goes to court and the court finds them binding, which is often the case. To avoid problems and remain friends, contracting parties should agree to terms before signing a contract or issuing an order.

There is no published documentation of trade practices for package printing. The trade customs most closely related to package printing are those developed by the Graphic Arts Council of North America (GACNA), but they apply primarily to commercial printing. Trade customs are not a satisfactory substitute for a good understanding, put into writing, between a packaging buyer and the printer. New TQM practices should produce far better results for the printer/converter and packaging buyer than the old trade practices. Acceptable variations in count, color and waste are much tighter than when the GACNA trade practices were drawn up as a result of both improved equipment, instruments and practices.

Packaging customers today are looking for a shipment containing no defects and no rejects. Packaging specs must be tighter than those for most commercial or publication work because packages from different print runs may be displayed side by side.

Ownership of color separations, films and plates has caused many arguments

TABLE 13-II. ITEMS TO CONSIDER IN PREPARING AN AGREEMENT TO PRINT OR PURCHASE PACKAGING

1. Quotation: How long will it remain firm?
2. Compensation for canceled orders
3. Payment for experimental or preliminary work.
4. Reworking or adaptation of copy or electronic files that differ from those originally described.
5. Alterations requested by the customer.
6. Ownership of creative work.
7. Ownership of mechanical art and other preparatory materials.
8. Type of proofs to be submitted, payment for proofs, payment for revised proofs.
9. Lost press time and costs due to customer delay in supplying materials or approving proofs, etc.
10. Acceptable differences between proofs and final product.
11. Over runs and under runs.
12. Payment terms.
13. Liability. Because defective packages can cause extensive damages to the packaged goods, this item needs extensive discussion. Legal counsel in advance of any problem may be a good investment.

and should be agreed upon in advance. This may not offer much of a problem to the modern customer who expects to develop a long-term relationship with the supplier, but if the customer wants to take repeat work elsewhere it can cause difficulties. Overrun and underrun customs for commercial printing are generally not suitable for package printing.

Issues not now covered by trade practices have been raised by rapid changes in electronic prepress. For example, it is advisable to have a written understanding about delays and costs originating from lost electronic files. Items that should be considered in preparing an agreement include those listed in table 13-II. The list is limited to printing characteristics, and many additional performance characteristics must be considered.

A complete report of GACNA trade practices appears in Ruggles's book on printing estimating.

FURTHER READING

Fibre Box Handbook. Fibre Box Association, Rolling Meadows, IL. (1992).

Printing Estimating, Third Edition. Philip K. Ruggles. Delmar Publishers, Albany, NY. (1991).

Getting It Printed. How to Work with Printers and Graphic Arts Services to Assure Quality, Stay on Schedule, and Control Costs. Mark Beach, Steve Shepro and Ken Russon. Coast to Coast Books, Portland, OR. (1987).

Glossary of Package Printing Terms and Defects

Abrasion. Scratches or marks formed by rubbing or grinding when a harder substance wears against a softer substance.

Abrasion Resistance. Ability to withstand abrasion. Also called scuff resistance or **rub resistance**.

Absolute Humidity. Weight of water vapor contained in a unit volume of air (usually grams per cubic meter).

Absorbency. The ability of a porous material to take up liquids or vapors.

Absorption. (1) Penetration of a liquid substance into a solid. (2) The failure to reflect light falling on an object.

Accelerate. To hasten. In printing, drying is accelerated by adding a low boiling solvent to a liquid ink or a drier to a sheetfed ink.

Acetate. One of a family of organic compounds, esters of acetic acid, such as ethyl acetate or propyl acetate used as solvents.

Abrasion on corrugated box.

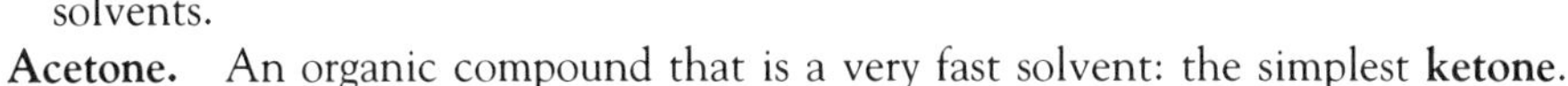

Acetone. An organic compound that is a very fast solvent: the simplest **ketone**.

Acid. A material that yields protons or hydrogen ions and reduces the **pH** when it is dissolved in water.

Acid Dyes. A group of anionic dyes that give bright colors. They are soluble in water and usually insoluble in organic solvents.

Across Web. The direction perpendicular to the direction of web travel. Also **cross web**.

Acrylic. An organic compound derived from acrylic acid, containing a double bond that is readily polymerized. Also a **polymer** produced from such a compound.

Actinic Ray. Electromagnetic radiation that causes chemical change.

Activate. To put into a state of increased chemical activity.

Additive. Material added to a formulation in small amounts to improve its properties.

Additive Colors. Colors such as red, green and blue that produce white light when combined, as in television.

Adhere. To stick together.

Adhesion. The force, either chemical or electrostatic, that holds two materials together.

Adhesive. A material that causes two materials to adhere.

Adsorption. The attraction of a liquid or gas to an interface owing to surface forces.

Afterimage. A visual sensation that, owing to fatigue of the retina, occurs after the stimulus has ceased.

Agglomerate. A cluster of undispersed particles, as in ink.

Airbrush. An air powered sprayer capable of producing subtle gradations of tone used to produce or retouch **artwork**.

Alcohol. One of a family of organic compounds containing a hydroxyl group, such as isopropyl alcohol (isopropanol) and ethyl alcohol (ethanol), used as solvents in printing operations.

Aliphatic Hydrocarbon. One of a group of organic compounds obtained from petroleum used as solvents and washes in printing. Aliphatic hydrocarbons contain no chemically unsaturated bonds.

Aliphatic Solvent. A hydrocarbon solvent such as **hexane** or **naphtha**, free of unsaturated carbon-to-carbon bonds.

Alkali. A caustic chemical that neutralizes **acid** and increases the **pH** of water.

Alkalinity. The concentration of alkali or the **pH**.

Alkali Resistance. The ability of a package to resist alkalies.

Alphabet Pulps. Thermomechanical pulp (TMP), chemithermomechanical pulp (CTMP), refiner groundwood or refiner mechanical pulp (RMP), etc. All alphabet pulps contain **lignin**.

Aluminum. A metal widely used as a foil in packaging, for lithographic plates and other industrial applications. Powdered aluminum is used to make "silver" inks.

Aluminum Foil. A sheet of aluminum metal rolled to a thickness of less than 0.006 inch (0.150 mm).

Ambient Conditions. The temperature, relative humidity and pressure of the surrounding air.

Amorphous. Having no crystalline structure.

Anchor Coat. A coating applied to improve the **adhesion** of subsequent coats. A **primer** coat.

Angle of Wipe. The angle formed between the doctor blade and the anilox roll or gravure cylinder.

Anhydrous. Free of water.

Aniline Dyes. Basic dyes, derived from coal tar, that have extreme brightness. **Acid dyes** are less brilliant but more lightfast.

Aniline Printing. An early name for printing from rubber plates using aniline dyes. The term is now obsolete.
Anilox Blade. See **doctor blade**.
Anilox Roll. A ceramic or chromed steel roll engraved to meter fluid ink.
ANSI. The American National Standards Institute.
Antifoaming Agent. An **additive** to ink that prevents foaming.
Antioxidant. A chemical that retards oxidation, notably in air-drying inks.
Anti-Skid Varnish. A coating formulated to reduce slippage of packages during stacking and shipping.
Anvil Cylinder. An impression cylinder, usually of urethane, against which a rotary die cuts through or presses the substrate.
Aperture. The size of the diaphragm opening admitting light into a camera.
Applicator Roll. Any roll used to apply a liquid: coating roll, print roll, lacquer roll or varnish roll.
Aromatic Hydrocarbon. A compound of hydrogen and carbon, such as benzene or toluene, containing a benzene ring.
Aromatic Solvents. Hydrocarbon solvents containing a benzene-ring structure, such as benzene and toluene.
Art. Drawings, paintings and photographs for reproduction.
Artwork. The original design, including drawings and text, produced by the artist and intended for reproduction.
ASCII. An acronym for American Standard Code for Information Interchange, a standard code used to help interface computers and digital equipment.
Ash. The mineral residue left after burning a sample of packaging material.
A.S.T.M. The American Society for Testing & Materials.
Azeotrope. A mixture of solvents that exhibits a constant maximum or minimum boiling point.

Backlash. Looseness in a gear mechanism that permits movement of one gear without movement by the others.
Back Printing. Printing on the underside of a transparent film with the printing viewed through the film. Also called **reverse printing**.
Back-Up Blade. A blade that supports the doctor blade.
Backup Roll. An **impression cylinder**.
Ballard Shell. A thin copper layer that carries a gravure image. It is plated non-adhesively over a copper base on the cylinder.
Barrier. A film or coating that limits the **migration** or **penetration** of one material into another.
Base. (1) A full strength ink or toner or the major ingredient in a clear lacquer, varnish or ink. (2) A flat sheet of paper or film that provides the support for a photosensitive coating or emulsion. (3) A material that increases the **pH** of water when it is dissolved.
Base Cylinder. The steel cylinder that carries the copper plating in gravure printing, or the coating of copper and chrome or of ceramic for an **anilox roll**.
Base Ink. A high-solids, single-pigment ink used for blending in package printing.
Basis Weight. The weight of a ream (usually 500 sheets) of a paper of specified size. Book reams are usually 500 sheets of 25 × 38 inches; bond and forms reams are 500 sheets of 17 × 22 inches. Film and paperboard are usually designated as weight per 1,000 square feet. Also basic weight. (See **grammage**.)
Bearers. (1) Metal spacers used to separate the platens when vulcanizing rubber flexo plates. (2) In lithography, the rings used to separate

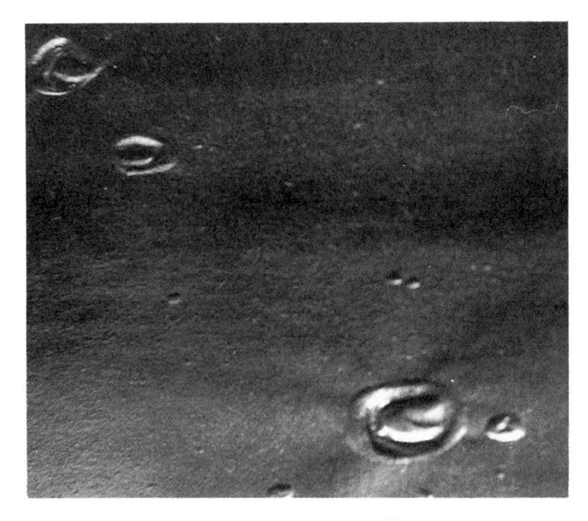

Blister.

Blocking.

the faces of the plate and blanket cylinders. (3) In flexography, raised areas of printing plates in non-printing areas to support the edges of the plate.

Binder. The adhesive component of inks or paper coating formulations.

Bit. In computers, the basic unit of digital information, a contraction of "binary digit."

Black & White. All elements of the design from which black and white art and printing plates are made. Also **B/W**.

Blade Coating. A method of applying a pigmented coating to paper or board in which a doctor blade meters and smooths the coating.

Blanket. (1) A fabric coated with an elastomer that is wrapped around a cylinder to transfer ink from an image carrier to the substrate; used principally in lithography and occasionally in letterpress and gravure. (2) In flexography, a sheet of dimensionally stable material on which plates are mounted. It is wrapped fully around the plate cylinder.

Bleed. (1) Staining of the non-image area with color from the print owing to solubility of the ink colorant. (2) Staining by fats or other packaged materials owing to solubility of the ink colorant. (3) The printed area that extends beyond the trim. (4) The overlap between two printed colors.

Blister. A small raised area caused by expansion of trapped moisture or other gas or fluid.

Blocking. Adhesion between a printed or varnished sheet and another material, such as the next sheet in a pile or roll.

Block Test. A test to determine the conditions under which **blocking** may occur.

Bloom. Material **migrating** or exuding to the surface.

Blushing. Haziness in the print usually caused by moisture trapped between the ink film and the surface of the substrate.

Board. A thick sheet of paper, usually greater than 0.010 inch thick. The distinction between **paper**, **bristol** and board is not distinct.

Body. The viscosity and flow characteristics of an ink, varnish or coating material.

Body Stock. The paper or board stock to which a paper coating is applied.

Body Type. Type or copy in the body of the text as opposed to headings or displays.

Bond. (1) A type of paper originally used for bonds and stock certificates; now a paper used for business records. (2) To attach materials by adhesives.

Bottle. A rigid or semirigid container with a narrow neck, typically made of glass or plastic.

Box. A rigid container, usually with a flat bottom and straight sides.

Boxboard. Paperboard of sufficient stiffness to be used for paperboard boxes. Commonly used types are: news, filled news, chip, straw, jute, patent coated, clay coated.

B.R.D.A. Boxboard Research and Development Association.

Breaking Length. The length of a strip of paper that weighs an amount equal to the tensile strength of the paper.

Brightness. (1) A paper property defined as the percent **reflectance** at a standard wavelength (457 nm). (2) The amount of diffuse light reflected from a surface. (3) The intensity of a light source. (4) The degree of **saturation** of a color—as a bright red.

Brilliance. The **intensity, chroma** or **strength** of a color. (See **color strength.**)

Bristol. A sheet of paper of thickness intermediate between that of **paper** and of **board,** usually between 0.006 and 0.010 inch.

Broke. Paper waste in the paper or board mill. It is recycled to the **furnish.** (See **waste.**)

Broken Type. A defect in type usually caused in art preparation. In letterpress, the defect is sometimes caused by breaking of the plate or matrix.

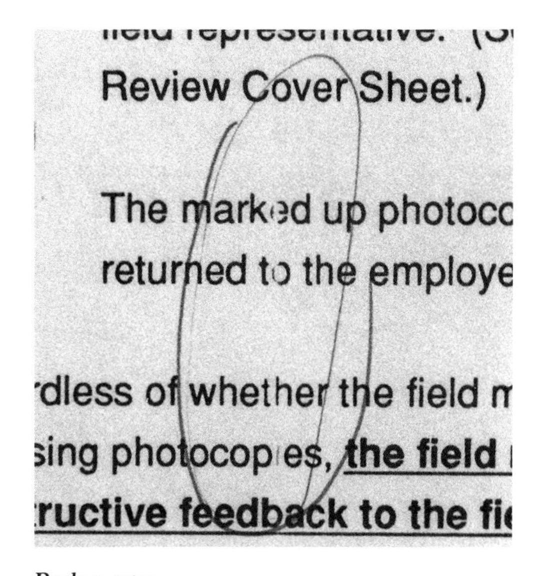

Broken type.

Bromide. A black-and-white print on paper or **continuous-tone** copy used as input on an electronic engraver.

Bronze. (1) A metallic sheen characteristic of some ink films, notably cyans. (2) A powder used for "gold" inks.

Buffer. A material that neutralizes both acids and bases and thereby controls the **pH** of water or other liquids.

Bulk. (1) The thickness of a sheet or the thickness relative to the **basis weight.** (2) Lack of compactness.

Buna-N. A synthetic rubber with good solvent resistance, made of butadiene and acrylonitrile.

Bursting Strength. The ability of paperboard or other substrate to resist pressure applied by the **burst tester.**

Burst Tester. A machine that measures the bursting strength or **mullen** of paper or board by stretching it until it bursts.

Butt. To join without overlapping or space between.

Butt Register. Printing of two or more colors that meet seamlessly without the need for **masks, spreads** or **chokes.**

Byte. In computers, a unit of digital information equivalent to one character, usually eight bits.

C1S. Paper coated on one side commonly used for printing labels.

C2S. Paper coated on two sides.

Cab. Short for **cabriolet,** the carrier that holds negatives taped into position for gravure printing. The word **flat** is used in lithography.

CAD/CAM. Computer aided design or drawing or drafting/computer aided makeup or manufacture. The design and manipulation of images with a computer.

Calender. A machine whose rolls are used to compress or smooth out paper or other materials.

Calender Stack. A group of rolls through which the paper is passed during calendering.

Calibration. The process of determining and controlling the accuracy of test equipment.

Caliper. The thickness of a sheet of material.

Camera-Ready Copy. Copy that is ready for camera reproduction.

Can. A cylindrical package made of metal, plastic or paperboard.

Cap. The cover of a bottle.

Capillary Action. The action of pores or capillaries on a liquid. It depends on the surface tension of the liquid, the wetability of the solid and the diameter of the pores. It causes liquids to rise in capillaries and porous materials.

Carbon Black. The most common black pigment. It is made by controlled burning of petroleum.

Carbon Paper Test. A procedure used in flexographic printing to determine where to make adjustments in plate makeready to assure even inking of a printing plate.

Carbon Tissue. A light-sensitive material once used to carry the image in etching of gravure cylinders.

Carcinogen. A material that causes cancer.

Cardboard. A term applied loosely to any paperboard by those unfamiliar with the industry. Correctly, a thin, stiff pasteboard sometimes used for signs or playing cards. The term is applied more broadly in British than in American usage.

Carrier Sheet. A plastic or metal sheet to which flexographic plates can be adhered before it is mounted on the cylinder. (See **blanket.**)

Case. A box or receptacle or a filled box.

Cast Coating. A process of coating paper or board in which the coating is given a high gloss by drying against a highly polished drum.

Catalyst. A substance that initiates or accelerates a chemical reaction without itself being consumed.

Caustic. A chemical that causes burns, usually alkaline.

Cellophane. An early packaging film made of plasticized **nitro-cellulose**.

Cellosolve. The trademark of Union Carbide Corporation for the monoethyl ether of ethylene glycol, a solvent used in inks to retard drying.

Cellulose. A natural carbohydrate polymer, insoluble in ordinary solvents, which is the major component of wood and of paper and board.

Cellulose Acetate. An ester of cellulose and acetic acid used for making films and fibers.

Centipoise. A unit of viscosity. Pure water has a viscosity of exactly 1 centipoise.

Central Impression. A CI press is one in which the web is supported by a single large impression cylinder while colors are applied from several printing stations around the cylinder.

Chalking. The dusting off of ink that is not firmly bound to the paper.

Chambered (or Chamber) Blade System. An enclosed anilox inking system in which one blade is used to retain the ink in a well while the other blade meters the ink.

Character. Each individual letter, symbol or punctuation mark that makes up a full **typeface**.

Chill Roll. (1) A roll cooled internally with water or brine, used to cool a web prior to rewinding. (2) In flexography, a roll following a dryer to return the printed web to its original dimensions.

Chipboard. A paperboard made of undeinked waste paper.

Choke. An image whose edges have been pulled in slightly from those of the original to improve the fit of the image on the print. Also called a **skinny**.

Chroma. A term of the Munsell Color Space Model that specifies the strength or saturation of a color. (See **color strength**.)

C.I.E. Commission Internationale de l'Éclairage, a standards-setting organization for color measurement.

Circumferential Register. See **running register** and **cutoff misregister**.

Clay. A material composed chiefly of aluminum silicates used as a filler and coating for paper and paperboard. Different clays have different properties. **Kaolin** is commonly used for paper coating.

Cling. The tendency of materials to adhere, usually owing to **static electricity**, as in blocking, except they can be separated without any visible damage.

Closure. A piece attached to a package, such as a cap or crown, to close the package.

Coating Binder or Adhesive. The material added to a paper or board coating formulation that binds the pigment to the surface of the paper or board.

Coefficient of Friction (COF). The measure of the slip properties between two surfaces.
Coextrusion. Production of films with more than one layer by forcing different polymers through a common die.
Cold Flow. Steady deformation under stress. Also called **creep**.
Collotype. A planographic printing process that uses a flat, gelatin-based plate with different water receptivity to separate image area from non-image area. Sometimes used for display and specialty printing.
Color. The impression, on the eye, of electromagnetic radiation with wavelengths between about 400 and 800 nanometers.
Colorant. The **pigment** or **dye** that produces color in an ink.
Color Balance. The combination of yellow, cyan and magenta needed to produce a **neutral gray**.
Color Bars. Small patches of color solids, overprints, tints and other targets used to help evaluate ink density, trap, etc. Also called **color control strip**.
Color Correction. (1) Changes required in color separation materials to produce the desired colors in the print. (2) Changes in color of ink to produce the desired print color.
Colorimeter. An instrument designed to measure color in a manner similar to that seen by the human eye.
Color Key. A trademark of 3M Company for its patented overlay proofing system.
Color Overlay. (1) Transparent sheets, usually acetate, bearing the colors to be printed, that are placed over a black-and-white drawing, as a guide for reproduction. (2) A term sometimes used at press-side referring to the number of colors that overprint each other.
Color Purity. Lack of grayness in a color. One of three properties (hue, strength and purity) that define a color.
Color Scanner. An instrument that generates **color separations** electronically.
Color Separation. The process of preparing **color separations** either photographically or electronically.
Color Separations. Film intermediates or digitized data that control the cyan, magenta, yellow and black images to be placed on the respective image carriers.
Color Sequence. The order in which colored inks are printed.
Color Strength. The intensity of coloring of an ink; **chroma**. One of three properties (hue, strength and purity) that define a color.
Color Temperature. The temperature, in degrees Kelvin, to which a black body must be heated to match a given white illuminant. Low color temperatures are reddish; higher temperatures are bluer.
Color Transparency. A full-color reproduction of the image or design on a photographic film.
Combination Press. A press that comprises more than one type of press in the line; e.g. a central impression press with outboard stacks, or a gravure or offset press with flexographic stations.
Composite Art. Black and white art where all colors are drawn on one piece of copy, sometimes keylined to facilitate color separation by the photoengraver.
Compressible Blanket. A blanket that is compressible because it contains foamed rubber or polymer.
Computer Aided Design/Computer Aided Manufacture. See **CAD/CAM**.
Consistency. (1) The body or viscosity of an ink or varnish. (2) The uniformity of reproduction.
Contact. A camera reproduction of an original negative to form a **positive** or a positive to produce a **negative**.

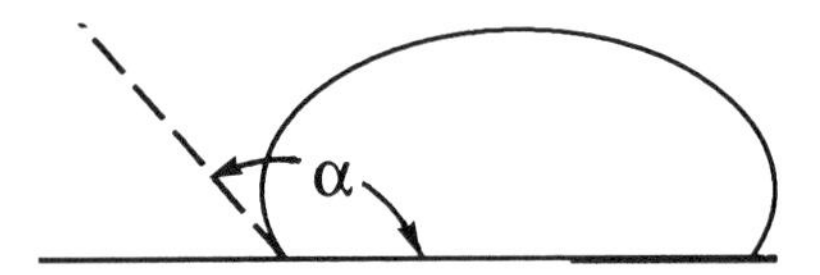

Contact angle of a liquid drop on a surface.

Contact Angle. (1) The angle formed between a drop of liquid and the surface of a package. A high contact angle indicates low wettability. (2) The angle of contact between the doctor blade and the gravure cylinder or anilox roll.

Contact Duplicate. A camera reproduction of an original positive or negative produced with materials that produce a positive from a positive and a negative from a negative.

Contact Film. Film specifically designed for producing conventional contacts: blue-sensitive continuous-tone film with a relatively high maximum density, excellent resolution, and a special antihalation back that allows exposure through the base of the film without loss in quality.

Contact Frame. A device that holds the image and the photosensitive material together for contact photography.

Contact Screen. See **halftone screen**.

Containerboard. Corrugated board.

Continuous-Tone Image. A picture or print with varying tones of color formed by the grain pattern of a photographic emulsion; made without halftone screen, as opposed to line copy in which all images are of a single density.

Converter. A manufacturer who produces printed packages or other materials from rolls of paper, film or foil.

Converter Waste. Waste produced from paper, paperboard or other substrate during conversion. It may or may not be printed. (See **waste**.)

Copolymer. A polymer produced from two or more **monomers**.

Copper Plating. The process by which a copper coat is applied to the base cylinder for gravure printing.

Copy. (1) The material furnished for graphic arts reproduction. It is the photographer's original. It comes in many forms: positive, negative, reflection, transmission, continuous or halftone, etc. (2) The printed product. (3) Textual material as distinguished from photographic or creative art.

Copy Viewer. A **light box** or other structure for viewing copy and prints under controlled conditions.

Type	Flutes per Length	Approx. Height*	Take-Up Factor**	
A-flute	33±3/ft. 110±10/m	.184 in. .467 cm	1.54	
B-flute	47±3/ft. 155±10/m	.097 in. .246 cm	1.32	
C-flute	39±3/ft. 130±10/m	.142 in. .361 cm	1.43	
E-flute	90±4/ft. 295±15/m	.062 in. .157 cm	1.27	
F-flute	96±4/ft. 315±15/m	.045 in. .114 cm	1.23	

* Not including thickness of facings.
**Amount of corrugating medium needed per unit length after fluting.

Corrugated board.

Corrugated Board. A composite paper product made by applying a **liner** to each side of a fluted or corrugated inner sheet or **medium**. The liner is usually made from kraft board, and the corrugated **medium** is made from neutral sulfite or recycled board. Variations include singleface board made from a corrugated sheet plus a single liner, doublewall made from two corrugated sheets and three liners, and triplewall made from three corrugated sheets and four liners. Also **containerboard**.

Corrugated Box. A box made of corrugated board.

Corrugating Medium. A board made of a stiff fiber, usually semichemical or recycled board, to be corrugated and used as the middle layer in a corrugated board.

Corrugation Marks. (1) Marks resembling ropes or chains in the surface of the paper owing to adjacent hard and soft spots in the web. (2) On printed boxes, the lighter and darker areas created by differential pressure of the printing plate on the substrate.

Cosolvent. One of two or more solvents in a mixture used to make an ink or other solution.

Crawl. The forming of droplets from a wet film of ink, owing to failure of the ink to wet the surface.

Crazing. A pattern of cracks that sometimes develops in heavy coatings.

Creep. Deformation that occurs in cured or uncured plastics under stress. (See **cold flow.**)

Cromalin. The trademark of E. I. duPont de Nemours & Co. for a patented single-sheet proofing material.

Crop Marks. Marks made on the outer edges of artwork to designate the area to be omitted from the print.

Cross Web. See **across web**

Crown. (1) The difference in diameter between the center of a roll and its ends. (2) The closure of a glass bottle that is crimped around the rib of glass at the neck.

Cure. A chemical reaction that converts a liquid to a solid, a wet film to a dry film, or raw rubber into vulcanized rubber.

Cut. (1) To dilute or thin an ink, lacquer or varnish. (2) An engraving. (3) A paper defect.

Cutoff. See **repeat.**

Cutoff Misregister. Misregister in the longitudinal or web direction.

Cyan. A shade of ink that absorbs red while transmitting blue and green.

Cylinder Gap. On lithographic presses, the gap or space in the blanket or plate cylinders where the plate or blanket are attached.

Cylinder Machine. A paper manufacturing machine on which the paper or board is formed against cylindrical screens.

Paper cut.

Dampening Solution. The aqueous solution in the water fountain of a lithographic press. **Fountain solution.**

Darkroom. The area in the camera department reserved for working with light-sensitive materials, as opposed to the **gallery.**

Deck. A print station with plate, impression cylinders and inking rolls. (Term used mostly in flexo.)

Deinking. Removal of ink from printed stock prior to recycling.

Delamination. (1) The partial or complete separation of the layers of a laminate. (2) Separation of a sheet or web of paper.

Densitometer. An instrument that measures the absorbency of light by an ink film or photographic film. The reflection densitometer measures the absorbency of a print while the transmission densitometer measures the absorbency by a film.

Density. (1) The ability of a printed image or a photographic image to absorb light. It is measured in a logarithmic scale. Thus a print with a density of 2 absorbs 10 times as much light as a density of 1. (2) With printing ink, the weight per unit volume.

Desktop Publishing. The generation and manipulation of images and type with a desktop computer, including digitization, color separation and color correction.

Developer. The solution or the chemical in the solution that converts a latent photographic image into a visible image.

Developing Agent. The chemical compound within the developer that converts a latent photographic image into a visible image.

Diazo. (1) Any of several photosensitive chemicals that are used to make an image. (2) Sometimes, the print or plate that is made by such a process.

Die. A sharp cutting form used to cut a desired shape from a sheet or film.

Diecutting. The process of using sharp steel rules to cut special shapes for labels, boxes and containers and other packaging and printing from printed sheets or webs.

Diffusion Transfer. A process of producing positive copy on paper, film or lithographic plates by physically transferring the image formed by a special light-sensitive material.

Dilatency. A reversible increase in **viscosity** when an ink or dispersion is stirred or worked. The opposite of **thixotropy**.

Diluent. A liquid with little solvent power that is used to thin or **cut** an ink or varnish.

Dimensional Stability. Resistance to dimensional change from changes in **ambient conditions**.

DIN. Deutches Institut fuer Normung. The German Standards Organization.

DIN Cup. An **efflux cup** that conforms to DIN standards for measuring viscosity.

Direct Positive. A photographic print obtained without the aid of a negative.

Doctor Blade. A thin flexible steel or plastic blade used (a) to remove excess ink from a gravure cylinder or anilox roll or (b) in blade coating of paper, to remove excess material and smooth the coating.

Dot. The individual element of a **halftone**.

Dot Etching. A process of reducing the size of halftone dots on a color separation film, either chemically or electronically, to change the density of the halftone print.

Doughnut or hickey.

Dot Gain. The increase in size of halftone dots in any of the steps from film to press sheet.

Dot Loss. The decrease in size of halftone dots in any of the steps from film to press sheet. Also **dot gain reduction**.

Doublewall. See **corrugated board**.

Doughnut. (1) A printed halftone dot that is missing its center. (2) A spot in a solid print that is surrounded by unprinted substrate—a hickey.

Drawdown. An ink testing procedure in which a dab of ink is spread onto the substrate with a wire-wound rod or an ink knife or spatula.

Drier. A chemical agent added to sheetfed inks that initiates drying.

Driving Side. The side of a web press on which the main gear trains are located. Also called the **gear side** or **drive side**; opposite of the **operating side**.

Dropout. A part of the original that does not reproduce, often on purpose.

Dryback. The change in ink color and gloss that occurs as the ink dries.

Dryer. A heater used to dry web inks, often called an **oven**.

Dry Offset. (1) In letterpress, dry offset means letterpress printing to an offset blanket. (2) A process of lithography in which special silicone-coated plates eliminate the necessity for dampening with water.

Drying Oil. An oil used for making inks and coatings that is changed to a solid by the action of the air.

Dummy. A **layout** of the design used to show placement of print and illustrations.

Duotone. A print made by superimposing prints from two halftone images.

Durometer. An instrument for measuring rubber hardness, used in flexo to measure plate hardness and in offset to measure roll hardness. Also the hardness of the rubber.

Dye. A soluble colorant, as opposed to a pigment which is insoluble.

Efflux Cup. A **viscometer** that consists of a cup with a calibrated hole in the bottom such as the **DIN**, **Zahn** or **Shell cup**. The reading is the number of seconds required for the liquid to flow from the cup in a continuous, unbroken stream.
Ektachrome. A trademark of Eastman Kodak Co. for a family of patented color transparency films.
Ektacolor. A trademark of Eastman Kodak Co. for a family of patented color negative films.
Elasticity. Resilience. The property of a substance that enables it to return to its original size or shape after being stretched or deformed.
Electrodeposition. **Electroplating**.
Electron. The most elementary of negative particles and a constituent of all atoms.
Electroplating. The electrolytic process by which copper or other metal is deposited on another metal. Also **electrodeposition**.
Electrostatic Assist. (ESA) A method of applying a high voltage to improve ink transfer in rotogravure printing.
Electrostatic Printing. Any of several **nonimpact printing** processes, including **xerography** and **laser printing** that form an image by attracting a **toner** to charged areas on the surface of a drum.
Elmendorf Test. A test for the resistance of paper or board to tearing.
Embossing. Production of a raised image in a substrate.
Emulsification. Mixing of two liquids that do not dissolve in each other to form tiny, stable droplets.
Emulsion. (1) An intimate mixture of two immiscible (insoluble) liquids. An emulsifying agent is required if the emulsion is to remain stable. (2) The image-bearing coating applied to photographic film.
Enamel. The finish or coating on a gloss-coated paper or board.
Engraved Roll. A transfer roll with cells engraved mechanically or by laser. (See **anilox roll**.)
Engraving. (1) A pattern incised or cut into a surface. (2) The process of forming an image in a gravure cylinder or anilox roll by physical or mechanical means. (3) A metal plate with a raised image.
EPA. The Environmental Protection Agency of the U.S. government or of many states or municipalities.
Epoxy Resin. A polymer or plastic used for strong, fast-setting adhesives.
ESA. **Electrostatic assist**.
Ester. An organic compound, such as ethyl acetate or isopropyl acetate, derived from an alcohol and an acid and used as a solvent.
Etch. To eat away a solid, such as metal or glass, to produce an image. Specifically, the process used to produce gravure images with chemicals.
Explosive Range. The range of vapor-to-air ratios of flammable products that will explode if ignited.
Extender. A colorless pigment put into ink to increase its body.
Extrusion. The production of a continuous sheet of film or other shapes by forcing hot plastic through a die or orifice.

Facings. Sheets of linerboard used as the flat members of corrugated board.
Facsimile Transmission. The transmission of images that have been digitized and transmitted electronically to produce a recorded likeness of the original. Also called **fax**.
Fading. A decrease in ink color density with time, light or heat.
Fade-Ometer. An instrument used to measure the lightfastness of printed ink, paper and other materials.

Fast. (1) Resistance to **fading**. (2) A characteristic of a low-boiling solvent that accelerates the drying of a solventbase ink.

Fatty. See **spread**.

Fax. See **facsimile transmission**.

F.D.A. The U. S. Food and Drug Administration.

Feathering. A ragged pattern that develops at the edges of lines or solids when ink is applied to a porous paper or board.

Fillers. Minerals added to paper to increase opacity or smoothness or to ink to alter color strength or flow.

Film. An unsupported, nonfibrous, thin, flexible material not exceeding 0.010 inch. Thicker materials are called **sheeting**.

Film Speed. The sensitivity of photographic film to light: the higher the sensitivity, the higher the speed.

Filter. (1) A thin transparent film that absorbs unwanted colors or wavelengths. (2) A mechanical device placed in a fluid ink system to remove solid contaminants.

Fineness of Grind. The degree of grinding or dispersion of pigment in a printing ink.

Fingerprinting. A study of the printing characteristics of a press on any particular ink and substrate.

Finish. The gloss or flatness of a print or surface.

First-Down Color. The first film of ink applied in multi-color printing.

Fit. The ability of two images to be printed in register.

Fixer. A chemical used to stop the development of a photographic image.

Flag. A small piece of paper or board inserted at the edge of a roll or skid of stock to indicate the location of a splice, imperfection, etc.

Flammable. Capable of burning. Coined as an unambiguous synonym for "inflammable."

Flare. Stray, non-image light that exposes a film. It arises from unwanted reflections, dust or unshielded lights.

Flash Point. The lowest temperature at which a substance can be ignited under standard test conditions.

Flat. (1) Lacking in contrast and definition of tone. (2) **Matte**. Opposite of **glossy**. (3) A sheet of goldenrod paper or other material with the film negatives or positives in place for plate exposure.

Flat Bed Press. A printing press on which the substrate is held flat during printing.

Flatbed Scanner. A scanner in which the original is mounted on a flat scanning table rather than around a drum.

Flat Die. A die on a flat base, used in reciprocating die cutters.

Flatfield Scanner. See **flatbed scanner**.

Flexible Packaging. Paper, film or foil printed and converted into pouches or bags for consumer or industrial uses.

Flexography. A method of direct rotary printing using resilient raised image printing plates, affixed to variable or fixed repeat plate cylinders, inked by a roll or doctor-blade-wiped engraved metal or ceramic coated roll, carrying fluid or paste-type inks to virtually any substrate.

Fling. See **flying**.

Flocculation. The clumping together of pigments in an ink.

Floppy Disk. A flat magnetic disk for recording digital electronic information.

Flow. The property of an ink that causes it to level as would a true liquid. Also **flowout**.

Fluid Ink. An ink of low viscosity used for flexography or gravure. Also called liquid ink.

Fluorescence. The emission of light of different wavelengths than that which falls on the object; for example, optical bleaches in paper convert ultraviolet rays into blue/white colors.
Flush Left (or Right). Type set to line up at the left (or right) of the column.
Flying. Ink thrown off the press by the inking rollers. Also **fling**.
Foil. An unsupported thin metal membrane less than 0.006 inch thick.
Folding Carton. A package made of boxboard that is folded flat during shipment and set up or opened at the packaging line.
Font. A complete assortment of all the different characters of a particular style and size of type.
Formation. The uniformity of distribution of fibers in a paper or board.
Fountain. A pan or trough that supplies ink to the press. The water pan on a litho press is also called a fountain.
Fountain Roll. The roll that picks up the ink or coating from the fountain and supplies it to a transfer roll or metering roll.
Fountain Solution. The dampening material in a lithographic water pan. **Dampening solution.**
Four-Color Process. Printing with yellow, magenta, cyan and black ink using halftone screens to create all other colors.
FPA. The Flexible Packaging Association.
Freeness. The ability of a pulp to drain quickly on the paper machine or on a test machine. Unrefined pulps have high freeness.
Free Sheet. A paper or board free of groundwood.
F-Stop. Also called f-number. The focal length of a lens divided by the diameter of the diaphragm opening. Large diaphragm openings give low f-stop numbers.
FTA. (Also F. T. A.) The Flexographic Technical Association, Inc.
Furnish. The mixture of pulp, water and additives fed to the paper machine to manufacture paper.

GAA. (Also G. A. A.) The Gravure Association of America.
Gallery. The area in the camera department for work under ordinary illumination, as opposed to the darkroom.
Galley Proof. A proof of text copy before it is assembled into the **paste-up**.
Gap. The opening in a lithographic plate or blanket cylinder into which the plate or blanket is fastened.
Gas Chromatograph. An instrument used to determine the composition of volatile materials.
GATF. The Graphic Arts Technical Foundation.
GCR. **Gray component replacement**.
Gear Side. **Driving side** of a press; opposite the **operating side**.
Gelatin. An animal byproduct, film-forming protein used for graphic arts films.
Ghosting. Appearance of a faint image of a design, appearing in areas not intended to print that part of the image.
Glassine. A type of translucent paper.
Gloss. The ability of a surface to reflect light at special angles, as opposed to diffuse reflection which is at all angles.
Gloss Meter. An instrument for measuring gloss.
Glue. Any adhesive, but specifically a natural adhesive prepared from animal protein.
Glue Line. (1) A line of adhesive between two surfaces to be adhered. (2) The space on a package for placing the adhesive.

Grain. (1) Clumping of silver grains in photographic film. Large grains may actually reduce the resolving power of the photographic film. (2) The alignment of paper or board fibers in a sheet or web.

Grain Direction. The orientation of fibers in a paper or board in relation to the web direction or the sheet direction.

Grammage. The weight per unit area of a paper or board, specified in grams per square meter. (See **basis weight**.)

Graphic Arts. The design and printing of graphic and alphanumeric material.

Gravure. The printing process using fluid inks in which the image area is depressed in a plate or cylinder and the non-image area is wiped with a doctor blade. It is usually printed from rotary cylinders, in which case gravure is synonymous with rotogravure.

Gravurescope. A microscope designed to study the cells on an engraved cylinder and to measure their depth and width.

Gray Balance. The values of yellow, magenta and cyan that are needed to reproduce a **neutral gray** when printed at normal density.

Gray Component Replacement. (GCR) A process of color separation in which the primary and secondary colors are the same as in normal process color printing, but black replaces the gray component created by the third color so that tertiary colors are composed of two colors and black. (See **undercolor removal**.)

Gray Scale. A strip of photopaper or film ranging in steps from clear to solid black (or other color) used to measure the exposure of a film or plate.

Groundwood. A paper pulp produced by grinding wood or pulping without removing the lignin, the natural adhesive in the wood. (See **free sheet**.)

Groundwood Papers. A general term applied to papers made of wood pulp that has been mechanically defibered.

Guillotine. A cutting machine in which a long knife descends vertically on the paper or other material.

Halftone. A print or photographic image formed by a pattern of discrete dots of various sizes that creates the illusion of **continuous tone** when viewed from a distance.

Halftone Gravure. The process of producing a gravure cylinder from material prepared for offset reproduction.

Halftone Screen. A grid composed of tiny squares produced in glass or plastic that breaks a **continuous-tone image** into the dots of a **halftone image**. An identical pattern produced electronically. Also called **contact screen**.

Halo. A printing defect of flexo and letterpress in which a heavy ridge occurs around the outside of the letter or image.

Hard Copy. A tangible image such as a printed proof or a printed sheet.

Hardness. Resistance to indentation.

Hardwood. Wood from deciduous trees (angiosperms) which include birch, maple, oak and poplar.

Haze. Appearance of milkiness in a transparent film.

Heat Seal. A method of uniting two surfaces by fusion of the base materials or coatings, one or both of which flow when heat is applied.

Helioklischograph. The trademark of Linotype-Hell AG. for an electromechanical machine that produces engraved cylinders.

He/Ne. A helium-neon laser.

Hexane. An aliphatic hydrocarbon containing six carbon atoms that is used as a solvent.

Hickey. A printing imperfection consisting of a small solid speck surrounded with a white halo, or the piece of foreign matter that produces it. (See **doughnut**.)

High Key. A design or picture in which the center of interest is of low color density.
Highlight. The lightest part of a picture; the lowest density of a positive or the highest density of a negative film.
Highlight Dots. The very small black dots in the highlight areas of the print or the tiny white dots in a film negative.
Holography. A method of producing a three-dimensional image in a film or foil by using interference patterns from a split laser beam.
Horizontal Camera. A camera that focuses in a horizontal direction and carries the copy and lens in a vertical direction.
Hue. The shade of an ink or print such as blue, red or green. One of three properties (hue, strength and purity) that define a color.
Hydrometer. A device for measuring the specific gravity of a liquid.
Hydrophilic. Having a strong affinity for water (**oleophobic**).
Hydrophobic. Lacking affinity or attraction for water (**oleophilic**).
Hygrometer. A device for measuring the relative humidity of the atmosphere.
Hygroscopic. The tendency of a material to take up moisture from its surroundings.
Hysteresis. (1) Failure of paper or board to return to its original dimensions when it has been wet or dried, then returned to its original humidity. (2) A retardation of the effect when the forces acting upon a body are changed.

I.D. The inside diameter of a core or other cylinder.
Image Area. (1) The area of the printing plate that transfers ink to the substrate. (2) The maximum area that can be printed on a press, usually slightly smaller than the maximum size sheet or web width that can pass through the press.
Image Carrier. The plate, cylinder, screen or magnetic tape that carries information suitable for reproducing an image on a substrate.
Impression. (1) The pressure of the plate against the substrate. (2) In offset, the pressure of the plate against the blanket or the blanket against the substrate. (3) The image transferred from the printing plate to the substrate.
Impression Cylinder. Roll or cylinder that backs up or supports the substrate at the point of impression, serving as an anvil.
Inching. See **jog**.
Infeed. A mechanism designed to control the forward travel of the web or sheet into the press.
Infrared. (IR). Properly "infrared radiation." Invisible electromagnetic radiation with wavelengths longer than red light, from about 800 nm to 100,000 nm.
Infrared Drying. Drying of printing inks by the use of infrared radiation.

Impression too light.

Ink Holdout. Ability of paper or board to resist penetration of an ink or varnish.
Ink Jet Printing. A computerized method of forming an image by controlling the flow of a stream of ink droplets or by generating drops as demanded by the computerized image.
Ink Mist. Flying droplets from threads formed by **long inks**.
Ink Splitting. The manner in which an ink film ruptures as it exits the nip between two inked rollers or a roller and a substrate or blanket.
In-line Press. A press composed of separate printing units mounted horizontally.
Intaglio. A method of printing from an engraved or etched design in which the ink is held in the engraved cells. Intaglio includes gravure and steel plate engraving.
Intensity. (See **color strength**.)

Ion. An electrically charged chemical fragment or molecule.
I.P.A. International Prepress Association.
IPA. Isopropyl alcohol used as a solvent in flexo and litho.
ISO. The International Standards Organization, an affiliate of the United Nations.
ISO 9000–9004. Five standards generated by the members of ISO relating to quality management of a service or manufactured product. (Equivalent to ANSI/ASQC Q-90 to Q-94 in the United States and BS 5750 in the United Kingdom.)

JAN. Joint Army-Navy specifications.
Jar. A wide-mouth cylindrical container, usually of glass, ceramic or plastic material.
Jet. An intense black. Also see **ink jet**.
Jog. (1) To align sheets of paper into a compact pile. (2) To operate a press intermittently for very short increments. Also **inching**.
Journals. The end shafts on which a roll rotates.
Joystick. A manual control device for a computer or electronic color imaging system.
Jumbo Roll. A roll of web material larger than standard.
Justify. To space words in such a way so that all lines are of equal length and line up vertically on both the left and right.

Kaolin. A type of clay used for coating paper and board.
KD Box. A knocked-down box. A folding carton or corrugated box that is shipped flat and erected at the customer's plant.
Kelvin. (1) The temperature scale of absolute temperature; the Celsius scale plus 273 degrees. (2) The unit of measurement for color of illumination, e.g., 5000 Kelvins is the color emitted by a radiator heated to 5000 degrees Kelvin.
Kerning. Reducing parts of type to bring letters closer together.
Ketone. One of a class of organic solvents including **acetone** and methyl ethyl ketone (**MEK**).
Keyline Art. Black and white production art for designs containing two or more colors, and with the colors indicated by non-reproducing outlines.
Key Plate. The printing plate that carries the detail and to which the other plates are registered.
K Film. A trademark of E. I. duPont de Nemours & Co. for a patented polymer-coated cellophane.
Kiss Impression. The lightest impression that will transfer ink from the transfer roll to the flexo plate and from the plate to the substrate. Insufficient pressure produces uneven solids and excess pressure produces halos and dot gain.
Kodachrome. A trademark of Eastman Kodak Co. for patented color transparency films.
Kodacolor. A trademark of Eastman Kodak Co. for patented color negative films.
Kraft. (1) A method of producing paper pulp by treating wood chips with alkali in the presence of sulfide ion. (2) The pulp produced by that method. (3) A strong, stiff paper or board produced from kraft pulp, also called **sulfate**.

Label. A printed sheet adhered to a package and carrying information about the contents.
Lacquer. (1) A coating solution made by dissolving a natural or synthetic polymer in a solvent. (2) The film produced by coating the solution onto a package.
Laminate. (1) To unite layers of materials with adhesives. (2) The product made by bonding two or more layers of material.

Laser. An acronym for light amplification by stimulated emission of radiation. An intense narrow-band light beam used in computer controlled imaging and in other processes.

Laser Printing. A method of printing in which an electrostatic image is produced with a laser beam on a semiconductor drum, and the image is developed and printed with a **toner**.

Layout. Preliminary design showing position, sizes, color and other details of the final design.

Leading. The vertical spacing between lines of type.

LEL. Lower explosion limit; the lowest concentration of combustible vapor in air that will explode.

Length. The property of an ink that allows it to be drawn into long strings.

Letterpress. A process of printing from rigid, raised images using a paste ink.

Letterset. A term sometimes used for offset letterpress printing in which the letterpress plate prints against an offset blanket instead of directly against the substrate.

Light. Electromagnetic radiation that is perceived by the eye. Near-visible radiation is sometimes called "light," as in ultraviolet light.

Light Box. A **viewing box.**

Lightfastness. The ability of paper, substrate or print to resist fading or color change on exposure to light.

Lightness. Brightness of a black-and-white or colored image regardless of its hue or saturation.

Lignin. A natural polymeric material that binds cellulose together in plants. It darkens with age or exposure to light and is removed during chemical pulping and bleaching.

Line Copy. Copy made up of solids and lines in contrast to halftones or shadings made up of dots. Also **line art**.

Linen Tester. A magnifying lens mounted on a small stand, used to inspect copy, negatives and printing.

Liner. One of the outer, flat components of corrugated board.

Linerboard. Paperboard used for liners in corrugated board.

Line Screen. A number defining the fineness of a halftone screen ranging from 25 lines per inch for very coarse to 300 lines per inch or more for very fine. (100 lines per inch = 39.4 lines per centimeter.)

Lint. Loose fibers.

Liquid Ink. See **fluid ink**.

Lithography. A method of printing from a flat plate on which the image to be printed is ink-receptive and the nonprinting area is ink-repellent.

Litho Laminate. A printed sheet that is attached to a singleface corrugated sheet to make a corrugated box. Even if printed by gravure or flexo, such sheets are often referred to as litho laminates.

Log. The roll of paper wound at the end of the paper machine.

Logo. The abbreviation or trade jargon for **logotype**.

Logotype. A name, symbol or mark to identify a company. Also a trademark.

Long Ink. See **length**.

Loupe. A **linen tester**.

Lux. The metric unit of illumination.

M. Roman numeral for 1,000; used for 1,000 copies or sheets.

Machine Direction. The direction of paper parallel to its forward direction on the paper machine. **Grain** direction.

Magenta. A shade of ink that absorbs green while transmitting blue and red.

Makeready. (1) The preparation of the printing press, including mounting the plates or cylinders and adjustments of folders and finishing equipment, before starting the printing run. (2) Mounting of printing plates on a printing cylinder, in register, off the press, especially for flexo.

Manuscript. The original written words from which **typeset** matter is prepared.

Masking. The process of making photographic images from the original for the purpose of color correction, contrast reduction, tonal adjustment or detail enhancement.

Masstone. The color reflected by bulk ink or a very heavy film.

Matchprint. A trademark of 3M Company for a patented integral color proof.

Mealiness or snowflaking in gravure printing.

Matte. A paper finish of very low **gloss**.

Mealiness. Missing halftone dots, especially in gravure printing. **Snowflaking.**

Mechanical Pulp. Pulp produced by grinding or refining wood without removing the **lignin**. (See **groundwood** and **alphabet pulps**.)

Medium. The paperboard material that is corrugated and used as the inner layer of corrugated board.

MEK. See **methyl ethyl ketone.**

Metamerism. The occurrence in which colors match under one light source but do not match under another.

Metering Blade. See **doctor blade**.

Metering Roll. (1) In flexography, a rubber roll that presses against the anilox roll to remove excess ink. (2) In offset litho, a roll that presses against a roll carrying ink or water to control the amount of ink fed to the press.

Methyl Ethyl Ketone. An organic solvent closely related to **acetone** but not quite as fast. Also **MEK**.

Meyer Rod. A wire-wound rod used for making **drawdowns** of ink or coating.

Mezzotint. An irregular, random dot halftone.

Micron. One-millionth of a meter; 25.4 microns = .001 inch.

Midtones. The tonal range between **highlights** and **shadows**.

Migration. Penetration of a liquid or color from one layer to another.

Mileage. The area covered by a given weight or volume of ink or coating.

Modem. Short for modulator/demodulator, a device that converts computer data into high-frequency signals that can be sent over telephone lines and reconverted to computer data.

Moiré.

Moiré. An undesirable checkerboard or plaid pattern observed in overprints of halftone images when the screen angles are improper.

Moisture Vapor Transmission Rate (MVTR). See **water vapor transmission rate**.

Monomer. A chemical that can be reacted to form a high molecular weight material such as a **resin** or **polymer**.

Monotone. Black and white reproduction or printing in one color only.

Mottle. An unwanted uneven or cloudy area in print.

Mouse. A handheld device for manipulating the image on a computer screen.

Mullen. The bursting strength of board as measured by the Mullen burst tester.

MVT. The water (moisture) vapor transmission rate. Also MVTR.
Mylar. A trademark of E. I. duPont de Nemours & Co. for their polyethylene terephthalate. (See **polyester.**)

Nanometer. One billionth of a meter, one thousandth of a micron or 10 Angstrom units. The unit of measurement of the wavelength of light.
Naphtha. A solvent that contains a mixture of aliphatic hydrocarbons. It swells natural rubber but has little effect on Buna-N synthetic rubber.
NAPIM. National Association of Printing Ink Manufacturers.
Negative. A photographic medium in which the tonal values are reversed from those in the original.
Neoprene. A synthetic rubber prepared from chlorinated butadiene that has good resistance to many solvents.
Neutral Gray. A gray overprint of three process colors that shows no shade or hue.
Newtonian. The behavior of a liquid in which the flow is proportional to the shearing force. The ratio of shearing stress to the rate of shear equals the coefficient of viscosity which is constant for a newtonian liquid.
Nitrocellulose. Cellulose nitrate. A film-forming polymer derived from cotton or paper pulp and widely used in flexographic and gravure inks. Also called **pyroxylin.**
Nonflammable. Not flammable. Incapable of supporting combustion.
Non-Impact Printing. Any of several methods of applying ink without printing pressure such as **laser printing** or **ink jet printing**.
Non-Newtonian. Flow behavior of many liquids, including printing inks, in which the flow is not proportional to the shearing force; flow in which the viscosity changes with stirring or agitation.
Nylon. A synthetic polymer prepared from an amine and an organic acid, used as a molding resin and for film and fibers.

O.D. The outside diameter of a cylinder or roll.
Offpress Proof. A simulation of the printed job produced directly from digital information or photographic films.
Offset. (1) Any printing process in which the plate or image carrier places an image on a blanket which prints the substrate. Since lithography usually prints in this manner, the term offset is often used instead of offset lithography. (2) To transfer a printed image from one surface to another. This may be printing from an offset blanket or the transfer of ink from the front of a printed sheet or web to the back of another. (See **setoff**.)
Offsetting. See **setoff**.
Oleophilic. Having a strong affinity for oil (**hydrophobic**).
Oleophobic. Lacking affinity for oil (**hydrophilic**).
Opacity. The ability to block the transmission of light.
Opaque. To paint out spots and areas in a negative not wanted on the plate.
Operating Side. The side of a press on which the printing unit adjustments are located; opposite of **driving side** or **gear side**.
Organic. A chemical compound containing carbon.
Original. The image or design to be reproduced by printing. Also called **copy**.
Out. See **up**.
Oven. A common name for the **dryer** on a web press.
Overlay. A transparent sheet attached to copy used to indicate changes, color separation, etc.
Overprint. An ink or varnish printed over an ink.

Pantone Matching System. A trademark of Pantone, Inc. for a system of solid ink color mixing based on eight colors plus white and black. Sometimes referred to as PMS colors, a term no longer used by Pantone.

Paper. A fibrous, felted sheet or web usually made of cellulose normally no thicker than 0.006 inch.

Paperboard. A fibrous, felted sheet or web usually made of cellulose normally thicker than 0.010 inch or 10 points (0.3 mm.)

Paste Ink. An ink of high viscosity or consistency commonly used in lithography, letterpress or screen printing.

Paste-Up. The artwork mounted on a stiff paperboard and ready for **photography**.

Pel. A picture element. See **pixel**.

Penetration. The absorption of ink or other liquid by a substrate.

pH. The potential of the hydrogen ion; a measure of acidity or alkalinity.

Phenolic. The common name for phenol-formaldehyde polymers, a group of polymers used as molding resins.

Photocomposition. The process of setting type copy photographically as opposed to manually or electronically.

Photoengraving. A template produced with photography used as a matrix or stereotype to produce a relief plate.

Photography. The production of images on a sensitized surface by use of light or radiant energy.

Photopolymer. A **polymer** that is chemically altered by exposure to light; used in film, printing plates and inks.

Photostat. (1) A photographic reproduction on paper. Photostats may be positive or negative. Also called a **stat**. (2) A process for making positive paper prints of line copy and halftones.

pH Value. The acidity of an aqueous solution. The lower the number, the higher the acidity. 0–7 is acid, 7–14 is alkaline, and 7.0 is neutral.

PIA. Printing Industries of America.

Pica. (1) A unit of type measure; $1/6$ of an inch. One pica equals 12 points. (2) Printing Industries of the Carolinas.

Picking. A lifting of a small piece of paper or a breaking of the surface of the paper during printing.

Pigment. An insoluble coloring material, including white.

Piling. Build-up of ink on the press rollers or offset blanket that creates its own image.

Picking.

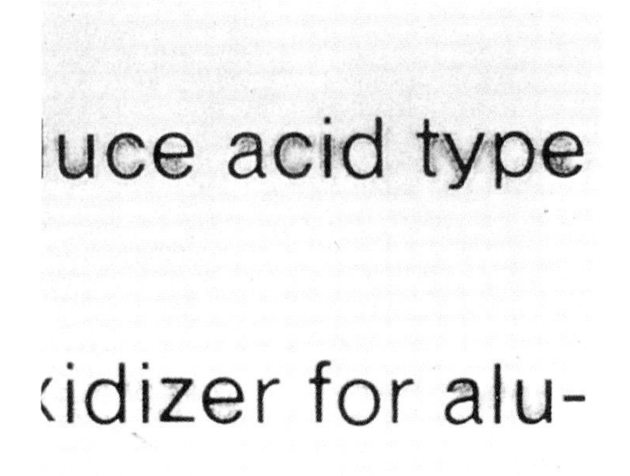

Piling.

Pin Register. A system of accurately positioned holes and pins on copy, film, plates and presses to insure proper **register** of colors.
PIV. Pulsating invariable variator. A speed control applicable to web presses to synchronize the line speed with the speed of the web.
Pixel. A picture element in electronic systems. The smallest tonal element in a digital imaging or display system. Sometimes called a **pel**. A halftone dot may be composed of several pixels.
Planography. Printing from a flat surface as in **lithography** and **collotype.**
Plasticizer. A liquid or solid material that makes films and other products more flexible.
Plate Cylinder. The cylinder on the printing press that carries the printing plate.
Platemaking. The production of a finished printing plate.
Plugging. Buildup of ink, lint or other foreign particles on a flexo plate to form an undesirable printing surface.
PMS. A contraction of Pantone Matching System.
PMT. A trademark of Eastman Kodak Co. for their patented photomechanical transfer or photostatic material comprising a photographic paper used for making prints of artwork.
Point. A unit of type measurement, 0.0138 inch; 12 points to the pica, 72 points to the inch.
Polyamide. A polymer, such as nylon, containing the amide group. Polyamides are used as packaging films, molding resins and as binders in some printing inks.
Polycarbonate. A polymer produced by polymerizing a phenol with a carbonate ester. Polycarbonates are molding resins used in making bottles.
Polyester. A polymer containing the ester group, usually polyethylene terephthalate or PET. Polyesters are used as packaging films and for bottles.
Polyethylene. A polymer produced by polymerizing ethylene, a monomer containing two carbon atoms, used as a packaging film or molding resin.
Polymer. A compound formed by combining many small, monomeric molecules into one high-molecular-weight compound.
Polyolefin. A polymer produced by polymerizing an olefin, commonly polyethylene or polypropylene, used as a packaging film.
Polypropylene. A polymer produced by polymerizing propylene, a monomer containing three carbon atoms; used as a packaging film or molding resin.
Polystyrene. A polymer produced by polymerizing styrene and used as a packaging material.
Polyvinyl Acetate. A polymer produced by polymerizing vinyl acetate. Often copolymerized with vinyl chloride to produce a resin used for making vinyl films.
Polyvinyl Alcohol. A polymer produced by hydrolyzing polyvinyl acetate and used as a laminate in some clear films, sometimes abbreviated PVOH.
Polyvinylidene Chloride. A polymer produced by polymerizing vinylidene chloride and used as a packaging film.
P.O.P. Point of purchase. A point-of-purchase display, usually incorporated into a package, is used to promote the product at the retail store.
Pop Test. Slang for the **mullen test.**
Porosity. The presence of pores or minute openings in natural or synthetic materials.
Porous. Containing minute openings that can take up liquid or gas.
Post-Consumer Waste. Waste paper or other material that has been discarded by the ultimate consumer. (See **waste.**)
Premount. (1) To place printing plates on the plate cylinder before the cylinder is placed on the press. (2) The product of premounting—the premounted plate or plates.
Prepolymer. A polymer of low molecular weight to be further polymerized.

Preprinted Liner. Linerboard that is printed and rewound before being combined into corrugated board.

Press Proof. Printed image produced on a proof press to illustrate what the final printed material will look like.

Primer Coat. See **anchor coat.**

Printability. The ability of a substrate to accept ink and produce an acceptable printed image. (Contrast with **runnability.**)

Printer-Slotter. A machine that prints fiberboard sheets and then scores and slots them to complete the manufacture of box blanks.

Printing. The production of multiple images on a substrate at high speed.

Process Color. A method of reproducing color that involves use of four halftone color separations and cyan, magenta, yellow and black ink.

Prog. Short for **progressive proof.**

Progressive Proofs. A set of press proofs, including proofs of the individual colors and of two-, three- and four-color overprints.

Proof. A prototype of the printed job used for in-house quality control or customer approval.

Purity. See **color purity.**

PVA. See **Polyvinyl alcohol.**

Pyroxylin. Nitrocellulose, especially the more soluble grades.

Quality. (1) An attribute, property or characteristic. (2) The degree of excellence.

Railroad Tracks. (1) A streak developed by oscillation of a nicked gravure doctor blade. (2) In narrow web printing, marks caused by gear backlash.

Raw Stock. The paper or board to which coating is applied.

Reducer. A material added to reduce the color strength of an ink or the viscosity of an ink or varnish.

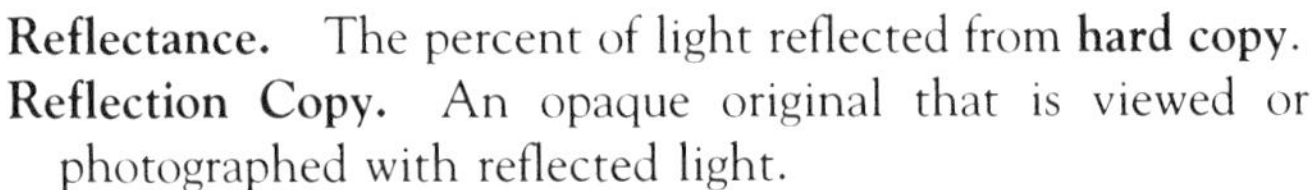

Reflectance. The percent of light reflected from **hard copy.**

Reflection Copy. An opaque original that is viewed or photographed with reflected light.

Misregister.

Reflection Densitometer. A densitometer for measuring the optical density of an image viewed by reflected light. (See **densitometer.**)

Reflection Image. An image such as a print or photograph that is viewed by reflected light rather than by transmitted light.

Register. The degree of alignment between two images in a proof or print. Also referred to as registration. (See **fit.**)

Relative Humidity. The percentage of water vapor in air compared to the amount of water required to saturate the air at that temperature.

Relief. In flexography and letterpress, the amount of relief in the printing plate. The distance between the surface of the type and the surface of the base of the plate. Also **undercut.**

Relief Printing. Printing from a plate in which the image areas are raised from the non-image areas, such as flexo and letterpress.

Repeat. The printing length of a plate cylinder. The length of an image that can be printed; on most presses this is nearly the same as the circumference of the plate cylinder. (Also cutoff.)

Resin. A natural or synthetic material used as a **binder** or film-forming material in inks or varnishes.
Retouching. The art of making selective manual or electronic corrections to images.
Reverse. An image printed as a void in a solid inked area.
Reverse Printing. Printing on the underside of a transparent film.
Rheology. The study of the flow and deformation of matter, notably printing inks.
Right Reading. Copy reading normally, not a mirror image.
Rigid Box. A box made of boxboard that is produced already set up.
Roll-out. A color sample made manually by rolling a sample of ink onto the substrate.
Rosin. A resin produced by pine trees.
Rotogravure. Gravure printing from a cylinder. Since most gravure printing is done in this manner, rotogravure is often called simply gravure.
Rub Resistance. Abrasion resistance.
Rubylith. A hand-cut masking film.
Runnability. The ability of a roll of paper or substrate to run without web breaks. (Contrast with **printability**.)
Running Register. The common name for the **circumferential register control**.

Sans Serif. A typeface that has no serifs.
Saran. A trademark of Dow Chemical Co. for polyvinylidene chloride.
Saturation. See **color strength**.
SBS. Solid bleached sulfate or board made from this stock.
Scale. A paper defect caused by flakes of dried paper coating embedded into the paper during **supercalendering**.
Score. (1) A groove or crease or a partial cut in a material to facilitate bending or folding. (2) To make such a groove or crease.
Screen Angle. The angle of rotation (counterclockwise) of a halftone screen from its normal, horizontal position or the angle between two screens.
Screen Printing. The process of printing in which the image is produced by pressing ink through a stencil supported by a screen. Formerly called **silk screen**.
Scuffing. See **abrasion**.
Scumming. Deposit of ink in the non-printing areas of a gravure or litho image.

Scumming.

Secondary Colors. Colors obtained by mixing any two primary colors in equal proportions. Subtractive secondary colors are red, green and blue.
Secondary Fibers. Fibers derived from wastepaper for reuse in paper or board manufacture.
Semichemical. A process of producing groundwood pulp or paper in which the lignin is softened chemically before pulping.
Sensitometer. An instrument for measuring the light sensitivity of photographic material.
Separations. A set of four halftone photographic films each of which represents one of the colors required for process color printing.
Serif. The short crossline at the end of the main stroke of many letters in some type faces.
Serigraphy. A name sometimes applied to **screen printing.**
Setoff. A printing defect in which wet ink transfers from the face of a print to the back of the print adjacent to it in the web or pile.
Setup Box. A rigid box made of paperboard that is manufactured in the shape of a box by the boxmaker and shipped in that form. (Contrast with a **KD box.**)
Shading. The addition of a color or tone to suggest three-dimensionality in a picture.
Shadow. The darkest part of a halftone image.
Shadow Dots. The large dots in the dark areas of a halftone print, interspersed with small white dots. Also, the small dots in a negative of the shadow area.
Sheeting. An unsupported, nonfibrous flexible material thicker than about 0.010 inch.

Setoff.

Shellac. An alcohol-soluble natural resin widely used in flexo inks.
Shell Cup. An efflux cup for measuring viscosity of liquid inks.
Shore Hardness. Hardness as measured by a Shore **durometer**. The numbers are usually reported as "Shore A" for soft materials such as rubber and "Shore D" for harder materials such as many thermosetting resins.
Short Ink. An ink that cannot be drawn out into long strings. (See **length.**)
Side-Lay Misregister. Misregister in the lateral or cross-web direction.
Silk Screen Printing. The early name for **screen printing.**
Sizing. Treatment of paper or other porous materials to improve resistance to water and other liquids.
Skid. (1) A platform supporting a pile of cut sheets. (2) The cut sheets in the pile.
Skinny. See **choke.**

Shell Cup.

Slime Hole. A hole in the paper caused by slime or fungus in the paper mill.
Slime Spot. A fragile spot in the paper resulting from slime or fungus growing in the paper mill.
Sling Psychrometer. A device with wet- and dry-bulb thermometers used for measuring relative humidity.
Slip Compound. A wax or other ink additive that lubricates the ink film after drying.
Slitter. A machine to cut a wide roll into narrower rolls.
Slitter-Rewinder. A machine that slits from a master roll and rewinds the strips into rolls of specified length.
Slurring. The elongation of halftone dots during printing, with resultant loss of sharpness in the print. Also called slur.
Smoothness. The microscopic evenness or flatness of paper or board required for good printing, especially by gravure or letterpress.
Snowflaking. White spots in the printing caused by missing dots in gravure and by emulsification in litho. (See **mealiness.**)
Soft Copy. An image on a video display terminal.
Softwood. Wood from coniferous trees (gymnosperms) that include pine, hemlock and spruce.

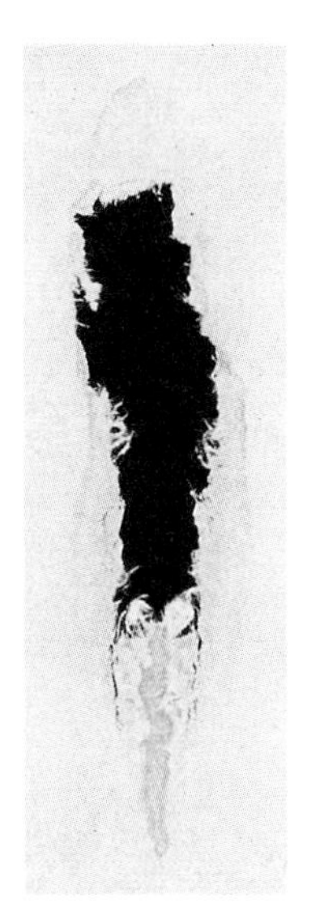

Slime hole.

Solid Fiber Box. A box made of solid fiberboard produced by laminating two or more plies of containerboard.
Solute. A material dissolved by a solvent.
Solvent. A liquid that dissolves or disperses other substances.
SPAI. Screen Printing Association, International.
SPC. Statistical process control. Control of manufacturing operations by using statistically significant information.
Specific Gravity. The weight of a unit volume, usually pounds per gallon.
Spectrophotometer. An instrument to measure color by determining the reflectance of a material at all wavelengths.
Spectrum. The electromagnetic spectrum of the entire range of wavelengths from radio waves to gamma rays. It includes visible light. A spectrum of visible light includes radiation between infrared and ultraviolet.
Splice. (1) To join two webs by pasting the end of one to the beginning of the next. (2) The joint itself.
Splitting. See **ink splitting.**
Spread. An image whose edges have been slightly enlarged in order to improve the fit of the image. Also called a **fatty**. The opposite of a **choke** or **skinny**.
SQC. Statistical quality control. Same as **SPC.**
Stack Press. A printing press in which the printing units are arranged vertically. (Contrast with **CI press** and **in-line press.**)
Star Target. A circular test image containing alternating light and dark wedges that meet at a point; used to monitor press performance.
Stat. Photostat. A photographic copy of artwork or photography.
Static. Electric charges on the surface of an insulating material.
Step & Repeat. Preparation of a plate or a film with many copies of the same image.
Stickyback. Double-faced adhesive used for mounting elastomeric printing plates to the plate cylinder.
Stock. The pulp used to make paper or paperboard.

Streaks and striations in flexography.

Streaking. Streaks in the print usually caused by poor wiping of the gravure cylinder or excess water in litho ink.

Stretch/shrink Factors. Calculations of dimensional changes that occur in mounting of photopolymer plates and the molding and mounting of rubber plates.

Striation. A streaky pattern of parallel lines, usually in the direction of the web caused by uneven inking across the web.

Sublimation. The process in which a solid is volatilized by heat and then converted back into a solid without passing through a liquid phase.

Substance. The weight in pounds of a ream of paper. **Basis weight.**

Substrate. The material to be printed: paper, paperboard, film or foil, etc.

Subtractive Primaries. Cyan, magenta and yellow; the hues used for process color printing.

Sulfate. See **kraft**. Often spelled sulphate.

Sulfite. A process of cooking wood under neutral or acid conditions in the presence of sulfite ion. Also the pulp and the paper or board produced from this pulp. Often spelled sulphite.

Supercalendering. A process in which paper is smoothed between nips that alternate between steel and hard-filled cotton rolls.

Surface Tension. A force existing at the surface of a liquid and the interface of a liquid with a solid or other liquid that tends to bring the liquid into the form of a droplet.

SUS. Solid unbleached sulfate or board made from this stock.

Swatch. A small piece of material cut for a sample.

SWOP. Specifications for Web Offset Publications. Specifications for the control and production of web offset publication printing, many of which are also applied to package printing.

Tack. The resistance of a film of liquid to **splitting** forces.

Tag. A stiff printed sheet attached to a package with string or wire and carrying information about the contents.

TAPPI. Technical Association of the Pulp and Paper Industry.

Teflon. A trademark of E. I. duPont de Nemours & Co. for a polymer of fluorinated ethylene. It is resistant to heat, solvents and chemical attack.

Thermocouple. A bimetallic device to measure temperature by converting heat to measurable electric potential.

Thermoplastic. A polymer capable of being resoftened after it is formed.

Thermoset. A polymer that does not soften when heated.

Thinner. A **solvent** or **diluent** added to the ink to reduce its **viscosity**.

Thixotropy. A reversible decrease in viscosity when an ink or dispersion is stirred or worked. The opposite of **dilatency**.

Tinctorial Strength. The power of a pigment or dye to impart color to an ink.

Tint. (1) A light shade of a color. (2) A halftone area that contains dots of uniform size. (3) A printing defect in which emulsified lithographic ink appears in the non-image area. (4) The mixture of a color with white.

Tinting. (1) A light scum that appears on unprinted areas of the plate and paper during offset printing. (2) Addition of a light overall base color to a substrate prior to printing.

TIR. Total indicated runout, a combined measurement of the lack of roundness of a cylinder plus the distance off center. Cylinders that are out-of round or off center run poorly on high-speed presses.

Tone. A color value or quality. Also a tint or shade of color. Sometimes short for **halftone**.

Toner. (1) A highly concentrated pigment or dye used to make black inks appear darker. (2) The colorant in electrostatic imaging.

Toxicity. The ability of a substance to damage living tissue.

TQM. Total Quality Management. Management of a company using **SQC** and other techniques to eliminate waste and increase productivity and quality.

Transfer Key. A trademark of 3M Company for a patented integral color proof.

Transmission Densitometer. An instrument for measuring the optical density of transparencies. (See **densitometer**.)

Transparency. A photographic positive on a clear or transparent support, viewed by transmitted light. The term is usually applied to full color transparencies.

Trap. The ability of a wet or dried printed ink film to accept the succeeding ink film.

Trapping. The process of correcting misregistration between colored inks by spreading the lighter color over the darker color.

Triplewall. See **corrugated board.**

Turning Bars. Stationary bars on a web press arranged to turn the web over so that the bottom becomes the top.

Type. A block of metal bearing a relief character for producing an inked print. Also the image produced from that character.

Typeface. All the type of a single design.

Typeset. To put manuscript into type form.

Ultraviolet. Electromagnetic radiation shorter than visible light, usually around 10–400 nm. Ultraviolet radiation initiates many chemical reactions, including those used for proofing materials, plates, inks and varnishes.

Undercolor Removal. Reduction of the amount of yellow, magenta and cyan dot percentage in three color overprint areas to reduce the amount of ink applied to the print.

Undercut. See **relief**.

Undertone. The color reflected by a very thin film of ink.

Up. After a number, this indicates the number of complete images printed for each revolution of the press. One-up means that the plate prints one image for each revolution, two-up means two, etc. (Also **out.**)

Vacuum Frame. A device that holds the photographic negative or positive in close contact with photosensitive materials such as proofing stock or printing plates.

Varnish. (1) An unpigmented resinous coating applied over the print. (2) The base of the printing ink to which are added **pigments** and **additives**.

Vehicle. The liquid component of a printing ink; varnish plus additives.

Vertical Process Camera. A camera that focuses in a vertical direction and carries the copy and lenses in a horizontal frame.

Vignette. An illustration in which the background gradually fades away until it blends into the unprinted substrate.

Vinyl. (1). Any of a group of polymers made from vinyl monomers such as vinyl chloride or vinyl acetate. (2) Pattern of a cutting and creasing die to serve as an overlay to verify alignment with the printing.

Viscometer. An instrument for measuring viscosity.

Viscosity. The resistance of a fluid to flow.

Volatility. The ability of a liquid to evaporate.

Volcano. A printing defect of heavy ink films in which evaporating solvent bursts through a dried ink surface.

Vulcanization. Curing of raw rubber or other elastomer to increase its strength and stability.

Wash marks.

Whiskers. (Courtesy of Gravure Association of America.)

Wash Drawing. A drawing, such as watercolor, that contains low pigment content.

Wash Marks. Streaks in lithographic printing caused by excess water in the system.

Waste. Printed or unprinted substrate that cannot be used or has served its purpose. Different problems are encountered in recycling **broke, converter waste** and **post-consumer waste.**

Water Vapor Transmission Rate. The weight of water vapor transmitted through a packaging material under standard conditions; reported in grams per square meter per hour.

Wavelength. The distance between two peaks on a wave curve. This is a fundamental property of light and is mathematically the speed of light divided by the frequency.

Web. Thin material, such as paper, foil or film, to be wound, unwound or printed from a roll.

Web Guide. A device that keeps the web traveling true through the press.

Web Press. A press that prints webs of paper.

Wet Strength. The tensile or burst properties of paper or paperboard when wet.

Wettability. Ability of a surface to accept a thin film of a liquid.

Whiskers. Hairy edges of shadow areas (especially gravure) due to static electricity.

Wire-wound Rod. A stainless steel rod wound with stainless steel wire used for drawing ink onto a substrate. (Also Meyer Rod.)

Wraparound Plate. A thin relief plate used in letterpress printing.

Wrapper. A printed sheet that covers a package and carries information about the contents. The distinction between a label and a wrapper is not always clear. The wrapper on a 3-piece can is usually called a label; the wrapper on a candy bar is usually called flexible packaging; the wrapper on a bar of soap or a package of gum is usually called a wrapper.

Wrong-reading Image. An image that is reversed, a mirror image, such as the image on a letterpress or flexo plate or a gravure cylinder. A right-reading image viewed from the wrong side.

Xerography. A method of printing in which an **electrostatic** image is produced by reflecting the copy onto a charged semiconductor drum and developing the electrostatic image and printing it with a dry **toner**.

Yellow. A shade of ink that absorbs blue while transmitting red and green.

Zahn Cup. An efflux cup for measuring viscosity of liquid inks.

FURTHER READING

Pocket Pal, Fifteenth Edition. Michael H. Bruno. International Paper Co., Memphis, TN. (1992).

Complete Color Glossary. Miles Southworth, Thad McIlroy and Donna Southworth. The Color Resource, Livonia, NY. (1992).

GATF Glossary of Graphic Arts Terms. Compiled by Pamela Groff. Graphic Arts Technical Foundation, Pittsburgh, PA. (1991).

Gravure: Process and Technology. E. Kendall Gillett, III, Chairman. Gravure Education Association and Gravure Association of America, Rochester, NY. (1991).

Flexography: Principles and Practices, Fourth Edition. Frank N. Siconolfi, Chairman. Flexographic Technical Association, Ronkonkoma, NY. (1991).

Packaging in Today's Society, Third Edition. Robert J. Kelsey. Technomic Publishing Co., Lancaster, PA. (1989).

Glossary of Packaging Terms. Sixth Edition. Institute of Packaging Professionals, Reston, VA. (1988).

Getting It Printed. Mark Beach, Steve Shepro and Ken Russon. Coast to Coast Books, Portland, OR. (1987).

Introduction to Flexo Folder-Gluers, Joel J. Shulman. Jelmar Publishing Co., Inc. Plainview, NY. (1986).

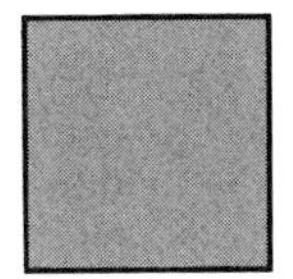

Index